T0338385

ACTIVE DISTURBANCE REJECTION CONTROL FOR NONLINEAR SYSTEMS

ACTIVE DISTURBANCE REJECTION CONTROL FOR NONLINEAR SYSTEMS

AN INTRODUCTION

Bao-Zhu Guo

Academy of Mathematics and Systems Science, Academia Sinica, People's Republic of China and University of the Witwatersrand, South Africa

Zhi-Liang Zhao

Shaanxi Normal University, Xi'an, People's Republic of China

WILEY

This edition first published 2016 © 2016 John Wiley & Sons (Asia) Pte Ltd.

Registered office
John Wiley & Sons Singapore Pte. Ltd., 1 Fusionopolis Walk, #07-01 Solaris South Tower, Singapore 138628.

For details of our global editorial offices, for customer services and for information about how to apply for permission to reuse the copyright material in this book please see our website at www.wiley.com.

All Rights Reserved. No part of this publication may be reproduced, stored in a retrieval system or transmitted, in any form or by any means, electronic, mechanical, photocopying, recording, scanning, or otherwise, except as expressly permitted by law, without either the prior written permission of the Publisher, or authorization through payment of the appropriate photocopy fee to the Copyright Clearance Center. Requests for permission should be addressed to the Publisher, John Wiley & Sons Singapore Pte. Ltd., 1 Fusionopolis Walk, #07-01 Solaris South Tower, Singapore 138628, tel: 65-66438000, fax: 65-66438008, email: enquiry@wiley.com.

Wiley also publishes its books in a variety of electronic formats. Some content that appears in print may not be available in electronic books.

Designations used by companies to distinguish their products are often claimed as trademarks. All brand names and product names used in this book are trade names, service marks, trademarks or registered trademarks of their respective owners. The Publisher is not associated with any product or vendor mentioned in this book. This publication is designed to provide accurate and authoritative information in regard to the subject matter covered. It is sold on the understanding that the Publisher is not engaged in rendering professional services. If professional advice or other expert assistance is required, the services of a competent professional should be sought.

Limit of Liability/Disclaimer of Warranty: While the publisher and author have used their best efforts in preparing this book, they make no representations or warranties with respect to the accuracy or completeness of the contents of this book and specifically disclaim any implied warranties of merchantability or fitness for a particular purpose. It is sold on the understanding that the publisher is not engaged in rendering professional services and neither the publisher nor the author shall be liable for damages arising here from. If professional advice or other expert assistance is required, the services of a competent professional should be sought.

Library of Congress Cataloging-in-Publication Data

Names: Guo, Bao-Zhu, 1962- author. | Zhao, Zhi-Liang, author.
Title: Active disturbance rejection control for nonlinear systems : an
    introduction / Bao-Zhu Guo and Zhi-Liang Zhao.
Description: Singapore ; Hoboken, NJ : John Wiley & Sons, 2016. | Includes
    bibliographical references and index.
Identifiers: LCCN 2016018239 (print) | LCCN 2016025091 (ebook) | ISBN
    9781119239925 (cloth) | ISBN 9781119239956 (pdf) | ISBN 9781119239949
    (epub)
Subjects: LCSH: Damping (Mechanics) | Automatic control. | Nonlinear systems.
Classification: LCC TA355 .G79 2016 (print) | LCC TA355 (ebook) | DDC
    620.3/7–dc23
LC record available at https://lccn.loc.gov/2016018239

A catalogue record for this book is available from the British Library.

Set in 10/12pt, TimesLTStd by SPi Global, Chennai, India.
Printed and bound in Singapore by Markono Print Media Pte Ltd

1 2016

Contents

Preface

The modern control theory came to mathematics via N. Wiener's book *Cybernetics or the Science of Control and Communication in the Animal and the Machine*, published in 1948. In 1954, H.S. Tsien published the book *Engineering Cybernetics*, which brought the control theory to engineering.

Roughly speaking, the modern control theory has been developed through three stages. The first stage is the classical control or automatic principle of compensation developed from the 1940s to the 1960s. During this period the single-input and single-output linear time-invariant systems were studied by the frequency domain approach. The second stage was from the 1960s to the 1980s, during which time the multi-input and multi-output systems were studied by the state space or time domain approach. The state-space approach relies heavily upon the mathematical models of the systems. After the 1980s, many control theories were developed to cope with uncertainties in systems. Several powerful methods were developed, including the internal model principle for output regulation (which started in the 1970s), as well as adaptive control, roust control, high-gain feedback control, and sliding mode control (which started even earlier). In particular, the robust control theory was well-established by both the frequency domain approach and the time domain approach. A common feature for these methods was the worst case scenario regarding the disturbance. A different way of dealing with uncertainty may be found in adaptive control where the unknown parameters are estimated under the "exciting persistent condition" and in output regulation where a special class of external disturbance is estimated through the observer and internal model and is compensated for in the feedback-loop.

During the late 1980s and 1990s, Jingqing Han of the Chinese Academy of Sciences proposed a powerful unconventional control approach to deal with vast uncertainty in nonlinear systems. This new control technology was later called the active disturbance rejection control (ADRC). The uncertainties dealt with by the ADRC can be very complicated. They could include the coupling of the external disturbances, the system unmodeled dynamics, the zero dynamics with unknown model, and the superadded unknown part of control input. The key idea of the ADRC considers the "total disturbance" as a signal of time, which can be estimated by the output of the system. Basically, the ADRC consists of three main parts. The first part is the tracking differentiator (TD) that is relatively independent and is actually thoroughly discussed in the control theory. The aim of the TD is to extract the derivatives of the reference signal and is also considered as transient profile for output tracking. The second part of the ADRC is the extended state observer (ESO) which is a crucial part of the ADRC. In ESO, both the state and the "total disturbance" are estimated by the output of the system. This remarkable feature makes the ADRC a very different way of dealing with uncertainty. The ESO is the generalization of the traditional state observer where only the state of the system is estimated. The final part of the ADRC is the extended state observer-based feedback control. Since the uncertainty is estimated in the ESO and is compensated for in the feedback loop, the barriers between the time invariant and time varying, linear and nonlinear have been broken down by considering the time-varying part and the nonlinear part as uncertainty. At the same time, the control energy is significantly reduced. More importantly, in this way, the closed-loop systems look like linear time-invariant systems, for which a reliable result can be applied.

In the past two decades, the ADRC has been successfully applied to many engineering control problems such as hysteresis compensation, high pointing accuracy and rotation speed, noncircular machining, fault diagnosis, high-performance motion control, chemical processes, vibrational MEMS gyroscopes, tension and velocity regulations in Web processing lines and DC–DC power converters by many researchers in different contexts. In all applications of process control and motion control, compared with the huge amount of literature on control theory in dealing with the uncertainty such as system unmodeled dynamics, external disturbance rejection, and unknown parameters, the ADRC has shown its remarkable PID nature of an almost independent mathematical model, no matter the high accuracy control of micrometre grade or the integrated control on a very large scale.

On the other hand, although many successful engineering applications have been developed, the theoretical research on these applications lags behind. This book serves as an introduction to the ADRC from a theoretical perspective in a self-contained way. In Chapter 1, some basic background is introduced on nonlinear uncertain systems that can be dealt with by the ADRC. Chapter 2 presents convergence of the different types of tracking differentiators proposed by Han in his original papers. Chapter 3 is devoted to convergence of the extended state observer for various nonlinear systems. Chapters 2 and 3 can be considered as independent sections of the book. Chapter 4 looks at convergence of the closed-loop based on the TD and ESO. This can be considered as a separation principle of the ADRC for uncertain nonlinear systems. The numerical simulations are presented from here to where to illustrate the applicability of the ADRC. Finally, in Chapter 5, the ESO and stabilization for lower triangular systems are discussed.

Most of the material in this book is from the authors' published papers on this topic. However, the idea for the book comes from ideas published in Han's many original numerical experiments, engineering applications that appeared publicly in Chinese, and the numerous works done by the ADRC group, in particular the group led by Dr. Zhiqiang Gao at Cleveland State University in the United States.

The authors are deeply indebted to those who helped with the works presented in this book. These include Zhiqiang Gao, Yi Huang, and Wenchao Xue. This book is dedicated to Bao-Zhu Guo's memory of Professor Jingqing Han who passed away in 2008.

This book is partially supported by the National Natural Science Foundation of China (No. 61403242) and the Nature Science Research Program of Shaanxi Province-Key Program (No. 2016JZ023).

Bao-Zhu Guo and Zhi-Liang Zhao
October, 2015

1

Introduction

In this chapter, we introduce some necessary background about the active disturbance rejection control (ADRC). Some notation and preliminary results are also presented.

1.1 Problem Statement

In most control industries, it is hard to establish accurate mathematical models to describe the systems precisely. In addition, there are some terms that are not explicitly known in mathematical equations and, on the other hand, some unknown external disturbances exist around the system environment. The uncertainty, which includes internal uncertainty and external disturbance, is ubiquitous in practical control systems. This is perhaps the main reason why the proportional–integral–derivative (PID) control approach has dominated the control industry for almost a century because PID control does not utilize any mathematical model for system control. The birth and large-scale deployment of the PID control technology can be traced back to the period of the 1920s–1940s in response to the demands of industrial automation before World War II. Its dominance is evident even today across various sectors of the entire industry. It has been reported that 98% of the control loops in the pulp and paper industries are controlled by single-input single-output PI controllers [18]. In process control applications, more than 95% of the controllers are of the PID type [9].

Let us look at the structure of PID control first. For a control system, let the control input be $u(t)$ and let the output be $y(t)$. The control objective is to make the output $y(t)$ track a reference signal $v(t)$. Let $e(t) = y(t) - v(t)$ be the tracking error. Then PID control law is represented as follows:

$$u(t) = k_0 e(t) + k_1 \int_0^t e(\tau) d\tau + k_2 \dot{e}(t), \tag{1.1.1}$$

where k_0, k_1, and k_2 are tuning parameters. The PID control is a typical error-based control method, rather than a model-based method, which is seen from Figure 1.1.1 for its advantage

Active Disturbance Rejection Control for Nonlinear Systems: An Introduction, First Edition.
Bao-Zhu Guo and Zhi-Liang Zhao.
© 2016 John Wiley & Sons Singapore Pte. Ltd. Published 2016 by John Wiley & Sons, Ltd.

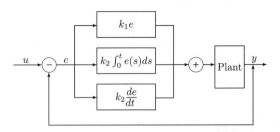

Figure 1.1.1   PID control topology.

of easy design. The nature of independent mathematical model and easy design perhaps have explained the partiality of control engineers to PID.

However, it is undeniable that PID is increasingly overwhelmed by the new demands in this era of modern industries where an unending efficiency is pursued for systems working in more complicated environments. In these circumstances, a new control technology named active disturbance rejection control (ADRC) was proposed by Jingqing Han in the 1980s and 1990s to deal with the control systems with vast uncertainty [58, 59, 60, 62, 63]. As indicated in Han's seminal work [58], the initial motivation for the ADRC is to improve the control capability and performance limited by PID control in two ways. One is by changing the linear PID (1.1.1) to nonlinear PID and the other is to make use of "derivative" in PID more efficiently because it is commonly recognized that, in PID, the "D" part can significantly improve the capability and transient performance of the control systems. However, the derivative of error is not easily measured and the classical differentiation most often magnifies the noise, which makes the PID control actually PI control in applications, that is, in (1.1.1), $k_2 = 0$.

In automatic principle of compensation, the differential signal for a given reference signal $v(t)$ is approximated by $y(t)$ in the following process:

$$\hat{y}(s) = \frac{s}{Ts + 1}\hat{v}(s) = \frac{1}{T}\left(\hat{v}(s) - \frac{1}{Ts + 1}\hat{v}(s)\right), \qquad (1.1.2)$$

where $\hat{L}(s)$ represents the Laplace transform of $L(t)$, T is a constant, and $\frac{1}{Ts+1}\hat{v}(s)$ represents the inertial element with respect to T (see Figure 1.1.2).

The time domain realization of (1.1.2) is

$$y(t) = \frac{1}{T}(v(t) - v(t - T)). \qquad (1.1.3)$$

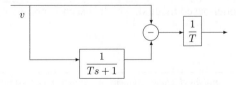

Figure 1.1.2   Classical differentiation topology.

If $v(t)$ is contaminated by a high-frequency noise $n(t)$ with zero expectation, the inertial element can filter the noise ([62], pp. 50–51):

$$y(t) = \frac{1}{T}(v(t) + n(t) - v(t - T)) \approx \dot{v}(t) + \frac{1}{T}n(t). \qquad (1.1.4)$$

That is, the output signal contains the magnified noise $\frac{1}{T}n(t)$. If T is small, the differential signal may be overwhelmed by the magnified noise.

To overcome this difficulty, Han proposed a noise tolerant tracking differentiator:

$$\hat{y}(s) = \frac{1}{T_2 - T_1}\left(\frac{1}{T_1 s + 1} - \frac{1}{T_2 s + 1}\right)\hat{v}(s), \qquad (1.1.5)$$

whose state-space realization is

$$\begin{cases} \dot{x}_1(t) = x_2(t), \\ \dot{x}_2(t) = -\dfrac{1}{T_1 T_2}(x_1(t) - v(t)) - \dfrac{T_2 - T_1}{T_1 T_2}x_2(t), \\ y(t) = x_2(t). \end{cases} \qquad (1.1.6)$$

The smaller T_1/T_2 is, the quicker $x_1(t)$ tracks $v(t)$. The abstract form of (1.1.6) is formulated by Han as follows:

$$\begin{cases} \dot{x}_1(t) = x_2(t), \\ \dot{x}_2(t) = r^2 f\left(x_1(t) - v(t), \dfrac{x_2(t)}{r}\right), \end{cases} \qquad (1.1.7)$$

where r is the tuning parameter and $f(\cdot)$ is an appropriate nonlinear function. Although a convergence of (1.1.7) is first reported in [59], it is lately shown to be true only for the constant signal $v(t)$. Nevertheless, the effectiveness of a tracking differentiator (1.1.7) has been witnessed by many numerical experiments and control practices [64, 147, 152, 153]. The convergence proof for (1.1.7) is finally established in [55 and 52]. In Chapter 2, we analyze this differentiator, and some illustrative numerical simulations and applications are also presented.

The second key part of the ADRC is the extended state observer (ESO). The ESO is an extension of the state observer in control theory. In control theory, a state observer is a system that provides an estimate of the internal state of a given real system from its input and output. For the linear system of the following:

$$\begin{cases} \dot{x}(t) = Ax(t) + Bu(t), \\ y(t) = Cx(t), \end{cases} \qquad (1.1.8)$$

where $x(t) \in \mathbb{R}^n (n \geq 1)$ is the state, $u(t) \in \mathbb{R}^m$ is the control (input), and $y(t) \in \mathbb{R}^l$ is the output (measurement). When $n = 1$, the whole state is measured and the state observer is unwanted. If $n > 1$, the Luenberger observer can be designed in the following way to recover the whole state by input and output:

$$\dot{\hat{x}}(t) = A\hat{x}(t) + Bu(t) + L(y(t) - C\hat{x}(t)), \qquad (1.1.9)$$

where the matrix L is chosen so that $A - LC$ is Hurwitz. It is readily shown that the observer error $x(t) - \hat{x}(t) \to 0$ as $t \to \infty$. The existence of the gain matrix L is guaranteed by the detectability of system (1.1.8). If it is further assumed that system (1.1.8) is stabilizable, then there exists a matrix K such that the closed-loop system under the state feedback $u(t) = Kx(t)$ is asymptotically stable: $x(t) \to 0$ as $t \to \infty$. In other words, $A + BK$ is Hurwitz. When the observer (1.1.9) exists, then under the observer-based feedback control $u(t) = K\hat{x}(t)$, the closed-loop system becomes

$$\begin{cases} \dot{x}(t) = (A + BK)x(t), \\ \dot{\hat{x}}(t) = (A - LC + BK)\hat{x}(t) + LCx(t). \end{cases} \tag{1.1.10}$$

It can be shown that $(x(t), \hat{x}(t)) \to 0$ as $t \to \infty$ and, moreover, the eigenvalues of (1.1.10) are composed of $\sigma(A + BK) \cup \sigma(A - LC)$, which is called the separation principle for the linear system (1.1.8). In other words, the matrices K and L can be chosen separately.

The observer design is a relatively independent topic in control theory. There are huge works attributed to observer design for nonlinear systems; see, for instance, the nonlinear observer with linearizable error dynamics in [87 and 88], the high-gain observer in [84], the sliding mode observer in [24, 26, and 130], the state observer for a system with uncertainty [22], and the high-gain finite-time observer in [103, 109, and 116]. For more details of the state observer we refer to recent monograph [14].

A breakthrough in observer design is the extended state observer, which was proposed by Han in the 1990s to be used not only to estimate the state but also the "total disturbance" that comes from unmodeled system dynamics, unknown coefficient of control and external disturbance. Actually, uncertainty is ubiquitous in a control system itself and the external environment, such as unmodeled system dynamics, external disturbance, and inaccuracy in control coefficient. The ubiquitous uncertainty in systems explains why the PID control technology is so popular in industry control because PID control is based mainly on the output error not on the systems' mathematical models. Since the ESO, the "total disturbance" and the state of the system are estimated simultaneously, we can design an output feedback control that is not critically reliant on the mathematical models. Let us start from an nth order SISO nonlinear control systems given by

$$\begin{cases} x^{(n)}(t) = f(t, x(t), \dot{x}(t), \ldots, x^{(n-1)}(t)) + w(t) + u(t), \\ y(t) = x(t), \end{cases}$$

which can be rewritten as

$$\begin{cases} \dot{x}_1(t) = x_2(t), \\ \dot{x}_2(t) = x_3(t), \\ \quad \vdots \\ \dot{x}_n(t) = f(t, x_1(t), x_2(t), \ldots, x_n(t)) + w(t) + u(t), \\ y(t) = x_1(t), \end{cases} \tag{1.1.11}$$

where $u(t) \in C(\mathbb{R}, \mathbb{R})$ is the control (input), $y(t)$ is the output (measurement), $f \in C(\mathbb{R}^n, \mathbb{R})$ is the system function, which is possibly unknown, and $w \in C(\mathbb{R}, \mathbb{R})$ is unknown external disturbance; $f(\cdot, t) + w(t)$ is called the "total disturbance" or "extended state" and

$\alpha_i \in \mathbb{R}, i = 1, 2, \ldots, n+1$ are the tuning parameters. The ESO designed in [60] is as follows:

$$\begin{cases} \dot{\hat{x}}_1(t) = \hat{x}_2(t) - \alpha_1 g_1(\hat{x}_1(t) - y(t)), \\ \dot{\hat{x}}_2(t) = \hat{x}_3(t) - \alpha_2 g_2(\hat{x}_1(t) - y(t)), \\ \quad \vdots \\ \dot{\hat{x}}_n(t) = \hat{x}_{n+1}(t) - \alpha_n g_n(\hat{x}_1(t) - y(t)) + u(t), \\ \dot{\hat{x}}_{n+1}(t) = -\alpha_{n+1} g_{n+1}(\hat{x}_1(t) - y(t)). \end{cases} \quad (1.1.12)$$

By appropriately choosing the nonlinear functions $g_i \in C(\mathbb{R}, \mathbb{R})$ and tuning the parameters α_i, we expect that the states $\hat{x}_i(t), i = 1, 2, \ldots, n+1$ of the ESO (1.1.12) can approximately recover the states $x_i(t), i = 1, 2, \ldots, n$ and the extended state $f(\cdot, t) + w(t)$, that is,

$$\hat{x}_i(t) \approx x_i(t), i = 1, 2, \ldots, n, \hat{x}_{n+1}(t) \approx f(\cdot, t) + w(t).$$

In Chapter 3, we have a principle of choosing the nonlinear functions $g_i(\cdot)$ and tuning the gain parameters α_i. The convergence of the ESO is established. We also present some numerical results to show visually the estimations of state and extended state. In particular, if the functions $g_i(\cdot)$ in (1.1.12) are linear, the ESO is referred to as the linear extended state observer (LESO). The LESO is also called the extended high-gain observer in [35].

The final key part of the ADRC is the TD and the ESO-based feedback control. In the feedback loop, a key component is to compensate (cancel) the "total disturbance" by making use of its estimate obtained from the ESO. The topology of the active disturbance rejection control is blocked in Figure 1.1.3.

Now we can describe the whole picture of the ADRC for a control system with vast uncertainty that includes the external disturbance and unmodeled dynamics. The control purpose is to design an output feedback control law that drives the output of the system to track a given reference signal $v(t)$. Generally speaking, the derivatives of the reference $v(t)$ cannot be measured accurately due to noise. The first step of the ADRC is to design a tracking differentiator

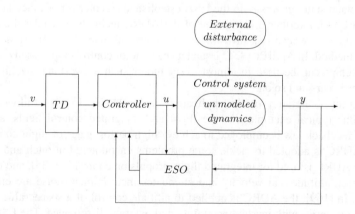

Figure 1.1.3 Topology of active disturbance rejection control.

(TD) to recover the derivatives of $v(t)$ without magnifying measured noise. Tracking differentiator also serves as transient profile for output tracking. The second step is to estimate, through the ESO, the system state and the "total disturbance" in real time by the input and output of the original system. The last step is to design an ESO-based feedback control that is used to compensate the "total disturbance" and track the estimated derivatives of $v(t)$. The whole ADRC design process and convergence are analyzed in Chapter 4.

The distinctive feature of the ADRC lies in its estimation/cancelation nature. In control theory, most approaches like high-gain control (HGC) and sliding mode control (SMC) are based on the worst case scenario, but there are some approaches that use the same idea of the ADRC to deal with the uncertainty. One popular approach is the internal model principle (IMP) [33, 34, 77, 99] and a less popular approach is the external model principle (EMP) [66, 104, 129, 149]. In the internal model principle and external model principle, the dynamic of the system is exactly known and the "external disturbance" is considered as a signal generated by the exogenous system, which follows exactly known dynamics. The unknown parts are initial states. However, in some complicated environments, it is very difficult to obtain the exact mathematical model of the exogenous system, which generates "external disturbance". In the ADRC configuration, we do not need a mathematical model of external disturbance and even most parts of the mathematical model of the control system itself can be unknown. This is discussed in Section 4.5.

The systems dealt with by the ADRC can also be coped with by high-gain control [128] and sometimes by sliding mode control [94, 130, 131]. However, control law by these approaches is designed in the worst case of uncertainty, which may cause unnecessary energy waste and may even be unrealizable in many engineering practices. In Section 4.6, three control methods are compared numerically by a simple example.

1.2 Overview of Engineering Applications

Nowadays, the ADRC is widely used in many engineering practices. It is reported in [166] that the ADRC control has been tested in the Parker Hannifin Parflex hose extrusion plant and across multiple production lines for over eight months. The product performance capability index (Cpk) is improved by 30% and the energy consumption is reduced by over 50%.

The Cleveland state university in the USA established a center for advanced control technologies (CACT) for further investigation of the ADRC technology. Under the cooperation of CACT and an American risk investment, the industrial giant Texas Instruments (TI) has adopted this method. In April 2013, TI issued its new motor control chips based on the ADRC. The control chips can be used in almost every motor such as washing machines, medical devices, electric cars and so on.

There is a lot of literature on the application of the ADRC. In what follows, we briefly overview some typical examples. In the flight and integrated control fields, an ESO and non-smooth feedback law is employed to achieve high performance of flight control [72]. In [126], the ADRC is adopted to tackle some problems encountered in pitch and roll altitude control. The ADRC is used for integrated flight-propulsion control in [135], and the coupling effects between altitude and velocity and attenuates measurement noise are eliminated by this method. In [169], the ADRC is applied to altitude control of a spacecraft model that is nonlinear in dynamics with inertia uncertainty and external disturbance. The ESO is applied to estimate the disturbance and the sliding mode control is designed based on the ESO to achieve the control purpose. The safe landing of unmanned aerial vehicles (UAVs) under various wind conditions has been a challenging task for decades. In [143], by using the ADRC method, an auto-landing control system consisting of a throttle control subsystem and an altitude control subsystem has been designed. It is indicated that this method can estimate directly in real time the UAV's internal and external disturbances and then compensate in the feedback. The simulation results show that this auto-landing control system can land the UAV safely under wide range wind disturbances (e.g., wind turbulence, wind shear). The application of the ADRC on this aspect can be found in monograph [139].

In the energy conversion and power plant control fields, [28] presents a controller for maximum wind energy capture of a wind power system by employing the ADRC method. The uncertainties in the torque of turbine and friction are both considered as an unknown disturbance to the system. The ESO is used to estimate the unknown disturbance. The maximum energy capture is achieved through the design of a tracking-differentiator. It is pointed out that this method has the merits of feasibility, adaptability, and robustness compared to the other methods. The paper [102] summarizes some methods for capturing the largest wind energy. It is indicated that the ADRC method captures the largest wind energy. The ADRC is used for a thermal power plant, which is characterized by nonlinearity, changing parameters, unknown disturbances, large time-delays, large inertia, and highly coupled dynamics among various control loops in [167]. In [121], the ADRC method is developed to cope with the highly nonlinear dynamics of the converter and the disturbances. The ADRC method is used for a thermal power generation unit in [69]. It is reported that the real-time dynamic linearization is implemented by disturbance estimation via the ESO and disturbance compensation via the control law, instead of differential geometry-based feedback linearization and direct feedback linearization theory, which need an accurate mathematical model of the plant. The decoupling for an MIMO coordinated system of boiler–turbine unit is also easily implemented by employing the ADRC. The simulation results on STAR-90 show that the ADRC coordinated control scheme can effectively solve problems of strong nonlinearity, uncertainty, coupling, and large time delays. It can also significantly improve the control performance of a coordinated control system. To eliminate the total disturbance effect on the active power filter (APF) performance, the ADRC is adopted in [95]. It is reported that the ADRC control has the merits of strong robustness, stability, and adaptability in dealing with the internal perturbation and external disturbance. In [151], the ADRC is used to regulate the frequency error for a three-area interconnected power system. As the interconnected power system transmits the power from one area to another, the system frequency will inevitably deviate from a scheduled frequency, resulting in a frequency error. A control system is essential to correct the deviation in the presence of external disturbances and structural uncertainties to ensure the safe and smooth operation of the power system. It is reported in [151] that the ADRC can extract the information of the disturbance from input and output data of the system and actively compensate for the disturbance in real time. Considering the difficulty of developing an accurate mathematical model for active power filters (APF), [168] uses the ADRC to parallel APF systems. It is reported that the analog signal detected in the ADRC controller is less than other control strategies. In [27], the ADRC is applied to an electrical power-assist steering system (EPAS) in automobiles to reduce the steering torque exerted by a driver so as to achieve good steering feel in the presence of external disturbances and system uncertainties. With the proposed ADRC, the driver can turn the steering wheel with the desired steering torque, which is independent of load torques, and tends to vary, depending on driving conditions.

As to motor and vehicle control, in [127], the ADRC is used to ensure high dynamic performance of a magnet synchronous motor (PMSM) servo system. It is concluded that the proposed topology produces better dynamic performance, such as smaller overshoot and faster transient time, than the conventional PID controller in its overall operating conditions. A matrix converter (MC) is superior to a drive induction motor since it has more attractive advantages than a conventional pulse width modulation (PWM) inverter such as the absence of a large dc-link capacitor, unity input power factor, and bidirectional power flow. However, due to the direct conversion characteristic of an MC, the drive performance of an induction motor is easily influenced by input voltage disturbances of the MC, and the stability of an induction motor drive system fed by an MC would be affected by a sudden change of load as well. In [105], the ADRC is applied to the MC fed induction motor drive system to solve the problems successfully. In [31], the ADRC is developed to ensure high dynamic performance of induction motors. In [123], the ADRC is developed to implement high-precision motion control of permanent-magnet synchronous motors. Simulations and experimental results show that the ADRC achieves a better position response and is robust to parameter variation and load disturbance. Furthermore, the ADRC is designed directly in discrete time with a simple structure and fast computation, which makes it widely applicable to all other types of drives. In [96], an ESO-based controller is designed for the permanent-magnet synchronous motor speed-regulator, where the ESO is employed to estimate both the states and the disturbances simultaneously, so that the composite speed controller can have a corresponding part to compensate the disturbances. Lateral locomotion control is a key technology for intelligent vehicles and is significant to vehicle safety itself. In [115], the ADRC is used for the lateral locomotion control. Simulation results show that, within the large velocity scale, the ADRC controllers can assist the intelligent vehicle to accomplish smooth and high precision on lateral locomotion, as well as remaining robust to system parameter perturbations and disturbances. In [146], the ADRC is applied to the anti-lock braking system (ABS) with regenerative braking of electric vehicles. Simulation results indicate that this method can regulate the slip rate at expired value in all conditions and, at the same time, it can restore the kinetic energy of a vehicle to an electrical source. In [142], the ADRC is applied to the regenerative retarding of a vehicle equipped with a new energy recovery retarder. Considering the railway restriction and comfort requirement, the ADRC is applied to the operation curve tracking of the maglev train in [100].

There is also a lot of literature on the ADRC's application in ship control. In [113], the ADRC is applied to the ship tracking control by considering the strong nonlinearity, uncertainty, and typical underactuated properties, as well as the restraints of the rudder. The simulation results show that the designed controller can achieve high precision on ship tracking control and has strong robustness to ship parameter perturbations and environment disturbances. In [108], the ADRC is used on the ship's main engine for optimal control under unmatched uncertainty. The simulation results show that the controller has strong robustness to parameter perturbations of the ship and environmental disturbances.

In robot control [73], the ESO is used to estimate and compensate the nonlinear dynamics of the manipulator and the external disturbances for a complex robot systems motion control. [120] applies the ADRC to the lateral control of tracked robots on stairs. The simulation results show that this algorithm can keep the robot smooth and precise in lateral control and effectively overcome the disturbance. In [114], the ADRC is applied to the rock drill robot joint hydraulic drive system. The simulation results show that the ADRC controller has ideal robustness to the system parameters' disturbances and the large load disturbance and a rapid and smooth control process and high steady precise performances can be implemented.

As to gyroscopes, [162] applies the ADRC to control two vibrating axes (or modes) of vibrational MEMS gyroscopes in the presence of the mismatch of natural frequencies between two axes, mechanical-thermal noises, quadrature errors, and parameter variations. The simulation results on a Z-axis MEMS gyroscope show that the controller is very effective by driving the output of the drive axis to a desired trajectory, forcing the vibration of the sense axis to zero for a force-to-rebalance operation, and precisely estimating the rotation rate. In [29], the ADRC is used for both vibrating axes (drive and sense) of vibrational gyroscopes, in both simulation and hardware tests on a vibrational piezoelectric beam gyroscope. The proposed controller proves to be robust against structural uncertainties and it also facilitates accurate sensing of time-varying rotation rates. [154] uses the ADRC and fuzzy control method for stabilizing circuits in platform inertial navigation systems (INS) based on fiber optic gyroscopes (FOGs).

1.3 Preliminaries

In this section, we first present a canonical form of active disturbance rejection control (ADRC). To make the book self-contained, we also present some notation and results about Lyapunov stability, asymptotical stability, finite-time stability, and weighted homogeneity.

1.3.1 Canonical Form of ADRC

As pointed out in the previous section, the ADRC can deal with nonlinear systems with vast uncertainty. However, for the sake of clarity, we first limit ourselves to a class of nonlinear systems that are canonical forms of the ADRC. Let us start with some engineering control systems.

Firstly, we consider micro-electro-mechanical systems (MEMS). The mechanical structure of the MEMS gyroscope can be understood as a proof mass attached to a rigid frame by springs and dampers, as shown in Figure 1.3.1. As the mass is driven to resonance along the drive (X) axis and the rigid frame is rotating along the rotation axis, a Coriolis acceleration will be produced along the sense (Y) axis, which is perpendicular to both drive and rotation axes. The Coriolis acceleration is proportional to the amplitude of the output of the drive axis and the unknown rotation rate. Therefore, we can estimate the rotation rate through measuring the vibration of the sense axis. To measure accurately the rotation rate, the vibration magnitude of the drive axis has to be regulated to a fixed level. Therefore, the controller of the drive axis is mainly used to drive the drive axis to resonance and to regulate the output amplitude.

The vibrational MEMS gyroscope can be modeled as follows:

$$\begin{cases} \ddot{x}(t) + 2\zeta\omega_n^2 x(t) + \omega_{xy}y(t) - 2\Omega\dot{y}(t) = \dfrac{k}{m}u(t) + N_x(t), \\ \ddot{y}(t) + 2\zeta\omega_n\dot{y}(t) + \omega_{xy}x(t) + 2\Omega\dot{x}(t) = N_y(t), \end{cases} \tag{1.3.1}$$

where $x(t)$ and $y(t)$ are the outputs of the drive and sense axes, $2\Omega\dot{x}(t)$ and $2\Omega\dot{y}(t)$ are the Coriolis accelerations, Ω is the rotation rate, ω_n is the natural frequency of the drive and sense axes, $\omega_{xy}y(t)$ and $\omega_{xy}x(t)$ are quadrature errors caused by spring couplings between two axes,

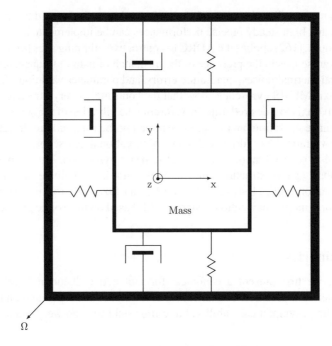

Figure 1.3.1   Mass–spring–damper structure of an MEMS gyroscope system.

ζ is the damping coefficient, m is the mass of the MEMS gyroscope, k is the control gain, and $u(t)$ is the control input for the drive axis. The $N_x(t)$ and $N_y(t)$ are external disturbances. We can rewrite system (1.3.1) as

$$\begin{cases} \dot{x}_1(t) = x_2(t), \\ \dot{x}_2(t) = f(x_1(t), x_2(t), Y(t), N_x(t)) + bu(t), \\ \dot{Y}(t) = F_0(x_1(t), x_2(t), Y(t), N_y(t)), \end{cases} \qquad (1.3.2)$$

where $x_1(t) = x(t)$, $x_2(t) = \dot{x}(t)$, $Y(t) = (y(t), \dot{y}(t))^\top$,

$$f(x_1(t), x_2(t), Y(t), N_x(t)) = -2\zeta\omega_n^2 x_1(t) - \omega_{xy}y(t) + 2\Omega\dot{y}(t) + N_x(t),$$

and

$$F_0(x_1(t), x_2(t), Y(t), N_y(t)) = \begin{pmatrix} 0 \\ -2\zeta\omega_n\dot{y}(t) - \omega_{xy}x(t) - 2\Omega\dot{x}(t) + N_y(t) \end{pmatrix}.$$

Obviously, both the nonlinear functions $f(\cdot)$ and $F_0(\cdot)$ contain external disturbances. However, the external disturbances, $f(\cdot)$ and $F(\cdot)$, cannot always be accurately measured due to the possible deviation of parameters ζ, Ω, ω_n, and ω_{xy} away from their real values.

Next, consider an hydraulic system where an inertia load is driven by a servo-valve-controlled hydraulic rotary actuator. A schematic structure is presented on the right of Figure 1.3.2. The objective is to drive the inertia load to track a given smooth motion trajectory by the position

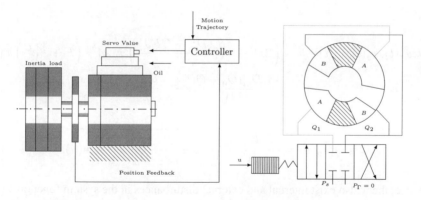

Figure 1.3.2   Architecture of the hydraulic system.

measurement. The motion dynamics of the inertia load can be described by the following equation:

$$J\ddot{x}(t) = P_L(t)D_m - F\dot{x}(t), \tag{1.3.3}$$

where J and $x(t)$ represent the moment of inertia and the angular displacement of the load, respectively, D_m is the radian displacement of the actuator, F represents the friction coefficient, $P_L(t) = P_1(t) - P_2(t)$ is the load pressure of the hydraulic actuator, and $P_1(t)$ and $P_2(t)$ are the pressures inside the two chambers of the actuator. The dynamics of load pressure can be written as

$$\frac{V_t}{4\beta_e}\dot{P}_L(t) = -D_m\dot{x}(t) - C_tP_L(t) + Q_0 + Q(t) + Q_L(t), \tag{1.3.4}$$

where V_t is the total control volume of the actuator, β_e is the effective oil bulk modulus, C_t is the coefficient of the total internal leakage of the actuator due to pressure, Q_0 is a constant modeling error and $Q(t)$ is the time-varying modeling error caused by complicated internal leakage, parameter deviations, unmodeled pressure dynamics, modeling error caused by the following flow equation, and so on, $Q_L(t) = (Q_1(t) + Q_2(t))/2$ is the load flow, Q_1 is the supplied flow rate to the forward chamber, and Q_2 is the return flow rate of the return chamber. $Q_L(t)$ is related to the spool valve displacement of the servovalve x_v by

$$Q_L(t) = k_qx_v(t)\sqrt{P_s - \text{sign}(x_v(t))P_L(t)}, \quad k_q = C_d\omega\sqrt{1/\rho}, \tag{1.3.5}$$

where C_d is the discharge coefficient, ω is the spool valve area gradient, ρ is the density of oil, and P_s is the supply pressure of the fluid with respect to the return pressure P_r. The control applied to the servovalve is directly proportional to the spool position, that is, $x_v(t) = k_iu(t)$, where k_i is a positive constant; $u(t)$ is the control input voltage.

Let $x_1(t) = x(t)$, $x_2(t) = \dot{x}(t)$, and $x_3(t) = \frac{D_m}{J}P_L(t) - F_2(x) + f(t, x_1(t), x_2(t))$. Then the control system can be rewritten as

$$\begin{cases} \dot{x}_1(t) = x_2(t), \\ \dot{x}_2(t) = x_3(t), \\ \dot{x}_3(t) = \varphi(t, x_1(t), x_2(t), x_3(t)) + b(x_2(t), x_3(t), u(t))u(t), \end{cases} \tag{1.3.6}$$

where

$$
\begin{aligned}
\varphi(t, x_1(t), x_2(t), x_3(t)) = {} & -\left(2F^2 + \frac{4\beta_e(D_m^2 + CFJ)}{JV}\right)x_2(t) - \left(F + \frac{4\beta_e C}{V}\right)x_3(t) \\
& + \frac{4\beta_e D_m(Q_0 + Q(t))}{JV}
\end{aligned}
$$

(1.3.7)

and

$$
b(x_2, x_3, u) = \frac{4\beta_e K_q D_m}{JV}\sqrt{P_S - \frac{J(Fx_2 + x_3)}{D_m}}\,\mathrm{sign}(u).
$$

(1.3.8)

In practice, there also exist internal and external disturbances in the system function $\varphi(\cdot)$.

Finally, we consider the dynamic of autonomous underwater vehicles (AUVs). The AUV can be modeled as follows:

$$
\begin{cases}
\dot{\mathbf{x}}(t) = J(\mathbf{x}(t))\mathbf{v}(t), \\
M\dot{\mathbf{v}}(t) + C(\mathbf{v}(t))\mathbf{v}(t) + D(\mathbf{v}(t))\mathbf{v}(t) + \mathbf{d}(t) = \mathbf{u}(t), \\
\mathbf{y}(t) = \mathbf{x}(t),
\end{cases}
$$

(1.3.9)

where

$$
\mathbf{x}(t) = (x(t), y(t), z(t), \phi(t), \theta(t), \psi(t))^\top
$$

(1.3.10)

is the vehicle location and orientation in the earth-fixed frame, $\mathbf{v}(t)$ is the vector of the vehicle's velocities expressed in the body-fixed frame, and $\mathbf{y}(t)$ is the output. The positive definite inertia matrix $M = M_{RB} + M_A$ includes the inertia M_{RB} of the vehicle as a rigid body and the added inertia M_A due to the acceleration of the wave. The matrix $C(\mathbf{v}) \in \mathbb{R}^{6\times6}$, which is skew-symmetrical groups of the coriolis and centripetal force. The hydrodynamic damping term $D(\mathbf{v}) \in \mathbb{R}^{6\times6}$ takes into account the dissipation due to the friction exerted by the fluid surrounding the AUV. The vector $g(\mathbf{x}) \in \mathbb{R}^6$ is the combined gravitation and buoyancy forces in the body-fixed frame, $\mathbf{d}(t) \in \mathbb{R}^6$ is the external disturbance, and $J(\mathbf{x})$ is the kinematic transformation matrix expressing the transformation from the body-fixed frame to the earth-fixed frame:

$$
J(\mathbf{x}) = \begin{pmatrix} J_1(\mathbf{x}) & 0 \\ 0 & J_2(\mathbf{x}) \end{pmatrix}
$$

(1.3.11)

with

$$
J_1(\mathbf{x}) = \begin{pmatrix} \cos\psi\cos\theta & -\sin\psi\cos\theta & \sin\psi\sin\phi + \cos\psi\cos\phi\sin\theta \\ \sin\psi\cos\theta & \cos\psi + \sin\phi\sin\theta\sin\psi & -\cos\psi\sin\phi + \sin\theta\sin\psi\cos\phi \\ -\sin\theta & \sin\phi\cos\theta & \cos\phi\cos\theta \end{pmatrix},
$$

$$
J_2(\mathbf{x}) = \begin{pmatrix} 1 & \sin\phi\tan\theta & \cos\phi\tan\theta \\ 0 & \cos\phi & -\sin\phi \\ 0 & \dfrac{\sin\phi}{\cos\theta} & \dfrac{\cos\phi}{\cos\theta} \end{pmatrix}.
$$

(1.3.12)

The control purpose is to make the output $\mathbf{y}(t)$ track the desired trajectory $\mathbf{x}_d(t)$. Let $\mathbf{x}_1(t) = \mathbf{x}(t)$ and $\mathbf{x}_2(t) = J(\mathbf{x}(t))\mathbf{v}(t)$. Then the dynamics (1.3.9) can be written as

$$\begin{cases} \dot{\mathbf{x}}_1(t) = \mathbf{x}_2(t), \\ \dot{\mathbf{x}}_2(t) = F(\mathbf{x}_1(t), \mathbf{x}_2(t), d(t)) + J(\mathbf{x}_1(t))M^{-1}\mathbf{u}(t), \end{cases} \tag{1.3.13}$$

where

$$F(\mathbf{x}_1, \mathbf{x}_2, \mathbf{d}) = -J(\mathbf{x}_1)M^{-1}C(J^{-1}(\mathbf{x}_1)\mathbf{x}_2)J^{-1}(\mathbf{x}_1)\mathbf{x}_2$$

$$- J(\mathbf{x}_1)M^{-1}D(J^{-1}(\mathbf{x}_1)\mathbf{x}_2)J^{-1}(\mathbf{x}_1)\mathbf{x}_2 - J(\mathbf{x}_1)M^{-1}\mathbf{d}, \tag{1.3.14}$$

and there are external disturbance and parameter uncertainty in $F(\cdot)$.

It is seen that all these systems, MEMS gyroscope (1.3.2), hydraulic system (1.3.6), and (1.3.13), are the special cases of the following nonlinear systems with vast uncertainty:

$$\begin{cases} \dot{x}_{i1}(t) = x_{i2}(t), \\ \dot{x}_{i2}(t) = x_{i3}(t), \\ \quad \vdots \\ \dot{x}_{in_i}(t) = f_i(x(t), \zeta(t), w(t)) + b(x(t), \zeta(t), w(t))u_i(t), \\ \dot{\zeta}(t) = f_{i0}(x(t), \zeta(t), w(t)), \quad i = 1, 2, \ldots, r, \end{cases} \tag{1.3.15}$$

where $(x^\top(t), \zeta^\top(t))^\top = ((x_{11}(t), \ldots, x_{1n_1}(t), x_{21}(t), \ldots, x_{rn_r}(t))^\top, \zeta^\top(t)) \in \mathbb{R}^{nm+l}$ is the system state, $y(t) = (x_{11}(t), \ldots, x_{r1}(t))^\top \in \mathbb{R}^r$ is the output (measurement), $u(t) = (u_1(t), \ldots, u_r(t))^\top \in \mathbb{R}^r$ is the input (control), and $w(t) \in \mathbb{R}^k$ is the external disturbance. The system functions $f_i \in C(\mathbb{R}^{n_1+\cdots+n_r+l+k}, \mathbb{R})$ and $f_{i0} \in C(\mathbb{R}^{n_1+\cdots+n_r+l+k}, \mathbb{R}^k)$ are completely unknown or partially unknown. In addition, some uncertainties are allowed in functions $b_i \in C(\mathbb{R}^{n_1+\cdots+n_r+l+k}, \mathbb{R})$. In fact, except for the above examples, there are many other control systems that can be modeled as (1.3.15). In this book, we consider system (1.3.15) as the control canonical form of ADRC.

To discuss further the canonical form of ADRC, we introduce some background about linear MIMO systems as follows:

$$\begin{cases} \dot{x}(t) = Ax(t) + Bu(t), \quad x(0) = x_0, \\ y(t) = Cx(t), \end{cases} \tag{1.3.16}$$

where $x(t) \in \mathbb{R}^n$ is the state, x_0 is the initial state, $u(t) \in \mathbb{R}^m$ is the input (control), $y(t) \in \mathbb{R}^p$ is the output (measurement), $A \in \mathbb{R}^{n \times n}$ is the system matrix, $B \in \mathbb{R}^{n \times m}$ is the control matrix, and $C \in \mathbb{R}^{p \times n}$ is the output matrix.

The concept of relative degree is useful for understanding the control structure of system (1.3.16).

Definition 1.3.1 *For system (1.3.16), let*

$$d_i = \begin{cases} \mu_i, & C_i A^k B = 0, \quad k = 0, 1, \ldots, \mu_i - 2(\mu_i \leq n), \quad C_i A^{\mu_i-1} B \neq 0, \\ n-1, & C_i A^k B = 0, \quad k = 0, 1, \ldots, n-1, \end{cases}$$

where C_i is the ith row of C, $i = 1, 2, \ldots, p$. Then $\{d_1, d_2, \ldots, d_p\}$ is called the relative degree of system (1.3.16) (or triple (A, B, C)).

Let

$$z_i(t) = \begin{pmatrix} z_{i1}(t) \\ z_{i2}(t) \\ \vdots \\ z_{id_i}(t) \end{pmatrix} = E_i x(t) = \begin{pmatrix} C_i x(t) \\ C_i A x(t) \\ \vdots \\ C_i A^{d_i - 1} x(t) \end{pmatrix}, \quad i = 1, 2, \ldots, m, \quad (1.3.17)$$

and assume that the following matrix E is full rank matrix, that is, $\mathrm{rank}(E) = d_1 + d_2 + \cdots + d_m$,

$$E = (E_1, E_2, \ldots, E_m)^\top. \quad (1.3.18)$$

Then there exists matrix $F \in \mathbb{R}^{s \times n}$ with rank $s = n - d_1 - \cdots - d_m$ such that $(E, F)^\top$ is invertible. Let $z(t) = (z_1(t), \ldots, z_m(t))^\top$ with

$$\begin{pmatrix} z_1(t) \\ z_2(t) \\ \vdots \\ z_m(t) \\ \zeta(t) \end{pmatrix} = Tx(t) = \begin{pmatrix} E \\ F \end{pmatrix} x(t). \quad (1.3.19)$$

It is obvious that the above transformation is invertible and under this transformation,

$$\begin{cases} \dot{z}_{i1}(t) = z_{i2}(t), \\ \dot{z}_{i2}(t) = z_{i3}(t), \\ \qquad \vdots \\ \dot{z}_{id_i}(t) = c_i A^{d_i} T^{-1}(z(t), \zeta(t))^\top + C_i A^{d_i} B u(t), i = 1, 2, \ldots, m, \\ \dot{\zeta}(t) = AFT^{-1}(z(t), \zeta(t))^\top + FBu(t). \end{cases} \quad (1.3.20)$$

Furthermore, if $FB = 0$, then system (1.3.20) is a special case of (1.3.15).

The following nonlinear system can also be transformed into a special case of (1.3.15) by a geometric method under some conditions:

$$\begin{cases} \dot{x}(t) = f(x(t)) + \sum_{i=1}^m g_i(x(t)) u_i(t), \quad x(0) = x_0 \in \mathbb{R}^n, \\ y(t) = (y_1(t), y_2(t), \ldots, y_m(t))^\top = h(x(t)), \end{cases} \quad (1.3.21)$$

where $x(t) \in \mathbb{R}^n$ is the system state, $u(t) = (u_1(t), u_2(t), \ldots, u_m(t))^\top \in \mathbb{R}^m$ is the control input, and $y(t) \in \mathbb{R}^m$ is the output, $f \in C(\mathbb{R}^n, \mathbb{R}^n)$ is the system function, $g_i \in C(\mathbb{R}^n, \mathbb{R}^n)$ $(i = 1, 2, \ldots, m)$ are control functions.

Now, we introduce the Lie derivative and Lie bracket in geometry.

Definition 1.3.2 *Suppose that* $h(x) = (h_1(x), h_2(x), \ldots, h_m(x))^\top \in C^1(\mathbb{R}^n, \mathbb{R}^m)$, $f(x) = (f_1(x), f_2(x), \ldots, f_n(x))^\top \in C(\mathbb{R}^n, \mathbb{R}^n)$. *The Lie derivative* $L_f h(x) : \mathbb{R}^n \to \mathbb{R}$ *of function* $h(x)$ *along vector field* $f(x)$ *is defined as*

$$L_f h(x) = \left(\frac{\partial h(x)}{\partial x_1}, \frac{\partial h(x)}{\partial x_2}, \ldots, \frac{\partial h(x)}{\partial x_n} \right) \begin{pmatrix} f_1(x) \\ f_2(x) \\ \vdots \\ f_n(x) \end{pmatrix} = \sum_{i=1}^{n} \frac{\partial h(x)}{\partial x_i} f_i(x). \qquad (1.3.22)$$

If $L_f h \in C^1(\mathbb{R}^n, \mathbb{R})$, *then the Lie derivative of* $L_f h(x)$ *along the vector field* $f(x)$ *is denoted by* $L_f^2 h(x)$ *that is,* $L_f^2 h(x) = (L_f(L_f h))(x)$. *Generally, we denote* $L_f^0 h(x) = h(x)$ *and* $L_f^i h(x) = L_f(L_f^{(i-1)} h(x))$, $i = 1, 2, \ldots, n$. *Similarly,* $L_g L_f h(x)$ *is the symbol of* $L_g(L_f h(x))$, *where* $g(x) = (g_1(x), g_2(x), \ldots, g_n(x))^\top \in C^1(\mathbb{R}^n, \mathbb{R}^n)$ *is another vector field.*

The Lie bracket of vector fields $f(x)$ *and* $g(x)$ *is a vector field denoted by* $[f, g](x)$ *given as*

$$[f, g](x) \triangleq \begin{pmatrix} \frac{\partial g_1(x)}{\partial x_1} & \frac{\partial g_1(x)}{\partial x_2} & \cdots & \frac{\partial g_1(x)}{\partial x_n} \\ \frac{\partial g_2(x)}{\partial x_1} & \frac{\partial g_2(x)}{\partial x_2} & \cdots & \frac{\partial g_2(x)}{\partial x_n} \\ \vdots & \vdots & \ddots & \vdots \\ \frac{\partial g_n(x)}{\partial x_1} & \frac{\partial g_n(x)}{\partial x_2} & \cdots & \frac{\partial g_n(x)}{\partial x_n} \end{pmatrix} \begin{pmatrix} f_1(x) \\ f_2(x) \\ \vdots \\ f_n(x) \end{pmatrix}$$

$$- \begin{pmatrix} \frac{\partial f_1(x)}{\partial x_1} & \frac{\partial f_1(x)}{\partial x_2} & \cdots & \frac{\partial f_1(x)}{\partial x_n} \\ \frac{\partial f_2(x)}{\partial x_1} & \frac{\partial f_2(x)}{\partial x_2} & \cdots & \frac{\partial f_2(x)}{\partial x_n} \\ \vdots & \vdots & \ddots & \vdots \\ \frac{\partial f_n(x)}{\partial x_1} & \frac{\partial f_n(x)}{\partial x_2} & \cdots & \frac{\partial f_n(x)}{\partial x_n} \end{pmatrix} \begin{pmatrix} g_1(x) \\ g_2(x) \\ \vdots \\ g_n(x) \end{pmatrix}. \qquad (1.3.23)$$

Generally, we denote $ad_f^k g(x) \triangleq [f, ad_f^{k-1} g](x)$ *and* $ad_f^0 g(x) = g(x)$.

For the Lie derivatives and Lie brackets, we have the following basic properties.

Lemma 1.3.1 *For the vector fields* $f, g \in C^1(\mathbb{R}^n, \mathbb{R}^n)$ *and functions* $\alpha, \beta \in C(\mathbb{R}^n, \mathbb{R})$, $\lambda \in C^1(\mathbb{R}^n, \mathbb{R})$, *the following conclusions hold true.*

(i) $L_{\alpha f} \lambda(x) = \alpha(x) L_f \lambda(x)$.
(ii) If $\alpha, \beta \in C^1(\mathbb{R}^n, \mathbb{R})$, *then*

$$[\alpha f, \beta g](x) = \alpha(x)\beta(x)[f, g](x) + \alpha(x)(L_f \beta(x))g(x) - \beta(x)(L_g \alpha(x))f(x).$$

(iii) $L_{[f,g]} \lambda(x) = L_f L_g \lambda(x) - L_g L_f \lambda(x)$.

For the given smooth vector fields $f_i \in C^1(\mathbb{R}^n, \mathbb{R}^n)$, $i = 1, 2, \ldots, d$, the vector space (depending on x) spanned by $f_1(x), f_2(x), \ldots, f_d(x)$ is called the distribution of vector fields $f_i(x)$, $i = 1, 2, \ldots, d$. We use the symbol $\Delta(x)$ to denote the distribution, that is,

$$\Delta(x) = \text{span}\{f_1(x), f_2(x), \ldots, f_d(x)\}. \tag{1.3.24}$$

The distribution $\Delta(x)$ is called involutive if the Lie bracket $[f_i, f_j](x)$ of any pair of vector fields $f_i(\cdot)$ and $f_j(\cdot)$ is a vector field that belongs to $\Delta(x)$, that is, there exist functions $a_k \in C(\mathbb{R}^n, \mathbb{R})$, $k = 1, 2, \ldots, d$, such that

$$[f_i, g_j](x) = \sum_{i=1}^{n} a_k(x) f_k(x). \tag{1.3.25}$$

In order to transform system (1.3.21) into the canonical form, we introduce the Frobenius theorem.

Suppose that one is interested in solving the following differential equation:

$$\begin{cases} L_{f_1}\varphi(x) = f_{11}(x)\dfrac{\partial\varphi(x)}{\partial x_1} + f_{12}(x)\dfrac{\partial\varphi(x)}{\partial x_2} + \cdots + f_{1n}(x)\dfrac{\partial\varphi(x)}{\partial x_n} = 0, \\[2mm] L_{f_2}\varphi(x) = f_{21}(x)\dfrac{\partial\varphi(x)}{\partial x_1} + f_{22}(x)\dfrac{\partial\varphi(x)}{\partial x_2} + \cdots + f_{2n}(x)\dfrac{\partial\varphi(x)}{\partial x_n} = 0, \\[2mm] \quad\vdots \\[2mm] L_{f_d}\varphi(x) = f_{d1}(x)\dfrac{\partial\varphi(x)}{\partial x_1} + f_{d2}(x)\dfrac{\partial\varphi(x)}{\partial x_2} + \cdots + f_{dn}(x)\dfrac{\partial\varphi(x)}{\partial x_n} = 0, \end{cases} \tag{1.3.26}$$

where $f_1, f_2, \ldots, f_d \in C^1(U \subset \mathbb{R}^n, \mathbb{R}^n)$ are vector fields that span a distribution $\Delta(x)$ for integer $d < n$, and $f_{ij}(\cdot)$ is the jth component of vector field $f_i(\cdot)$. The system of partial differential equations (1.3.26) or the d-dimensional distribution $\Delta(\cdot)$ is said to be *completely integrable* if there exist $n - d$ independent smooth functions $\varphi_i \in C^1(\mathbb{R}^n, \mathbb{R})$, $i = 1, 2, \ldots, n - d$, satisfying differential equations (1.3.26) on U. By "independent", we mean that the row vector group composed by gradients $\nabla\varphi_1(x), \nabla\varphi_2(x), \ldots, \nabla\varphi_{n-d}(x)$ are independent at every $x \in U$.

Lemma 1.3.2 *A distribution is completely integrable if and only if it is involutive.*

Now we give the definition of relative degree for nonlinear systems (1.3.21).

Definition 1.3.3 *Let $U \subset \mathbb{R}^n$ be a neighborhood near the initial state of system (1.3.21). If there exist positive integers r_i, $i = 1, 2, \ldots, m$ such that*

$$L_{g_j} L_f^k h_i(x) = 0 \quad \forall\ x \in U, \quad 0 \leq k \leq r_i - 2, \ i = 1, 2, \ldots, m, j = 1, 2, \ldots, m, \tag{1.3.27}$$

and the following matrix function $A(x)$ is invertible at x_0, then we say that system (1.3.21) has the relative degree $\{r_1, r_2, \ldots, r_m\}$ at initial state x_0:

$$A(x) = \begin{pmatrix} L_{g_1} L_f^{r_1-1} h_1(x) & L_{g_2} L_f^{r_1-1} h_1(x) & \cdots & L_{g_m} L_f^{r_1-1} h_1(x) \\ L_{g_1} L_f^{r_2-1} h_2(x) & L_{g_2} L_f^{r_2-1} h_2(x) & \cdots & L_{g_m} L_f^{r_2-1} h_2(x) \\ \vdots & \vdots & \ddots & \vdots \\ L_{g_1} L_f^{r_m-1} h_m(x) & L_{g_2} L_f^{r_m-1} h_m(x) & \cdots & L_{g_m} L_f^{r_m-1} h_m(x) \end{pmatrix}. \qquad (1.3.28)$$

Lemma 1.3.3 *Suppose that system (1.3.21) has a (vector) relative degree $\{r_1, \ldots, r_m\}$ at the initial state x_0. Then*

$$r_1 + r_2 + \cdots + r_m \leq n.$$

Set, for $i = 1, 2, \ldots, m$,

$$\begin{cases} \xi_{i1}(x) = \phi_{i1}(x) = h_i(x), \\ \xi_{i2}(x) = \phi_{i2}(x) = L_f h_i(x), \\ \quad \vdots \\ \xi_{ir_i}(x) = L_f^{r_i-1} h_i(x). \end{cases} \qquad (1.3.29)$$

We assume without loss of generality that $r = r_1 + r_2 + \cdots + r_i < n$ and there exist $n - r$ functions $\phi_{r+1}, \phi_{r+2}, \ldots, \phi_n \in C(\mathbb{R}^n, \mathbb{R})$ such that the mapping

$$\Phi(x) = (\phi_{11}(x), \phi_{12}(x), \ldots, \phi_{1r_i}(x), \ldots, \phi_{mr_m}(x), \phi_{r+1}(x), \ldots, \phi_n(x))^\top$$

has a Jacobian matrix that is nonsingular at x_0. Moreover, if the distribution

$$G(x) = \operatorname{span}\{g_1(x), g_2(x), \ldots, g_m(x)\}$$

is involutive near x_0, then $\phi_{r+1}(x), \ldots, \phi_n(x)$ satisfy

$$L_{g_j} \phi_i(x) = 0, \quad j = 1, 2, \ldots, m, \quad j = r + 1, \ldots, n, \quad x \in U,$$

where $U \subset \mathbb{R}^n$ is a neighborhood of initial state x_0.

Set

$$\xi_i(x) = \begin{pmatrix} \xi_{i1}(x) \\ \xi_{i2}(x) \\ \vdots \\ \xi_{ir_i}(x) \end{pmatrix} = \begin{pmatrix} \phi_{i1}(x) \\ \phi_{i2}(x) \\ \vdots \\ \phi_{ir_i}(x) \end{pmatrix}, \quad i = 1, 2 \ldots, m, \qquad (1.3.30)$$

$$\xi(x) = (\xi_1(x), \xi_2(x), \ldots, \xi_m(x))^\top, \qquad (1.3.31)$$

$$\eta(x) = \begin{pmatrix} \eta_1(x) \\ \eta_2(x) \\ \vdots \\ \eta_{n-r}(x) \end{pmatrix} = \begin{pmatrix} \phi_{r+1}(x) \\ \phi_{r+2}(x) \\ \vdots \\ \phi_n(x) \end{pmatrix}, \tag{1.3.32}$$

and

$$b_{ij}(\xi, \eta) = L_{g_i} L_f^{r_i-1} h_i(\Phi^{-1}(\xi, \eta)), \ \forall \ 1 \le i, j \le m,$$

$$\psi_i(\xi, \eta) = L_f^{r_i} h_i(\Phi^{-1}(\xi, \eta)), \ \forall \ 1 \le i \le m, \tag{1.3.33}$$

$$F_0(\xi, \eta) = (L_f \phi_{r+1}(\Phi^{-1}(\xi, \eta)), L_f \phi_{r+2}(\Phi^{-1}(\xi, \eta)), \dots, L_f \phi_n(\Phi^{-1}(\xi, \eta)))^{\top}.$$

Then the general nonlinear affine system (1.3.21) has the form

$$\begin{cases} \dot{\xi}_{i1}(t) = \xi_{i2}(t), \\ \dot{\xi}_{i2}(t) = \xi_{i3}(t), \\ \qquad \vdots \\ \dot{\xi}_{ir_i}(t) = \psi(\xi(t), \eta(t)) + \sum_{j=1}^{m} b_{ij}(\xi(t), \eta(t)) u_j(t), \\ \dot{\eta}(t) = F_0(\xi(t), \eta(t)), \\ y_i(t) = \xi_{i1}(t), \end{cases} \tag{1.3.34}$$

which is also a special case of the canonical form of ADRC (1.3.15).

1.3.2 Stability for Nonlinear Systems

In this section, we give some basic notation and results about stability for nonlinear systems. In this book, the stability means Lyapunov stability, which is named after Aleksandr Lyapunov, a Russian mathematician who published his doctoral thesis, *The General Problem of Stability of Motion*, in 1892. It was becoming more interests during the Cold War period when it was found to be applicable to the stability of aerospace guidance systems, which typically contains strong nonlinearities that are not treatable by other methods.

In this book, we use $\| \cdot \|$ to denote the Euclidian norm of $\mathbb{R}^n$: $\| (\nu_1, \nu_2, \dots, \nu_n) \| = \left(\sum_{i=1}^{n} |\nu_i|^2 \right)^{1/2}$ and $\| \cdot \|_{\infty}$ the infinite norm of $\mathbb{R}^n$: $\| (\nu_1, \dots, \nu_n) \|_{\infty} = \max_{i=1, \dots, n} |\nu_i|$. It is well known that the two norms are equivalent; however, for the simplicity, we use different norms in different circumstances.

Consider the nonlinear system of the following:

$$\dot{x}(t) = f(t, x(t)), \tag{1.3.35}$$

where $f(t, x) = (f_1(t, x), f_2(t, x), \dots, f_n(t, x))^{\top} \in C([0; \infty) \times \mathbb{R}^n, \mathbb{R}^n)$ with $f_i(t, x)$ being locally Lipschitz continuous with respect to x (i.e., $|f_i(t, x_1) - f_i(t, x_2)| \le L_t \|$

$x_1 - x_2 \parallel$, for some $L_t > 0$ and all $x_1, x_2 \in \mathbb{R}^n$) and $f_i(0) = 0$, $i = 1, 2, \ldots, n$. It is obvious that $x(t) \equiv 0$ is a trivial solution of system (1.3.35). The trivial solution is also said to be an equilibrium state of the system. To represent the dependence of a system solution with initial state, we denote, in this section, the solution of system (1.3.35) with the initial state $x(0) = x_0 \in \mathbb{R}^n$ as $x(t; x_0)$.

Definition 1.3.4 *If for any positive constant $\epsilon > 0$ there exists $\sigma > 0$ such that for any $x_0 \in \mathbb{R}^n$ satisfying $\parallel x_0 \parallel < \sigma$, the solution $x(t; x_0)$ of (1.3.35) satisfies $\parallel x(t; x_0) \parallel < \epsilon, \forall\ t \geq 0$, then the zero equilibrium of system (1.3.35) is said to be stable in the sense of Lyapunov.*

Definition 1.3.5 *A domain $\Omega \subset \mathbb{R}^n (0 \in \Omega^\circ)$ is said to be the attracting basin of the zero equilibrium state of (1.3.35), if for any $x_0 \in \Omega$, the solution with initial value x_0 tends to zero as time goes to ∞, that is, $\lim_{t \to \infty} \parallel x(t; x_0) \parallel = 0$, and for any $x \in \mathbb{R}^n \setminus \Omega$, $\lim_{t \to \infty} \parallel x(t; x_0) \parallel = 0$ is not valid any more. We say that the zero equilibrium of system (1.3.35) is attractive on Ω. Furthermore, if $\Omega = \mathbb{R}^n$, we say that the zero equilibrium of system (1.3.35) is globally attractive.*

Definition 1.3.6 *The zero equilibrium of system (1.3.35) is said to be asymptotically stable on attracting basin Ω if it is stable and attracting on Ω. If $\Omega = \mathbb{R}^n$, we say that the equilibrium is globally asymptotically stable.*

We point out that there is no implication relation between stability and attractiveness. For example, consider the following system:

$$\begin{cases} \dot{x}_1(t) = x_2(t), & x_1(0) = x_{10}, \\ \dot{x}_2(t) = x_1(t), & x_2(0) = x_{20}. \end{cases} \tag{1.3.36}$$

The solution of system (1.3.36) is

$$\begin{cases} x_1(t; x_{10}, x_{20}) = x_{10} \cos t - x_{20} \sin t, \\ x_2(t; x_{10}, x_{20}) = x_{10} \sin t + x_{20} \cos t. \end{cases} \tag{1.3.37}$$

A straightforward computation shows that

$$x_1^2(t; x_{10}, x_{20}) + x_2^2(t; x_{10}, x_{20}) = x_{10}^2 + x_{20}^2.$$

It is obvious that the zero equilibrium state of system (1.3.36) is stable, but not attractive. Also there exists an example where the zero equilibrium is attractive but not stable. Consider the following system:

$$\begin{cases} \dot{x}(t) = f(x(t)) + y(t), \\ \dot{y}(t) = -x(t), \end{cases} \qquad f(x) = \begin{cases} -4x, & x > 0; \\ 2x, & -1 \leq x \leq 0; \\ -x - 3, & x \leq -1. \end{cases} \tag{1.3.38}$$

If $x > 0$, the general solution of the system (1.3.38) is

$$\begin{cases} x(t) = c_1(2 - \sqrt{3})e^{(-2+\sqrt{3})t} + c_2(2 + \sqrt{3})e^{(-2-\sqrt{3})t}, \\ y(t) = c_1 e^{(-2+\sqrt{3})t} + c_2 e^{(-2-\sqrt{3})t}. \end{cases} \tag{1.3.39}$$

For $x \in [-1, 0]$, the general solution is

$$\begin{cases} x(t) = c_1 e^t + c_2 t e^t, \\ y(t) = (-c_1 + c_2)e^t - c_2 t e^t. \end{cases}$$

(1.3.40)

For $x < -1$, the general solution is

$$\begin{cases} x(t) = 1/2\, c_1 e^{-\frac{t}{2}} \left(\cos \frac{\sqrt{3}}{2}t + \sqrt{3}\sin \frac{\sqrt{3}}{2}t \right) + 1/2\, c_2 e^{-\frac{t}{2}} \left(\cos \frac{\sqrt{3}}{2}t - \sqrt{3}\sin \frac{\sqrt{3}}{2}t \right), \\ y(t) = c_1 e^{-\frac{t}{2}} \cos \dfrac{\sqrt{3}}{2}t + c_2 e^{-\frac{t}{2}} \sin \dfrac{\sqrt{3}}{2}t + 3. \end{cases}$$

(1.3.41)

The trajectories of system (1.3.38) are plotted in Figure 1.3.3.

By the general solution and Figure 1.3.3, we can obtain $\lim_{t\to\infty} x(t) = \lim_{t\to\infty} y(t) = 0$ for each solution $(x(t), y(t))$ of system (1.3.38). Consider the solution of system (1.3.38) with the initial value

$$\begin{pmatrix} x(0) \\ y(0) \end{pmatrix} = \begin{pmatrix} x_0 \\ y_0 \end{pmatrix} = \begin{pmatrix} -e^{-t_0} \\ e^{-t_0} \end{pmatrix},$$

(1.3.42)

where $t_0 > 0$ is a positive constant. The solution of system (1.3.38) with initial value $(x_0, y_0)^\top$ in interval $[0, t_0]$ is

$$x(t; x_0) = -e^{t-t_0}, \quad y(t) = e^{t-t_0}, \quad t \le t_0,$$

(1.3.43)

which satisfies $x(t_0) = -1$ and $y(t_0) = 1$. A simple computation shows that

$$\lim_{t_0\to\infty} x_0 = \lim_{t_0\to\infty} -e^{-t_0} = 0, \quad \lim_{t_0\to\infty} y_0 = \lim_{t_0\to\infty} e^{-t_0} = 0.$$

(1.3.44)

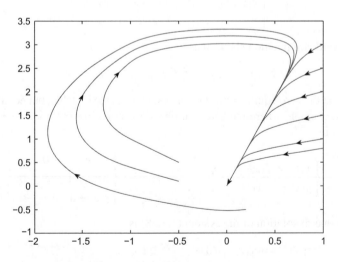

Figure 1.3.3   Orbit distribution of system (1.3.38).

This implies that when t_0 is large enough, $\| (x_0, y_0)^\top \|$ can be as small as expected. However, system (1.3.38) is not stable because no matter how small the norm of the initial state is, $\| (x(t_0; x_0), y(t_0; y_0))^\top \| = \sqrt{2}$ can not be always arbitrarily small.

The class $\mathcal{K}$ and $\mathcal{K}_\infty$ functions and Lyapunov functions are important in stability analysis.

Definition 1.3.7 *The function $\varphi \in C([0, a), [0, \infty))$ is said to be a class $\mathcal{K}$ function if $\varphi(r)$ is strictly increasing on $[0, a)$ and $\varphi(0) = 0$. Furthermore, if $a = +\infty$ and $\lim_{r \to +\infty} \varphi(r) = \infty$, then $\varphi(r)$ is a class $\mathcal{K}_\infty$ function.*

Definition 1.3.8 *Let $\Omega \subset \mathbb{R}^n$ and $0 \in \Omega^\circ$. Function $V \in C(\Omega, [0, \infty))$ is said to be positive definite $(-V(x)$ is said to be negative definite) if, for any $x \in \Omega, V(x) \geq 0$ and $V(x) = 0$ if and only if $x = 0$. Furthermore, if $\Omega = \mathbb{R}^n$ and $\lim_{\|x\| \to +\infty} V(x) = +\infty$, then $V(x)$ is said to be a radially unbounded positive definite function. In stability analysis, the positive definite function is also said to be a Lyapunov function.*

Theorem 1.3.1 *Suppose that $V \in C(\Omega, [0, \infty))$ is a positive definite function on Ω, where $\Omega \subset \mathbb{R}^n (0 \in \Omega^\circ)$ is a connected domain and $B_r = \{x \in \mathbb{R}^n : \| x \| \leq r\} \subset \Omega$ for $r > 0$. Then there exist class $\mathcal{K}$ functions $\kappa_1, \kappa_2 \in C([0, r), \bar{\mathbb{R}}^+)$ such that*

$$\kappa_1(\| x \|) \leq V(x) \leq \kappa_2(\| x \|), \quad \forall x \in B_r.$$

Furthermore, if $V(x)$ is radially unbounded, then $\kappa_1(\cdot)$ and $\kappa_2(\cdot)$ are class $\mathcal{K}_\infty$ functions.

Proof. Let $\kappa(\tau) = \inf_{\tau \leq \|x\| \leq r} W(x), \ \tau \in [0, r)$. It is easy to verify that for any $\tau \in (0, r)$, $\kappa(\tau) > 0, \kappa(0) = 0$, and $\kappa(\tau)$ is continuous on $(0, r)$.

Let $\kappa_1(\tau) = \frac{\tau \kappa(\tau)}{r}$. A direct computation shows that $\kappa_1(0) = 0$. For any $\tau_1, \tau_2 \in [0, r)$, if $\tau_1 < \tau_2$ then

$$\kappa_1(\tau_1) = \frac{\tau_1 \kappa(\tau_1)}{r} \leq \frac{\tau_1 \kappa(\tau_2)}{r} < \frac{\tau_2 \kappa(\tau_2)}{r} = \kappa_1(\tau_2).$$

Let $\tilde{\kappa}(\tau) = \max_{\|x\| \leq \tau} |V(\tau)|, \ \tau \in [0, r)$. Also, we can verify that $\tilde{\kappa}(\tau)$ is continuous on $[0, r)$, for any $\tau \in (0, r), \tilde{\kappa}(\tau) > 0$. Therefore $\kappa_1(\tau)$ is a class $\mathcal{K}$ function.

Let $\kappa_2(\tau) = \tilde{\kappa}(\tau) + \tau, \tau \in [0, r)$. A simple computation shows that $\kappa_1(0) = 0$, for any $\tau \in (0, \tau), \kappa_2(\tau) > 0$, and for any $\tau_1, \tau_2 \in [0, r)$, if $\tau_1 < \tau_2$; then

$$\kappa_2(\tau_1) = \tilde{\kappa}(\tau_1) + \tau_1 \leq \tilde{\kappa}(\tau_2) + \tau_1 < \tilde{\kappa}(\tau_2) + \tau_2 = \kappa_2(\tau_2).$$

Therefore κ_2 is a $\mathcal{K}$ function.

Finally, for any $x \in B_r$,

$$\kappa_1(\| x \|) \leq \kappa(\| x \|) = \inf_{\xi \in B_r, \|x\| \leq \|\xi\| \leq r} V(\xi) \leq V(x) \leq \max_{\xi \in B_r, 0 \leq \|\xi\| \leq \|x\|} V(\xi)$$

$$= \tilde{\kappa}(\| x \|) \leq \kappa_2(\| x \|).$$

This completes the proof of the theorem. $\square$

The next result is the Lyapunov theorem on stability for an autonomous system:

$$\dot{x}(t) = f(x(t)), \quad f(x) = (f_1(x), f_2(x), \ldots, f_n(x))^\top, \quad f_i \in C(\mathbb{R}^n, \mathbb{R}), i = 1, 2, \ldots, n.$$
(1.3.45)

Theorem 1.3.2 *Let $f(0) \equiv 0$ in (1.3.45) and hence the zero state is an equilibrium state of system (1.3.45). Let $\Omega = B_r(0) \subset \mathbb{R}^n$ and $V \in C^1(\Omega, \mathbb{R})$ is a positive definite Lyapunov function:*

1. If for every $x \in \Omega$, the Lie derivative of $V(x)$:

$$L_f V(x) = \left. \frac{dV(x)}{dt} \right|_{along\ (1.3.45)} = \sum_{i=1}^n \frac{\partial V(x)}{\partial x_i} f_i(x) \le 0,$$
(1.3.46)

then the zero equilibrium state of system (1.3.45) is Lyapunov stable.
2. If $-\left. \frac{dV(x(t))}{dt} \right|_{(1.3.45)}$ is positive definite on Ω, then the zero equilibrium of system (1.3.45) is asymptotically stable.

Proof.
1. For the positive definite Lyapunov function $V(x)$, it follows from Theorem 1.3.1 that there exist class $\mathcal{K}$ functions $\kappa_1, \kappa_2 \in \mathbb{C}([0, r), [0, \infty))$ such that

$$\kappa_1(\| x \|) \le V(x) \le \kappa_2(\| x \|).$$

Let $x(t; x_0)$ be the solution of system (1.3.45) with initial condition $x(0) = x_0$. For any $\epsilon > 0$, let $\delta = \kappa_1(\kappa_2^{-1}(\epsilon))$. It follows from (1.3.46) that $V(x(t; x_0)) \le V(x(t; x_0)) = V(x_0)$ for any $t > 0$. For any $x_0 \in \mathbb{R}^n$, if $\| x_0 \| < 0$, then

$$\| x(t; x_0) \| \le \kappa_1^{-1}(V(x(t; x_0))) \le \kappa_1^{-1}(V(x_0)) \le \kappa_1^{-1}(\kappa_2(\| x_0 \|)) < \epsilon.$$
(1.3.47)

The Lyapunov stability of zero equilibrium of system (1.3.45) is proved.
2. Let

$$W(x(t; x_0)) = \left. \frac{dV(x(t; x_0))}{dt} \right|_{along\ (1.3.45)} = \sum_{i=1}^n \frac{\partial V(x(t; x_0))}{\partial x_i} f_i(x(t; x_0)).$$
(1.3.48)

From the positive definiteness of $W(x)$, there exist class $\mathcal{K}$ functions $\tilde{\kappa}_1(\cdot)$ and $\tilde{\kappa}_2(\cdot)$ such that

$$\tilde{\kappa}_1(\| x \|) \le W(x) \le \tilde{\kappa}_2(\| x \|), \quad x \in \mathbb{R}^n.$$

Therefore if $V(x(t; x_0)) > \sigma$ for some $\sigma > 0$, then

$$\left. \frac{dV(x(t; x_0))}{dt} \right|_{along\ (1.3.45)} \le -W(x(t; x_0)) \le -\tilde{\kappa}_1(\| x(t; x_0) \|) \le -\tilde{\kappa}_1(\kappa_1^{-1}(V(x(t; x_0))))$$
$$\le -\tilde{\kappa}_1(\kappa_1^{-1}(\sigma)) < 0.$$

This implies that

$$\lim_{t \to \infty} V(x(t; x_0)) = 0,$$

which together with the Lyapunov stability proved in (1) yields the asymptotical stability. This completes the proof of the theorem.   □

The following Theorem 1.3.3 is the inverse Lyapunov theorem 1.3.2.

Theorem 1.3.3 *Suppose that the zero equilibrium of system (1.3.35) is asymptotically stable and the attracting basin is $\Omega \subset \mathbb{R}^n$, where Ω is a connected domain and $0 \in \Omega^\circ$. If $f \in C(\Omega, \mathbb{R}^n)$ is locally Lipschitz continuous, then there exist Lyapunov functions $V \in C^1(\Omega, [0, \infty))$ and $W \in C(\Omega, [0, \infty))$ such that*

$$\left. \frac{dV(x)}{dt} \right|_{\text{along}(1.3.35)} \leq -W(x), \quad \forall\ x \in \Omega,\ \lim_{x \to \partial\Omega} V(x) = +\infty.$$

The Theorem 1.3.3 is a special case of Theorem 1.3.11, which is proved in Section 1.3.6.

The well-known Lasalle invariance principle is a powerful tool for verifying stability for autonomous systems.

Theorem 1.3.4 *Suppose that in system (1.3.45), $f(0) = 0$ and $\Omega = B_r(0) \subset \mathbb{R}^n$ is a connected domain. The function $V \in C^1(\Omega, [0, \infty))$ is positive definite and satisfies*

$$L_f V(x) \leq 0, \quad \forall x \in \Omega. \tag{1.3.49}$$

In addition, there is no nonzero solution of system (1.3.35) staying in the following set $L_f V^{-1}(0)$:

$$L_f V^{-1}(0) = \{x \in \Omega : L_f V(x) = 0\}.$$

Then the zero equilibrium of (1.3.35) is asymptotically stable.

As a preliminary of proving Theorem 1.3.4, we give Lemma 1.3.4.

Lemma 1.3.4 *Let $x(t; x_0)$ be the solution of system (1.3.45) and let x^* be the limit point of $x(t; x_0)$, that is, there exists series t_k: $t_k \to \infty$ as k goes to ∞ such that $\lim_{k \to \infty} x(t_k; x_0) = x^*$. Then any point in $E = \{x(t; x^*); t \geq 0\}$ is the limit point of $x(t; x_0)$.*

Proof of Theorem 1.3.4. Let $\Omega(x_0) = \{x^* \in \mathbb{R}^n|\ x^*$ is the limit point of $x(t; x_0)\}$. It follows from (1.3.49) that the equilibrium of system (1.3.45) is Lyapunov stable and $\Omega(x_0)$ is a bounded nonempty set. We show that $\Omega(x_0) = \{0\}$. If this is not true, then there exists a sequence t_n: $t_n \to \infty$ such that $\lim_{n \to \infty} x(t_n; x_0) = x^* \neq 0$. Again, by using (1.3.49), we can obtain that $V(x(t; x_0))$ is nonincreasing as t increases. This together with the positive definiteness and continuity of $V(x)$ gives

$$\lim_{t \to \infty} V(x(t; x_0)) = V(x^*) > 0. \tag{1.3.50}$$

Now consider the solution of (1.3.45) starting from x^*. From (1.3.49), we have $V(x(t; x_*)) < V(x^*)$ for all $t > 0$. If for any $t \geq 0$

$$V(x(t; x^*)) \equiv V(x^*), \tag{1.3.51}$$

then
$$\frac{dV(x(t;x^*))}{dt} = 0, \tag{1.3.52}$$

which implies that $\{x(t;x^*)| \ t \geq 0\} \subset L_f V^{-1}(0)$. This is a contradiction. Hence there exists $t_1 > 0$ such that $V(x(t_1;x^*)) < V(x^*)$. It follows from Lemma 1.3.4 that there exists a sequence $\{t_n^*\}$ such that
$$\lim_{n\to\infty} x(t_n^*;x_0) = x(t_1;x^*),$$

which yields
$$\lim_{n\to\infty} V(x(t_n^*;x_0)) = V(x(t_1;x^*)) < V(x^*).$$

This contradicts (1.3.50). The result is thus concluded. $\qquad\square$

1.3.3   Stability of Linear Systems

Let $A \in \mathbb{R}^{n\times n}$. Consider the linear system of the following:
$$\dot{x}(t) = Ax(t), \quad x(0) = x_0. \tag{1.3.53}$$

First of all, we introduce the Kronecker product and straightening operator of the matrices.

Definition 1.3.9 *Let*

$$A = \begin{pmatrix} a_{11} & a_{12} & \cdots & a_{1n} \\ a_{21} & a_{22} & \cdots & a_{2n} \\ \vdots & \vdots & \ddots & \vdots \\ a_{m1} & a_{m2} & \cdots & a_{mn} \end{pmatrix}_{m\times n}, \quad B = \begin{pmatrix} b_{11} & b_{12} & \cdots & b_{1l} \\ b_{21} & b_{22} & \cdots & a_{2l} \\ \vdots & \vdots & \ddots & \vdots \\ b_{s1} & b_{s2} & \cdots & b_{sl} \end{pmatrix}_{s\times l}. \tag{1.3.54}$$

The Kronecker product of A and B is an $(ml) \times (ns)$ matrix, which is defined as follows:

$$A \otimes B = \begin{pmatrix} a_{11}B & a_{12}B & \cdots & a_{1n}B \\ a_{21}B & a_{22}B & \cdots & a_{2n}B \\ \vdots & \vdots & \ddots & \vdots \\ a_{m1}B & a_{m2}B & \cdots & a_{mn}B \end{pmatrix}_{(ml)\times(ns)}. \tag{1.3.55}$$

The straightening operator is a $1 \times (nm)$ matrix given by

$$\overrightarrow{A} = (a_{11}, \ldots, a_{1n}, a_{21}, \ldots a_{2n}, \ldots, a_{n1}, \ldots, a_{nn})^{\top}. \tag{1.3.56}$$

We can verify that the Kronecker product and straightening operator have the following properties.

Property 1.3.1

(i) *If $m = n$ and $s = l$, then*

$$\det(A \otimes B) = (\det(A))^m (\det(B))^s.$$

(ii) *The Kronecker product $A \otimes B$ is invertible if and only if both A and B are invertible, and*

$$(A \otimes B)^{-1} = A^{-1} \otimes B^{-1}.$$

(iii) *Let E_{ij} be an $m \times n$ matrix with the ith row and jth column component being one and other entries being identical to zero. Let $e_i \in \mathbb{R}^{1 \times m}$ (or $e_i \in \mathbb{R}^{1 \times n}$) with the ith component being one and other entries being zero. Then*

$$A = \sum_{i=1}^{m} \sum_{j=1}^{n} a_{ij} E_{ij}, \quad A e_i = (a_{1i}, a_{2i}, \dots, a_{mi})^\top,$$
(1.3.57)

$$e_i^\top A = (a_{i1}, a_{i2}, \dots, a_{in}), E_{ij} = e_i e_j^\top, \quad \overrightarrow{E_{ij}} = e_i \otimes e_j.$$

(iv) *Let $A \in \mathbb{R}^{n \times m}$, $B \in \mathbb{R}^{m \times s}$, and $C \in \mathbb{R}^{s \times l}$. Then $\overrightarrow{ABC} = (A \otimes C^\top)\overrightarrow{B}$.*

Let $A, C \in \mathbb{R}^{n \times n}$. The Lyapunov equation determined by unknown matrix X with respect to A and C is defined by

$$A^\top X + XA = C.$$
(1.3.58)

The following Property 1.3.2 is about solvability of the Lyapunov equation (1.3.58).

Property 1.3.2 *Let $A, C \in \mathbb{R}^{n \times n}$. The following conclusions are equivalent:*

(i) *There exists a unique matrix $X \in \mathbb{R}^{n \times n}$ satisfying (1.3.58).*
(ii) *There exists a unique vector $x \in \mathbb{R}^{n^2}$ satisfying the linear equation*

$$(A^\top \otimes I_{n \times n} + I_{n \times n} \otimes A^\top) x = \overrightarrow{C}.$$
(1.3.59)

(iii) *The matrix $A^\top \otimes I_{n \times n} + I_{n \times n} \otimes A^\top$ is invertible, that is, $\text{rank}(A^\top \otimes I_{n \times n} + I_{n \times n} \otimes A^\top) = n^2$.*
(iv) *$\prod_{i,j=1}^{n}(\lambda_i + \lambda_j) \neq 0$.*

Based on Property 1.3.2, we have immediately Theorem 1.3.5.

Theorem 1.3.5 *If A is a Hurwitz matrix, that is, all the eigenvalues of A have the negative real part, then for any positive definite symmetrical matrix $C \in \mathbb{R}^{n \times n}$ there is a unique positive definite symmetrical matrix solution $X \in \mathbb{R}^{n \times n}$ to the Lyapunov equation:*

$$A^\top X + XA = -C.$$

Let $V(x) = x^\top Bx$. Then $V(x)$ can be written as

$$V(x) = \frac{1}{\det(\Delta)} \begin{vmatrix} 0 & X \\ \overrightarrow{C} & \Delta \end{vmatrix}, \tag{1.3.60}$$

where $\Delta = A^\top \otimes I_{n\times n} + I \otimes A^\top$, $X = (X_1, X_2, \ldots, X_n)$, and

$$X_1 = (x_1^2, 2x_1x_2, \ldots, 2x_1x_n), \quad X_2 = (0, x_2^2, \ldots, 2x_2x_n), \quad X_n = (0, 0, \ldots, x_n^2).$$

Proof. Let $B = (b_{ij})$ satisfy

$$(A^\top \otimes I + I \otimes A^\top)\overrightarrow{B} = \overrightarrow{C}. \tag{1.3.61}$$

It follows from the Gramer law that

$$b_{ij} = \frac{\det(\Delta_{ij})}{\det(\Delta)}, \quad i, j = 1, 2, \ldots, n,$$

where Δ_{ij} is the matrix where the $(i-1)n_j$th column (the number of b_{ij}'s coefficient column) in Δ is replaced by $\overrightarrow{C}$ and other columns are the same as in Δ. Then

$$V(x) = x^\top Bx = \sum_{i,j=1}^{n} b_{ij}x_ix_j. \tag{1.3.62}$$

On the other hand, a direct computation shows that

$$\frac{1}{\det(\Delta)} \begin{vmatrix} 0 & X \\ \overrightarrow{C} & \Delta \end{vmatrix} = \sum_{i,j=1}^{n} \frac{\det(\Delta_{ij})}{\det(\Delta)} x_ix_j = \sum_{i,j=1}^{n} b_{ij}x_ix_j. \tag{1.3.63}$$

$\square$

The following stability theorem can be directly obtained as an application of Theorem 1.3.5.

Theorem 1.3.6 *If A is a Hurwitz matrix, then the zero equilibrium of system (1.3.53) is globally asymptotical stable.*

Proof. Since A is Hurwitz, it follows from Theorem 1.3.5 that there exists a positive definite symmetrical matrix P_A such that

$$A^\top P_A + P_A A = -I_{n\times n},$$

where $I_{n\times n}$ is the $n \times n$ identity matrix. Let $V(\nu) = \nu^\top P_A \nu$ for all $\nu \in \mathbb{R}^n$. A direct computation shows that

$$\left. \frac{dV(x(t;x_0))}{dt} \right|_{\text{along } (1.3.53)} = (x(t;x_0))^\top (A^\top P_A + P_A A)x(t;x_0) = -\parallel x(t;x_0)\parallel^2.$$

$$\tag{1.3.64}$$

Theorem 1.3.6 then follows by setting $W(\nu) = \parallel \nu \parallel^2$ for all $\nu \in \mathbb{R}^n$.

$\square$

1.3.4   Finite-Time Stability of Continuous System

The finite-time stability for continuous systems has many investigations. Here we list some preliminary results.

Definition 1.3.10 *Let $\Omega \in \mathbb{R}^n$ be a connected domain, $0 \in \Omega^\circ$, and $f(t, \cdot) \in C(\Omega, \mathbb{R}^n)$, $f(t, 0) \equiv 0$, that is, the zero state is the equilibrium state of the system (1.3.35). The zero state is finite stable on the attracting basin Ω if it is Lyapunov stable and, for every $x_0 \in \Omega$, there exits a positive constant $T(x_0) > 0$ such that the solution of (1.3.35) starting from x_0 satisfies*

$$\lim_{t \uparrow T(x_0)} x(t; x_0) = 0,$$

$$x(t) = 0, \ \forall \ t \in [T(x_0), \infty).$$

Furthermore, if $\Omega = \mathbb{R}^n$, then the zero equilibrium of system (1.3.35) is globally finite-time stable, where $T(\cdot) : \Omega \to \mathbb{R}$ is a positive-valued function defined on Ω, which is said to be a setting-time function.

Now we look at an example. Consider the differential equation

$$\dot{x}(t) = -|x(t)|^\alpha \text{sign}(x(t)), \quad x(0) = x_0, \alpha \in (0, 1). \tag{1.3.65}$$

If $x_0 > 0$, then the solution of (1.3.65) is

$$x(t; x_0) = \begin{cases} \left(x_0^{1-\alpha} - t\right)^{\frac{1}{1-\alpha}}, & t < x_0^{1-\alpha}, \\ 0 & t \geq x_0^{1-\alpha}, \end{cases} \tag{1.3.66}$$

while $x_0 < 0$, the solution is

$$x(t; x_0) = \begin{cases} \left(t - |x_0|^{1-\alpha}\right)^{\frac{1}{1-\alpha}}, & t < |x_0|^{1-\alpha}, \\ 0 & t \geq |x_0|^{1-\alpha}. \end{cases} \tag{1.3.67}$$

We can clearly see that the system (1.3.65) is finite-time stable.

For the zero equilibrium state of a nonlinear system, we can verify the finite-time stability by Theorem 1.3.7.

Theorem 1.3.7 *Suppose that there exists a positive definite function $V \in C^1(\Omega, [0, \infty))$ and positive constants $\alpha \in (0, 1)$ and $C > 0$ such that*

$$L_f V(x) \leq -CV^\alpha(x).$$

Then the zero equilibrium of system (1.3.35) is finite-time stable on $\Omega \in \mathbb{R}^n$ and the setting time $T(x_0)$ satisfies

$$T(x_0) \leq \frac{1}{C(1 - \alpha)} V^{1-\alpha}(x_0), \tag{1.3.68}$$

where x_0 is the initial state of the system.

Proof. Let $x(t; x_0)$ be the solution of the initial value problem following

$$\dot{x}(t) = f(x(t)), \quad x(0) = x_0. \tag{1.3.69}$$

Then

$$\frac{dV(x(t;x_0))}{dt} = L_f V(x(t;x_0)) \leq -CV^\alpha(x(t;x_0)). \tag{1.3.70}$$

Solve the following initial value problem:

$$\dot{z}(t) = -C|z(t)|^\alpha \mathrm{sign}(z(t)), \quad z(0) = V(x_0), \tag{1.3.71}$$

to obtain

$$z(t) = \begin{cases} \left(t + \dfrac{1}{C(1-\alpha)}(V(x_0))^{1-\alpha}\right)^{\frac{1}{1-\alpha}}, & t < \dfrac{1}{C(1-\alpha)}(V(x_0))^{1-\alpha}, \\ 0, & t \geq \dfrac{1}{C(1-\alpha)}(V(x_0))^{1-\alpha}. \end{cases} \tag{1.3.72}$$

This together with the comparison principle of the ordinary differential equations gives

$$V(x(t;x_0)) = 0, \quad \forall\, t \geq \frac{1}{C(1-\alpha)}(V(x_0))^{1-\alpha}. \tag{1.3.73}$$

This completes the proof of the theorem.

The most popular continuous finite-time stable systems are those weighted homogeneous systems.

Definition 1.3.11 *The function* $V : \mathbb{R}^n \to \mathbb{R}$ *is said to be d-degree weighted homogeneous with the weights* $\{r_i > 0\}_{i=1}^n$, *if there exist positive constant* $\lambda > 0$ *and* $x = (x_1, x_2, \dots, x_n) \in \mathbb{R}^n$ *such that*

$$V(\lambda^{r_1}x_1, \lambda^{r_2}x_2, \dots, \lambda^{r_n}x_n) = \lambda^d V(x_1, x_2, \dots, x_n). \tag{1.3.74}$$

A vector field $g : \mathbb{R}^n \to \mathbb{R}^n$ *is said to be d-degree weighted homogeneous with weights* $\{r_i > 0\}_{i=1}^n$ *if, for every* $i = 1, 2, \dots, n$, $\lambda > 0$, *and* $(x_1, x_2, \dots, x_n) \in \mathbb{R}^n$,

$$g_i(\lambda^{r_1}x_1, \lambda^{r_2}x_2, \dots, \lambda^{r_n}x_n) = \lambda^{d+r_i} g_i(x_1, x_2, \dots, x_n), \tag{1.3.75}$$

where $g_i : \mathbb{R}^n \to \mathbb{R}$ *is the ith component of* $g(\cdot)$.

If the vector field $g : \mathbb{R}^n \to \mathbb{R}^n$ *is d-degree weighted homogeneous with weights* $\{r_i > 0\}_{i=1}^n$, *then we say that the system*

$$\dot{x}(t) = g(x(t))$$

is d-degree weighted homogeneous with weights $\{r_i > 0\}_{i=1}^n$.

Example 1.3.1 *The following nonlinear system*

$$\begin{cases} \dot{x}_1(t) = x_2(t), \\ \dot{x}_2(t) = -|x_1(t)|^\alpha \mathrm{sign}(x_1(t)) - |x_2(t)|^\beta \mathrm{sign}(x_2(t)), \end{cases} \tag{1.3.76}$$

is weighted homogeneous if $\beta = \frac{2\alpha}{1+\alpha}$, $\alpha > 0$. *Actually, let* $r_1 = 1$, $r_2 = (\alpha+1)/2$, *and let*

$$f_1(x_1, x_2) = x_2, \quad f_2(x_1, x_2) = -|x_1|^\alpha \mathrm{sign}(x_1) - |x_2|^\beta \mathrm{sign}(x_2). \tag{1.3.77}$$

For any vector $(x_1, x_2) \in \mathbb{R}^2$ and positive constant $\lambda > 0$,

$$\begin{cases} f_1(\lambda^{r_1} x_1, \lambda^{r_2} x_2) = \lambda^{r_2} x_2 = \lambda^{\frac{\alpha-1}{2}+r_1} f_1(x_1, x_2), \\ f_2(\lambda^{r_1} x_1, \lambda^{r_2} x_2) = -\lambda^{\alpha r_1} |x_1|^\alpha \mathrm{sign}(x_1) - \lambda^{\beta r_2} |x_2|^\beta \mathrm{sign}(x_2) = \lambda^{\frac{\alpha-1}{2}+r_2} f_2(x_1, x_2). \end{cases} \tag{1.3.78}$$

Therefore system (1.3.76) is $\frac{\alpha-1}{2}$ degree homogeneous with weights $\{r_1, r_2\}$.

The following system (1.3.79) is also weighted homogeneous.

Example 1.3.2

$$\begin{cases} \dot{x}_1(t) = x_2(t) - |x_1(t)|^\theta \mathrm{sign}(x_1(t)), \\ \dot{x}_2(t) = -|x_1(t)|^{2\theta-1} \mathrm{sign}(x_1(t)), \end{cases} \tag{1.3.79}$$

where $\theta > 0$. Actually, let

$$f_1(x_1, x_2) = x_2 + |x_1|^\theta \mathrm{sign}(x_1), \quad f_2(x_1, x_2) = |x_1|^{2\theta-1} \mathrm{sign}(x_1).$$

Then for any vector $(x_1, x_2) \in \mathbb{R}^2$ and positive constant $\lambda > 0$,

$$\begin{cases} f_1(\lambda x_1, \lambda^\theta x_2) = \lambda^\theta x_2 + |x_1|^\theta \mathrm{sign}(x_1) = \lambda^{\theta-1+1} f_1(x_1, x_2), \\ f_2(\lambda x_1, \lambda^\theta x_2) = \lambda^{2\theta-1} |x_1|^{2\theta-1} \mathrm{sign}(x_1) = \lambda^{\theta-1+\theta} f_2(x_1, x_2). \end{cases} \tag{1.3.80}$$

This means that system (1.3.79) is $\theta - 1$ degree homogeneous with weights $\{1, \theta\}$.

Property 1.3.3 *Suppose that $V_1, V_2 : \mathbb{R}^n \to \mathbb{R}$ are continuous weighted homogeneous functions with the same weights $\{r_i > 0\}_{i=1}^n$, with degree $l_1 > 0$ and $l_2 > 0$ respectively. Assume that $V_1(x)$ is positive definite. Then for any $x \in \mathbb{R}^n$,*

$$\left(\min_{y \in V_1^{-1}(1)} V_2(y) \right) (V_1(x))^{l_2/l_1} \le V_2(x) \le \left(\max_{y \in V_1^{-1}(1)} V_2(y) \right) (V_1(x))^{l_2/l_1}, \tag{1.3.81}$$

where $V_1^{-1}(1) \triangleq \{x \in \mathbb{R}^n | V_1(x) = 1\}$.

Theorem 1.3.8 *If the matrix*

$$K = \begin{pmatrix} -k_1 & 1 & 0 & \cdots & 0 \\ -k_2 & 0 & 1 & \cdots & 0 \\ \vdots & \vdots & \vdots & \ddots & \vdots \\ -k_n & 0 & 0 & \cdots & 1 \\ -k_{n+1} & 0 & 0 & \cdots & 0 \end{pmatrix} \tag{1.3.82}$$

is Hurwitz, then there exists $\theta^* \in \left(\frac{n}{n+1}, 1\right)$ such that for any $\theta \in (\theta^*, 1)$, the system following is finite-time stable:

$$\begin{cases} \dot{x}_1(t) = x_2(t) - k_1[x_1(t)]^\theta, \\ \dot{x}_2(t) = x_3(t) - k_2[x_1(t)]^{2\theta-1}, \\ \quad \vdots \\ \dot{x}_n(t) = x_{n+1}(t) - k_n[x_1(t)]^{n\theta-(n-1)}, \\ \dot{x}_{n+1}(t) = -k_{n+1}[x_1(t)]^{(n+1)\theta-n}. \end{cases} \tag{1.3.83}$$

Proof. We can verify that system (1.3.83) is $(\theta - 1)$ degree homogeneous with weights $\{(i-1)\theta - (i-2)\}_{i=1}^{n+1}$. Let $q = \Pi_{i=1}^n((i-1)\theta - (i-2))$. Then

$$y(x) = \begin{pmatrix} [x_1]^{\frac{1}{q}} \\ [x_2]^{\frac{1}{\theta q}} \\ \vdots \\ [x_n]^{\frac{1}{((n-1)\theta-(n-2))q}} \\ [x_{n+1}]^{\frac{1}{(n\theta-(n-1))q}} \end{pmatrix} \tag{1.3.84}$$

and

$$V(\theta, x) = y^\top P y, \tag{1.3.85}$$

where P is the positive definite matrix solution to the Lyapunov equation $K^\top P + PK = -I_{(n+1)\times(n+1)}$. Let

$$S = \{x \in \mathbb{R}^{n+1} : \ V(1, x) = 1\}. \tag{1.3.86}$$

It is easy to verify that S is a compact set. Let $\theta = 1$. Then system (1.3.83) becomes $\dot{x}(t) = Kx(t)$, which is asymptotically stable and

$$\frac{dV(1, x(t))}{dt} = - \| x(t) \|^2 < -a < 0, \ \ a > 0, \ \ \forall \ x \in S.$$

By the continuity of $V(\theta, x)$ on θ, there exists $\theta^* \in \left(\frac{n}{n+1}, 1\right)$ such that, for any $\theta \in (\theta^*, 1)$,

$$\frac{dV(\theta, x(t))}{dt} < -\frac{a}{2} < 0, \ \forall \ x \in S.$$

In addition, we can verify that for any $\theta \in (\theta^*, 1)$, $V(\theta, x)$ is $1/q^2$ degree homogeneous with weights $\{(i-1)\theta - (i-2)\}_{i=1}^{n+1}$. This implies that $\frac{dV(\theta, x(t))}{dt}$ is negative definite. By Theorem 1.3.7, system (1.3.83) is finite-time stable. □

The following Theorem 1.3.9 is about finite stability for the weighted homogeneous systems.

Theorem 1.3.9 *Suppose that the vector field* $f \in C(\mathbb{R}^n, \mathbb{R}^n)$ *is* d*-degree homogeneous with weights* $\{r_i\}_{i=1}^n$, $f(0) = 0$.

(i) *If the zero equilibrium of the system*

$$\dot{x}(t) = f(x(t)) \tag{1.3.87}$$

is finite-time stable on attracting basin $\Omega \subset \mathbb{R}^n$, then it is asymptotically stable on Ω.

(ii) *If the zero equilibrium of system (1.3.87) is asymptotically stable on attracting basin Ω, and the degree $d < 0$, then the zero equilibrium of system (1.3.87) is finite-stable on Ω. Furthermore, let $U \subset \Omega$ be an open neighborhood of zero state. Then for any integer $k > \max\{d, r_1, r_2, \ldots, r_n\}$ there exists a positive definite function $V \in C^1(U, [0, \infty))$, which is k-degree homogeneous with weights $\{r_i > 0\}_{i=1}^n$. In addition, if $\Omega = \mathbb{R}^n$ then the Lyapunov function $V(x)$ is radially unbounded.*

Proof. We only need to prove (ii). For the sake of simplicity and without loss of generality, we may assume that $\Omega = \mathbb{R}^n$. Since the zero equilibrium state of system (1.3.87) is finite-time stable, it is asymptotically stable. By Theorem 1.3.3, there exists a positive definite Lyapunov function $\tilde{V} : \mathbb{R}^n \to \mathbb{R}$ such that $L_F \tilde{V}$ is negative definite on $\mathbb{R}^n$. Let $\alpha \in C^\infty(\mathbb{R}, \mathbb{R})$ be such that

$$\alpha(s) = \begin{cases} 0, & s \in (-\infty, 1], \\ 1, & s \in [2, +\infty), \end{cases} \quad \text{and} \ \forall \ s \in \mathbb{R}, \ \alpha'(s) > 0, \tag{1.3.88}$$

and

$$V(x) = \begin{cases} \displaystyle\int_0^{+\infty} \frac{1}{\mu^{k+1}} (\alpha \circ \tilde{V})(\mu^{r_1} x_1, \ldots, \mu^{r_n} x_n) d\mu, & x \in \mathbb{R}^n \setminus \{0\}, \\ 0, & x = 0. \end{cases} \tag{1.3.89}$$

Apparently, $V(x)$ is positive definite. For any $\lambda > 0$, $x \neq 0$,

$$V(\lambda^{r_1} x_1, \ldots, \lambda^{r_n} x_n) = \int_0^{+\infty} \frac{1}{\mu^{k+1}} (\alpha \circ \tilde{V})((\lambda\mu)^{r_1} x_1, \ldots, (\lambda\mu)^{r_n} x_n) d\mu$$

$$= \lambda^k \int_0^{+\infty} \frac{1}{(\lambda\mu)^{k+1}} (\alpha \circ \tilde{V})((\lambda\mu)^{r_1} x_1, \ldots, (\lambda\mu)^{r_n} x_n) d(\lambda\mu)$$

$$= \lambda^k V(x). \tag{1.3.90}$$

This shows that $V(x)$ is k-degree homogenous with weights $\{r_1, \ldots, r_n\}$. Furthermore, we can find that there exist $l, L > 0$ such that

$$\begin{aligned} \tilde{V}(\mu^{r_1} x_1, \ldots, \mu^{r_n} x_n) &\leq 1 \ \forall \ x \in \mathbb{R}^n, \tfrac{1}{2} \leq \| x \| \leq 2, \ \mu \leq l, \\ \tilde{V}(\mu^{r_1} x_1, \ldots, \mu^{r_n} x_n) &\geq 2 \ \forall \ x \in \mathbb{R}^n, \tfrac{1}{2} \leq \| x \| \leq 2, \ \mu \geq L. \end{aligned} \tag{1.3.91}$$

Therefore, for any $x \in \mathbb{R}^n$, $1/2 \leq \| x \| \leq 2$,

$$V(x) = \int_l^L \frac{1}{\mu^{k+1}} (\alpha \circ \tilde{V})(\mu^{r_1} x_1, \ldots, \mu^{r_n} x_n) d\mu + \frac{1}{kL^k}. \tag{1.3.92}$$

It is easy to see that $V(x)$ is of C^∞ on $\{x \in \mathbb{R} : 1/2 <\| x \|< 2\}$ and

$$\frac{\partial V(x)}{\partial x_i} = \int_l^L \frac{\mu^{r_i}}{\mu^{k+1}} \alpha'(\tilde{V}(y_1, \ldots, y_n)) \frac{\partial \tilde{V}(y_1, \ldots, y_n)}{\partial y_i} d\mu, \quad y_i = \mu^{r_i} x_i. \qquad (1.3.93)$$

It then follows that

$$\sum_{i=1}^n f_i(x) \frac{\partial V(x)}{\partial x_i} = \int_l^L \frac{1}{\mu^{d+k+1}} \alpha'(\tilde{V}(y_1, \ldots, y_n))$$

$$\times \left(\sum_{i=1}^n \left(f_i(y_1, \ldots, y_n) \frac{\partial V(y_1, \ldots, y_n)}{\partial y_i} \right) \right) d\mu, \quad y_i = \mu^{r_i} x_i.$$

$$(1.3.94)$$

Since $\alpha'(s) > 0$, $L_f \tilde{V}(x) < 0$ for $x \in \mathbb{R}^n$ and $\frac{1}{2} \le \| x \| \le 2$. A straightforward computation shows that $L_F V(x)$ is homogeneous of degree $k + d$ with weights $\{r_i\}_{i=1}^n$. This together with (1.3.94) yields that $L_f V(x)$ is negative definite. By Property 1.3.3,

$$L_F V(x) \le \left(\min_{y \in V^{-1}(1)} L_F V(y) \right) (V(x))^{\frac{k+d}{k}}.$$

Since $d < 0$, this together with Theorem 1.3.7 completes the proof of the theorem. $\qquad \square$

1.3.5   Stability of Discontinuous Systems

In this section, we investigate stability for system (1.3.35), where $f(t, x)$ is not continuous with respect to x. In this case, we consider system (1.3.35) as the following differential inclusion:

$$\dot{x}(t) \in F(t, x), \qquad (1.3.95)$$

where

$$F(t, x) = \mathbf{K}_x f(t, x) \triangleq \bigcap_{\delta>0} \bigcap_{\mu(N)=0} \overline{\text{co}}\{f(t, B_\delta(x) \setminus N)\}, \qquad (1.3.96)$$

where $\overline{\text{co}}(\cdot)$ denotes the convex closure of a set, $B_\delta(x) = \{v \in \mathbb{R}^n | \ \| v - x \|_\infty< r\}$, and $\mu(\cdot)$ is the Lebesgue measure of $\mathbb{R}^n$. If $f(t, x)$ is Lebesgue measurable and locally bounded, then there exist t and $f(t, x)$-dependent zero measure subset N_0^t of $\mathbb{R}^n$ such that for any $x \in \mathbb{R}^n$ and $N \subset \mathbb{R}^n : \ \mu(N) = 0$,

$$\mathbf{K}_x f(t, x) = \overline{\text{co}}\{v = \lim_{n \to \infty} f(t, x_n) : x_i \notin N_0^t \cup N, \lim_{i \to \infty} x_i = x\}. \qquad (1.3.97)$$

We say that $x(t)$ is a generalized solution (or a Filippov solution) of (1.3.35) if $x(t)$ is absolutely continuous on each compact subinterval $I \subset [0, \infty)$ and

$$\dot{x}(t) \in F(t, x(t)) \text{ almost everywhere on } I. \qquad (1.3.98)$$

The following Definition 1.3.12 defines stability for systems with discontinuous right-hand sides.

Definition 1.3.12 *Let $f(t, \cdot)$ be Lebesgue measurable and locally bounded, let $F(t, \cdot)$ be defined in (1.3.96), and $0 \in F(t, 0)$ for almost all $t \geq 0$. For any $x_0 \in \mathbb{R}^n$, the set of solution of (1.3.35) (or (1.3.95)) with initial condition $x(0) = x_0$ is denoted by $\mathcal{S}_{t;x_0}$. The zero equilibrium of system (1.3.35) (or differential inclusion (1.3.95)) is uniformly globally asymptotically stable if*

(i) *For any $\delta > 0$, $x_0 \in \mathbb{R}$, and $x(t; x_0) \in \mathcal{S}_{t;x_0}$, if $\| x_0 \|_\infty < \delta$, then for any $t > 0$, $\| x(t; x_0) \|_\infty < m(\delta)$, where $m \in C((0, +\infty), (0, +\infty))$ satisfies $\lim_{\delta \to 0^+} m(\delta) = 0$;*

(ii) *For any $R > 0$, $\epsilon > 0$, $x_0 \in \mathbb{R}^n$, and $x(t; x_0) \in \mathcal{S}_{t;x_0}$ if $\| x_0 \|_\infty \leq R$, then $\| x(t; x_0) \|_\infty < \epsilon$ for any $t > T(R, \epsilon)$, where $T(R, \epsilon)$ is an R and ϵ-dependent constant.*

The following Theorem 1.3.10 is an extension of Theorem 1.3.2.

Theorem 1.3.10 *Let $f(\cdot, x)$ be Lebesgue measurable and locally bounded, and let $F(t, x)$ be defined in (1.3.96), and $0 \in F(t, 0)$ for almost all $t \geq 0$. Assume that there exists a Lyapunov function $V(t, x)$ and the class $\mathcal{K}_\infty$ functions $\kappa_1(\cdot)$, $\kappa_2(\cdot)$, and $\kappa_3(\cdot)$ such that*

$$\kappa_1(\| \nu \|_\infty) \leq V(t, \nu) \leq \kappa_2(\| \nu \|_\infty) \quad \forall\, t \in [0, \infty), \quad \nu \in \mathbb{R}^n, \tag{1.3.99}$$

and for any $0 < t_1 \leq t_2$

$$V(t_2, x(t_2; x_0)) - V(t_1, x(t_1; x_0)) \leq \int_{t_1}^{t_2} \kappa_3(\| x(\tau; x_0) \|_\infty) d\tau, \tag{1.3.100}$$

then the zero equilibrium state of system (1.3.35) (or differential inclusion (1.3.95)) is uniformly globally asymptotically stable.

When $V(t, \cdot)$ is of C^1 class, the inequality (1.3.100) can be obtained by the following infinitesimal decreasing condition: there exists a class $\mathcal{K}_\infty$ function $\kappa(\cdot)$ such that for almost all $t \geq 0$, all $x \in \mathbb{R}^n$, and $\nu \in F(t, x)$,

$$\frac{\partial V(t, x)}{\partial t} + \langle \nabla_x V(t, x), \nu \rangle \leq -\kappa(\| x \|_\infty). \tag{1.3.101}$$

The proof of Theorem 1.3.10 is similar to Theorem 1.3.2, and the details are omitted. The following Theorem 1.3.11 is the converse of the Lyapunov theorem.

Theorem 1.3.11 *(Converse of second Lyapunov theorem) Let $F(t, x)$ be defined in (1.3.96) and $0 \in F(t, 0)$ for almost all $t \geq 0$. Assume that the zero equilibrium state of system (1.3.35) (or differential inclusion (1.3.95)) is uniformly globally asymptotically stable and there exists a zero measure set $N_0 \subset [0, \infty)$ such that*

- *$F(t, x)$ is a nonempty convex compact set for any $(t, x) \in ([0, \infty) \setminus N_0) \times \mathbb{R}^n$.*
- *For any $R > 0$, if $\| x \|_\infty \leq R$ and $t \in [0, R] \setminus N_0$ then $F(t, x) \subset \overline{B}_M(0)$ for some $M > 0$.*
- *For any $(t_0, x_0) \in ([0, \infty) \setminus N_0) \times \mathbb{R}^n$ and $\epsilon > 0$, there exists $\delta > 0$ such that, for any $(t, x) \in ([0, \infty) \setminus N_0)$, if $\| (t - t_0, x - x_0) \|_\infty < \delta$ then $F(t, x) \subset F(t_0, x_0) + B_\epsilon(0)$.*

Then for any $\lambda > 0$, there exist $V \in C^\infty([0, \infty) \times \mathbb{R}^n, [0, \infty))$ and the class $\mathcal{K}_\infty$ functions $\kappa_1(\cdot)$ *and* $\kappa_2(\cdot)$ *such that*

$$\kappa_1(\| \nu \|_\infty) \leq V(t, \nu) \leq \kappa_2(\| \nu \|_\infty) \quad \forall \, t \geq 0, \; \nu \in \mathbb{R}^n \tag{1.3.102}$$

and

$$\frac{\partial V(t, x)}{\partial t} + \langle \nabla_x V(t, x), \nu \rangle \leq -\lambda V(t, x), \; \forall \, t \in [0, \infty) \setminus N_0, x \in \mathbb{R}^n, \nu \in F(t, x). \tag{1.3.103}$$

The proof of Theorem 1.3.11 is presented in the next subsection.

1.3.6   Proof of Theorem 1.3.11

The proof of Theorem 1.3.11 is lengthy and we split the proof into three steps. In the first step, we show that the uniform global asymptotical stability also holds true for some perturbed system

$$\dot{x}(t) \in F_2(t, x(t)), \tag{1.3.104}$$

where $F(t, x) \subset F_2(t, x)$ for every x and almost all t, $F_2(t, x)$ is locally Lipschitz continuous in $[0, \infty) \times (\mathbb{R}^n \setminus \{0\})$, that is, for each $(t_0, x_0) \in [0, \infty) \times (\mathbb{R}^n \setminus \{0\})$, there exist $L > 0$ and $\delta > 0$ such that for any $(t_1, x_1), (t_2, x_2) \in B_\delta((t_0, x_0))$

$$\hbar(F_2(t_1, x_1), F_2(t_2, x_2)) \leq \|(t_1, x_1) - (t_2, x_2)\|_\infty, \tag{1.3.105}$$

where $\hbar(\cdot, \cdot)$ is the Hausdorff distance between nonempty compact subsets of $\mathbb{R}^n$:

$$\hbar(A, B) = \max \left\{ \sup_{a \in A} \mathrm{dis}(a, B), \; \sup_{b \in A} \mathrm{dis}(b, A) \right\} \tag{1.3.106}$$

with $\mathrm{dis}(a, B) = \inf_{b \in B} \|a - b\|_\infty$. In the second step, we construct a Lipschitz continuous function $V_L(t, x)$. In the final step, we smooth $V_L(t, x)$ to be a C^∞ function. All these steps are accomplished by a series of lemmas.

The following Lemma 1.3.5 gives an initial value continuous dependence on differential inclusion (1.3.95) for which the proof is omitted.

Lemma 1.3.5 *Suppose that $F(t, x)$ satisfies three conditions in Theorem 1.3.11. Let $y : [T_1, T_2] \to \mathbb{R}^n$ be the solution of (1.3.95) and $b > 0$. Assume that $F(t, x)$ is Lipschitz continuous in $[T_1, T_2]$, $\|x - y(t)\|_\infty < b$, that is, there exists some constant $K > 0$ such that, for any $t, \bar{t} \in [T_1, T_2]$ and any $x, \bar{x} \in \mathbb{R}^n$ with $\|x - y(t)\|_\infty \leq b$, $\|\bar{x} - y(\bar{t})\|_\infty \leq b$,*

$$\hbar(F(t, x), F(\bar{t}, \bar{x})) \leq K\|(t - \bar{t}, x - \bar{x})\|_\infty. \tag{1.3.107}$$

Let $(t_0, x_0) \in [T_1, T_2] \times \mathbb{R}^n$ satisfy $\|x_0 - y(t_0)\|_\infty \leq b$. Then there exists a solution $x(t)$ of (1.3.95) with $x(t_0) = x_0$ satisfying

$$\|x(t) - y(t)\|_\infty \leq \|x_0 - y(t_0)\| e^{K|t - t_0|} \tag{1.3.108}$$

as long as $\|x_0 - y(t_0)\| e^{K|t - t|} \leq b$.

To prove Theorem 1.3.11, we need firstly to regularize $F(t,x)$. To facilitate the construction of a Lyapunov function that is smooth up to $t = 0$, we extend $F(t,x)$ on $[-1,0) \times \mathbb{R}^n$ by setting $F(t,x) = \{-x\}$. It is easy to verify that if $F(t,x)$ is a nonempty convex compact set for any $(t,x) \in ([0,\infty) \setminus N_0) \times \mathbb{R}^n$, then, after the extension, it is also a nonempty convex compact set for any $(t,x) \in ([-1,\infty) \setminus N_0) \times \mathbb{R}^n$. We need the following lemma to smooth $F(t,x)$.

Lemma 1.3.6 *Let $F(t,x)$ be a nonempty, compact, and convex subset on $[-1,\infty) \times \mathbb{R}^n$, R be a positive constant, and let $\{t_j^1\}_{j=1}^\infty$, $\{t_j^2\}_{j=1}^\infty$, and $\{\delta_j\}_{j=1}^\infty$ be sequences of numbers satisfying*

$$-1 \leq t_j^1 \leq t_j^2 \leq R, \ \forall\, j \in \mathbb{N}^+ = \mathbb{N} \setminus \{0\}, \quad \lim_{j \to \infty} \delta_j = 0. \tag{1.3.109}$$

Let $\{x_j(t)\}_{j=1}^\infty$ be a sequence of absolutely continuous functions $x_j : [t_j^1, t_j^2] \to \overline{B}_R(0)$ such that for almost all $t \in [t_j^1, t_j^2]$

$$\dot{x}_j(t) \in \overline{\mathrm{co}}\{F(\overline{B}_{\delta_j}(t, x_j(t))) \cap (((-1,\infty) \setminus N_0) \times \mathbb{R}^n))\}. \tag{1.3.110}$$

Then there exist numbers $t_1, t_2 \in [-1, R]$, function $x : [t_1, t_2] \to \overline{B}_R(0)$, and sequence $j_k \to \infty$ such that $x(t)$ is a solution of (1.3.95), and $t_{j_k}^1 \to t_1$, $t_{j_k}^2 \to t_2$ as $k \to \infty$ such that

$$\lim_{k \to \infty} x_{j_k}(t_{j_k}^1) = x(t_1), \quad \lim_{k \to \infty} x_{j_k}(t_{j_k}^2) = x(t_2). \tag{1.3.111}$$

Proof. In what follows, we need to take frequently a convergent subsequence from a sequence of numbers or functions. For the sake of simplicity, we avoid using multiple indices and just assume that the given sequence itself is convergent. According to the second item of the conditions on $F(t,x)$, there exists $M > 0$ such that for any $t \in [-1, R+1] \setminus N_0$ and $\|x\|_\infty \leq R+1$, $F(t,x) \subset \overline{B}_M(0)$.

For any $j \geq 0$ and $t \in [-1, R]$, set

$$\tilde{x}_j(t) = \begin{cases} x_j(t_j^1), & \text{if } -1 \leq t \leq t_j^1, \\ x_j(t), & \text{if } t_j^1 \leq t \leq t_j^2, \\ x_j(t_j^2), & \text{if } t_j^2 \leq t \leq R. \end{cases} \tag{1.3.112}$$

Since for any $j \geq 0$ and $t \in [-1, R]$, $\|\tilde{x}_j(t)\|_\infty \leq R$, and for almost all $t \in [-1, R]$ $\|\dot{\tilde{x}}_i(t)\|_\infty \leq M$, we can obtain a sequence $\{\tilde{x}_j(t)\}_{j=1}^\infty$ that is bounded in the Sobolev space $H^1((-1, R), \mathbb{R}^n)$ and hence a subsequence (still denoted by itself) that is weakly convergent to a function $x(t)$. It follows that $x_j \to x$ in $C^0([-1, R], \mathbb{R}^n)$ and $\tilde{x}_j \rightharpoonup \dot{x}$ in $L^2((-1, R), \mathbb{R}^n)$. As a consequence, $\tilde{x}_j(t_j^1) \to x(t_1)$ and $\tilde{x}_j(t_j^2) \to x(t_2)$. To prove that $x(t)$ is a solution to (1.3.95), that is, $x(t) \in F(t, x(t))$ for almost all $t \in [t_1, t_2]$, we consider the functional $J(w)$ defined on $L^2((-1, R), \mathbb{R}^n)$ by

$$J(w) = \int_{t_1}^{t_2} \mathrm{dis}(w(t), F(t, x(t)))dt. \tag{1.3.113}$$

Since the non-negative map from t to $\operatorname{dis}(w(t), F(t, x(t))$ is measurable, it is easy to verify that the functional $J(w)$ is well defined, convex, and continuous in the strong topology of $L^2((-1, \mathbb{R}), \mathbb{R}^n)$. Since $\dot{\tilde{x}}_j \rightharpoonup \dot{x}$ in $L^2((-1, \mathbb{R}), \mathbb{R}^n)$, we can obtain

$$0 \leq J(\dot{x}(t)) \leq \liminf_{j \to \infty} J(\dot{\tilde{x}}_j(t)). \tag{1.3.114}$$

To prove $J(\dot{x}) = 0$, we only need to show that $\lim_{j \to \infty} J(\dot{\tilde{x}}_j) = 0$. Noting $\operatorname{dis}(\dot{\tilde{x}}_j(t), F(t, x(t))) \leq M$ for every j and almost all $t \in (t_1, t_2)$, by the Lebesgue theorem, we only need to prove that

$$\lim_{j \to \infty} \operatorname{dis}(\dot{\tilde{x}}_j(t), F(t, x(t))) = 0. \tag{1.3.115}$$

For almost all $t_0 \in (t_1, t_2) \setminus N_0$, there exists an integer $j_0 \geq 0$ such that, for any $j \geq j_0$ and $t_0 \in (t_1^j, t_2^j)$, $\dot{x}_j(t_0)$ exists and belongs to $\overline{co}\{F(B_{\delta_j}(t_0, x_j(0)) \cap (([-1, +\infty) \setminus N_0) \times \mathbb{R}^n))\}$. Let $\epsilon > 0$. By the third condition in Theorem 1.3.11, there exists $\delta > 0$ such that for any $(t, x) \in ([-1, +\infty) \setminus N_0) \times \mathbb{R}^n$, if $\|(t - t_0, x - x(t_0))\|_\infty \leq \delta$, then $F(t, x) \subset F(t_0, x(t_0)) + B_\epsilon(0)$. We may assume without loss of generality that for all $j \geq j_0$,

$$\delta_j < \frac{\delta}{2}, \quad \|x_j(t_0) - x(t_0)\| < \frac{\delta}{2}. \tag{1.3.116}$$

This implies that for all $j > j_0$, $\overline{B_{\delta_j}(t_0, x_j(t_0))} \subset \overline{B_\delta(t_0, x(t_0))}$ and

$$\overline{co}\{F(\overline{B_{\delta_j}(t_0, x_j(t_0))} \cap (([-1, \infty) \setminus N_0) \times \mathbb{R}^n))\} \subset F(t_0, x(t_0)) + \overline{B_\epsilon(0)}. \tag{1.3.117}$$

It follows that for any $j \geq j_0$, $\operatorname{dis}(\dot{\tilde{x}}_j(t_0), F(t_0, x(t_0))) \leq \epsilon$. This completes the proof of the lemma. $\qquad \square$

Now we use $\delta(t, x)$ to denote any continuous function defined on $[-1, \infty) \times \mathbb{R}^n \to \mathbb{R}^n$ such that, for any $(t, x) \in [-1, \infty) \times \mathbb{R}^n$, $\delta(t, x) \geq 0$ and $\delta(t, x) = 0$ if and only if $x = 0$. For such a given function $\delta(t, x)$, set

$$F_1(t, x) = \overline{co}\{F(\overline{B_{\delta(t,x)}} \cap E); \quad \forall (t, x) \in [-1, \infty) \times \mathbb{R}^n, \tag{1.3.118}$$

where $E = ([-1, \infty) \setminus N_0) \times \mathbb{R}^n$. We can verify that $F_1(t, x)$ also satisfies the condition of Theorem 1.3.11. Now we show that the following differential inclusion is globally asymptotically stable:

$$\dot{x}(t) \in F_1(t, x(t)), \quad t \geq -1. \tag{1.3.119}$$

From the uniform global asymptotical stability of (1.3.95), for any solution $x(t_0, x_0)$ of (1.3.95), there exists a class $\mathcal{KL}$ function $\beta : [0, \infty) \times [0, \infty) \to [0, \infty)$ such that $\|x(t_0 + h)\| \leq \beta(h, \|x_0\|)$. We say that $\beta(t, s)$ is the class $\mathcal{KL}$ function if, for any given t, $\beta(t, s)$ is the class $\mathcal{K}_\infty$ function with respect to s and, for any given s, $\beta(t, s)$ is decreasing with respect to t and $\lim_{t \to \infty} \beta(t, s) = 0$. Let $\varphi_i(h) = \beta(h, 2^i)$. We can prove that the sequence $\{\varphi_i(h)\}_{i=-\infty}^\infty$ of positive continuous decreasing functions on $[0, \infty)$ satisfies:

(i) For any $(t^0, x_0) \in [-1, +\infty) \times \mathbb{R}^n$ and any solution $x(t)$ of (1.3.95) with $x(t_0) = x_0$, if $\|x^0\|_\infty \leq 2^i$ then $\|x(t^0 + h)\|_\infty < \varphi_i(h)$ for any $h \geq 0$.

(ii) $\lim_{h \to \infty} \varphi_i(h) = 0$ for any i.

(iii) $\{\varphi_i(0)\}_{i=-\infty}^{\infty}$ is a nondecreasing sequence such that $\lim_{i \to -\infty} \varphi_i(0) = 0$ and $\lim_{i \to \infty} \varphi_i(0) = \infty$.

For integer $i \in \mathbb{Z}$, let $p_i \in \mathbb{Z}$ be the greatest natural number such that $\varphi_{p_i}(0) \leq 2^{i-1}$. Choose $T_i \geq 1$ so that $\varphi_i(T_i) \leq 2^{i-1}$ and set $\hat{T}_i = \max\{T_i, \max\{T_j : p_j = i\}\}$.

Lemma 1.3.7 *Let* $i \in \mathbb{Z}$ *and* $k \in \mathbb{N}^+ \cup \{-1\}$. *Then there exists constant* $\delta > 0$ *such that for any solution* $x(t)$ *of the following differential inclusion:*

$$\dot{x}(t) \in \overline{co}\{F(\overline{B_\delta(t, x)} \cap E)\} \qquad (1.3.120)$$

with $\|x(t_0)\| \leq 2^i$, $t_0 \in [k, k+1]$, *it has*

$$\|x(t_0 + h)\|_\infty \leq \varphi_i(h), \quad \forall \ h \in [0, \hat{T}_i]. \qquad (1.3.121)$$

Proof. Suppose that the conclusion is false. Then there exists a decreasing sequence of positive numbers $\{\delta_j\}_{j=1}^{\infty} : \lim_{j \to \infty} \delta_j = 0$ and a sequence of absolutely continuous functions $\{x_j(t)\}_{j=1}^{\infty}$ with $x_j : [t_j^0, t_j^1] \to \mathbb{R}^n$, $t_j^0 \leq t_j^1 \leq t_j^0 + \hat{T}_i$ such that

$$\begin{cases} \dot{x}_j(t) \in \overline{co}\{F(\overline{B_{\delta_j}(t, x_j(t))} \cap E)\} \text{ for almost all } t \in [t_j^0, t_j^1], \\ \|x_j(t_j^0)\|_\infty \leq 2^i, \ \|x_j(t)\|_\infty \leq \varphi_i(t - t_j^0) \text{ for all } t \in [t_j^0, t_j^1), \|x_j(t_j^1)\|_\infty = \varphi_i(t_j^1 - t_j^0). \end{cases}$$
$$(1.3.122)$$

By Lemma 1.3.6 and extracting a subsequence if necessary, we may also assume that for some $t^0, t^1 \in [k, k+1+\hat{T}_i]$ and some solution $x: [t^0, t^1] \to \mathbb{R}^n$ of (4.1):

$$\lim_{j \to \infty}(t_j^0, x_j(t_j^0)) = (t^0, x(t^0)), \quad \lim_{j \to \infty}(t_j^1, x_j(t_j^1)) = (t^1, x(t^1)). \qquad (1.3.123)$$

This yields $\|x(t^0)\|_\infty \leq 2^i$ and $\|x(t^1)\|_\infty = \varphi_i(t^1 - t^0)$, which contradicts the definition of $\varphi_i(t)$. This completes the proof of the lemma. $\qquad\square$

For any $(i, k) \in \mathbb{Z} \times (-1 \cup \mathbb{N}^+)$, the number $\delta > 0$ in Lemma 1.3.7 related to i and k is denoted by δ_i^k. Let $\delta : [-1, \infty) \times \mathbb{R}^n \to [0, \infty)$ be a Lipschitz continuous function with a Lipschitz constant one, and satisfy

$$\delta(t, x) = 0 \text{ if and only if } x = 0 \qquad (1.3.124)$$

and

$$\delta(t, x) < \min(\delta_i^k, \delta_{p_i}^k), \ \forall \ k \leq t, \ 2^{p_i} \leq \|x\| \leq \varphi_i(0). \qquad (1.3.125)$$

Lemma 1.3.8 *Let* $\delta(t)$ *satisfy (1.3.124) and (1.3.125), and let* $x(t)$ *be any solution of (1.3.119). Then for* $t_0 \geq -1$, $i \in \mathbb{Z}$ *and* $\|x(t_0)\| \leq 2^i$,

(a) $\|x(t_0 + h)\|_\infty < \varphi_i(0)$ *for any* $h \in [0, T_i]$.
(b) $\|x(t_0 + T_i)\|_\infty \leq 2^{i-1}$.

Proof. If (a) is false, then there exist $t_1, t_2\colon t_0 < t_1 < t_2 \leq t_0 + T_i$ such that

$$2^i = \|x(t_1)\|_\infty < \|x(t)\|_\infty < \|x(t_2)\|_\infty = \varphi_i(0), \quad t \in (t_1, t_2). \tag{1.3.126}$$

Set $k = [t_1]$. Since $2^{p_i} \leq \|x(t)\| \leq \varphi_i(0)$ for $t \in [t_1, t_2]$, $x(t)$ is also a solution of the following differential inclusion:

$$\dot{x}(t) \in \overline{\mathrm{co}}\{F(\overline{B_{\delta_i^k}(t, x)} \cap E)\}. \tag{1.3.127}$$

By Lemma 1.3.7, $\|x(t_2)\|_\infty \leq \varphi_i(t_2 - t_1) < \varphi_i(0)$, which contradicts $\|x(t_2)\|_\infty = \varphi_i(0)$. Therefore, (a) is valid.

Now we prove (b). Assume that $\|x(t_0 + T_i)\|_\infty > 2^{i-1}$. If $2^{p_i} \leq \|x(t_0 + h)\|_\infty \leq \varphi_i(0)$ for every $h \in [0, h_i]$, then $x(t)$ also satisfies (1.3.127) on $[t_0, t_0 + T_i]$ with $k = [t_0]$. By Lemma 1.3.7, $\|x(t_0 + T_i)\|_\infty < \varphi_i(T_i) \leq 2^{i-1}$, which is a contradiction. Therefore, there exist t_1 and t_2 such that $t_0 \leq t_1 < t_2 \leq t_0 + T_i$ and $2^{p_i} = \|x(t_1)\|_\infty < \|x(t)\|_\infty < \|x(t_2)\|_\infty = 2^{i-1}$ for $t \in (t_1, t_2)$. For $k = [t_1]$, consider $\delta(t, x(t)) < \delta_{p_i}^k$ for $t_1 < t < t_2$ and $t_2 - t_1 \leq T_i \leq \hat{T}_{p_i}$. Once again, by Lemma 1.3.7, $\|x(t_2)\|_\infty < \varphi_{p_i}(0) \leq 2^{i-1}$. The conclusion is obtained by the contradiction of the property of T_2. $\qquad\square$

By Lemma 1.3.8, we can obtain that $\|x(t_0 + h)\|_\infty < \varphi_{i-l}(0)$ for any $l \in \mathbb{N}^+$ and $h \geq \sum_{j=i-l+1}^{i} T_j$. This means that (1.3.119) is uniformly globally asymptotically stable.

We are now in a position to enlarge and regularize the differential inclusion $\dot{x}(t) \in F_2(t, x(t))$. To this purpose, we need some suitable partition of unity. Set

$$U = (-1, +\infty) \times (\mathbb{R}^n \setminus \{0\}), \tag{1.3.128}$$

and for any $(t, x) \in U$,

$$W(t, x) = \left\{ (s, y) \in U \mid \|(s - t, y - x)\| < \frac{1}{3}\delta(t, x) \right\}. \tag{1.3.129}$$

It is easy to see that the family $\{W(t, x)\}_{(t,x)\in U \cap E}$ is an open covering of U.

Let $\{\psi_i(t, x)\}_{i \in \mathbb{N}^+}$ be a C^∞-partition of unity on U subordinate to the open covering $\{W(t, x)\}_{(t,x)\in U \cap E}$ of U. It means that, firstly, each $\psi_i(t, x)$ is a nonnegative function of class C^∞ on $\mathbb{R}^{n+1}$, with support contained in $W(t_i, x_i)$ for $(t_i, x_i) \in U \cap E$; secondly, for any $(t, x) \in U$, $\sum_{i=1}^{\infty} \psi_i(t, x) = 1$; and lastly, for any $(t, x) \in U$, there exists a number $\rho > 0$ such that $\psi_i(t, x) \equiv 0$ on $B_\rho(t, x)$ for all $i \in \mathbb{N}^+$ except finitely many i's.

For any $(t, x) \in (-1, +\infty) \times \mathbb{R}^n$, set

$$F_2(t, x) = \begin{cases} \displaystyle\sum_{i=1}^{\infty} \psi_i(t, x)\overline{\mathrm{co}}\{F(\overline{B_{\frac{1}{3}\delta(t_i, x_i)}(t_i, x_i)} \cap E)\}, & x \neq 0, \\[2mm] F(t, 0), & x = 0. \end{cases} \tag{1.3.130}$$

Since the summation in (1.3.130) is finite on the compact subset of U, we see that $F_2(t, x)$ is locally Lipschitz continuous in the Hausdorff distance on U. It is clear that for $x = 0$, $F(t, x) \subset F_2(t, x)$. Let $x \neq 0$ and $t \in (-1, +\infty) \setminus N_0$, $i \in \mathbb{N}^+$ such that $\psi_i(t, x) > 0$. By the definition of $\psi(t, x)$, $(t, x) \in W(t_i, x_i)$. This together with (1.3.129) yields $\|(t - t_i, x - x_i)\|_\infty < \frac{1}{3}\delta(t_i, x_i)$. Hence

$$F(t, x) \subset F(\overline{B_{\frac{1}{3}\delta(t_i, x_i)}(t_i, x_i)} \cap E), \tag{1.3.131}$$

which implies that $F(t, x) \subset F_2(t, x)$. Therefore, for every $(t, x) \in ((-1, +\infty) \setminus N_0) \times \mathbb{R}^n$, $F(t, x) \subset F_2(t, x)$.

Furthermore, for any $(t, x) \in U$ and $i \in \mathbb{N}^+$ satisfying $\psi_i(t, x) > 0$, since $\delta(t, x)$ is Lipschitz continuous with the Lipschitz constant one, we can obtain

$$\delta(t_i, x_i) - \delta(t, x) \leq \|(t - t_i, x - x_i)\|_\infty \leq \frac{1}{3}\delta(t_i, x_i). \tag{1.3.132}$$

This yields

$$\overline{B_{\frac{1}{3}\delta(t_i, x_i)}(t_i, x_i)} \subset \overline{B_{\frac{2}{3}\delta(t_i, x_i)}(t, x)} \subset \overline{B_{\delta(t, x)}(t, x)}, \tag{1.3.133}$$

and hence $F_2(t, x) \subset \overline{co}\{F(\overline{B_{\delta(t, x)}(t, x)} \cap E)\} \subset F_1(t, x)$. This together with the uniform global asymptotical stability of $\dot{x}(t) \in F_1(t, x(t))$ deduces that $\dot{x}(t) \in F_2(t, x(t))$ is also uniformly globally asymptotically stable.

Secondly, we construct a local Lipschitz continuous Lyapunov function.

For any $(t_0, x_0) \in (-1, +\infty) \times \mathbb{R}^n$, let $\mathcal{S}_{t_0; x_0}$ be the set of solutions $x(t)$ of the differential inclusion $\dot{x}(t) \in F_2(t, x(t))$ with initial condition $x(t_0) = x_0$. For any $q \in \mathbb{N}^+$, $r \in [0, \infty)$, and $(t, x) \in (-1, +\infty) \times \mathbb{R}^n$, set

$$G_q(r) = \max\left\{0, \ r - \frac{1}{q}\right\} \tag{1.3.134}$$

and

$$V_q(t, x) = \sup_{\varphi \in \mathcal{S}_{t; x}} \sup_{\tau \geq 0} e^{2\lambda\tau} G_q(\|\varphi(t + \tau)\|_\infty), \tag{1.3.135}$$

where λ is the positive number appearing in Theorem 1.3.11.

From the uniform global asymptotical stability of $\dot{x}(t) \in F_2(t, x(t))$, we can infer that, for any $R > 0$ and $q \in \mathbb{N}^+$, there exist the class $\mathcal{K}_\infty$ function $m(R)$ and nondecreasing function $T(R, q)$ such that, as long as $\|x_0\|_\infty \leq R$, for each $(t_0, x_0) \in (-1, +\infty) \times \mathbb{R}^n$, $\varphi \in \mathcal{S}_{t_0; x_0}$, $\|\varphi(t_0 + \tau)\|_\infty < m(R)$ for all $\tau \geq 0$, and $\|\varphi(t_0 + \tau)\|_\infty < \frac{1}{q}$ for any $\tau > T(R, q)$.

The following Lemma 1.3.9 is a direct consequence of (1.3.134) and (1.3.135).

Lemma 1.3.9 *Let $R > 0$ and $(t, x) \in (-1, +\infty) \times \overline{B_R(0)}$. Then for any $q \in \mathbb{N}^+$,*

$$G_q(\|x\|_\infty) \leq V_q(t, x) \leq e^{2\lambda T(R, q)} m(R) < \infty. \tag{1.3.136}$$

Another important property of $V_q(t, x)$ is the local Lipschitz continuity.

Proposition 1.3.1 *Let $q \in \mathbb{N}^+$ and $R > 0$. Then there exists a positive constant $C_q(R)$ such that for any $t_1, t_2 \in [-R/(R + 1), R]$ and $x_1, x_2 \in \overline{B_R(0)}$,*

$$|V_q(t_1, x_1) - V_q(t_2, x_2)| \leq C_q(R)\|(t_1 - t_2, x_1 - x_2)\|_\infty. \tag{1.3.137}$$

We assume without loss of generality that, for every $q \in \mathbb{N}^+$, the function $C_q(R)$ is nondecreasing.

To prove Proposition 1.3.1, we need the following elementary lemma.

Lemma 1.3.10 *Let $V(x)$ be a function defined on a set $\mathbb{K} = \prod_{i=1}^{n}[a_i, b_i] \subset \mathbb{R}^n$. Assume that there exists a constant $L > 0$ such that for any $x_0 \in \mathbb{K}$ there exists $\eta_0 > 0$ satisfying*

$$|V(x) - V(x_0)| \leq \|x - x_0\|_{\infty}, \quad \forall\, x \in \overline{B_{\eta_0}(x_0)} \cap \mathbb{K}. \tag{1.3.138}$$

Then $V(x)$ is Lipschitz continuous on $\mathbb{K}$ with the Lipschitz constant nL.

Proof. Let

$$x^1 = (x_1^1, x_2^1, \ldots, x_n^1) \in \mathbb{K}, \quad x^2 = (x_1^2, x_2^2, \ldots, x_n^2) \in \mathbb{K}. \tag{1.3.139}$$

Then

$$|V(x^1) - V(x^2)| \leq \sum_{j=1}^{n} |V(x_1^1, \ldots, x_j^1, x_{j+1}^2, \ldots, x_n^2) - V(x_1^1, \ldots, x_{j-1}^1, x_j^2, \ldots, x_n^2)|. \tag{1.3.140}$$

By (1.3.138), it follows that, for any $j \in [1, n]$,

$$|V(x_1^1, \ldots, x_j^1, x_{j+1}^2, \ldots, x_n^2) - V(x_1^1, \ldots, x_{j-1}^1, x_j^2, \ldots, x_n^2)| \leq L|x_j^1 - x_j^2|. \tag{1.3.141}$$

This completes the proof of the lemma. $\qquad\square$

Proposition 1.3.1 can be obtained directly by Lemma 1.3.10 and the following Proposition 1.3.2.

Proposition 1.3.2 *Let $q \in \mathbb{N}^+$ and $R > 0$. Then there exists a positive constant $L > 0$ such that, for any $t_0 \in [-R/(R+1), R]$ and any $x_0 \in \overline{B_R(0)}$, there exists $\eta_0 > 0$ satisfying*

$$|V_q(t, x) - V_q(t_0, x_0)| \leq L\|(t - t_0, x - x_0)\|_{\infty}, \quad \forall\, (t, x) \in \overline{B_{\eta_0}(t_0, x_0)}. \tag{1.3.142}$$

Let $T = T(R+1, q)$. By local Lipschitz continuity of $F_2(t, x)$ on U, there exists $K > 0$ such that for any (t_1, x_1) and (t_2, x_2) satisfying

$$-\frac{R+1}{R+2} \leq t_i \leq R + T + 1, \quad \frac{1}{2}m^{-1}\left(\frac{1}{q}\right) \leq \|x_1\|_{\infty} \leq m(R+2), \quad i = 1, 2, \tag{1.3.143}$$

we have

$$\hbar(F_2(t_1, x_1), F_2(t_2, x_2)) \leq \|(t_1 - t_2, x_1 - x_2)\|_{\infty}. \tag{1.3.144}$$

Let $M \geq 1$ be a constant such that for every $(t, x) \in (-(R+1)/(R+2), R+T+1) \setminus N_0 \times \overline{B_{m(R+2)}(0)}$,

$$F_2(t, x) \subset \overline{B_M(0)}. \tag{1.3.145}$$

Let

$$L = e^{2\lambda T}((M+1)e^{K(T+1)} + 2\lambda m(R+1)) \tag{1.3.146}$$

and let $\bar{\eta}_0$ be a constant satisfying

$$\bar{\eta}_0 \in \left(0, \ \min\left\{\frac{R+1}{R+2} - \frac{R}{R+1}, e^{-K(T+1)}\min\left\{m(R+2) - m(R+1), \frac{1}{2}m^{-1}\left(\frac{1}{q}\right)\right\}\right\}\right) \tag{1.3.147}$$

and

$$b = \bar{\eta}_0 e^{K(T+1)}. \tag{1.3.148}$$

The proof of Proposition 1.3.2 is lengthy. Before giving the proof, we present the following Lemma 1.3.11, which is useful in the proof of Proposition 1.3.2.

Lemma 1.3.11 *Let* $(t_0, x_0) \in [-R/(R+1), R] \times \overline{B_R(0)}$ *and* $(t_1, x_1) \times \overline{B_{\bar{\eta}_0}(t_0, x_0)}$. *If* $V_q(t_1, x_1) > 0$, *then for any* $\varphi_1 \in S_{t_1; x_1}$ *satisfying* $\|\varphi_1(t_1 + \tau)\|_\infty > 1/q$ *for some* $\tau \in [0, T]$,

$$m^{-1}\left(\frac{1}{q}\right) < \|\varphi_1(t_1 + h)\|_\infty < m(R+1) \quad \forall \ h \in [0, \tau]. \tag{1.3.149}$$

Proof. By

$$\bar{\eta}_0 < \frac{R+1}{R+2} - \frac{R}{R+1} < 1, \tag{1.3.150}$$

it has

$$-\frac{R+1}{R+2} < t_1 \le t_1 + \tau < R + T + 1. \tag{1.3.151}$$

Let $\varphi_1(t)$ satisfy the conditions of the lemma. Since $\|x_1\|_\infty \le \|x_0\|_\infty + \eta_0 < R+1$,

$$\|\varphi_1(t_1 + h)\| < m(R+1), \quad \forall \ h \ge 0. \tag{1.3.152}$$

Since $\|\varphi_1(t_1 + \tau)\|_\infty > 1/q$, we have

$$\|\varphi_1(t_1 + h)\|_\infty > m^{-1}\left(\frac{1}{q}\right), \quad \forall \ h \in [0, \tau]. \tag{1.3.153}$$

The remaining proof of the lemma can be obtained from (1.3.147) to (1.3.149). $\square$

Proof of Proposition 1.3.2. Let $(t_0, x_0) \in [-R/(R+1), R] \times \overline{B_R(0)}$ be fixed. In what follows, we always assume that

$$\eta_0 \in \left(0, \frac{\eta_0}{2M+1}\right). \tag{1.3.154}$$

The proof is divided into two cases: $V_q(t_0, x_0) \ne 0$ and $V_q(t_0, x_0) = 0$.

Case 1: $V_q(t_0, x_0) \ne 0$. In this case, the proof is accomplished by the following two claims.
  Claim 1: If η_0 *is small enough, then* $V_q(t, x) \ne 0$ *for any* $(t, x) \in \overline{B_{\eta_0}(t_0, x_0)}$.
  Let $\varphi_0 \in S_{t_0, x_0}$ satisfy

$$V_q(t_0, x_0) - e^{2\lambda \tau} G_q(\|\varphi_0(t_0 + \tau)\|_\infty) < \frac{V_q(t_0, x_0)}{2}, \quad \tau \in (0, T]. \tag{1.3.155}$$

Then $\|\varphi_0(t_0 + \tau)\| > 1/q$. This together with Lemma 1.3.11 shows that

$$m^{-1}\left(\frac{1}{q}\right) < \|\varphi_0(t_0 + h)\|_\infty < m(R+1), \quad \forall \ h \in [0, \tau]. \tag{1.3.156}$$

We assume without loss of generality that $\eta_0 < \tau$. Then $\varphi_0(t)$ is defined on $[t_0 - \eta_0, t_0]$ and (1.3.156) holds true for $h \in [-\eta_0, \tau$ and $[t_0 - \eta_0, t_0 + \tau] \subset$

$[-(R+1)/(R+2), R+T+1]$. Let $(t,x) \in \overline{B_{\eta_0}}(t_0, x_0)$. Then $|t-t_0| \leq \eta_0$. By (1.3.145),

$$\|\varphi_0(t) - x\|_\infty \leq \|\varphi_0(t) - \varphi_0(t_0)\|_\infty \leq M|t-t_0| + \eta_0$$
$$\leq (M+1)\eta_0 < \overline{\eta}_0 < b. \tag{1.3.157}$$

By Lemmas 1.3.5 and 1.3.12, there exists $\psi \in \mathcal{S}_{t;x}$ such that

$$|\varphi_0(t+s) - \psi(t+s)\|_\infty \leq \|\varphi_0(t) - x\|_\infty e^{Ks} \tag{1.3.158}$$

as long as $m^{-1}(1/q) \leq \|\varphi_0(t+s)\|_\infty \leq m(R+1)$ and s is small enough so that $\|\varphi_0(t) - x\|_\infty e^{Ks} < b$.

It follows from (1.3.158) that if η_0 is sufficiently small, then

$$\|\psi(t_0 + \tau)\|_\infty \geq \|\varphi_0(t_0 + \tau)\|_\infty - (M+1)e^{K(\eta_0 + \tau)}$$
$$\geq \|\varphi_0(t_0 + \tau)\|_\infty - (M+1)e^{K(T+1)}\eta_0 > \frac{1}{q}.$$

This yields $V_q(t,x) > 0$.

Let η_0 be the same as in Claim 1 and let $(t_1, x_1), (t_2, x_2) \in \overline{B_{\eta_0}}(t_0, x_0)$. The inequality (1.3.142) is a consequence of the following Claim 2.

Claim 2: $|V_q(t_1, x_1) - V_q(t_2, x_2)| \leq L\|(t_1 - t_2, x_1 - x_2)\|_\infty$.

We may assume without loss of generality that $t_1 \leq t_2$. Firstly, we prove that

$$|V_q(t_1, x_1) - V_q(t_1, x_2)| \leq e^{(2\lambda + K)T}\|x_1 - x_2\|_\infty. \tag{1.3.159}$$

By the definition of $V_q(t,x)$, for every $\sigma \in (0, V_q(t_1, x_1))$, there exist $\varphi_1 \in \mathcal{S}_{t_1; x_1}$ and $\tau \in [0, T]$ such that

$$V_q(t_1, x_1) - \sigma < e^{2\lambda\tau} G_q(\|\varphi_1(t_1 + \tau)\|_\infty) \leq V_q(t_1, x_1). \tag{1.3.160}$$

Hence

$$V_q(t_1, x_1) - V_q(t_1, x_2) < e^{2\lambda\tau} G_q(\|\varphi_1(t_1 + \tau)\|_\infty) - V_q(t_1, x_2) + \sigma. \tag{1.3.161}$$

Since $\|x_1 - x_2\| \leq 2\eta_0 < \overline{\eta}_0 < b$, we infer from Lemmas 1.3.5, 1.3.11, and Claim 1 that there exists a solution $\psi_2 \in \mathcal{S}_{t_1, x_2}$ such that $\|\varphi_1(t) - \varphi_2(t)\|_\infty \leq \|x_1 - x_2\|_\infty e^{K|t - t_1|}$ as long as $m^{-1}(1/q) \leq \|\varphi_1(t)\|_\infty \leq m(R+1)$ and $\|x_1 - x_2\|_\infty e^{K|t_1 - t|} \leq b$. Since $V_q(t_1, x_2) \geq e^{2\lambda\tau} G_q(\|\varphi_2(t_1 + \tau)\|_\infty)$ and $G_q(\cdot)$ is Lipschitz continuous with Lipschitz constant one, for any $t \in [t_1, t_1 + \tau]$, we have

$$V_q(t_1, x_1) - V_q(t_1, x_2) \leq e^{2\lambda\tau}(G_q(\|\varphi_1(t_1 + \tau)\|_\infty) - G_q(\|\varphi_2(t_1 + \tau)\|_\infty)) + \sigma$$
$$\leq e^{(2\lambda + K)T}\|x_1 - x_2\|_\infty + \sigma. \tag{1.3.162}$$

Exchanging x_1 and x_2, we obtain (1.3.159) by the arbitrariness of σ.

Now we show that

$$V_q(t_2, x_2) - V_q(t_1, x_2) \leq Me^{(2\lambda + K)T}|t_2 - t_1|. \tag{1.3.163}$$

For any $\varphi \in \mathcal{S}_{t_1;x_2}$, set $x_3 = \varphi(t_2)$. It follows from (1.3.12) that

$$V_q(t_2, x_3) \le e^{-2\lambda(t_2-t_1)}V_q(t_1, x_2) \le V_q(t_1, x_2) \tag{1.3.164}$$

and hence

$$V_q(t_2, x_2) - V_q(t_1, x_2) \le V_q(t_2, x_2) - V_q(t_2, x_3). \tag{1.3.165}$$

Since

$$\|x_2 - x_3\|_\infty \le \int_{t_1}^{t_2} \|\dot\varphi(t)\|_\infty dt \le M|t_2 - t_1| \le 2\eta_0 M \tag{1.3.166}$$

and $2\eta_0 M < \bar\eta_0 < b$ by (1.3.154), we conclude, with a similar proof to that of (1.3.159), that

$$V_q(t_2, x_2) - V_q(t_2, x_3) \le e^{(2\lambda+K)T}\|x_2 - x_3\|_\infty \le Me^{(2\lambda+K)T}|t_2 - t_1|. \tag{1.3.167}$$

This together with (1.3.165) gives (1.3.163).

We show that

$$V_q(t_1, x_2) - V_q(t_2, x_2) \le (Me^{KT} + 2\lambda m(R+1))e^{2\lambda T}|t_1 - t_2|. \tag{1.3.168}$$

Actually, from the definition, for each $\sigma \in (0, V_q(t_1, x_2))$ there exists a solution $\psi \in \mathcal{S}_{t_1,x_2}$ and $\tau \in [0, T]$ such that $V_q(t_1, x_2) \le e^{2\lambda\tau}G_q(\|\psi(t_1 + \tau)\|_\infty) + \sigma$. The proof of (1.3.168) is accomplished with two cases.

(i) $t_1 + \tau > t_2$. In this case, set $x_4 = \psi(t_2)$. We can obtain that

$$\|x_4 - x_0\|_\infty \le \int_{t_1}^{t_2} \|\dot\psi(t)\|_\infty dt \le M|t_2 - t_1| \le 2M\eta_0. \tag{1.3.169}$$

Hence

$$\|x_4 - x_0\|_\infty < (2M+1)\eta_0 > \bar\eta_0. \tag{1.3.170}$$

It follows from Lemmas 1.3.11 and 1.3.5 that there exists a solution $\psi \in \mathcal{S}_{t_2;x_2}$ such that for any $t \in [t_2, t_1 + \tau]$,

$$\|\varphi(t) - \psi(t)\|_\infty \le \|x_4 - x_2\|_\infty e^{K|t-t_2|}. \tag{1.3.171}$$

By $V_q(t_2, x_2) \ge e^{2\lambda(t_1+\tau-t-2)}G_q(\|\varphi(t_1 + \tau)\|_\infty)$, it follows that

$$V_q(t_1, x_2) - V_q(t_2, x_2) \le e^{2\lambda\tau}G_q(\|\psi(t_1 + \tau)\|_\infty) - e^{2\lambda(\tau+t_1-t_2)}$$
$$G_q(\|\varphi(t_1 + \tau)\|_\infty) + \sigma$$
$$\le e^{2\lambda\tau}(|G_q(\|\psi(t_1 + \tau)\|_\infty) - G_q(\|\varphi(t_1 + \tau)\|_\infty)|$$
$$+ (1 - e^{-2\lambda|t_1-t_2|})G_q(\|\varphi(t_1 + \tau)\|_\infty) + \sigma). \tag{1.3.172}$$

In addition,

$$|G_q(\|\psi(t_1+\tau)\|_\infty) - G_q(\|\varphi(t_1+\tau)\|_\infty)| \le \|\psi(t_1+\tau) - \varphi(t_1+\tau)\|_\infty$$
$$\le \|x_4 - x_2\|_\infty e^{K|t_1+\tau-t_2|}$$
$$\le Me^{KT}|t_2 - t_1| \qquad (1.3.173)$$

and

$$(1 - e^{-2\lambda|t_1-t_2|})G_q(\|\varphi(t_1+\tau)\|_\infty) \le 2\lambda|t_1 - t_2|m(R+1). \qquad (1.3.174)$$

Therefore,

$$V_q(t_1, x_2) - V_q(t_2, x_2) \le e^{2\lambda T}(Me^{KT} + 2\lambda m(R+1))|t_1 - t_2| + \sigma. \quad (1.3.175)$$

(ii) $t_1 + \tau \le t_2$. In this case, by $V_q(t_2, x_2) \ge G_q(\|x_2\|_\infty)$, we obtain

$$V_q(t_1, x_1) - V_q(t_2, x_2) \le [e^{2\lambda\tau}G_q(\|\psi(t_1+\tau)\|_\infty) - G_q(\|x_2\|_\infty) + \sigma$$
$$\le e^{2\lambda\tau}|G_q(\|\psi(t_1+\tau)\|_\infty) - G_q(\|x_2\|_\infty)$$
$$+ (e^{2\lambda\tau-1})G_q(\|x_2\|_\infty) + \sigma. \qquad (1.3.176)$$

Since

$$|\|\psi(t_1+\tau)\|_\infty - \|x_2\|_\infty| \le \left\|\int_{t_1}^{t_2} \dot\psi(t)dt\right\|_\infty \le M\tau \le M|t_2 - t_1| \quad (1.3.177)$$

and

$$|e^{2\lambda\tau} - 1| \le 2\lambda\tau e^{2\lambda\tau} \le 2\lambda e^{2\lambda T}|t_2 - t_1|, \qquad (1.3.178)$$

we obtain

$$V_q(t_1, x_2) - V_q(t_2, x_2) \le e^{2\lambda T}(M + 2\lambda m(R+1))|t_1 - t_2| + \sigma$$
$$\le e^{2\lambda T}(Me^{KT} + 2\lambda m(R+1))|t_1 - t_2| + \sigma. \quad (1.3.179)$$

Therefore (1.3.175) holds in both cases and (1.3.168) is valid by the arbitrariness of σ.

Finally, by (1.3.159), (1.3.163), and (1.3.168),

$$|V_q(t_1, x_1) - V_q(t_2, x_2)|$$
$$\le |V_q(t_1, x_1) - V_q(t_1, x_2)| + |V_q(t_1, x_2) - V_q(t_2, x_2)|$$
$$\le e^{(2\lambda+K)T}\|x_1 - x_2\|_\infty + (Me^{KT} + 2\lambda m(R+1)e^{2\lambda T}|t_1 - t_2|)$$
$$\le L\|(t_1 - t_2, x_1 - x_2)\|_\infty. \qquad (1.3.180)$$

This completes the proof of Claim 2.

Case 2: $V_q(t_0, x_0) = 0$. By (1.3.154), $M\eta_0 < 1$. We claim that for any $(t, x) \in \overline{B_{\eta_0}(t_0, x_0)}$ and any $\varphi \in \mathcal{S}_{t;x}$, φ is defined on $[t - \eta_0, +\infty)$ (including t_0). Indeed, by (1.3.145), for any $s \in \text{dom}(\varphi) \cap [t - \eta_0, t]$, if $\|\varphi(s)\|_\infty \le m(R + 2)$, then

$$\|\varphi(s)\|_\infty < \|\varphi(s) - \varphi(t)\|_\infty + \|x\|_\infty \le M|s - t| + R + \eta_0$$

$$\le (M + 1)\eta_0 + R \le R + 1. \tag{1.3.181}$$

Since $R + 1 < m(R + 2)$, a direct computation shows that $[t - \eta_0, t] \subset \text{dom}(\varphi)$ and (1.3.181) holds true on $[t_0 - \eta, t]$. Pick any $(t, x) \in B_{\eta_0}(t_0, x_0)$. If $V_q(t, x) = 0$, then (1.3.142) is trivial. If $V_q(t, x) > 0$, then for any $\sigma \in (0, V_q(t, x))$, there exists a solution $\varphi \in \mathcal{S}_{t;x}$ and $\tau \in [0, T]$ such that

$$V_q(t, x) \le e^{2\lambda\tau} G_q(\|\varphi(t + \tau)\|_\infty) + \sigma. \tag{1.3.182}$$

Once again we divide the remaining proof into two cases.

Case (a): $t_0 < t + \tau$. In this case, since $\varphi(t)$ is defined on $[t - \eta_0, +\infty)$, it is also well-defined at t_0. Since $\|\varphi(t + \tau)\|_\infty > 1/q$, we have $m^{-1}(1/q) < \|\varphi(s)\|_\infty < m(R + 1)$ for any $s \in [t - \eta_0, t + \tau]$. Furthermore, by (1.3.154),

$$\|\varphi(t_0) - x_0\|_\infty \le \|\varphi(t_0) - \varphi(t)\|_\infty + \|x - x_0\|_\infty \le (M + 1)\eta_0 < \bar{\eta}_0. \tag{1.3.183}$$

It follows from Lemmas 1.3.5 and 1.3.11 that there exists a solution $\psi \in \mathcal{S}_{t_0;x_0}$ such that

$$\|\psi(t + \tau) - \varphi(t + \tau)\|_\infty \le \|\psi(t_0) - \varphi(t_0)\|_\infty e^{K|t + \tau - t_0|}$$

$$\le (\|x_0 - x\|_\infty + \|\varphi(t) - \varphi(t_0)\|_\infty)e^{K(T+1)}$$

$$\le (M + 1)e^{K(T+1)}\|(t - t_0, x - x_0)\|_\infty. \tag{1.3.184}$$

This yields from $V_q(t_0, x_0) = 0$ that $G_q(\|\psi(t + \tau)\|_\infty) = 0$. Therefore,

$$V_q(t, x) \le e^{2\lambda\tau}(G_q(\|\varphi(t + \tau)\|_\infty) - G_q(\|\psi(t + \tau)\|_\infty)) + \sigma$$

$$\le (M + 1)e^{2\lambda T + K(T+1)}\|(t - t_0, x - x_0)\|_\infty + \sigma. \tag{1.3.185}$$

Case (b): $t_0 > t + \tau$. In this case, since

$$G_q(\|\varphi(t + \tau)\|_\infty) = G_q(\|\varphi(t + \tau)\|_\infty) - G_q(\|x_0\|_\infty)$$

$$\le |G_q(\|\varphi(t + \tau)\|_\infty) - G_q(\|x\|_\infty)|$$

$$+ |G_q(\|x\|_\infty) - G_q(\|x_0\|_\infty)|$$

$$\le M\tau + \|x - x_0\|_\infty \le M|t - t_0| + \|x - x_0\|_\infty, \tag{1.3.186}$$

by (1.3.182) and (1.3.186), it follows that

$$V_q(t, x) \le e^{2\lambda T}(M + 1)\|(t - t_0, x - x_0)\|_\infty + \sigma. \tag{1.3.187}$$

To sum up, in any case, $0 \le V_q(t, x) \le L\|(t - t_0, x - x_0)\|_\infty + \sigma$. Therefore, (1.3.142) is valid by the arbitrariness of σ. This completes the proof of Proposition 1.3.2. $\qquad\square$

We are now in a position to construct a continuous Lyapunov function for the differential inclusion $\dot{x}(t) \in F_2(t, x(t))$. For any $(t, x) \in (-1, +\infty) \times \mathbb{R}^n$, set

$$V_L(t, x) = \sum_{q=1}^{\infty} \frac{2^{-q}}{1 + C_q(q)} r^{-2\lambda T(q,q)} V_q(t, x). \tag{1.3.188}$$

For any $r \geq 0$, set

$$a_L(r) = \sum_{q=1}^{+\infty} \frac{2^{-q} e^{-2\lambda T(q,q)}}{1 + C_q(q)} G_q(r). \tag{1.3.189}$$

Clearly, $a_L(r)$ is well-defined, increasing, Lipschitz continuous, and $\lim_{r \to +\infty} a_L(r) = +\infty$, that is, $a_L(r)$ belongs to the class $\mathcal{K}_\infty$. Furthermore,

$$a_L(\|x\|_\infty) \leq V_L(t, x), \ \forall \ (t, x) \in (-1, +\infty) \times \mathbb{R}^n. \tag{1.3.190}$$

Let

$$L(R) = \sum_{q=1}^{+\infty} 2^{-q} \frac{C_q(R)}{1 + C_q(q)} e^{-2\lambda T(q,q)}, \ \forall \ R > 0, q \in \mathbb{N}^+. \tag{1.3.191}$$

It is easy to verify that $L(R)$ is nondecreasing and

$$|V_L(t_1, x_1) - V_L(t_2, x_2)|_\infty \leq L(R)\|(t_1 - t_2, x_1 - x_2)\|_\infty. \tag{1.3.192}$$

By (1.3.136), it follows that for any $R > 0$ and $(t, x) \in (-1, +\infty) \times \overline{B_R(0)}$,

$$
\begin{aligned}
V_L(t, x) &\leq \sum_{i=1}^{+\infty} 2^{-q} \frac{e^{2\lambda(T(R,q) - T(q,q))m(R)}}{1 + C_q(q)} \\
&\leq \left[\sum_{q=1}^{[R]} 2^{-q} \frac{e^{2\lambda(T(R,q) - T(q,q))}}{1 + C_q(q)} + 1 \right] m(R) = \tilde{m}(R).
\end{aligned} \tag{1.3.193}
$$

It is easy to obtain that $\tilde{m}(R)$ is nondecreasing and $\lim_{R \to 0^+} \tilde{m}(R) = 0$. Hence, there exists a class $\mathcal{K}_\infty$ function $b_L(R)$ such that $\tilde{m}(R) \leq b_L(R)$. Therefore,

$$V_L(t, x) \leq b_L(\|x\|_\infty), \ \forall \ (t, x) \in (-1, +\infty) \times \mathbb{R}^n. \tag{1.3.194}$$

Lemma 1.3.12 *Let $(t_0, x_0) \in (-1, +\infty) \times \mathbb{R}^n$ and let $\psi \in \mathcal{S}_{t_0, x_0}$. Then for any $q \in \mathbb{N}^+$ and $h > 0$,*

$$V_q(t_0 + h, \psi(t_0 + h)) \leq e^{-2\lambda h} V_q(t_0, x_0). \tag{1.3.195}$$

As a direct consequence of Lemma 1.3.12, for any $(t_0, x_0) \in (-1, +\infty) \times \mathbb{R}^n$ and $\psi \in \mathcal{S}_{t_0, x_0}$,

$$V_L(t_0 + h, \psi(t_0 + h)) \leq e^{-2\lambda h} V_L(t_0, x_0), \forall \ h \geq 0. \tag{1.3.196}$$

For any $\psi \in \mathcal{S}_{t_0;x_0}$ and $\varphi \in \mathcal{S}_{t_0+h;\psi(t_0+h)}$, set

$$\overline{\varphi}(t) = \begin{cases} \psi(t), & t_0 \leq t \leq t_0 + h, \\ \varphi(t), & t_0 + h \leq t. \end{cases} \tag{1.3.197}$$

It is clear that $\overline{\varphi} \in \mathcal{S}_{t_0;x_0}$, and for any $\varphi \in \mathcal{S}_{t_0+h;\psi(t_0+h)}$,

$$V_q(t_0, x_0) \geq \sup_{\tau \geq 0} e^{2\lambda\tau} G_q(\|\overline{\varphi}(t_0 + \tau)\|_\infty) \geq e^{2\lambda h} \sup_{\tau \geq 0} e^{2\lambda\tau} G_q(\|\varphi(t_0 + h + \tau)\|_\infty). \tag{1.3.198}$$

It follows that

$$V_q(t_0, x_0) \geq e^{2\lambda h} V_q(t_0 + h, \psi(t_0 + h)), \tag{1.3.199}$$

and (1.3.195) holds true.

Corollary 1.3.1 *For almost all $(t_0, x_0) \in U$, $\nu \in F_2(t_0, x_0)$,*

$$\frac{\partial V_L(t_0, x_0)}{\partial t} + \langle \nabla_x V_L(t_0, x_0), \nu \rangle \leq -2\lambda V_L(t_0, x_0). \tag{1.3.200}$$

Proof. Since $V_L(t, x)$ is Lipschitz continuous on U, it is therefore differentiable for almost all $(t_0, x_0) \in U$. We show that for almost all $(t_0, x_0) \in U$ and $\nu \in F_2(t_0, x_0)$,

$$\limsup_{h \to 0^+} \frac{V_L(t_0 + h, x_0 + h\nu) - V_L(t_0, x_0)}{h} \leq -2\lambda V_L(t_0, x_0), \tag{1.3.201}$$

and for any $\nu \in F_2(t_0, x_0)$, there exists a solution of the differential inclusion $\dot{x}(t) \in F_2(t, x(t))$ satisfying $x(t_0) = t_0$ and $\dot{x}(t_0) = \nu$. Indeed, by the local Lipschitz continuity of $F_2(t, x)$, the projection on the convex compact set $F_2(s, y)$:

$$g(s, y) = \pi F_2(s, y)(\nu), \quad (s, y) \in U, \tag{1.3.202}$$

is continuous. Hence, there exists a solution $x(t)$ to the following initial value problem on the interval $[t_0, t_0 + \epsilon]$:

$$\begin{cases} \dot{x}(t) = g(t, x)(\in F_2(t, x(t))), \\ x(t_0) = x_0. \end{cases} \tag{1.3.203}$$

It is clear that $\dot{x}(t_0) = g(t_0, x_0) = \nu$. Hence there exists constant $K > 0$ such that for any (t_1, x_1) and (t_2, x_2) in some neighborhood of (t_0, x_0),

$$|V_L(t_1, x_1) - V_L(t_2, x_2)| \leq K\|(t_1 - t_2, x_1 - x_2)\|_\infty. \tag{1.3.204}$$

It follows that when h is sufficiently small,

$$\frac{V_L(t_0 + h, x_0 + h\nu) - V_L(t_0, x_0)}{h} = \frac{V_L(t_0 + h, x_0 + h\nu) - V_L(t_0 + h, x(t_0 + h))}{h}$$

$$+ \frac{V_L(t_0 + h, x(t_0 + h)) - V_L(t_0, x_0)}{h}$$

$$\leq K \left\| \frac{x(t_0 + h) - x_0}{h} - \nu \right\|_\infty + \frac{e^{-2\lambda h} - 1}{h} V_L(t_0, x_0). \tag{1.3.205}$$

Therefore,

$$\limsup_{h \to 0^+} \frac{V_L(t_0 + h, x_0 + h\nu)V_L(t_0, x_0)}{h} \leq -2\lambda V_L(t_0, x_0). \tag{1.3.206}$$

This completes the proof of (1.3.12).

Proof of Theorem 1.3.11 Let S be any compact set in $U = (-1, +\infty) \times (\mathbb{R}^n \setminus \{0\})$ and $\epsilon > 0$. We will show in what follows that there exists a function $\overline{V}(t, x)$ of class C^∞, with compact support in $(-1, \infty) \times \mathbb{R}^n$, such that

$$\|\overline{V}(t, x) - V_L(t, x)\| < \epsilon, \tag{1.3.207}$$

and for any $(t_0, x_0) \in S$, $v \in F_2(t_0, x_0)$,

$$\frac{\partial \overline{V}(t_0, x_0)}{\partial t} + \langle \nabla_x \overline{V}(t_0, x_0), v \rangle \leq -\frac{3}{2}\lambda V_L(t_0, x_0). \tag{1.3.208}$$

Let $\rho \in C^\infty(\mathbb{R}^{n+1}, \mathbb{R})$ be an mollifier given by

$$\rho(t, x) = \begin{cases} C_\rho \exp\left(-1/(1 - |(t, x)|^2)\right), \|(t, x)\|_{\mathbb{R}^{n+1}} < 1, \\ 0, \|(t, x)\|_{\mathbb{R}^{n+1}} \geq 1, \end{cases} \tag{1.3.209}$$

where C_ρ is chosen so that $\int_{\mathbb{R}^{n+1}} \rho(t, x)dt\, dx = 1$. Then $\rho(t, x)$ is non-negative. For any $\sigma > 0$, set $\rho_\delta(t, x) = (1/\delta^{n+1})\rho(t/\delta, x/\delta)$ and

$$V_\delta(t, x) = V_L * \rho_\delta(t, x) = \int_{\mathbb{R}^{n+1}} V_L(t - s, x - y)\rho_\delta(s, y)dsdy$$

$$= \int_{\|(\bar{s}, \bar{y})\|_\infty \leq 1} V_L(t - \delta\bar{s}, x - \delta\bar{y})\rho(\bar{s}, \bar{y})d\bar{s}d\bar{y}. \tag{1.3.210}$$

Therefore, $V_\delta(t, x)$ is well defined and is of class C^∞ on $(-1 + \delta, +\infty) \times \mathbb{R}^n$. In addition, $V_\delta(t, x) \to V_L(t, x)$ uniformly on S as $\delta \to 0$. If $\theta(t, x)$ is a function of class C^∞ with compact support in $(-1, +\infty) \times \mathbb{R}^n$ and is taking value one in the neighborhood of S, then the function $\overline{V}(t, x) = \theta \cdot V_\delta(t, x)$ has a compact support in $(-1, +\infty) \times \mathbb{R}^n$ and satisfies (1.3.207) if δ is small enough. To complete the proof of the lemma, it remains to show that there exists $\delta_0 > 0$ such that for any $\delta \in (0, \delta_0)$, $(t_0, x_0) \in S$ and $v \in F_2(t_0, x_0)$,

$$\frac{\partial V_\delta(t_0, x_0)}{\partial t} + \langle \nabla_x V_\delta(t_0, x_0), v \rangle \leq -\frac{3}{2}\lambda V_L(t_0, x_0). \tag{1.3.211}$$

Let $\delta_1 > 0$ be a small constant so that $S + \overline{B_{\delta_1}(0)} \subset U$ and let $L > 0$ be a positive constant so that for any pairs $(t_1, x_1), (t_2, x_2) \in S + B_{\delta_1}(0)$. The Hausdorff distance between the two pairs satisfies

$$\hbar(F_2(t_1, x_1), F_2(t_2, x_2)) + |V_L(t_1, x_1) - V_L(t_2, x_2)| \leq L\|(t_1 - t_2, x_1 - x_2)\|_\infty. \tag{1.3.212}$$

Then it follows that for almost all $(t, x) \in S + \overline{B_{\delta_1}(0)}$, $V_L(t, x)$ is differentiable at (t, x) and

$$\left\|\left(\frac{\partial V_L(t, x)}{\partial t}, \nabla_x V_L(t, x)\right)\right\| \leq L. \tag{1.3.213}$$

Let $\delta \in (0, \bar{\delta})$, $(t_0, x_0) \in S$, and $v \in F_2(t_0, x_0)$. Applying the Lebesgue dominant convergence theorem, we infer from (1.3.212) that

$$\frac{\partial V_\delta(t_0, x_0)}{\partial t} + \langle \nabla_x V_\delta(t_0, x_0), v \rangle$$

$$= \lim_{\eta \to 0} \int_{\|(\bar{s}, \bar{y})\|_\infty \leq 1} \frac{1}{\eta} (V_L(t_0 - \delta \bar{s} + \eta, x_0 - \delta \bar{y} + \eta v) - V_L(t_0 - \delta \bar{s}, x_0 - \delta \bar{y})) \rho(\bar{s}, \bar{y}) d\bar{s} d\bar{y}$$

$$= \int_{\|(\bar{s}, \bar{y})\|_\infty \leq 1} \left(\frac{\partial V_L(t_0 - \delta \bar{s}, x_0 - \delta \bar{y})}{\partial(t_0 - \delta \bar{s})} + \langle \nabla_x V_L(t_0 - \delta \bar{s}, x_0 - \delta \bar{y}), v \rangle \right) \rho(\bar{s}, \bar{y}) d\bar{s} d\bar{y}.$$

$$(1.3.214)$$

Let $g(s, y)$ be the map defined in (1.3.202). By (1.3.213), (1.3.214), and Corollary 1.3.1,

$$\frac{\partial V_\delta(t_0, x_0)}{\partial t} + \langle \nabla_x V_\delta(t_0, x_0), v \rangle$$

$$\int_{\|(\bar{s}, \bar{y})\|_\infty \leq 1} \left(\frac{\partial V_L}{\partial t} + \langle \nabla_x V_L, g \rangle \right) (t_0 - \delta \bar{s}, x_0 - \delta \bar{y}) \rho(\bar{s}, \bar{y}) d\bar{s} d\bar{y}$$

$$+ \int_{\|(\bar{s}, \bar{y})\|_\infty \leq 1} \langle \nabla_x V_L(t_0 - \delta \bar{s}, x_0 - \delta \bar{y}), v - g(t_0 - \delta \bar{s}, x_0 - \delta \bar{y}) \rangle \rho(s, \bar{y}) d\bar{s} d\bar{y}$$

$$\leq -2\lambda V_\delta(t_0, x_0) + \sqrt{n} L \int_{\|(\bar{s}, \bar{y})\| \leq 1} \|v - g(t_0 - \delta \bar{s}, x_0 - \delta \bar{y})\|_\infty \rho(\bar{s}, \bar{y}) d\bar{s} d\bar{y}. \quad (1.3.215)$$

This yields from (1.3.212) that for any $\|(\bar{s}, \bar{y})\|_\infty \leq 1$,

$$\|v - g(t_0 - \delta \bar{s}, x_0 - \delta \bar{y})\|_\infty \leq \hbar(F_2(t_0, x_0), F_2(t_0 - \delta \bar{s}, x_0 - \delta \bar{y})) \leq L\delta. \quad (1.3.216)$$

Hence

$$\frac{\partial V_\delta(t_0, x_0)}{\partial t} + \langle \nabla_x V_\delta(t_0, x_0), v \rangle \leq -2\lambda V_\delta(t_0, x_0) + \sqrt{n} L^2 \delta \leq -\frac{3}{2}\lambda V_L(t_0, x_0) \quad (1.3.217)$$

for sufficiently small δ.

Let $\{\psi_i(t, x)\}_{i=1}^\infty$ be a C^∞ partition of unity for U. For any $i \geq 1$, the support $S_i(t, x)$ of $\psi_i(t, x)$ is a compact set in U. For each $i \geq 1$, set

$$\begin{cases} q_i = \displaystyle\sup_{(t,x) \in S_i, v \in F_2(t,x)} \left| \frac{\partial \psi(t, x)}{\partial t} + \langle \nabla_x \psi_i(t, x), v \rangle \right| < +\infty, \\[2mm] \epsilon_i = \dfrac{\lambda}{2^{i+2}(1 + q_i)(\lambda + 1)} \displaystyle\min_{(t,x) \in S_i} V_L(t, x) > 0. \end{cases} \quad (1.3.218)$$

It follows from the fact presented in the beginning of the proof that there exist $V_i \in C^\infty((-1, +\infty) \times \mathbb{R}^n, \mathbb{R})$, $i = 1, 2, \ldots$, such that for any $(t, x) \in S_i$, $v \in F_2(t, x)$,

$$|V_L(t, x) - V_i(t, x)| < \epsilon_i, \quad \frac{\partial V_i(t, x)}{\partial t} + \langle \nabla_x V_i(t, x), v \rangle \leq -\frac{3}{2}\lambda V_L(t, x). \quad (1.3.219)$$

For any $(t, x) \in (-1, +\infty) \times \mathbb{R}^n$, let

$$\tilde{V}(t, x) = \begin{cases} \sum_{i=1}^{\infty} \psi_i(t, x)|V_i(t, x)|, & x \neq 0, \\ 0, & x = 0. \end{cases} \tag{1.3.220}$$

It is easy to verify that $\tilde{V}(t, x)$ is class C^∞ on U, and for any $(t, x) \in U$, $v \in F_2(t, x)$,

$$|\tilde{V}(t, x) - V_L(t, x)| \leq \frac{1}{4} V_L(t, x), \tag{1.3.221}$$

$$\frac{3}{4} a_L(\|x\|_\infty) + \langle \nabla_x \tilde{V}(t, x), v \rangle \leq \frac{5}{4} b_L(\|x\|_\infty), \tag{1.3.222}$$

where a_L is defined in (1.3.190) and b_L is defined in (1.3.194). A direct computation shows that

$$\frac{\partial \tilde{V}(t, x)}{\partial t} + \langle \nabla_x \tilde{V}(t, x), v \rangle \leq -\lambda \tilde{V}(t, x). \tag{1.3.223}$$

In the following, we smooth $\tilde{V}(t, x)$ up to $x = 0$. For this purpose, let $\nu : \mathbb{R} \to \mathbb{R}$ be C^∞, $\nu(r) = 0, \forall\ r \in (-1, 0]$, $\dot{\nu}(r) \geq 0, \forall\ r > 0$, $\lim_{r \to \infty} \nu(r) = \infty$, $\partial^\alpha (\nu \circ \tilde{V})(t, 0) = 0, \forall\ t > -1$, and $\alpha \in \mathbb{N}^{n+1}$.

For any $(t, x) \in (-1, \infty) \times \mathbb{R}^n$, let

$$V(t, x) = \nu(\tilde{V}(t, x)). \tag{1.3.224}$$

For any $(t, x) \in ([0, \infty) \setminus N_0) \times (\mathbb{R}^n \setminus \{0\})$, $v \in F(t, x)$. According to a conclusion proved in Step 1, $v \in F_2(t, x)$, and hence (1.3.223) is valid. This together with the fact that

$$\nu(r) = \int_0^r \dot{\nu}(s)ds \leq \int_0^r \dot{\nu}(r)ds = r\dot{\nu}(r), r \geq 0 \tag{1.3.225}$$

gives

$$\frac{\partial V(t, x)}{\partial t} + \langle \nabla_x V(t, x), v \rangle \leq -\lambda V(t, x). \tag{1.3.226}$$

This completes the proof of Theorem 1.3.11. $\qquad\qquad\qquad\square$

1.4   Remarks and Bibliographical Notes

Section 1.2 For TI issues, we refer to the report *LineStream Technologies signs licensing deal with Texas Instruments, The Plain Dealer, July 12, 2011.*

Section 1.3.1 The details of MEMS gyroscope and Figure 1.3.1 can be found in [161]. An hydraulic system is studied in [148]. Figure 1.3.2 is taken from [148]. Autonomous underwater vehicles (AUV) are modeled in [155]. The notation of the relative degree of nonlinear systems is taken from [79].

Section 1.3.2 For Lyapunov's doctoral thesis, we refer to [101]. A large number of publications appeared after Cold War for Lyapunov stability in the control and systems literature

[92, 80, 91]. There is plenty of literature on this topic in monographs, see for instance, [84, 12, 70, and 98].

Section 1.3.4 The finite-time stability for continuous systems was investigated more recently in [112, 17, 67, 15, 12, 16, 109, and 116].

Section 1.3.5 For the Filippov solution, we refer to the monograph [32].

Section 1.3.6 This section is refereed largely from [12].

2

The Tracking Differentiator (TD)

It is known that the powerful yet primitive proportional–integral–derivative (PID) control law developed in the period of the 1920s to the 1940s in the last century still plays a very important role in modern engineering control practices. However, because of the noise sensitivity, derivative control is not always physically implementable for most control systems. A noise-tolerate tracking differentiator was proposed in 1989 by Jingqing Han, which also serves as transient profile for output tracking in active disturbance rejection control. Subsequently, many engineering applications have been made. Han's TD can be described by the following Theorem 2.0.1.

Theorem 2.0.1 *(Han's TD)* *If any solution of the following system*

$$\begin{cases} \dot{x}_1(t) = x_2(t), \\ \dot{x}_2(t) = f(x_1(t), x_2(t)) \end{cases} \tag{2.0.1}$$

satisfies $\lim_{t \to \infty}(x_1(t), x_2(t)) = 0$, *then for any bounded integral function* $v(t)$ *and any constant* $T > 0$, *the solution of the following system*

$$\begin{cases} \dot{z}_{1R}(t) = z_{2R}(t), \\ \dot{z}_{2R}(t) = R^2 f\left(z_{1R}(t) - v(t), \dfrac{z_{2R}(t)}{R}\right) \end{cases} \tag{2.0.2}$$

satisfies

$$\lim_{R \to \infty} \int_0^T |z_{1R}(t) - v(t)| dt = 0. \tag{2.0.3}$$

In applications, the signal $v(t)$ may be only locally integratable like the piecewise continuous signal or bounded measurable $v(t)$, and hence its classical (pointwise) derivative may not exist but its $(i-1)$th generalized derivative, still denoted by $v^{(i-1)}(t)$, always exists in the sense of distribution, which is defined as a functional of $C_0^\infty(0, T)$ for any $T > 0$ as follows:

$$v^{(i-1)}(\varphi) = (-1)^{(i-1)} \int_0^T v(t)\varphi^{(i-1)}(t)dt, \tag{2.0.4}$$

Active Disturbance Rejection Control for Nonlinear Systems: An Introduction, First Edition.
Bao-Zhu Guo and Zhi-Liang Zhao.
© 2016 John Wiley & Sons Singapore Pte. Ltd. Published 2016 by John Wiley & Sons, Ltd.

where $\varphi \in C_0^\infty(0, T), i > 1$. The definition (2.0.4) is a standard definition of the generalized derivative. From this definition, we see that any order of the generalized derivative $v^{(i)}(t)$ always exists provided that $v(t)$ is bounded measurable. Suppose that (2.0.3) holds true. Then considering $z_{iR}(t)$ as a functional of $C_0^\infty(0, T)$, we have

$$
\lim_{R \to \infty} z_{iR}(\varphi) = \lim_{R \to \infty} \int_0^T z_{iR}(t)\varphi(t)dt = \lim_{R \to \infty} \int_0^T z_{1R}^{(i-1)}(t)\varphi(t)dt
$$

$$
= \lim_{R \to \infty} (-1)^{(i-1)} \int_0^T z_{1R}(t)\varphi^{(i-1)}(t)dt \qquad (2.0.5)
$$

$$
= (-1)^{(i-1)} \int_0^T v(t)\varphi^{(i-1)}(t)dt, \ \varphi \in C_0^\infty(0, T), \ i > 1.
$$

Comparing the right-hand sides of (2.0.4) and (2.0.5), we see that

$$
\lim_{R \to \infty} z_{iR}(t) = v^{(i-1)}(t)
$$

in the sense of distribution. Therefore, $x_{iR}(t)$ can be regarded as an approximation of the $(i-1)$th generalized derivative $v^{(i-1)}(t)$ of $v(t)$ in $[0, T]$.

There has been a lot of other research done on differentiation trackers such as the high-gain observer-based differentiator, the super-twisting second-order sliding mode algorithm, linear time-derivative tracker, robust exact differentiation, to name just a few. However, the tracking differentiator (2.0.2) has the advantages that (a) it has weak stability; (b) it requires a weak condition on the input; and (c) it has a small integration value of $|z_{1R}(t) - v(t)|$ in any bounded time interval rather than the small error of $|z_{1R}(t) - v(t)|$ after a finite transient time. In addition, this TD also has the advantage of smoothness compared with the obvious chattering problem encountered by sliding-mode-based differentiators. Moreover, it has been shown by linear cases that the tracking differentiator (2.0.2) is noise-tolerant.

Theorem 2.0.1 is never proved for its original form. In this chapter, we give convergence for this tracking differentiator under some additional conditions that the system (2.0.1) is required to be Lyapunov stable. In addition, we extend the conclusion of the second-order tracking differentiator to the high-order tracking differentiator. Especially in the case of the linear tracking differentiator, we prove a much stronger convergence result:

$$
\lim_{R \to \infty} |z_{1R}(t) - v(t)| = 0, \ \lim_{R \to \infty} |z_{2R}(t) - \dot{v}(t)| = 0. \qquad (2.0.6)
$$

For the general nonlinear tracking differentiator, we get weaker convergence result than (2.0.6), which is also stronger than (2.0.3):

$$
\lim_{R \to \infty} |z_{1R}(t) - v(t)| = 0. \qquad (2.0.7)
$$

From now on, we say that the TD (2.0.2) is strong convergent if it satisfies (2.0.6) and weak convergent if it satisfies (2.0.7). The context of this chapter is as follows. In Section 2.1, we focus on the linear tracking differentiator, which is the most simple tracking differentiator in use. In Section 2.2, we discuss the general nonlinear tracking differentiator. The finite-time stable system-based tracking differentiator is considered in Section 2.3. Finally, in Section 2.4, we give an application of the tracking differentiator to the online frequency estimation of the finite sum of the sinusoidal signals. Some numerical simulations are presented to illustrate the effectiveness of the estimation.

2.1 Linear Tracking Differentiator

In this section, we discuss the linear tracking differentiator as follows:

$$\begin{cases} \dot{z}_{1R}(t) = z_{2R}(t), \\ \dot{z}_{2R}(t) = -k_1 R^2((z_{1R}(t) - v(t)) - k_2 R z_{2R}(t), \end{cases} \tag{2.1.1}$$

where $k_1 > 0$ and $k_2 > 0$ are constants, and $R > 0$ is the tuning parameter. For linear TD (2.1.1), we can prove the strong convergence.

Theorem 2.1.1 *Suppose that $k_1, k_2 > 0$ and $v : [0, \infty) \to \mathbb{R}$ is a function satisfying $\sup_{t \in [0,\infty)}(|v(t)| + |\dot{v}(t)|) = M < \infty$ for constant $M > 0$. Then the linear tracking differentiator (2.1.1) is convergent in the sense that, for any $a > 0$,*

$$\lim_{R \to \infty} |z_{1R}(t) - v(t)| = 0, \quad \lim_{R \to \infty} |z_{2R}(t) - \dot{v}(t)| = 0$$

uniformly for $t \in [a, \infty)$.

Proof. Suppose that $(z_{1R}(t), z_{2R}(t))$ is the solution of system (2.1.13). Let $t = s/R$. Then

$$\begin{cases} \dfrac{d}{ds} z_{1R}\left(\dfrac{s}{R}\right) = \dfrac{1}{R} z'_{1R}\left(\dfrac{s}{R}\right) = \dfrac{1}{R} z_{2R}\left(\dfrac{s}{R}\right), \\ \dfrac{d}{ds} z_{2R}\left(\dfrac{s}{R}\right) = \dfrac{1}{R} z'_{2R}\left(\dfrac{s}{R}\right). \end{cases}$$

Let

$$\begin{cases} y_{1R}(s) = z_{1R}\left(\dfrac{s}{R}\right) - v\left(\dfrac{s}{R}\right), \\ y_{2R}(s) = \dfrac{1}{R} z_{2R}\left(\dfrac{s}{R}\right). \end{cases} \tag{2.1.2}$$

Then

$$\begin{cases} \dot{y}_{1R}(s) = y_{2R}(s) - \dfrac{\dot{v}(s/R)}{R}, \\ \dot{y}_{2R}(s) = -k_1 y_{1R}(s) - k_2 y_{2R}(s). \end{cases} \tag{2.1.3}$$

Therefore, we can write (2.1.3) as

$$\dot{Y}_R(t) = A Y_R(t) + \frac{\dot{v}(t/R)}{R} B, \tag{2.1.4}$$

with

$$A = \begin{pmatrix} 0 & 1 \\ -k_1 & -k_2 \end{pmatrix}, \quad B = \begin{pmatrix} 0 \\ 1 \end{pmatrix}. \tag{2.1.5}$$

Solve the linear differential equation (2.1.4) to obtain

$$Y_R(t) = e^{At} Y_R(0) + \int_0^t e^{A(t-s)} \frac{\dot{v}(s/R)}{R} B \, ds. \tag{2.1.6}$$

It then follows that

$$y_{1R}(t) = [e^{At}]_1 Y_R(0) + \int_0^t [e^{A(t-s)}]_{11} \frac{\dot{v}(s/R)}{R} ds, \tag{2.1.7}$$

where $[e^{At}]_1$ denotes the first row of the matrix e^{At} and $[e^{A(t-s)}]_{11}$ the first entry of $e^{A(t-s)}$. By (2.1.2) and (2.1.7), we have

$$z_{1R}(t) = [e^{RAt}]_1 Y_R(0) + \int_0^{Rt} \left[e^{A(Rt-s)} \right]_{11} \frac{\dot{v}(s/R)}{R} ds + v(t). \tag{2.1.8}$$

Differentiate $z_{1R}(t)$ with respect to t to give

$$z_{2R}(t) = \dot{z}_{1R}(t)$$

$$= [RAe^{RAt}]_1 Y_R(0) + \dot{v}(t) + \int_0^{Rt} \frac{d}{dt} \left(\left[e^{A(Rt-s)} \right]_{11} \right) \frac{\dot{v}(s/R)}{R} ds + \dot{v}(t)$$

$$= [RAe^{RAt}]_1 Y_R(0) + \dot{v}(t) - \int_0^{Rt} \frac{d}{ds} \left(\left[e^{A(Rt-s)} \right]_{11} \right) \dot{v}\left(\frac{s}{R}\right) ds + \dot{v}(t)$$

$$= [RAe^{RAt}]_1 Y_R(0) + \dot{v}(t) - \left[e^{A(Rt-s)} \right]_{11} \dot{v}\left(\frac{s}{R}\right) \Big|_0^{Rt}$$

$$+ \int_0^{Rt} \left[e^{A(Rt-s)} \right]_{11} \frac{\ddot{v}(s/R)}{R} ds + \dot{v}(t)$$

$$= [RAe^{RAt}]_1 Y_R(0) + [e^{RAt}]_{11} \dot{v}(0) + \int_0^{Rt} \left[e^{A(Rt-s)} \right]_{11} \frac{\ddot{v}(s/R)}{R} ds + \dot{v}(t). \tag{2.1.9}$$

It is easy to verify that A is Hurwitz. From this, we may assume without loss of generality that there exist constants $L, \omega > 0$ such that all entries of $e^{At} = \{e_{ij}(t)\}_{i,j=1}^2$ satisfy

$$|e_{ij}(t)| \leq Le^{-\omega t}, \ \forall \, t \geq 0, \ i,j = 1,2. \tag{2.1.10}$$

Since $|v^{(k)}(t)| \leq M$ for $k = 0,1$ and all $t \in [0,\infty)$, we have for every $t \in [0,\infty)$ that

$$\left| \int_0^{Rt} \left[e^{A(Rt-s)} \right]_{11} \frac{v^{(k)}(s/R)}{R} ds \right| = \left| \int_0^{Rt} e_{11}(Rt-s) \frac{v^{(k)}(s/R)}{R} ds \right|$$

$$\leq \frac{ML}{R} \int_0^{Rt} e^{-\omega(Rt-s)} ds \leq \frac{ML}{\omega R}.$$

This together with (2.1.8) and (2.1.9) gives

$$\lim_{R \to \infty} z_{1R}(t) = v(t) \text{ uniformly in } [a,\infty) \tag{2.1.11}$$

and

$$\lim_{R \to \infty} z_{2R}(t) = \dot{v}(t) \text{ uniformly in } [a,\infty). \tag{2.1.12}$$

This completes the proof of the theorem. □

In many control practices, we also need a high-order differential signal of a measured signal. To do this, we can use the above differentiator repeatedly. The linear high-order tracking differentiator can be designed as follows:

$$
\begin{cases}
\dot{z}_{1R}(t) = z_{2R}(t), \; z_{1R}(0) = z_{10}, \\
\dot{z}_{2R}(t) = z_{3R}(t), \; z_{2R}(0) = z_{20}, \\
\quad \vdots \\
\dot{z}_{(n-1)R}(t) = z_{nR}(t), \; z_{(n-1)R}(0) = z_{(n-1)0}, \\
\dot{z}_{nR}(t) = R^n \left(a_1(z_{1R}(t) - v(t)) + \dfrac{a_2 z_{2R}(t)}{R} + \cdots + \dfrac{a_n z_{nR}(t)}{R^{n-1}} \right), \; z_{nR}(0) = z_{n0}.
\end{cases}
$$
$$(2.1.13)$$

For the nth-order tracking differentiator (2.1.13), we can also prove the following strong convergence, which is similar with the second-order linear tracking differentiator.

Theorem 2.1.2 *Suppose that the matrix of the following*

$$
A = \begin{pmatrix}
0 & 1 & 0 & \cdots & 0 \\
0 & 0 & 1 & \cdots & 0 \\
\vdots & \vdots & \vdots & \ddots & \vdots \\
0 & 0 & 0 & \cdots & 1 \\
a_1 & a_2 & a_3 & \cdots & a_n
\end{pmatrix}
\tag{2.1.14}
$$

is Hurwitz and $v : [0, \infty) \to \mathbb{R}$ is a function satisfying $\sup_{t \in [0,\infty), 1 \leq k \leq n} |v^{(k)}(t)| = M < \infty$ for some constant $M > 0$. Then the linear tracking differentiator (2.1.13) is convergent in the sense that: for any $a > 0$, $z_{kR}(t)(k = 1, 2, \ldots, n)$ converges uniformly to $v^{(k-1)}(t)$ in $[a, \infty)$, where $(z_{10}, z_{20}, \ldots, z_{n0})$ is any given initial value.

Proof. Suppose that $(z_{1R}(t), z_{2R}(t), \ldots, z_{nR}(t))$ is the solution of system (2.1.13). Let $t = s/R$. Then

$$
\begin{cases}
\dfrac{d}{ds} z_{1R}\left(\dfrac{s}{R}\right) = \dfrac{1}{R} z'_{1R}\left(\dfrac{s}{R}\right) = \dfrac{1}{R} z_{2R}\left(\dfrac{s}{R}\right), \\
\dfrac{d}{ds} z_{2R}\left(\dfrac{s}{R}\right) = \dfrac{1}{R} z'_{2R}\left(\dfrac{s}{R}\right) = \dfrac{1}{R} z_{3R}\left(\dfrac{s}{R}\right), \\
\quad \vdots \\
\dfrac{d}{ds} z_{(n-1)R}\left(\dfrac{s}{R}\right) = \dfrac{1}{R} z'_{(n-1)R}\left(\dfrac{s}{R}\right) = \dfrac{1}{R} z_{nR}\left(\dfrac{s}{R}\right), \\
\dfrac{d}{ds} z_{nR}\left(\dfrac{s}{R}\right) = \dfrac{1}{R} z'_{nR}\left(\dfrac{s}{R}\right).
\end{cases}
$$

Let

$$
\begin{cases}
y_{1R}(s) = z_{1R}\left(\dfrac{s}{R}\right) - v\left(\dfrac{s}{R}\right), \\[2mm]
y_{2R}(s) = \dfrac{1}{R} z_{2R}\left(\dfrac{s}{R}\right), \\[2mm]
y_{3R}(s) = \dfrac{1}{R^2} z_{3R}\left(\dfrac{s}{R}\right), \\[2mm]
\qquad \vdots \\[2mm]
y_{nR}(s) = \dfrac{1}{R^{n-1}} z_{nR}\left(\dfrac{s}{R}\right).
\end{cases}
\tag{2.1.15}
$$

Then

$$
\begin{cases}
\dot{y}_{1R}(s) = y_{2R}(s) - \dfrac{\dot{v}(s/R)}{R}, \\[2mm]
\dot{y}_{2R}(s) = y_{3R}(s), \\[2mm]
\qquad \vdots \\[2mm]
\dot{y}_{(n-1)R}(s) = y_{nR}(s), \\[2mm]
\dot{y}_{nR}(s) = a_1 y_{1R}(s) + a_2 y_{2R}(s) + a_n y_{nR}(s).
\end{cases}
\tag{2.1.16}
$$

Therefore, we can write (2.1.16) as

$$
\dot{Y}_R(t) = A Y_R(t) + \left[\frac{\dot{v}(t/R)}{R}, 0, \ldots, 0\right]^{\top}.
\tag{2.1.17}
$$

Solve the linear differential equation (2.1.17) to obtain

$$
Y_R(t) = e^{At} Y_R(0) + \int_0^t e^{A(t-s)} \left[\frac{\dot{v}(s/R)}{R}, 0, \ldots, 0\right]^{\top} ds.
\tag{2.1.18}
$$

It then follows that

$$
y_{1R}(t) = [e^{At}]_1 Y_R(0) + \int_0^t \left[e^{A(t-s)}\right]_{11} \frac{\dot{v}(s/R)}{R} ds,
\tag{2.1.19}
$$

where once again, $[e^{At}]_1$ denotes the first row of the matrix e^{At} and $[e^{A(t-s)}]_{11}$ the first entry of $e^{A(t-s)}$.

By (2.1.15) and (2.1.19), we have

$$
z_{1R}(t) = [e^{RAt}]_1 Y_R(0) + \int_0^{Rt} \left[e^{A(Rt-s)}\right]_{11} \frac{\dot{v}(s/R)}{R} ds + v(t).
\tag{2.1.20}
$$

Differentiate $z_{1R}(t)$ with respect to t to give

$$z_{2R}(t) = \dot{z}_{1R}(t)$$

$$= [RAe^{RAt}]_1 Y_R(0) + \dot{v}(t) + \int_0^{Rt} \frac{d}{dt}\left(\left[e^{A(Rt-s)}\right]_{11}\right)\frac{\dot{v}(s/R)}{R}ds + \dot{v}(t)$$

$$= [RAe^{RAt}]_1 Y_R(0) + \dot{v}(t) - \int_0^{Rt} \frac{d}{ds}\left(\left[e^{A(Rt-s)}\right]_{11}\right)\dot{v}\left(\frac{s}{R}\right)ds + \dot{v}(t) \quad (2.1.21)$$

$$= [RAe^{RAt}]_1 Y_R(0) + \dot{v}(t) - \left[e^{A(Rt-s)}\right]_{11}\dot{v}\left(\frac{s}{R}\right)\Big|_0^{Rt}$$

$$+ \int_0^{Rt} \left[e^{A(Rt-s)}\right]_{11}\frac{\ddot{v}(s/R)}{R}ds + \dot{v}(t)$$

$$= [RAe^{RAt}]_1 Y_R(0) + [e^{RAt}]_{11}\dot{v}(0) + \int_0^{Rt} \left[e^{A(Rt-s)}\right]_{11}\frac{\ddot{v}(s/R)}{R}ds + \dot{v}(t).$$

Generally, we have by induction that

$$z_{kR}(t) = [(RA)^{k-1}e^{RAt}]_1 Y_R(0) + [(RA)^{k-2}e^{ARt}]_{11}\dot{v}(0) + \cdots + [e^{ARt}]_{11}v^{(k-1)}(0)$$

$$+ \int_0^{Rt} [e^{A(Rt-s)}]_{11}\frac{v^{(k)}(s/R)}{R}ds + v^{(k-1)}(t), \ 2 \le k \le n. \quad (2.1.22)$$

Since A is Hurwitz, we may assume without loss of generality that there exist constants L and $\omega > 0$ such that all entries of $e^{At} = \{e_{ij}(t)\}_{i,j=1}^n$ satisfy

$$|e_{ij}(t)| \le Le^{-\omega t}, \ \forall\, t \ge 0, \ i,j = 1,2\cdots,n. \quad (2.1.23)$$

Since $|v^{(k)}(t)| \le M$ for all $t \in [0,\infty)$, we have, for every $t \in [0,\infty)$, that

$$\left|\int_0^{Rt} \left[e^{A(Rt-s)}\right]_{11}\frac{v^{(k)}(s/R)}{R}ds\right| = \left|\int_0^{Rt} e_{11}(Rt-s)\frac{v^{(k)}(s/R)}{R}ds\right|$$

$$\le \frac{ML}{R}\int_0^{Rt} e^{-\omega(Rt-s)}ds \le \frac{ML}{\omega R}.$$

This together with (2.1.22) and (2.1.23) gives

$$\lim_{R\to\infty} z_{kR}(t) = v^{(k-1)}(t) \text{ uniformly in } [a,\infty) \text{ for any } 0 < a < T, 2 \le k \le n. \quad (2.1.24)$$

This completes the proof of the theorem. $\square$

2.2 Nonlinear Tracking Differentiator

In this subsection, we firstly analyze the second-order nonlinear tracking differentiator and then consider the high-order nonlinear tracking differentiator.

2.2.1 Second-Order Nonlinear Tracking Differentiator

We first give convergence of a second-order nonlinear tracking differentiator, which is the same as Han's TD but with additional assumption on the Lyapunov stability.

Theorem 2.2.1 *Let $f : \mathbb{R}^2 \to \mathbb{R}$ be a locally Lipschtz continuous function, $f(0,0) = 0$. Suppose that the equilibrium point $(0,0)$ of the following system is globally asymptotically stable:*

$$\begin{cases} \dot{x}_1(t) = x_2(t), \ x_1(0) = x_{10}, \\ \dot{x}_2(t) = f(x_1(t), x_2(t)), \ x_2(0) = x_{20}, \end{cases} \tag{2.2.1}$$

where (x_{10}, x_{20}) is any given initial value. If the signal $v(t)$ is differentiable and $A = \sup_{t \in [0,\infty)} |\dot{v}(t)| < \infty$, then the solution of the following tracking differentiator:

$$\begin{cases} \dot{z}_{1R}(t) = z_{2R}(t), \ z_{1R}(0) = z_{10}, \\ \dot{z}_{2R}(t) = R^2 f\left(z_{1R}(t) - v(t), \dfrac{z_{2R}(t)}{R}\right), \ z_{2R}(0) = z_{20}, \end{cases} \tag{2.2.2}$$

is convergent in the sense that: for every $a > 0$, $z_{1R}(t)$ is uniformly convergent to $v(t)$ on $[a, \infty)$ as $R \to \infty$, where (z_{10}, z_{20}) is any given initial value.

Proof. The proof will be split into five steps.

Step 1. *Transform system (2.2.2) into system (2.2.1) with a perturbation.*
Suppose that $(z_{1R}(t), z_{2R}(t))$ is the solution of system (2.2.2). Let $t = s/R$. Then

$$\begin{cases} \dfrac{d}{ds} z_{1R}\left(\dfrac{s}{R}\right) = \dfrac{1}{R} z_{1R}'\left(\dfrac{s}{R}\right) = \dfrac{1}{R} z_{2R}\left(\dfrac{s}{R}\right), \\ \dfrac{d}{ds} z_{2R}\left(\dfrac{s}{R}\right) = \dfrac{1}{R} z_{2R}'\left(\dfrac{s}{R}\right) = R f\left(z_{1R}\left(\dfrac{s}{R}\right) - v\left(\dfrac{s}{R}\right), \dfrac{z_{2R}(s/R)}{R}\right). \end{cases}$$

Let

$$\begin{cases} y_{1R}(s) = z_{1R}\left(\dfrac{s}{R}\right) - v\left(\dfrac{s}{R}\right), \\ y_{2R}(s) = \dfrac{1}{R} z_{2R}\left(\dfrac{s}{R}\right). \end{cases} \tag{2.2.3}$$

Then

$$\begin{cases} \dot{y}_{1R}(s) = y_{2R}(s) - \dfrac{\dot{v}(s/R)}{R}, \ y_{1R}(0) = z_{1R}(0) - v(0), \\ \dot{y}_{2R}(s) = f(y_{1R}(s), y_{2R}(s)), \ y_{2R}(0) = \dfrac{z_{2R}(0)}{R}. \end{cases} \tag{2.2.4}$$

Therefore, $Y_R(t) = (y_{1R}(t), y_{2R}(t))^\top$ is a solution to the system of the following:

$$\dot{Y}_R(t) = F(Y_R(t)) + G_R(t), \ Y_R(0) = Y_{R0} = \left(z_{1R}(0) - v(0), \dfrac{z_{2R}(0)}{R}\right)^\top, \tag{2.2.5}$$

where

$$F(Y_R(t)) = (y_{2R}(t), f(y_{1R}(t), y_{2R}(t)))^\top, \; G_R(t) = \left(-\frac{\dot{v}(t/R)}{R}, 0\right)^\top.$$

If $X(t) = (x_1(t), x_2(t))^\top$ is a solution to system (2.2.1), then (2.2.1) can be written as

$$\dot{X}(t) = F(X(t)). \tag{2.2.6}$$

It is seen that system (2.2.5) is a perturbed system of (2.2.6).

Step 2. *The existence of the Lyapunov function.*

Since $f(x_1, x_2)$ is locally Lipschtz continuous and system (2.2.1) is globally asymptotically stable, by the inverse Lyapunov Theorem 1.3.3, there is a smooth, positive definite function $V : \mathbb{R}^2 \to \mathbb{R}$ and a continuous, positive definite function $W : \mathbb{R}^2 \to \mathbb{R}$ such that

- $V(x_1, x_2) \to \infty$ as $\|(x_1, x_2)\| \to \infty$;
- $\dfrac{dV(x_1, x_2)}{dt} = x_2 \dfrac{\partial V(x_1, x_2)}{\partial x_1} + f(x_1, x_2) \dfrac{\partial V(x_1, x_2)}{\partial x_2} \le -W(x_1, x_2)$

  along the trajectory of (2.2.1);

- $\{(x_1, x_2) \in \mathbb{R}^2 | V(x_1, x_2) \le d\}$ is bounded closed in $\mathbb{R}^2$ for any given $d > 0$.

By the existence of the above continuous positive definite functions, it follows from Theorem 1.3.1 that there exist class $\mathcal{K}_\infty$ functions $K_i : [0, \infty) \to [0, \infty)$, $i = 1, 2, 3, 4$, such that

$$K_1(\|(x_1, x_2)\|) \le V(x_1, x_2) \le K_2(\|(x_1, x_2)\|),$$

$$K_3(\|(x_1, x_2)\|) \le W(x_1, x_2) \le K_4(\|(x_1, x_2)\|), \lim_{r \to \infty} K_i(r) = \infty, \; i = 1, 2, 3, 4.$$

Denote by $Y_R(t; 0, Y_{R0})$ the solution of (2.2.5).

Step 3. *For each $Y_{R0} \in \mathbb{R}^2$, there exists an $R_1 > 1$ such that when $R > R_1$,*

$$\{Y_R(t; 0, Y_{R0}) | \, t \in [0, \infty)\} \subset \{Y = (y_1, y_2) | \, V(Y) \le c\},$$

$$c = \max\{K_2(\|Y_{10}\|), 1\} > 0. \tag{2.2.7}$$

We prove this claim by contradiction. Firstly, since $\frac{\partial V(Y)}{\partial y_1}$ is continuous and the set $\{Y | \, c \le V(Y) \le c+1\}$ is bounded, we have

$$M = \sup_{Y \in \{Y | \, c \le V(Y) \le c+1\}} \left| \frac{\partial V(Y)}{\partial y_1} \right| < \infty.$$

Secondly,

$$W(Y) \ge K_3(\|Y\|) \ge K_3 K_2^{-1}(V(Y)) \ge K_3 K_2^{-1}(c) > 0,$$

$$\forall Y \in \{Y | \, c \le V(Y) \le c+1\}. \tag{2.2.8}$$

Suppose that the claim (2.2.7) is false. Notice that $V(Y_{R0}) \leq K_2(\| Y_{R0} \|) \leq K_2$ $(\| Y_{10} \|) \leq c$. For R_1 given by

$$R_1 = \max\left\{1, \frac{AM}{K_3 K_2^{-1}(c)}\right\}, \qquad (2.2.9)$$

there exist $R > R_1$ and $0 \leq t_1^R < t_2^R < \infty$ such that

$$Y_R(t_1^R; 0, Y_{R0}) \in \{Y | V(Y) = c\}, \ Y_R(t_2^R; 0, Y_{R0}) \in \{Y | V(Y) > c\} \quad (2.2.10)$$

and

$$\{Y_R(t; 0, Y_{R0}) | t \in [t_1^R, t_2^R]\} \subset \{Y | \ c \leq V(Y) \leq c + 1\}. \qquad (2.2.11)$$

Combining (2.2.8) and (2.2.11) yields

$$\inf_{t \in [t_1^R, t_2^R]} W(Y_R(t; 0, Y_{R0})) \geq K_3 K_2^{-1}(c). \qquad (2.2.12)$$

Therefore, for $t \in [t_1^R, t_2^R]$,

$$\frac{dV(Y_R(t; 0, Y_{R0}))}{dt} = \left. \frac{dV(Y_R(t; 0, Y_{R0}))}{dt} \right|_{\text{along (2.2.5)}}$$

$$\leq -W(Y_R(t; 0, Y_{R0})) + \frac{AM}{R} \leq -K_3 K_2^{-1}(c)$$

$$+ AM \frac{K_3 K_2^{-1}(c)}{AM}$$

$$= 0,$$

which shows that $V(Y_R(t; 0, Y_{R0}))$ is nonincreasing in $[t_1^R, t_2^R]$, and hence

$$V(Y_R(t_2^R; 0, Y_{R0})) \leq V(Y_R(t_1^R; 0, Y_{R0})) = c.$$

This contradicts (2.2.10) and hence (2.2.7) is valid.

For any given $\epsilon > 0$, since $V(Y)$ is continuous, there exists a $\delta \in (0, \epsilon)$ such that

$$0 \leq V(Y) \leq K_1(\epsilon), \ \forall \ \| Y \| \leq \delta. \qquad (2.2.13)$$

Now, for every $Y \in \{Y \| V(Y)| \geq \delta\}$,

$$W(Y) \geq K_3(\| Y \|) \geq K_3 K_2^{-1}(V(Y)) \geq K_3 K_2^{-1}(\delta) > 0. \qquad (2.2.14)$$

By Step 3, for every $R > R_1$, $\{Y_R(t; 0, Y_{R0}) | t \in [0, \infty)\} \subset \{Y | V(Y) \leq c\}$, and hence

$$H = \sup_{t \in [0, \infty)} \left| \frac{\partial V(Y_R(t; 0, Y_{R0}))}{\partial y_1} \right| \leq \sup_{Y \in \{Y | V(Y) \leq c\}} \left| \frac{\partial V(Y)}{\partial y_1} \right| < \infty.$$

Step 4. *There is an $R_2 \geq R_1$ such that for every $R > R_2$ there exists a $T_R \in \left[0, \frac{2c}{K_3 K_2^{-1}(\delta)}\right]$ such that $|Y_R(T_R; 0, Y_{R0})| < \delta$.*

Suppose that the claim is false. Then for

$$R_2 = \max\left\{R_1, \frac{2HA}{K_3 K_2^{-1}(\delta)}\right\} \tag{2.2.15}$$

there exists an $R > R_2$ such that $|Y_R(t; 0, Y_{R0})| \geq \delta$ for any $t \in \left[0, \frac{2c}{K_3 K_2^{-1}(\delta)}\right]$. This together with (2.2.14) concludes that for any $R > R_2$ and all $t \in \left[0, \frac{2c}{K_3 K_2^{-1}(\delta)}\right]$,

$$\frac{dV(Y_R(t; 0, Y_{R0}))}{dt} = \left.\frac{dV(Y_R(t; 0, Y_{R0}))}{dt}\right|_{\text{along (2.2.5)}}$$

$$\leq -W(Y_R(t; 0, Y_{R0})) + \left|\frac{\partial V(Y_R(t; 0, Y_{R0}))}{\partial y_1} \frac{v'(t/R)}{R}\right|$$

$$\leq -\frac{K_3 K_2^{-1}(\delta)}{2} < 0.$$

Perform integration over $\left[0, \frac{2c}{K_3 K_2^{-1}(\delta)}\right]$ to give

$$V\left(Y_R\left(\frac{2c}{K_3 K_2^{-1}(\delta)}; 0, Y_{R0}\right)\right) = \int_0^{\frac{2c}{K_3 K_2^{-1}(\delta)}} \frac{dV(Y_R(t; 0, Y_{R0}))}{dt} dt + V(Y_{R0})$$

$$\leq -\frac{K_3 K_2^{-1}(\delta)}{2} \frac{2c}{K_3 K_2^{-1}(\delta)} + V(Y_{R0})$$

$$\leq 0.$$

This is a contradiction because for each $t \in \left[0, \frac{2c}{K_3 K_2^{-1}(\delta)}\right]$, $\parallel Y_R(t; 0, Y_{R0}) \parallel \geq \delta$. The claim follows.

Step 5. *For any $R > R_2$, if there exists a $t_0^R \in [0, \infty)$ such that*

$$Y_R(t_0^R; 0, Y_{R0}) \in \{Y | \parallel Y \parallel \leq \delta\},$$

then

$$\{Y_R(t; 0, Y_{R0}) | t \in (t_0^R, \infty]\} \subset \{Y | \parallel Y \parallel \leq \epsilon\}. \tag{2.2.16}$$

Suppose that (2.2.16) is not valid. Then there is a $t_2^R > t_1^R \geq t_0^R$ such that

$$\parallel Y_R(t_1^R; 0, Y_{R0}) \parallel = \delta, \parallel Y_R(t_2^R; 0, Y_{R0}) \parallel > \epsilon,$$

$$\{Y_R(t; 0, Y_{R0}) | t \in [t_1^R, t_2^R]\} \subset \{Y | \parallel Y \parallel \geq \delta\}. \tag{2.2.17}$$

This together with (2.2.14) concludes that for $t \in [t_1^R, t_2^R]$,

$$K_1(\| Y_R(t_2^R; 0, Y_{R0}) \|) \leq V(Y(t_2^R; 0, Y_{R0}))$$

$$= \int_{t_1^R}^{t_2^R} \frac{dV(Y(t; 0, Y_{R0}))}{dt} dt + V(Y_R(t_1^R; 0, Y_{R0}))$$

$$\leq \int_{t_1^R}^{t_2^R} -\frac{K_3 K_2^{-1}(\delta)}{2} dt + V(Y_R(t_1^R; 0, Y_{R0}))$$

$$\leq V(Y_R(t_1^R; 0, Y_{R0})). \tag{2.2.18}$$

By (2.2.13) and $|Y_R(t_1^R; 0, Y_{R0})| = \delta$, we have

$$V(Y_R(t_1^R; 0, Y_{R0})) \leq K_1(\epsilon).$$

This together with (2.2.18) gives

$$K_1(|Y_R(t_2^R; 0, Y_{R0})|) \leq K_1(\epsilon). \tag{2.2.19}$$

Since the wedge function $K_1(\cdot)$ is increasing, (2.2.19) implies that $|Y_R(t_2^R; 0, Y_{R0})| \leq \epsilon$, which contradicts with the middle inequality of (2.2.17). The claim (2.2.16) follows.

Finally, for each $a > 0$, by results of Step 4 and Step 5, for $R > \max \left\{ R_2, \frac{2c}{a K_3 K_2^{-1}(\delta)} \right\}$ and $t \in [a, \infty)$,

$$|z_{1R}(t) - v(t)| = |y_{1R}(Rt)| \leq Y_R(Rt) \leq \epsilon.$$

Hence $z_{1R}(t)$ converges uniformly to $v(t)$ in $[a, \infty)$ as $R \to \infty$. This completes the proof of the theorem. □

2.2.2 High-Order Nonlinear Tracking Differentiator

In order to obtain approximations of high-order derivatives of a given signal, we need the high-order tracking differentiator. The following Theorem 2.2.2 is about the convergence of the high-order nonlinear tracking differentiator.

Theorem 2.2.2 *Let $f : \mathbb{R}^n \to \mathbb{R}$ be a locally Lipschitz continuous function, $f(0,0) = 0$. Suppose that the zero equilibrium point of the following system is globally asymptotically stable:*

$$\begin{cases} \dot{x}_1(t) = x_2(t), & x_1(0) = x_{10}, \\ \dot{x}_2(t) = x_3(t), & x_2(0) = x_{20}, \\ \quad \vdots \\ \dot{x}_n(t) = f(x_1(t), x_2(t), \dots, x_n(t)), & x_n(0) = x_{n0}, \end{cases} \tag{2.2.20}$$

where $(x_{10}, x_{20}, \ldots, x_{n0})^\top$ is a given initial value. If the signal $v(t)$ is differentiable and $A = \sup_{t \in [0,\infty)} |v^{(n+1)}(t)| < \infty$, then the solution of the following tracking differentiator:

$$
\begin{cases}
\dot{z}_{1R}(t) = z_{2R}(t), \ z_{1R}(0) = z_{10}, \\
\dot{z}_{2R}(t) = z_{R3}(t), \ z_{R2}(0) = z_{20}, \\
\dot{z}_{nR}(t) = R^n f\left(z_{1R}(t) - v(t), \dfrac{z_{2R}(t)}{R}, \ldots, \dfrac{z_{nR}(t)}{R^{n-1}} \right), \ z_{nR}(0) = z_{n0},
\end{cases}
\tag{2.2.21}
$$

is convergent in the sense that: for every $a > 0$, $z_{1R}(t)$ is uniformly convergent to $v(t)$ on $[a, \infty)$ as $R \to \infty$, where $(z_{10}, z_{20}, \ldots, z_{n0})$ is any given initial value.

Proof. The proof will be split into the following steps.

Step 1. *Transform system (2.1.15) into system (2.2.21) with a perturbation.*
Suppose that $(z_{1R}(t), z_{2R}(t), \ldots, z_{nR}(t))$ is the solution of system (2.1.15). Let $t = s/R$. Then

$$
\begin{cases}
\dfrac{d}{ds} z_{1R}\left(\dfrac{s}{R}\right) = \dfrac{1}{R} z_{2R}\left(\dfrac{s}{R}\right), \\
\dfrac{d}{ds} z_{2R}\left(\dfrac{s}{R}\right) = \dfrac{1}{R} z_{3R}\left(\dfrac{s}{R}\right), \\
\dfrac{d}{ds} z_{nR}\left(\dfrac{s}{R}\right) = R^{n-1} f\left(z_{1R}\left(\dfrac{s}{R}\right) - v\left(\dfrac{s}{R}\right), \dfrac{z_{2R}(s/R)}{R}, \ldots, \dfrac{z_{nR}\left(\frac{s}{R}\right)}{R^{n-1}} \right).
\end{cases}
$$

Let

$$
\begin{cases}
y_{1R}(s) = z_{1R}\left(\dfrac{s}{R}\right) - v\left(\dfrac{s}{R}\right), \\
y_{iR}(s) = \dfrac{1}{R} z_{iR}\left(\dfrac{s}{R^{i-1}}\right), \ i = 1, 2, \ldots, n.
\end{cases}
\tag{2.2.22}
$$

Then

$$
\begin{cases}
\dot{y}_{1R}(s) = y_{2R}(s) - \dfrac{\dot{v}(s/R)}{R}, \ y_{1R}(0) = z_{1R}(0) - v(0), \\
\dot{y}_{2R}(s) = y_{3R}(s), \ y_{2R}(0) = \dfrac{z_{2R}(0)}{R}, \\
\dot{y}_{nR}(s) = f(y_{1R}(s), y_{2R}(s)), \ldots, y_{nR}(s)), \ y_{nR}(0) = \dfrac{z_{2R}(0)}{R^{n-1}}.
\end{cases}
\tag{2.2.23}
$$

Therefore, $Y_R(t) = (y_{1R}(t), y_{2R}(t), \ldots, y_{nR}(t))^\top$ is a solution to the system

$$
\dot{Y}_R(t) = F(Y_R(t)) + G_R(t), Y_R(0) = Y_{R0}
$$

$$
= \left(z_{1R}(0) - v(0), \dfrac{z_{2R}(0)}{R}, \ldots, \dfrac{z_{nR}(0)}{R^{n-1}} \right)^\top.
\tag{2.2.24}
$$

where

$$F(Y_R(t)) = (y_{2R}(t), y_{3R}(t), \ldots, f(y_{1R}(t), y_{2R}(t), \ldots, y_{nR}(y)))^\top,$$

$$G_R(t) = \left(-\frac{\dot{v}(t/R)}{R}, 0, \ldots, 0\right)^\top.$$

If $X(t) = (x_1(t), x_2(t), \ldots, x_n(t))^\top$ is a solution to system (2.2.21), then (2.2.21) can be written as

$$\dot{X}(t) = F(X(t)). \tag{2.2.25}$$

It is seen that system (2.2.24) is a perturbed system of (2.2.25).

Step 2. *The existence of the Lyapunov function.*

Since $f(\cdot)$ is locally Lipschitz continuous and system (2.2.20) is globally asymptotically stable, by the inverse Lyapunov theorem 1.3.3, there is a smooth, positive definite function $V : \mathbb{R}^n \to \mathbb{R}$ and a continuous, positive definite function $W : \mathbb{R}^n \to \mathbb{R}$ such that

- $V(x_1, x_2, \ldots, x_n) \to \infty$ as $\| (x_1, x_2, \ldots, x_n) \| \to \infty$;

- $\dfrac{dV}{dt} = \displaystyle\sum_{i=1}^{n-1} x_{i+1} \frac{\partial V}{\partial x_i} + f(x_1, x_2, \ldots, x_n) \frac{\partial V}{\partial x_n} \leq -W(x_1, x_2, \ldots, x_n)$

  along the trajectory of (2.2.20);

- $\{(x_1, x_2, \ldots, x_n) \in \mathbb{R}^n |\ V(x_1, x_2, \ldots, x_n) \leq d\}$ is bounded closed in
  $\mathbb{R}^n$ for any given $d > 0$.

It follows from Theorem 1.3.1 that there exist class $\mathcal{K}_\infty$ functions $K_i : [0, \infty) \to [0, \infty), i = 1, 2, 3, 4$, such that

$$K_1(\| (x_1, x_2, \ldots, x_n) \|) \leq V(x_1, x_2, \ldots, x_n) \leq K_2(\| (x_1, x_2, \ldots, x_n) \|),$$

$$\lim_{r \to \infty} K_i(r) = \infty, i = 1, 2,$$

$$K_3(\| (x_1, x_2, \ldots, x_n) \|) \leq W(x_1, x_2, \ldots, x_n) \leq K_4(\| (x_1, x_2, \ldots, x_n) \|).$$

Denote by $Y_R(t; 0, Y_{R0})$ the solution of (2.2.24).

Step 3. *For each $Y_{R0} \in \mathbb{R}^n$, there exists an $R_1 > 1$ such that, when $R > R_1$,*

$$\{Y_R(t; 0, Y_{R0}) |\ t \in [0, \infty)\} \subset \{Y = (y_1, y_2, \ldots, y_n) |\ V(Y) \leq c\},$$

$$c = \max\{K_2(\| Y_{10} \|), 1\} > 0. \tag{2.2.26}$$

We prove this claim by contradiction. Firstly, since $\dfrac{\partial V(Y)}{\partial y_1}$ is continuous and the set $\{Y |\ c \leq V(Y) \leq c + 1\}$ is bounded, we have

$$M = \sup_{Y \in \{Y|\ c \leq V(Y) \leq c+1\}} \left| \frac{\partial V(Y)}{\partial y_1} \right| < \infty.$$

Secondly,

$$W(Y) \geq K_3(\| Y \|) \geq K_3 K_2^{-1}(V(Y)) \geq K_3 K_2^{-1}(c) > 0,$$
$$\forall Y \in \{Y | c \leq V(Y) \leq c+1\}. \tag{2.2.27}$$

Suppose that the claim (2.2.26) is false. Notice that $V(Y_{R0}) \leq K_2(|Y_{R0}|) \leq K_2(|Y_{10}|) \leq c$. For R_1 given by

$$R_1 = \max\left\{1, \frac{AM}{K_3 K_2^{-1}(c)}\right\}, \tag{2.2.28}$$

there exist an $R > R_1$ and $0 \leq t_1^R < t_2^R < \infty$ such that

$$Y_R(t_1^R; 0, Y_{R0}) \in \{Y | V(Y) = c\}, \quad Y_R(t_2^R; 0, Y_{R0}) \in \{Y | V(Y) > c\}, \tag{2.2.29}$$

and

$$\{Y_R(t; 0, Y_{R0}) | t \in [t_1^R, t_2^R]\} \subset \{Y | c \leq V(Y) \leq c+1\}. \tag{2.2.30}$$

Combining (2.2.27) and (2.2.30) yields

$$\inf_{t \in [t_1^R, t_2^R]} W(Y_R(t; 0, Y_{R0})) \geq K_3 K_2^{-1}(c). \tag{2.2.31}$$

Therefore, for $t \in [t_1^R, t_2^R]$,

$$\frac{dV(Y_R(t; 0, Y_{R0}))}{dt} = \frac{dV(Y_R(t; 0, Y_{R0}))}{dt}\bigg|_{\text{along } (2.2.24)}$$
$$\leq -W(Y_R(t; 0, Y_{R0})) + \frac{AM}{R}$$
$$\leq -K_3 K_2^{-1}(c) + AM \frac{K_3 K_2^{-1}(c)}{AM}$$
$$= 0,$$

which shows that $V(Y_R(t; 0, Y_{R0}))$ is nonincreasing in $[t_1^R, t_2^R]$, and hence

$$V(Y_R(t_2^R; 0, Y_{R0})) \leq V(Y_R(t_1^R; 0, Y_{R0})) = c.$$

This contradicts (2.2.29), and hence (2.2.26) is valid.

Step 4. *There is an $R_2 \geq R_1$ such that for each $R > R_2$ there exists a $T_R \in \left[0, \frac{2c}{K_3 K_2^{-1}(\delta)}\right]$ such that $|Y_R(T_R; 0, Y_{R0})| < \delta$.*

Actually, for any given $\epsilon > 0$, since $V(Y)$ is continuous, there exists a $\delta \in (0, \epsilon)$ such that

$$0 \leq V(Y) \leq K_1(\epsilon), \quad \forall |Y| \leq \delta. \tag{2.2.32}$$

Now, for each $Y \in \{Y \| V(Y) | \geq \delta\}$,

$$W(Y) \geq K_3(|Y|) \geq K_3 K_2^{-1}(V(Y)) \geq K_3 K_2^{-1}(\delta) > 0. \tag{2.2.33}$$

By Step 3, for every $R > R_1$, $\{Y_R(t; 0, Y_{R0}) | t \in [0, \infty)\} \subset \{Y | V(Y) \le c\}$, and hence

$$H = \sup_{t \in [0, \infty)} \left| \frac{\partial V(Y_R(t; 0, Y_{R0}))}{\partial y_1} \right| \le \sup_{Y \in \{Y | V(Y) \le c\}} \left| \frac{\partial V(Y)}{\partial y_1} \right| < \infty.$$

Suppose that the claim is false. Then for

$$R_2 = \max \left\{ R_1, \frac{2HA}{K_3 K_2^{-1}(\delta)} \right\} \tag{2.2.34}$$

there exists an $R > R_2$ such that $|Y_R(t; 0, Y_{R0})| \ge \delta$ for any $t \in \left[0, \frac{2c}{K_3 K_2^{-1}(\delta)} \right]$. This together with (2.2.33) concludes that for any $R > R_2$ and all $t \in \left[0, \frac{2c}{K_3 K_2^{-1}(\delta)} \right]$,

$$\frac{dV(Y_R(t; 0, Y_{R0}))}{dt} = \left. \frac{dV(Y_R(t; 0, Y_{R0}))}{dt} \right|_{\text{along } (2.2.24)}$$

$$\le - W(Y_R(t; 0, Y_{R0})) + \left| \frac{\partial V(Y_R(t; 0, Y_{R0}))}{\partial y_1} \frac{v'(t/R)}{R} \right|$$

$$\le - \frac{K_3 K_2^{-1}(\delta)}{2} < 0.$$

Perform integration over $\left[0, \frac{2c}{K_3 K_2^{-1}(\delta)} \right]$ to give

$$V \left(Y_R \left(\frac{2c}{K_3 K_2^{-1}(\delta)}; 0, Y_{R0} \right) \right) = \int_0^{\frac{2c}{K_3 K_2^{-1}(\delta)}} \frac{dV(Y_R(t; 0, Y_{R0}))}{dt} dt + V(Y_{R0})$$

$$\le - \frac{K_3 K_2^{-1}(\delta)}{2} \frac{2c}{K_3 K_2^{-1}(\delta)} + V(Y_{R0})$$

$$\le 0.$$

This is a contradiction because for each $t \in \left[0, \frac{2c}{K_3 K_2^{-1}(\delta)} \right]$, $|Y_R(t; 0, Y_{R0})| \ge \delta$. The claim follows.

Step 5. *For each $R > R_2$, if there exists a $t_0^R \in [0, \infty)$ such that*

$$Y_R(t_0^R; 0, Y_{R0}) \in \{Y | \ |Y| \le \delta\},$$

then

$$\{Y_R(t; 0, Y_{R0}) | t \in (t_0^R, \infty]\} \subset \{Y | \ |Y| \le \epsilon\}. \tag{2.2.35}$$

Suppose (2.2.35) is not valid. Then there is a $t_2^R > t_1^R \ge t_0^R$ such that

$$|Y_R(t_1^R; 0, Y_{R0})| = \delta, \ |Y_R(t_2^R; 0, Y_{R0})| > \epsilon,$$
$$\{Y_R(t; 0, Y_{R0}) | t \in [t_1^R, t_2^R]\} \subset \{Y | |Y| \ge \delta\}. \tag{2.2.36}$$

This together with (2.2.33) concludes that, for $t \in [t_1^R, t_2^R]$,

$$K_1(\| Y_R(t_2^R; 0, Y_{R0}) \|) \leq V(Y(t_2^R; 0, Y_{R0}))$$

$$= \int_{t_1^R}^{t_2^R} \frac{dV(Y(t; 0, Y_{R0}))}{dt} dt + V(Y_R(t_1^R; 0, Y_{R0}))$$

$$\leq \int_{t_1^R}^{t_2^R} -\frac{K_3 K_2^{-1}(\delta)}{2} dt + V(Y_R(t_1^R; 0, Y_{R0}))$$

$$\leq V(Y_R(t_1^R; 0, Y_{R0})). \tag{2.2.37}$$

By (2.2.32) and $\| Y_R(t_1^R; 0, Y_{R0}) \| = \delta$, we have

$$V(Y_R(t_1^R; 0, Y_{R0})) \leq K_1(\epsilon).$$

This together with (2.2.37) gives

$$K_1(\| Y_R(t_2^R; 0, Y_{R0}) \|) \leq K_1(\epsilon). \tag{2.2.38}$$

Since the wedge function $K_1(\cdot)$ is increasing, (2.2.38) implies that $|Y_R(t_2^R; 0, Y_{R0})| \leq \epsilon$, which contradicts the middle inequality of (2.2.36). The claim (2.2.35) follows.

Finally, for each $a > 0$, from the results of Step 4 and Step 5, for $R > \max\left\{ R_2, \frac{2c}{a K_3 K_2^{-1}(\delta)} \right\}$ and $t \in [a, \infty)$,

$$|z_{1R}(t) - v(t)| = |y_{1R}(Rt)| \leq \| Y_R(Rt) \| \leq \epsilon.$$

Hence $z_{1R}(t)$ converges uniformly to $v(t)$ in $[a, \infty)$ as $R \to \infty$. This completes the proof of the theorem.   □

2.3   Finite-Time Stable System-Based Tracking Differentiator

In this section, we study the following tracking differentiator:

$$\begin{cases} \dot{z}_{1R}(t) = z_{2R}(t), \ z_{1R}(0) = z_{10}, \\ \dot{z}_{2R}(t) = z_{3R}(t), \ z_{2R}(0) = z_{20}, \\ \quad \vdots \\ \dot{z}_{nR}(t) = R^n f\left(z_{1R}(t) - v(t), \frac{z_{2R}(t)}{R}, \dots, \frac{z_{nR}(t)}{R^{n-1}} \right), \ z_{nR}(0) = z_{n0}, \end{cases} \tag{2.3.1}$$

based on continuous finite-time stable systems. We first generalize stability for the perturbed finite-time stable systems and then apply it to the proof of strong and weak convergence of the finite-time stable tracking differentiators without assuming the Lipschitz continuous for Lyapunov function. A second-order finite-time stable differentiator is constructed using homogeneity. All required conditions are verified.

2.3.1   Convergence of Finite-Time Stable System-Based TD

The main purpose of this section is to prove the strong and weak convergence of (2.3.1). Before going on, we give some preliminary results on finite-time stability.

Lemma 2.3.1   *For the following system*

$$\dot{x}(t) = f(x(t)), \qquad x(0) = x_0 \in \mathbb{R}^n \tag{2.3.2}$$

suppose that there exists a continuous, positive definite function $V : \mathbb{R}^n \to \mathbb{R}$, constants $c > 0, \alpha \in (0, 1)$ such that

$$L_f V(x) = \sum_{i=1}^{n} \frac{\partial V(x)}{\partial x_i} f_i(x) \le - c(V(x))^\alpha, \ x = (x_1, x_2, \ldots, x_n) \neq 0, \tag{2.3.3}$$

where $f_i(x)$ denotes the ith component of $f(x)$. Then (2.3.2) is globally finite-time stable. Furthermore, there exists $\sigma > 0$ such that for any $x_0 \in \mathbb{R}^n$, $\| x_0 \| \le \sigma$, the following inequality holds:

$$\| x_0 \| \le \frac{1}{c(1 - \alpha)} (V(x_0))^{1-\alpha}. \tag{2.3.4}$$

Proof. The finite-time stability follows directly from Theorem 1.3.7. The finite-time stability together with the continuity of $f(x)$ shows that there exists $\sigma > 0$ such that if $\| x_0 \| \le \sigma$, then the solution $x(t; x_0)$ of (2.3.2) satisfies

$$\| f(x(t; x_0)) \| \le 1 \ \forall \, t > 0. \tag{2.3.5}$$

Integrating (2.3.2) from 0 to $T(x_0)$ gives

$$0 = x(T(x_0); x_0) = x_0 + \int_0^{T(x_0)} f(x(t; x_0)) dt, \tag{2.3.6}$$

where $T(x_0)$ is the setting time, which gives

$$\| x_0 \| = \left\| \int_0^{T(x_0)} f(x(t; x_0)) dt \right\| \le T(x_0). \tag{2.3.7}$$

This together with (1.3.68) completes the proof of the lemma.   □

Lemma 2.3.2   *Consider the following perturbed system of (2.3.2):*

$$\dot{y}(t) = f(y(t)) + g(t, y(t)), \ y(0) = y_0. \tag{2.3.8}$$

If there exists a continuous, positive definite, and radially unbounded function $V : \mathbb{R}^n \to \mathbb{R}$ with all continuous partial derivatives in its variables, and constants $c > 0$ and $\alpha \in (0, 1)$ such that (2.3.3) holds, then there exists $\delta_0 > 0$ such that, for every continuous function $g : \mathbb{R}^{n+1} \to \mathbb{R}^n$ with

$$\delta = \sup_{(t,x) \in \mathbb{R}^{n+1}} \| g(t, x) \| \le \delta_0, \tag{2.3.9}$$

the solution of (2.3.8) is bounded and

$$\| y(t) \| \le L\delta^{\frac{1-\alpha}{\alpha}}, \ \forall\, t \in [T, \infty), \tag{2.3.10}$$

where L and T are δ-independent positive constants.

Proof. We split the proof into two steps.

Step 1. *There exists a $\delta_1 > 0$ such that for each $\delta < \delta_1$, where δ is defined in (2.3.9), the solution of (2.3.8) is bounded.*

Let $b = \max\{1, V(y_0)\}$, $\delta_1 = cb^\alpha/M$ where $M = \sup_{x \in \{x:\ V(x) \le b+1\}} \| \nabla_x V \|$. For a radially unbounded positive definite function $V(x)$, there are strictly increasing functions $\kappa_1, \kappa_2 : [0, \infty) \to [0, \infty)$ such that $\lim_{r \to \infty} \kappa_i(r) = \infty$ and $\kappa_1(\| x \|) \le V(x) \le \kappa_2(\| x \|)$ for all $x \in \mathbb{R}^n$. For $x \in \{x :\ V(x) \le b+1\}$, $\kappa_1(\| x \|) \le b+1$, $\| x \| \le \kappa^{-1}(b+1)$, and hence the set $\{x :\ V(x) \le b+1\}$ is bounded. This together with the continuity of $\nabla_x V(x)$ concludes that $M < \infty$, so δ_1 is a positive number. If the claim is not true, since $y(t)$ is continuous and $V(y(0)) \le b$, for $\delta < \delta_1$, there exist $t_1, t_2 : 0 < t_1 < t_2$ such that the solution of (2.3.8) satisfies

$$V(y(t_1)) = b, V(y(t_2)) > b, y(t) \in \{x :\ b \le V(x) \le b+1\}, \forall\, t \in [t_1, t_2]. \tag{2.3.11}$$

Finding the derivative of $V(y(t))$ in $[t_1, t_2]$ gives

$$\dot{V}(y(t)) = L_f V(y(t)) + \langle \nabla_x V, g(t, y(t)) \rangle \le -c(V(y(t)))^\alpha + cb^\alpha \le 0, \tag{2.3.12}$$

which contradicts (2.3.11). So the claim is true.

Step 2. *There exists $\delta_0 : 0 < \delta_0 < \delta_1$ such that for every $\delta < \delta_0$, where δ is defined in (2.3.9), the solution of (2.3.8) satisfies $\| y(t) \| \le \sigma$ for all $t \in [T, \infty)$ and some $T > 0$, where σ is the same as in Lemma 2.3.1.*

Let $d = \kappa_1(\sigma) > 0$, $\delta_0 = \min\{\delta_1, cd^\alpha/2M\}$, $\mathcal{A} = \left\{x :\ V(x) \le \left(\frac{2M\delta}{c}\right)^{1/\alpha}\right\}$, where M and κ_1 are the same as in Step 1. Then the derivative of $V(x)$ along the solution of (2.3.8) is found to satisfy

$$\dot{V}(y(t)) \le -c(V(y(t)))^\alpha + M\delta < -\frac{1}{2}c(V(y(t)))^\alpha, \text{ if } y(t) \notin \mathcal{A}. \tag{2.3.13}$$

Consider the following scalar differential equation:

$$\dot{z}(t) = -k(z(t))^\alpha, \ z(0) > 0, \ k > 0. \tag{2.3.14}$$

Its solution can be found as

$$z(t) = \begin{cases} \left((z(0))^{1-\alpha} - k(1-\alpha)t\right)^{\frac{1}{1-\alpha}}, & t < \dfrac{1}{k(1-\alpha)}(z(0))^{1-\alpha}, \\[2mm] 0, & t \ge \dfrac{1}{k(1-\alpha)}(z(0))^{1-\alpha}. \end{cases} \tag{2.3.15}$$

If for all $t \ge 0$, $y(t) \in \mathcal{A}^c$, the complementary set of $\mathcal{A}$, then by the comparison principle in ordinary differential equation, $V(y(t)) \le z(t)$ with $z(0) = V(y(0))$, $k = c/2$, which is a contradiction since, on the one hand, $y(t) \in \mathcal{A}^c$ yields $V(y(t)) > 0$ and, on the other hand, $V(y(t)) \leq z(t)$ and $z(t)$ is identical to zero for $t \geq \frac{1}{k(1-\alpha)}(z(0))^{1-\alpha}$ from (2.3.15). Therefore, there exists a constant $T > 0$ such that $y(T) \in \mathcal{A}$. Since, for $y(t) \in \mathcal{A}^c$, $\dot{V}(y(t)) < 0$, it must have $y(t) \in \mathcal{A}$, for all $t \geq T$. However, for any $x \in \mathcal{A}, \parallel x \parallel \leq \kappa_1^{-1}(V(x)) \leq \kappa_1^{-1}(d) \leq \sigma$. This is the claim in Step 2.

By Lemma 2.3.1, and Steps 1 and 2, we obtain

$$\parallel y(t) \parallel \leq \frac{1}{c(1-\alpha)}(V(y(t)))^{1-\alpha} \leq \frac{1}{c(1-\alpha)}\left(\frac{2M\delta}{c}\right)^{\frac{1-\alpha}{\alpha}} \tag{2.3.16}$$

for all $t \geq T$. This completes the proof of the lemma. $\square$

Theorem 2.3.1 *Suppose that*

(i) $\sup_{t \in [0,\infty)} |v^{(i)}(t)| < \infty, i = 1, 2, \ldots, n;$
(ii) *the nonlinear function $f(x)$ in (2.3.1) satisfies*

$$|f(x) - f(\overline{x})| \leq \sum_{j=1}^{n} k_j \parallel x_j - \overline{x}_j \parallel^{\theta_j} \text{ for some } k_j > 0, \ \theta_j \in (0,1]; \tag{2.3.17}$$

(iii) *there exists a continuous, positive definite function $V : \mathbb{R}^n \to \mathbb{R}$, with all continuous partial derivatives in its variables, satisfying*

$$L_h V(x) \leq -c(V(x))^{\alpha}, \tag{2.3.18}$$

with $c > 0, \alpha \in (0,1), \gamma = \frac{1-\alpha}{\alpha}$, and $h(x)$ is the vector field: $h(x) = (x_2, x_3, \ldots, x_{n-1}, f(x))^{\top}$. Then for any initial value of (2.3.1) and constant $a > 0$, there exists $R_0 > 0$ such that, for every $R > R_0$,

$$|z_{iR}(t) - v^{(i-1)}(t)| \leq L\left(\frac{1}{R}\right)^{\theta\gamma - i + 1}, \ \forall\, t > a, \tag{2.3.19}$$

where L is some positive constant, $\theta = \min\{\theta_2, \theta_3, \ldots, \theta_n\}$, and $z_{iR}(t)$ is the solution of (2.3.1), $i = 1, 2, \ldots, n$.

Proof. Let $e_R(t) = (e_{1R}(t), e_{2R}(t), \ldots, e_{nR}(t))^{\top}$, where

$$e_{iR}(t) = \frac{z_{iR}\left(\frac{t}{R}\right) - v^{(i-1)}\left(\frac{t}{R}\right)}{R^{i-1}}, \ i = 1, 2, \ldots, n. \tag{2.3.20}$$

Then $e_{iR}(t)$ satisfies the following system of differential equations:

$$\begin{cases} \dot{e}_{1R}(t) = e_{2R}(t), \\ \dot{e}_{2R}(t) = e_{3R}(t), \\ \quad \vdots \\ \dot{e}_{nR}(t) = f\left(e_{1R}(t), e_{2R}(t) + \frac{\dot{v}\left(\frac{t}{R}\right)}{R}, \ldots, e_{nR}(t) + \frac{v^{(n-1)}\left(\frac{t}{R}\right)}{R^{n-1}}\right) - \frac{v^{(n)}\left(\frac{t}{R}\right)}{R^n}, \end{cases}$$
$$\tag{2.3.21}$$

which can be written as a perturbed system of the following finite-time stable system:

$$
\begin{cases}
\dot{e}_{1R}(t) = e_{2R}(t), \\
\dot{e}_{2R}(t) = e_{3R}(t), \\
\quad \vdots \\
\dot{e}_{nR}(t) = f\left(e_{1R}(t), e_{2R}(t), \ldots, e_{nR}(t)\right) + \Delta(t),
\end{cases}
\tag{2.3.22}
$$

with

$$
\Delta(t) = f\left(e_{1R}(t), e_{2R}(t) + \frac{\dot{v}\left(\frac{t}{R}\right)}{R}, \ldots, e_{nR}(t) + \frac{v^{(n-1)}\left(\frac{t}{R}\right)}{R^{n-1}}\right)
$$

$$
- \frac{v^n\left(\frac{t}{R}\right)}{R^n} - f(e_{1R}(t), e_{2R}(t), \ldots, e_{nR}(t)).
\tag{2.3.23}
$$

By conditions (i) and (ii), there exists a constant $B > 0$ such that

$$
|\Delta(t)| \le B\left(\frac{1}{R}\right)^{\theta}, \ \forall \, t \ge 0, \ R > 1.
\tag{2.3.24}
$$

By Lemma 2.3.2, there exist constants $R_0 > 1, T > 0, L > 0$ such that, for each $R > R_0$ and $t > T$,

$$
\| e_R(t) \| = \| (e_{1R}(t), e_{2R}(t), \ldots, e_{nR}(t))^{\top} \| \le L\left(\frac{1}{R}\right)^{\gamma}.
\tag{2.3.25}
$$

By (2.3.20), it follows that, for all $R > \left\{R_0, \frac{T}{a}\right\}$ and $t > a$,

$$
|z_{iR}(t) - v^{(i-1)}(t)| = R^{i-1}|e_{iR}(Rt)| \le R^{i-1} \| e_R(Rt) \| \le L\left(\frac{1}{R}\right)^{\gamma\theta - i + 1}.
\tag{2.3.26}
$$

This ends the proof of the theorem. $\square$

Remark 2.3.1 *It is noted that in (2.3.19) that if $\theta\gamma - n + 1 > 0$ then $z_{iR}(t) \to v^{(i-1)}(t)$ uniformly on (a, ∞) as $R \to \infty$, while $\theta\gamma - n + 1 < 0$, then $z_{1R}(t) \to v(t)$ uniformly on (a, ∞) as $R \to \infty$. Since constant a can be sufficiently small, we have*

$$
\lim_{R\to\infty} \int_0^T |z_{1R}(t) - v(t)| dt = 0.
\tag{2.3.27}
$$

Hence $z_{iR}(t) \to v^{(i-1)}(t)$ in the sense of distribution.

Theorem 2.3.1 can be extended to piecewise continuous signal $v(t)$.

Theorem 2.3.2 *Suppose that there exist $0 = t_0 < t_1 < t_2 < \cdots < t_m$ such that $v(t)$ is n times differentiable in $(t_j, t_{j+1}), (t_m, \infty)$, and is left and right differentiable at t_j. Assume that*

$$
\max_{1 \le i \le n, 0 \le j \le m-1, 1 \le k \le m} \left\{ \sup_{t \in (t_j, t_{j+1}) \cup (t_m, \infty)} \left\{ v^{(i)}(t), v_-^{(i)}(t_k), v_+^{(i)}(t_k) \right\} \right\} < \infty
$$

and the function $f(\cdot)$ in (2.3.1) satisfies conditions (ii) and (iii) of Theorem 2.3.1. Then for any initial value of (2.3.1) and $a \in (0, \min_{0 \leq j \leq m-1}(t_{j+1} - t_j))$, there exists $R_0 > 0$ such that, for any $R > R_0$ and $t \in (t_j + a, t_{j+1})$ or $t > t_m + a$, it has

$$|z_{iR}(t) - v^{(i-1)}(t)| \leq L\left(\frac{1}{R}\right)^{\theta\gamma - i + 1}, \quad \forall\, i = 1, 2, \ldots, n, \tag{2.3.28}$$

for some constant $L > 0$, where we use $v_-(t)$ to denote the left derivative and $v_+(t)$ for the right one.

Proof. Let

$$\begin{cases} e_{iR}(t) = \dfrac{z_{iR}\left(\frac{t}{R}\right) - v^{(i-1)}\left(\frac{t}{R}\right)}{R^{i-1}}, & t_j R < t < t_{j+1}R,\ t > t_m R, \\[3mm] e_R(t_j R) = \dfrac{z_{iR-}(t_j) - v_+^{(i-1)}(t_j)}{R^{i-1}}, & 1 \leq i \leq n,\ 0 \leq j \leq m. \end{cases} \tag{2.3.29}$$

Then $\{e_{iR}(t)\}$ satisfies the following system of impulsive differential equations:

$$\begin{cases} \begin{cases} \dot{e}_{1R}(t) = e_{2R}(t), \\ \dot{e}_{2R}(t) = e_{3R}(t), \\ \quad\vdots \\ \dot{e}_{nR}(t) = f\left(e_{1R}(t), e_{2R}(t) + \dfrac{\dot{v}\left(\frac{t}{R}\right)}{R}, \ldots, e_{nR}(t) + \dfrac{v^{(n-1)}\left(\frac{t}{R}\right)}{R^{n-1}}\right) - \dfrac{v^{(n)}\left(\frac{t}{R}\right)}{R}, \\ t_j R < t < t_{j+1}R,\ 0 \leq j \leq m - 1,\ t > t_m R, \end{cases} \\[3mm] e_R(0) = \left(z_{10}, \dfrac{z_{20}}{R}, \ldots, \dfrac{z_{n0}}{R^{n-1}}\right), \quad e_R(t_j R) = \dfrac{z_{iR-}(t_j) - v_+^{(i-1)}(t_j)}{R^{i-1}}. \end{cases} \tag{2.3.30}$$

The result follows by successively applying Theorem 2.3.1 to (2.3.30) in intervals $[0, t_1 R)$, $\ldots$, $[t_j R, t_{j+1} R)$, $\ldots$, $[t_m R, \infty)$, respectively. $\qquad\square$

Theorem 2.3.3 *Suppose that there exist $0 = t_0 < t_1 < t_2 < \cdots < t_m$ such that $v(t)$ is differentiable in $(t_j, t_{j+1}), (t_m, \infty)$, and is left and right differentiable at t_j. Assume that*

$$\max\left\{ \sup_{t \in (t_j, t_{j+1}) \cup (t_m, \infty)} \{\dot{v}(t), \dot{v}_-(t_k), \dot{v}_+(t_k)\} \right\} < \infty$$

and the function $f(\cdot)$ in (2.3.1) satisfies conditions (ii) and (iii) in Theorem 2.3.1. Then for any initial value of (2.3.1) and $a \in (0, \min_{0 \leq j \leq m-1}(t_{j+1} - t_j))$, there exists $R_0 > 0$ such that for any $R > R_0$, $t \in (t_j + a, t_{j+1})$ or $t > t_m + a$,

$$|z_{1R}(t) - v(t)| \leq L\left(\frac{1}{R}\right)^{\theta\gamma - i + 1}, \quad \forall\, i = 1, 2, \ldots, n, \tag{2.3.31}$$

where $L > 0$ is a constant and $\gamma = \frac{1-\alpha}{\alpha}$.

Proof. Let

$$
\begin{cases}
e_{1R}(t) = z_{1R}\left(\dfrac{t}{R}\right) - v\left(\dfrac{t}{R}\right), \quad e_{iR}(t) = \dfrac{z_{iR}\left(\frac{t}{R}\right)}{R^{i-1}}, \quad t_j R < t < t_{j+1}R, \\[3mm]
e_{iR}(t_j R) = \dfrac{z_{iR-}(t_j) - v_+^{(i-1)}(t_j)}{R^{i-1}}, \quad 2 \le i \le n, \ 0 \le j \le m.
\end{cases}
\tag{2.3.32}
$$

Then $\{e_{iR}(t)\}$ satisfies the following system of impulsive differential equations:

$$
\begin{cases}
\dot{e}_{1R}(t) = e_{2R}(t) - \dfrac{\dot{v}\left(\frac{t}{R}\right)}{R}, \\[3mm]
\dot{e}_{2R}(t) = e_{3R}(t), \\
\quad\vdots \\
\dot{e}_{nR}(t) = f(e_{1R}(t), e_{2R}(t), \ldots, e_{nR}(t)), \\
\quad t_j R < t < t_{j+1}R, \ 0 \le j \le m-1, \ t > t_m R, \\[2mm]
e_{1R}(t_j R) = \dfrac{z_{1R-}(t_j) - \dot{v}_-(t_j R)}{R}, \quad e_{iR}(t_j R) = \dfrac{z_{iR-}(t_j)}{R^{i-1}}, \quad 2 \le i \le n, \ 0 \le j \le m.
\end{cases}
\tag{2.3.33}
$$

Also, (2.3.33) is a perturbed system of a finite-time stable system. Similar to the proof of Theorem 2.3.2, we can get the required result. The details are omitted. $\qquad\square$

Remark 2.3.2 *For any constants $B > b > 0$, it follows from Theorem 2.3.3 that $\int_b^B |z_{1R}(t) - v(t)|dt \to 0$ as $R \to \infty$, that is to say, the weak convergence is also true. Therefore, we can understand $z_{iR}(t)$ as an approximation of the weak derivative $v^{(i-1)}(t)$ of $v(t)$ on (b, B) even if the derivative may not exist in the classical sense at some points.*

2.3.2   A Second-Order Finite-Time Stable Tracking Differentiator

In this section, we construct a second-order finite-time stable differentiator with the help of homogeneity.

Consider the second-order system:

$$
\dot{x}(t) = f(x(t)) = (f_1(x(t)), f_2(x(t)))^\top,
\tag{2.3.1}
$$

where $x(t) = (x_1(t), x_2(t))$ and

$$
\begin{cases}
f_1(x_1, x_2) = x_2, \\
f_2(x_1, x_2) = -k_1[x_1]^\alpha - k_2[x_2]^\beta,
\end{cases}
\tag{2.3.2}
$$

with $[r]^\alpha = \text{sign}(r)|r|^\alpha$ and

$$
k_1, \ k_2 > 0, \quad \alpha = \frac{b-1}{a}, \quad \beta = \frac{b-1}{b}, \quad a = b+1, \ b > 1.
\tag{2.3.3}
$$

It is seen that, for any $\lambda > 0$,

$$
\begin{cases}
f_1\left(\lambda^a x_1, \lambda^b x_2\right) = \lambda^b x_2 = \lambda^{-1+a} f_1(x_1, x_2), \\
f_2(\lambda^a x_1, \lambda^b x_2) = -k_1 \lambda^{a\alpha}[x_1]^\alpha - k_2 \lambda^{b\beta}[x_2]^\beta = \lambda^{-1+b} f_2(x_1, x_2).
\end{cases}
\tag{2.3.4}
$$

Therefore, the vector field $f(x)$ is homogeneous of degree -1 with respect to weights a, b.

Let $W : \mathbb{R}^2 \to \mathbb{R}$ be given by

$$W(x_1, x_2) = \frac{1}{2k_1}x_2^2 + \frac{|x_1|^{1+\alpha}}{1+\alpha}. \tag{2.3.5}$$

A direct computation shows that

$$L_f W(x) = -\frac{k_2}{k_1}|x_2|^{\beta+1} \leq 0. \tag{2.3.6}$$

By Lasalle's invariance principle of Theorem 1.3.4, system (2.3.1) is globally asymptotically stable. For some positive numbers $l > \max\{1, a, b\}$ and $c > 0$, there is a continuous, positive definite, and radially unbounded Lyapunov function $V : \mathbb{R}^2 \to \mathbb{R}$ such that $\nabla_x V(x)$ is continuous on $\mathbb{R}^n$ and

$$L_f V(x) \leq -c(V(x))^{\frac{l-1}{l}}. \tag{2.3.7}$$

Now we show that $f_2(x)$ satisfies condition (2.3.17) in Theorem 2.3.1. Let $\phi : [a, \infty) \to \mathbb{R}$ be defined by $\phi(x) = x^\theta - a^\theta - (x - a)^\theta$ for some $a > 0, \theta \in (0, 1)$. Then $\phi(x)$ is decreasing since $\dot{\phi}(x) = \theta(x^{\theta-1} - (x-a)^{\theta-1}) < 0$ for all $x > a$. It follows that

$$x^\theta - y^\theta \leq (x - y)^\theta, \ \forall \, x > y > 0. \tag{2.3.8}$$

Furthermore, since $\phi''(x) = \theta(\theta - 1)x^{\theta-2} < 0$ for all $x > 0$, $\phi(x)$ is convex on $(0, \infty)$. By Jessen's inequality,

$$x^\theta + y^\theta \leq 2^{1-\theta}(x + y)^\theta, \ \forall \, x, y > 0. \tag{2.3.9}$$

Combining inequalities (2.3.8) and (2.3.9), we obtain

$$|f_2(x_1, x_2) - f_2(y_1, y_2)| \leq k_1 2^{1-\alpha}|x_1 - y_1|^\alpha + k_2 2^{1-\beta}|x_2 - y_2|^\beta. \tag{2.3.10}$$

That is, the condition (2.3.17) of Theorem 2.3.1 is satisfied for $f_2(x_1, x_2)$. Applying Theorem 2.3.1 to (2.3.1), we have the following Theorem 2.3.4.

Theorem 2.3.4 *If the signal $v(t)$ satisfies $\sup_{t\in[0,\infty)} |v^{(i)}(t)| < \infty$ for $i = 1, 2$, then the following second-order finite-time stable differentiator:*

$$\begin{cases} \dot{z}_{1R}(t) = z_{2R}(t), \\ \dot{z}_{2R}(t) = R^2\left(-k_1[z_{1R}(t) - v(t)]^\alpha - k_2\left[\dfrac{z_{2R}(t)}{R}\right]^\beta\right), \end{cases} \tag{2.3.11}$$

is convergent in the sense that for any initial value of (2.3.11) and $a > 0$, there exists a constant $R_0 > 0$ such that for all $R > R_0$ and $t > a$,

$$|z_{1R}(t) - v(t)| \leq M_1\left(\frac{1}{R}\right)^{\beta\frac{1-\gamma}{\gamma}}, \ |z_{2R}(t) - \dot{v}(t)| \leq M_2\left(\frac{1}{R}\right)^{\beta\frac{1-\gamma}{\gamma}-1}, \tag{2.3.12}$$

where $\gamma = \frac{l-1}{l}$, $l > \max\{1, a, b\}$ and the parameters in (2.3.11) are selected to satisfy (2.3.3). If $\beta\frac{1-\gamma}{\gamma} > 1$ then $z_{2R}(t) \to \dot{v}(t)$ in classical sense, while $\beta\frac{1-\gamma}{\gamma} \leq 1$, $z_{2R}(t) \to \dot{v}(t)$ in the sense of a weak derivative (weak convergence).

Let $a = 4, b = 3, k_1 = k_2 = 1, \alpha = \frac{1}{2}, \beta = \frac{2}{3}$. Then it is easy to verify that (2.3.11) is weakly convergent. Unfortunately, it seems hard to find parameters so that (2.3.11) is strongly convergent. Nevertheless, in the next section, we give the numerical simulation of differentiator (2.3.11) using these parameters.

2.4 Illustrative Examples and Applications

In this section, we first give numerical experiments to compare the linear tracking differentiator with a finite-time stable systems-based tracking differentiator and a robust exact differentiator. Secondly, we use a tracking differentiator to frequency online estimation and boundary displacement feedback control for stabilizing a one-dimensional wave equation.

2.4.1 Comparison of Three Differentiators

In this section, we compare the following three differentiators numerically.

DI. Robust exact differentiator using the sliding mode technique:

$$\begin{cases} \dot{z}_{1R}(t) = z_{2R}(t) - R[z_{1R}(t) - v(t)]^{\frac{1}{2}}, \\ \dot{z}_{2R}(t) = -R \, \text{sign}(z_{1R}(t) - v(t)). \end{cases} \tag{2.4.1}$$

DII. Linear tracking differentiator:

$$\begin{cases} \dot{z}_{1R}(t) = z_{2R}(t), \\ \dot{z}_{2R}(t) = R^2 \left(-(z_{1R}(t) - v(t)) - \frac{z_{2R}(t)}{R} \right). \end{cases} \tag{2.4.2}$$

DIII. Finite-time stable tracking differentiator (2.3.11):

$$\begin{cases} \dot{z}_{1R}(t) = z_{2R}(t), \\ \dot{z}_{2R}(t) = R^2 \left(-[z_{1R}(t) - v(t)]^{\frac{1}{2}} - \left[\frac{z_{2R}(t)}{R} \right]^{\frac{2}{3}} \right). \end{cases} \tag{2.4.3}$$

The Matlab program of the Eular method is adopted in the investigation. We choose the same zero initial value, step $h = 0.001$, respectively, and $v(t) = \sin t$ in all simulations.

The results by differentiator DI are plotted in Figure 2.4.1, where in Figure 2.4.1(a) $R = 20$ is taken, while in Figure 2.4.1(b) $R = 10$. Figure 2.4.1(c) is the magnification of Figure 2.4.1(b). The results by differentiator DII are plotted in Figure 2.4.2, where in Figure 2.4.2(a) $R = 20$ is taken, while in Figure 2.4.2(b) $R = 10$. The results by differentiator DIII are plotted in Figure 2.4.3, where in Figure 2.4.3(a) $R = 20$ is taken, while in Figure 2.4.3(b) $R = 10$. Figure 2.4.3(c) plots the results of differentiator DIII with a delayed signal $v(t)$ of delay 0.02 and $R = 20$. Figure 2.4.4 plots the numerical results of differentiator

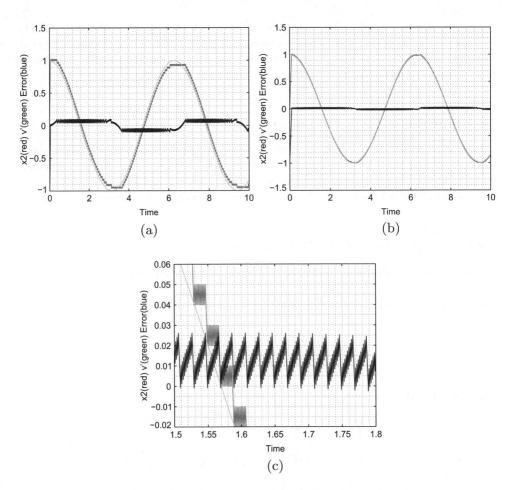

Figure 2.4.1   Derivative tracking for $v(t) = \sin t$ by DI.

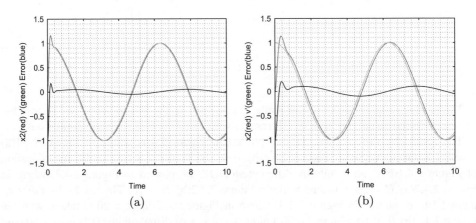

Figure 2.4.2   Derivative tracking for $v(t) = \sin t$ by DII.

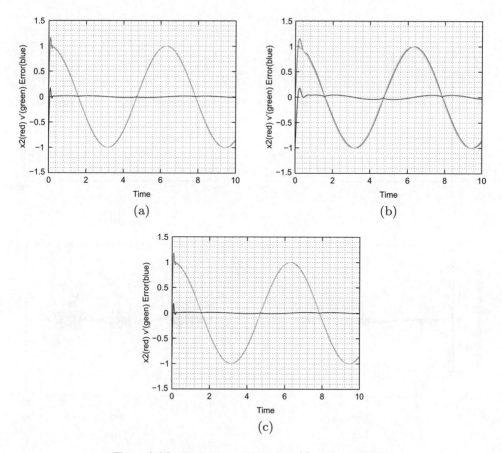

Figure 2.4.3 Derivative tracking for $v(t) = \sin t$ by DIII.

DIII with singular $v(t)$ disturbed by its 1% noise, while the noise in Figure 2.4.4(a) is uniform noise and $R = 10$, in Figure 2.4.4(b) the noise is Gaussian noise and $R = 10$, in Figure 2.4.4(c) the noise is uniform noise and $R = 20$, and in Figure 2.4.4(d) the noise is Gaussian noise and $R = 20$.

From Figures 2.4.1, 2.4.2, and 2.4.3, we see that our finite-time stable tracking differentiator DIII is smoother than differentiator DI, in which the discontinuous of function produces a big problem of chattering. In addition, differentiator DIII tracks faster than linear differentiator DII. Moreover, it seems that the finite-time stable tracking differentiator DIII is more accurate than linear DII. Finally, the finite-time stable tracking differentiator is tolerant to small time delays and noise.

From Figures 2.4.3 and 2.4.4, we see that the turning parameter R in differentiator DIII plays a significant role in convergence and noise tolerance: the larger R is, the more accurate the tracking effect would be, but the more sensitive the noise would be. This suggests that the choice of parameter R in DIII is a tradeoff between tracking accuracy and noise tolerance in practice.

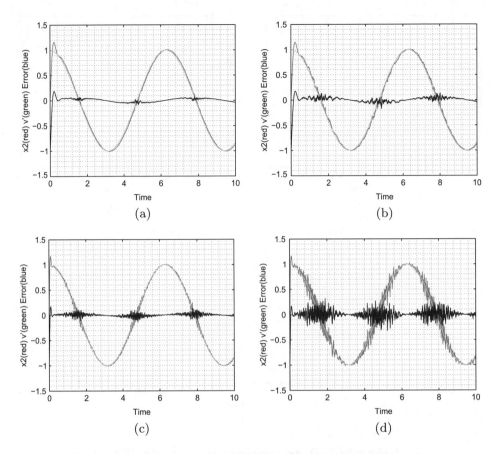

Figure 2.4.4 Derivative tracking of DIII for $v(t) = \sin t$ disturbed by noise.

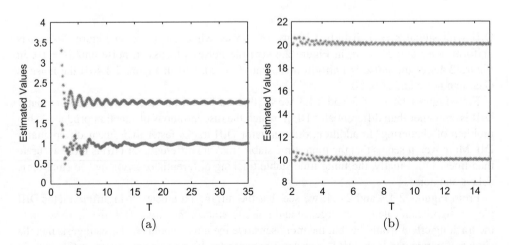

Figure 2.4.5 Estimation of two frequencies by the linear tracking differentiator (2.4.13).

2.4.2   Applications to Frequency Online Estimation

In this subsection, we apply the tracking differentiator to frequency estimation. We consider the finite sum of sinusoidal signals $v(t) = \sum_{i=1}^{n} A_i \sin(\omega_i t + \phi_i)$, where the $\omega_i > 0$ are different frequencies. The aim is to estimate all frequencies ω_i by using the tracking differentiator. The even-order derivatives of $v(t)$ with respect to t up to $2n - 2$ are found to be

$$\ddot{v}(t) = \sum_{i=1}^{n} \theta_i A_i \sin(\omega_i t + \phi_i),$$

$$v^{(4)}(t) = \sum_{i=1}^{n} \theta_i^2 A_i \sin(\omega_i t + \phi_i),$$

$$\cdots$$

$$v^{(2n-2)}(t) = \sum_{i=1}^{n} \theta_i^{n-1} A_i \sin(\omega_i t + \phi_i),$$

where $\theta_i = -\omega_i^2$. That is,

$$\begin{pmatrix} v(t) \\ \ddot{v}(t) \\ \vdots \\ v^{(2n-2)}(t) \end{pmatrix} = \Lambda \begin{pmatrix} A_1 \sin(\omega_1 t + \phi_1) \\ A_2 \sin(\omega_2 t + \phi_2) \\ \vdots \\ A_n \sin(\omega_n t + \phi_n) \end{pmatrix}, \tag{2.4.4}$$

where

$$\Lambda = \begin{pmatrix} 1 & 1 & \cdots & 1 \\ \theta_1 & \theta_2 & \cdots & \theta_n \\ \cdots & \cdots & \ddots & \cdots \\ \theta_1^{n-1} & \theta_2^{n-1} & \cdots & \theta_n^{n-1} \end{pmatrix}. \tag{2.4.5}$$

Since Λ is invertible, one has

$$\begin{pmatrix} A_1 \sin(\omega_1 t + \phi_1) \\ A_2 \sin(\omega_2 t + \phi_2) \\ \vdots \\ A_n \sin(\omega_n t + \phi_n) \end{pmatrix} = \Lambda^{-1} \begin{pmatrix} v(t) \\ \ddot{v}(t) \\ \vdots \\ v^{(2n-2)(t)} \end{pmatrix}. \tag{2.4.6}$$

Denote Λ^{-1} by

$$\Lambda^{-1} = \begin{pmatrix} \lambda_{11} & \lambda_{12} & \cdots & \lambda_{1n} \\ \lambda_{21} & \lambda_{22} & \cdots & \lambda_{2n} \\ \cdots & \cdots & \ddots & \cdots \\ \lambda_{n1} & \lambda_{n2} & \cdots & \lambda_{nn} \end{pmatrix}. \tag{2.4.7}$$

Since for any $b \geq 0$

$$\omega_k^2 = \lim_{t \to \infty} \frac{\int_b^t ((A_k \sin(\omega_k t + \phi_k))')^2 dt}{\int_b^t (A_k \sin(\omega_k t + \phi_k))^2 dt}, \quad k = 1, 2, \ldots, n,$$

it follows from (2.4.6) that

$$-\theta_k = \lim_{t \to \infty} \frac{\sum_{i,j=1}^{n} \lambda_{ki} \lambda_{kj} \int_b^t v^{(2i-1)}(s) v^{(2j-1)}(s) ds}{\sum_{i,j=1}^{n} \lambda_{ki} \lambda_{kj} \int_b^t v^{(2i-2)}(s) v^{(2j-2)}(s) ds}, \quad k = 1, 2, \ldots, n.$$

Let T be a sufficiently large number and let

$$\begin{cases} a_{ij} = \int_b^T v^{(2i-1)}(s) v^{(2j-1)}(s) ds, \\ b_{ij} = \int_b^T v^{(2i-2)}(s) v^{(2j-2)}(s) ds, \quad i, j = 1, 2, \ldots, n. \end{cases} \quad (2.4.8)$$

By solving the following high-order equations of n unknown elements $\theta_i, i = 1, 2, \ldots, n$, we can obtain an approximate values of θ_i:

$$\begin{cases} \sum_{i,j=1}^{n} \lambda_{1i} \lambda_{1j} b_{ij} \theta_1 + \sum_{i,j=1}^{n} \lambda_{1i} \lambda_{1j} a_{1j} = 0, \\ \sum_{i,j=1}^{n} \lambda_{2i} \lambda_{2j} b_{ij} \theta_2 + \sum_{i,j=1}^{n} \lambda_{2i} \lambda_{2j} a_{2j} = 0, \\ \cdots\cdots\cdots\cdots\cdots\cdots\cdots\cdots\cdots\cdots\cdots \\ \sum_{i,j=1}^{n} \lambda_{ni} \lambda_{nj} b_{nj} \theta_1 + \sum_{i,j=1}^{n} \lambda_{ni} \lambda_{nj} a_{nj} = 0, \end{cases} \quad (2.4.9)$$

where $\theta_i = -\omega_i^2$, λ_{ij} are rational functions of θ_i, and the values of a_{ij} and b_{ij} in (2.4.8) can be approximated by the high-order tracking differentiator (2.2.21).

Example 2.4.1 *Let us investigate the following two different frequencies:*

$$v(t) = A_1 \sin(\omega_1 t + \phi_1) + A_2 \sin(\omega_2 t + \phi_2). \quad (2.4.10)$$

This is the case of $n = 2$. The equation (2.4.9) now becomes

$$\begin{cases} (a_{11} + 2b_{12}) \theta_1 \theta_2 - b_{22} \theta_1 - b_{22} \theta_2 - a_{22} = 0, \\ b_{11} \theta_1 \theta_2 + a_{11} \theta_2 + a_{11} \theta_1 - 2a_{12} - b_{22} = 0 \end{cases} \quad (2.4.11)$$

and (2.4.8) becomes, in this case,

$$
\begin{cases}
a_{11} = \int_b^T z_{2R}^2(t)dt, \ a_{12} = a_{21} = \int_b^T z_{2R}(t)z_{4R}(t)dt, \ a_{22} = \int_b^T z_{4R}^2(t)dt, \\
b_{11} = \int_b^T z_{1R}^2(t)dt, \ b_{12} = b_{21} = \int_b^T z_{1R}(t)z_{3R}(t)dt, \ b_{22} = \int_b^T z_{3R}^2(t)dt,
\end{cases}
\tag{2.4.12}
$$

where instead of $v^{(i-1)}(t)$, we used directly $z_{iR}(t)$ to be an approximation of $v^{(i-1)}(t)$, $i = 1, 2, 3, 4$, by the linear tracking differentiator (2.1.13):

$$
\begin{cases}
\dot{z}_{1R}(t) = z_{2R}(t), z_{1R}(0) = z_{10}, \\
\dot{z}_{2R}(t) = z_{3R}(t), z_{2R}(0) = z_{20}, \\
\dot{z}_{3R}(t) = z_{4R}(t), z_{3R}(0) = z_{30}, \\
\dot{z}_{4R}(t) = -24R^4(z_{1R}(t) - v(t)) - 50R^3 z_{2R}(t) \\
\qquad\qquad - 35R^2 z_{3R}(t) - 10R z_{4R}(t), z_{4R}(0) = z_{40},
\end{cases}
\tag{2.4.13}
$$

Now matrix A becomes

$$
\begin{pmatrix}
0 & 1 & 0 & 0 \\
0 & 0 & 1 & 0 \\
0 & 0 & 0 & 1 \\
-24 & -50 & -35 & -10
\end{pmatrix},
\tag{2.4.14}
$$

which has eigenvalues $\{-1, -2, -3, -4\}$, so it is Hurwitz. Hence the tracking differentiator (2.4.13) is well defined. Note that in (2.4.11), θ_1 and θ_2 are symmetrical. If we cancel θ_1 from (2.4.11), we get a quadratic equation of θ_2. Therefore, if there are two real solutions to (2.4.11), they must be (θ_1, θ_2).

Let $A_1 = 1, A_2 = 2, \omega_1 = 1, \omega_2 = 2, \phi_1 = \phi_2 = 0$ in (2.4.10), $b = 1, T : 2 \to 35$ with the step equal to 0.1 in (2.4.12) and $z_{10} = z_{20} = z_{30} = z_{40} = 0, R = 20$ in (2.4.13). The numerical results for frequency estimate by (2.4.11) to (2.4.13) are plotted in Figure 2.4.1. Figure 2.4.2 is the simulation for $A_1 = A_2 = 1, \omega_1 = 10, \omega_2 = 20, \phi_1 = \phi_2 = 0, z_{10} = z_{20} = z_{30} = z_{40} = 0, R = 20, b = 1, T : 2 \to 15$ with steps equal to 0.01. It is seen that the estimates are quite satisfactory.

Example 2.4.2 *In this example, we use a nonlinear second-order tracking differentiator to estimate the frequency of the signal $v(t) = A \sin(\omega t + \phi)$. This is the case of $n = 1$. In this case,*

$$
\omega = \lim_{T \to \infty} \sqrt{\frac{\int_b^T \dot{v}^2(t)dt}{\int_b^T v^2(t)dt}}
$$

The nonlinear tracking differentiator that we use here is

$$
\begin{cases}
\dot{z}_{1R}(t) = z_{2R}(t), \ z_{1R}(0) = z_{10}, \\
\dot{z}_{2R}(t) = -R^2 \text{sign}(z_{1R}(t) - v(t))|z_{1R}(t) - v(t)|^{1.3} - R z_{2R}(t), \ z_{2R}(0) = z_{20}.
\end{cases}
\tag{2.4.15}
$$

In order for the tracking differentiator (2.4.15) to satisfy all conditions of Theorem 2.2.2 as (2.2.21), we only need to prove that the equilibrium point (0,0) of the following systems is globally asymptotically stable:

$$\begin{cases} \dot{x}_1(t) = x_2(t), \\ \dot{x}_2(t) = -\text{sign}(x_1(t))|x_1(t)|^{1.3} - x_2(t). \end{cases} \tag{2.4.16}$$

In fact, let the Lyapunov function $v(x_1, x_2)$ be defined by

$$V(x_1, x_2) = \frac{|x_1|^{2.3}}{1.3} + \frac{x_2^2}{2}. \tag{2.4.17}$$

Then

$$\frac{dV(x_1(t), x_2(t))}{dt}\bigg|_{\text{along } (2.4.16)} = -x_2^2(t) \le 0.$$

Note that the set

$$\left\{ (x_1(t), x_2(t)) \,\bigg|\, \frac{dV(x_1(t), x_2(t))}{dt}\bigg|_{\text{along } (2.4.16)} = 0 \right\}$$

does not contain any nonzero trajectory of system (2.4.16). By Theorem 1.3.4, the equilibrium point (0,0) of the system (2.4.16) is globally asymptotically stable. Hence the tracking differentiator (2.4.15) satisfies all conditions of Theorem 2.2.2.

In Figure 2.4.6, we plot the numerical results for $A = 1, z_{10} = z_{20} = 0, \omega = 2, b = 10, R = 100, T : 10 \to 25$ with step $= 0.0001$. It is obviously convergent.

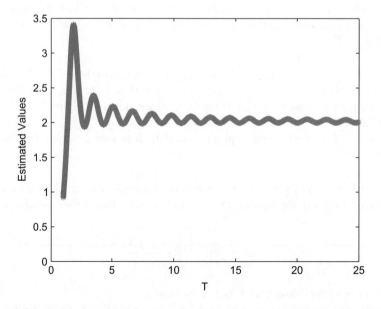

Figure 2.4.6 Estimation of one frequency by the nonlinear tracking differentiator (2.4.15).

2.4.3 Application to the Boundary Stabilization of Wave Equation

In the following, we apply the tracking differentiator to boundary stabilization for a one-dimensional wave equation using the displacement signal. Consider the following one-dimensional wave equation:

$$\begin{cases} w_{tt}(x,t) - w_{xx}(x,t) = 0, \ 0 < x < 1, \ t > 0, \\ w(0,t) = 0, \ w_x(1,t) = u(t), \ t \geq 0, \\ w(x,0) = w_0(x), \ w_t(x,0) = w_1(x), \ 0 \leq x \leq 1, \end{cases} \tag{2.4.18}$$

which describes the vibration of string, where $w(x,t)$ is the amplitude, $w_t(x,t)$ is the velocity, $w_x(x,t)$ is the vertical force, $(w_0(x), w_1(x))$ is the initial value, and $u(t)$ is the boundary control input. The control objective is practically to stabilize the wave equation (2.4.18). In the following, we will show that with the boundary feedback $u(t) = -w_t(1,t)$, the system (2.4.18) is exponentially stable in the sense that the energy $E(t)$ decays exponentially:

$$E(t) \leq Me^{-\omega t}E(0)$$

for some positive constants $M, \omega > 0$, where

$$E(t) = \frac{1}{2}\int_0^1 [w_x^2(x,t) + w_t^2(x,t)]dx \tag{2.4.19}$$

is the energy of the vibrating string. For this purpose, let the energy multiplier

$$\beta(t) = \int_0^1 xw_x(x,t)w_t(x,t)dt. \tag{2.4.20}$$

By a direct computation, we obtain

$$|\beta(t)| \leq \int_0^1 |w_x(x,t)w_t(x,t)|dt \leq \frac{1}{2}\left(\int_0^1 w_x^2(x,t)dt + \int_0^1 w_t^2(x,t)dt\right) = \frac{1}{2}E(t) \tag{2.4.21}$$

and

$$\begin{aligned}
\dot{\beta}(t) &= \int_0^1 x(w_{xt}(x,t)w_t(x,t) + w_x(x,t)w_{tt}(x,t))dx \\
&= \int_0^1 x(w_{xt}(x,t)w_t(x,t) + w_x(x,t)w_{xx}(x,t))dx \\
&= \frac{1}{2}\left(\int_0^1 x\ dw_t^2(x,t) + \int_0^1 x\ dw^2(x,t)\right) \\
&= \frac{1}{2}(xw_t^2(x,t)\big|_0^1 + xw_x^2(x,t)\big|_0^1) - \frac{1}{2}\left(\int_0^1 w_t^2(x,t)dx + \int_0^1 w_x^2(x,t)dx\right) \\
&= w_t^2(1,t) - \frac{1}{2}E(t).
\end{aligned} \tag{2.4.22}$$

Set

$$\widetilde{E}(t) = E(t) + \beta(t). \tag{2.4.23}$$

It is easy to obtain

$$\frac{1}{2}E(t) \le \widetilde{E}(t) \le \frac{3}{2}E(t). \tag{2.4.24}$$

Finding the derivative of $\widetilde{E}(t)$ with respect to t yields

$$\dot{\widetilde{E}}(t) = -w_t^2(1,t) - \frac{1}{2}E(t) + w_t^2(1,t) \le -\frac{1}{3}\widetilde{E}(t). \tag{2.4.25}$$

It follows that

$$E(t) \le \frac{3}{2}\widetilde{E}(t) \le \frac{3}{2}e^{-\frac{t}{3}}\widetilde{E}(0) \le \frac{3}{2}e^{-\frac{t}{3}}E(0). \tag{2.4.26}$$

It should be pointed out that the feedback controller $u(t) = -w_t(1,t)$ requires the information of the velocity $w_t(1,t)$. If our measurement is the amplitude $w(1,t)$ instead of the velocity $w_t(1,t)$, the direct proportional control cannot make the system stable. In this case, we can use the differentiator to track $w_t(1,t)$ from $w(1,t)$ and then design the stabilizing controller for system (2.4.18).

We use our constructed finite-time stable tracking differentiator (2.4.3) to design the stabilizing controller $u(t) = -z_{2R}(t)$, where

$$\begin{cases} \dot{z}_{1R}(t) = z_{2R}(t), \\ \dot{z}_{2R}(t) = R^2\left(-[z_{1R}(t) - w(1,t)]^{1/2} - \left[\dfrac{z_{2R}(t)}{R}\right]^{2/3}\right). \end{cases} \tag{2.4.27}$$

The closed-loop system of (2.4.18) is

$$\begin{cases} w_{tt}(x,t) - w_{xx}(x,t) = 0, \ 0 < x < 1, \ t > 0, \\ w(0,t) = 0, \ w_x(1,t) = -z_{2R}(t), \ t \ge 0, \\ w(x,0) = w_0(x), \ w_t(x,0) = w_1(x), \ 0 \le x \le 1. \end{cases} \tag{2.4.28}$$

We use the numerical method to study the stability of system (2.4.28).

The finite difference method is adopted in simulation. Let h and k be steps along the x and t axes, respectively. We set

$$R = 50, w_0(x) = \sin\ x, w_1(x) = \cos\ x$$

in (2.4.27) and (2.4.28), and $h = 0.01, k = 0.005$.

The numerical result for amplitude $w(x,t)$ from Equations (2.4.27) and (2.4.28) is plotted in Figure 2.4.7(a). The velocity $w_t(x,t)$ is obtained from $w(x,t)$ by difference interpolation with time step $= 0.05$ and the result is plotted in Figure 2.4.7(b). It is seen that both are convergent satisfactory.

The corresponding $z_{1R}(t)$ and $z_{2R}(t)$ from (2.4.27) with the same parameters as that used for Figure 2.4.7 are plotted in Figure 2.4.8. It is seen that $z_{1R}(t)$ is convergent also but $z_{2R}(t)$ is divergent. This suggests that for stability of the system (2.4.28), one cannot couple the controller (2.4.27) together as an overall system.

Figure 2.4.9 demonstrates both $w(x,t)$ and $w_t(x,t)$ from (2.4.27) and (2.4.28) with the disturbed measurement $w(1,t) + 0.005\sin(t)$ instead of $w(1,t)$ in Figure 2.4.7. All the other parameters are the same as in Figure 2.4.7. This shows that $w(x,t)$ and $w_t(x,t)$ are still convergent, which means that the stabilizing controller (2.4.27) is tolerant to small output disturbance.

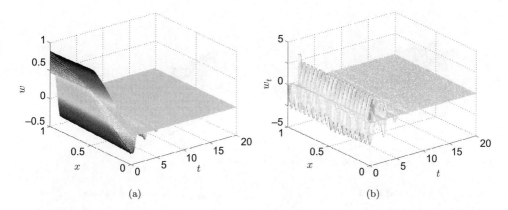

Figure 2.4.7   Numerical simulations for $w(x,t)$ and $w_t(x,t)$ from (2.4.27) and (2.4.28).

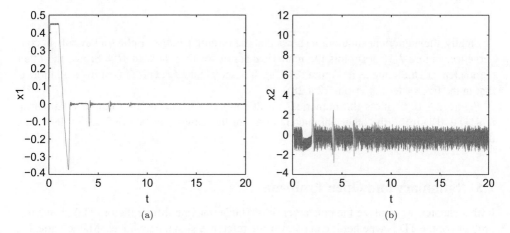

Figure 2.4.8   Numerical simulations for $z_{1R}(t)$ and $z_{2R}(t)$ in (2.4.27).

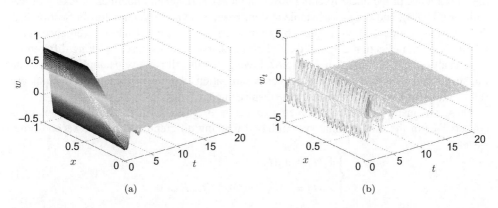

Figure 2.4.9   Numerical simulations for $w(x,t)$ and $w_t(x,t)$ from (2.4.27) and (2.4.28) with disturbed measurements.

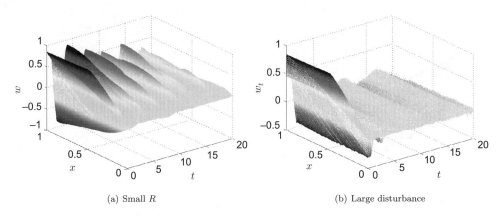

(a) Small R (b) Large disturbance

Figure 2.4.10   Numerical simulation for $w(x, t)$ from (2.4.27) and (2.4.28) with small R or large output disturbance.

Finally, the numerical simulation shows that the turning parameter plays a key role in convergence. Figure 2.4.10(a) plots the numerical result for $w(x, t)$ with $R = 8$; and the other parameters are the same as in Figure 2.4.7(a). It is seen that $w(x, t)$ is almost in oscillation as that in the free system ($a = 0$ in (2.4.28)).

Figure 2.4.10(b) plots the numerical result for $w(x, t)$ with a large output disturbance $w(1, t) + 0.1 \sin(t)$; the other parameters used are the same as in Figure 2.4.7(a). It is seen that $w(x, t)$ is divergent.

2.5   Summary and Open Problems

In this chapter, we analyze the convergence of Han's tracking differentiator (TD). The basic problem of the TD is whether it can recover the reference signal derivatives. Mathematically, it means the TD's states converge to the derivatives of the reference. In this chapter, we give the convergence of the linear tracking differentiator (LTD), general nonlinear tracking differentiator (GNTD), and finite-time stable system-based tracking differentiator. In Section 2.1, the mechanism of the LTD is analyzed by giving a convergence proof. The convergence of the GNTD is given in Section 2.2, and the finite-time stable systems-based tracking differentiator is analyzed in Section 2.3. In Section 2.4, we give some illustrative examples and use the tracking differentiators to the online frequency estimation for sinusoid signals and to boundary stabilization for a wave equation with the displacement.

There are many theoretical problems to be investigated further. The first problem is convergence of the time-optimal system-based tracking differentiator. For the second-order system,

$$\begin{cases} \dot{x}_1(t) = x_2(t), \\ \dot{x}_2(t) = u(t), \ |u(t)| \le R, \ \ R > 0. \end{cases} \tag{2.5.1}$$

The time-optimal feedback control is

$$u(x_1(t), x_2(t)) = -R \operatorname{sign}\left(x_1(t) + \frac{x_2(t)|x_2(t)|}{2R} \right). \qquad (2.5.2)$$

It is this state feedback control $u(t)$ that drives the system from the initial state to the zero state in the shortest time. Based on the time-optimal feedback control, we can construct the following TD:

$$\begin{cases} \dot{z}_{1R}(t) = x_{2R}(t), \\ \dot{x}_{2R}(t) = -R \operatorname{sign}\left(z_{1R}(t) - v(t) + \frac{z_{2R}(t)|z_{2R}(t)|}{2R} \right). \end{cases} \qquad (2.5.3)$$

The numerical results show that the above TD (2.5.3) is convergent and fast. However, the convergence of the TD (2.5.3) remains open.

The second problem is whether a tracking differentiator that is based on an attractive system only is also convergent. In this chapter, the zero equilibrium state of the free system of the TD (TD with a reference signal of 0 and tuning parameter of one) is assumed to be asymptotically stable, that is, it is Lyapunov stable and attracting. We have pointed out in Chapter 1 that there are systems that are attracting but are not Lyapunov stable. It is interesting to know that an attractive-based TD is also convergent.

Another problem is that except for LTD, convergence for other tracking differentiators is in the sense of distribution, which is weaker than uniform convergence. It is interesting to know whether a nonlinear tracking differentiator is also convergent uniformly.

In Section 2.4, we apply the TD to boundary stabilization for a wave equation by the displacement. The numerical results show that this method is effective. However, the mathematical convergence remains open.

2.6   Remarks and Bibliographical Notes

The important role of PID played in modern engineering can be referred to in [63] and [117]. The TD is firstly proposed in [58]. Although the first effort was made in [59], the proof there is only true for a constant signal and the proof for a general signal through the approximation of step functions is baseless, which was finally publicly indicated in [55]. A theoretical analysis for linear TD with noise can be found in [41]. The practice application of TD can be found in [30, 121, 124], and [123]. The detail of the distribution can be found in [1]. There are other analogs of TD. A high-gain observer-based differentiator is presented in [23]; the super-twisting second-order sliding mode algorithm differentiator is discussed in [26]; a linear time-derivative tracker is studied in [78]; and a robust exact differentiation can be found in [94] and [93]. The comparison of a TD with different differentiation trackers is presented in [144].

Section 2.2: This section is taken from paper [55] and is reproduced by permission of the Taylor and Francis license.

Section 2.3: This section is taken from paper [52] and is reproduced by permission of the IEEE license.

The finite-time stable system-based tracking differentiator was first studied in [134], where a strict condition is used and it is assumed that the Lyapunov function $V : \mathbb{R}^n \to \mathbb{R}$ satisfies the following:

I. $\sum_{i=1}^{n-1} \frac{\partial V(x)}{\partial x_i} x_{i+1} + \frac{\partial V(x)}{\partial x_n} f(x) \leq -cV(x)^\theta$ in $\mathbb{R}^n$ for some $c > 0$, $\theta \in (0,1)$.

II. $V(x)$ is Lipschitz continuous or the gradient $\nabla_x V(x)$ is bounded in $\mathbb{R}^n$.

The condition I above is to guarantee that the free system of (2.3.1) (i.e., $R = 1, v \equiv 0$) is globally finite-time stable. The second condition is quite strong and is actually not necessary. The example of a finite-time stable differentiator given in [134] is as follows:

$$\begin{cases} \dot{x}_1(t) = x_2(t), \\ \epsilon^2 \dot{x}_2(t) = -\mathrm{sat}_{\epsilon_b} \left\{ \mathrm{sign}(\phi_\alpha(x_1(t) - v(t), \epsilon x_2(t))) | \phi_\alpha(x_1(t) - v(t), \epsilon x_2(t))|^{\frac{\alpha}{2-\alpha}} \right\} \\ \qquad\quad -\mathrm{sat}_{\epsilon_b} \left\{ \mathrm{sign}(x_2(t)) | \epsilon x_2(t)|^\alpha \right\} \end{cases}$$

(2.6.1)

with $\phi_\alpha(x, y) = x + \frac{\mathrm{sign}(y)|y|^{2-\alpha}}{2-\alpha}$, $\mathrm{sat}_{\epsilon_b}(x) = x$ for $|x| < \epsilon_b$, $\mathrm{sat}_{\epsilon_b}(x) = \epsilon_b \,\mathrm{sign}(x)$ for $|x| \geq \epsilon_b$. The main conclusion given in [134] is that for some signal $v(t)$, there exists $\gamma > 0$ with $\rho\gamma > 2$ and $\rho = \alpha/(2 - \alpha)$, such that

$$x_i(t) - v^{(i-1)}(t) = O(\epsilon^{\rho\gamma - i + 1}), \; i = 1, 2,$$

where $t > T$ and T is a positive number.

There is no direct verification of the conditions I and II (finite-time stability and Lipschitz continuous Lyapunov function) in [134] for system (2.6.1), but instead is refered to [17]. Although the finite-time stability of (2.6.1) for $\epsilon = 1, v(t) \equiv 0$ is studied in [17], the Lyapunov function is not available in [17]. So it is not clear how to verify the global Lipschitz condition for the Laypunov function required in [134] for the system (2.6.1), which is far from simple. For instance, a very simple Lyapunov function like $V(x_1, x_2) = x_1^2 + x_2^2$ does not satisfy the global Lipschitz condition.

Moreover, the choice of the parameter γ is also important. According to the proof in [134], $\gamma = \frac{1-\theta}{\theta}$, where θ is the power exponent in its assumption II. However, since the Lyapunov function is not given explicitly in both [134] and [17], we are not clear why the required parameter condition $\rho\gamma > 2$ is satisfied. Notice that the finite-time stability of (2.6.1) is concluded from the following system:

$$\begin{cases} \dot{x}_1(t) = x_2(t), \\ \epsilon^2 \dot{x}_2(t) = -\mathrm{sign}(\phi_\alpha(x_1(t) - v(t), \epsilon x_2(t))) | \phi_\alpha(x_1(t) - v(t), \epsilon x_2(t))|^{\frac{\alpha}{2-\alpha}} \\ \qquad\quad -\mathrm{sign}(x_2(t)) | \epsilon x_2(t)|^\alpha, \end{cases}$$

(2.6.2)

which is equal to (2.6.1) in some neighborhood of zero in $\mathbb{R}^2$. For system (2.6.2), the Lyapunov function satisfying its assumption II is given in [17] with $\theta = \frac{2}{3-\alpha}$, $\gamma = \frac{1-\theta}{\theta}$ and $\alpha \in (0,1)$. By a simple computation, we obtain $\gamma \in \left(0, \frac{1}{2}\right)$ and $\rho = \frac{\alpha}{2-\alpha} \in (0,1)$, so it seems impossible to choose γ satisfying $\rho\gamma > 2$, even for system (2.6.2).

The stability results for the perturbed finite-time stable systems have been studied in [15]. Lemma 2.3.2 in Subsection 2.3.1 is a generalization of Theorem 5.2 of [15] by replacing the global Lipschitz continuity with continuity only for the Lyapunov function $V(x)$. In the proof of Theorem 5.2 in [15], the inequality (2.3.13)is obtained by the global Lipschitz continuity of the Lyapunov function. Such a Lyapunov function is hard to construct in applications. Here we derive it from the boundedness of the solution claimed in Step 1 in the proof of Lemma 2.3.2.

The finite-time stability of system (2.3.2) in Subsection 12 is also studied in [12] on page 191. However, here we are more interested in inequality (2.3.7) since it means that our condition (2.3.18) in Theorem 2.3.1 is valid.

Section 2.4. A different approach by the nonlinear observer for frequency estimation of the finite sum of the sinusoidal signals was also discussed in [141]. However, due to different approaches, it is hard to compare the effectiveness of these two approaches, but the TD-based approach is at least as simple as the nonlinear observer. The first far more simple global convergent frequency estimator for a single sinusoidal was presented in [74].

Section 2.4.2. The Lyapunov function $V(x_1, x_2)$ defined by (2.4.17) does not satisfy the condition required in [134] and is different to the super-twisting observer in [26], where the function is not Lipschitz continuous.

3

Extended State Observer

For a control system, we are usually not able to determine the physical state of the system by direct observation. What we know is the measurement or output of the system, which is partial information of the state. How to understand the state from the input and output of a system becomes the central issue in control theory. This is realized by the design of a state observer, that is, a system to provide an estimate of the internal state of the system. The observer makes the output feedback possible once the state feedback is available. A celebrated Luenberger observer is named after Luenberger's PhD dissertation, where Luenberger introduced a new method for construction of state observers. Usually, for a linear system, the higher the observer gain is, the quicker the linear Luenberger observer converges to the system state. However, high observer gain leads to a peaking phenomenon in which initial error can be amplified by the high gain. To overcome this difficulty, different nonlinear high-gain observers are proposed that may converge quickly without the peaking phenomenon. A typical successful method is the sliding mode-based observer. Some of these methods are advantageous in terms of robustness, but most of them have problems with things like uncertainty estimation, adaptability, and anti-chattering.

In his seminal work, Han proposed the following extended state observer (ESO):

$$\begin{cases} \dot{\hat{x}}_1(t) = \hat{x}_2(t) - \alpha_1 g_1(\hat{x}_1(t) - y(t)), \\ \dot{\hat{x}}_2(t) = \hat{x}_3(t) - \alpha_2 g_2(\hat{x}_1(t) - y(t)), \\ \qquad \vdots \\ \dot{\hat{x}}_n(t) = \hat{x}_{n+1}(t) - \alpha_n g_n(\hat{x}_1(t) - y(t)) + u(t), \\ \dot{\hat{x}}_{n+1}(t) = -\alpha_{n+1} g_{n+1}(\hat{x}_1(t) - y(t)), \end{cases} \tag{3.0.1}$$

for an n-dimensional SISO nonlinear system

$$\begin{cases} x^{(n)}(t) = f(t, x(t), \dot{x}(t), \ldots, x^{(n-1)}(t)) + w(t) + u(t), \\ y(t) = x(t), \end{cases} \tag{3.0.2}$$

Active Disturbance Rejection Control for Nonlinear Systems: An Introduction, First Edition.
Bao-Zhu Guo and Zhi-Liang Zhao.
© 2016 John Wiley & Sons Singapore Pte. Ltd. Published 2016 by John Wiley & Sons, Ltd.

where $u \in C(\mathbb{R}, \mathbb{R})$ is the input (control), $y(t)$ the output (measurement), $f \in C(\mathbb{R}^{n+1}, \mathbb{R})$ a possibly unknown system function, and $w \in C(\mathbb{R}, \mathbb{R})$ the uncertain external disturbance; $f(t, \cdot) + w(t)$ is called the "total disturbance", $(x_{10}, x_{20}, \ldots, x_{n0})$ is the initial state, and $\alpha_i, i = 1, 2, \ldots, n+1$ are regulable gain constants. The main idea of the extended state observer is that for appropriately chosen functions $g_i \in C(\mathbb{R}, \mathbb{R})$, the states $\hat{x}_i(t), i = 1, 2, \ldots, n$, and $\hat{x}_{n+1}(t)$ of the ESO can be, through regulating α_i, used to recover the corresponding states $x_i(t)$ for $i = 1, 2, \ldots, n$, and the total disturbance $f(t, \cdot) + w(t)$, respectively. The last remarkable fact is the source in which the external state observer is rooted. The numerical studies and many other studies over the years have shown that for some nonlinear functions $g_i(\cdot)$ and parameters α_i, the observer (3.0.1) performs very satisfactorily with regard to adaptability, robustness, and anti-chattering.

Unfortunately, although huge applications have been carried out in engineering environments since then, the choice of functions $g_i(\cdot)$ is essentially experiential. In order to apply conveniently in practice, we first introduce a linear extended state observer (LESO) in what follows, a special case of (3.0.1) in (3.1.2).

The LESO (3.1.2) is essentially similar to the "extended high-gain observer" in literature. The main part of the ESO is not only to estimate the state $(x_1(t), x_2(t), \ldots, x_n(t))$ but also the total disturbance $f(t, \cdot) + w(t)$ in which $f(t, \cdot)$ is also not known in many cases. Through the ESO (state plus extended state observer), we are able to compensate (cancel) the total disturbance in the feedback loop. From this point of view, the study of the ESO becomes significant both theoretically and practically. In this chapter, we firstly focus on a linear ESO and then on a nonlinear ESO for SISO systems. Secondly, we discuss the ESO with time-varying tuning gain. We then consider the ESO for MIMO systems. In the last section, we present a brief summary of this chapter and some theoretical aspects of the ESO are also proposed.

3.1 Linear Extended State Observer for SISO Systems

In this section, we introduce a linear ESO (LESO) for SISO systems with vast uncertainty, which is a simple ESO and is easy to design in practice. The first part is about convergence of LESO and the second part includes some examples as illustrations.

Consider the following nonlinear system with vast uncertainty:

$$\begin{cases} \dot{x}_1(t) = x_2(t), x_1(0) = x_{10}, \\ \dot{x}_2(t) = x_3(t), x_2(0) = x_{20}, \\ \quad \vdots \\ \dot{x}_n(t) = f(t, x_1(t), x_2(t), \ldots, x_n(t)) + w(t) + u(t), \quad x_n(0) = x_{n0}, \\ y(t) = x_1(t), \end{cases} \quad (3.1.1)$$

where $u \in C(\mathbb{R}, \mathbb{R})$ is the input (control), $y(t)$ the output (measurement), $f \in C(\mathbb{R}^{n+1}, \mathbb{R})$ a possibly unknown system function, and $w \in C(\mathbb{R}, \mathbb{R})$ the uncertain external disturbance; $x_{n+1}(t) \triangleq f(t, \cdot) + w(t)$ is the "total disturbance" or "extended state" and $(x_{10}, x_{20}, \ldots, x_{n0})$

is the initial state. The LESO for system (3.1.1) is designed as follows:

$$\begin{cases} \dot{\hat{x}}_1(t) = \hat{x}_2(t) + \dfrac{\alpha_1}{\epsilon}(y(t) - \hat{x}_1(t)), \\[2mm] \dot{\hat{x}}_2(t) = \hat{x}_3(t) + \dfrac{\alpha_2}{\epsilon^2}(y(t) - \hat{x}_1(t)), \\[2mm] \quad\vdots \\[2mm] \dot{\hat{x}}_n(t) = \hat{x}_{n+1}(t) + \dfrac{\alpha_n}{\epsilon^n}(y(t) - \hat{x}_1(t)) + u(t), \\[2mm] \dot{\hat{x}}_{n+1}(t) = \dfrac{\alpha_{n+1}}{\epsilon^{n+1}}(y(t) - \hat{x}_1(t)), \end{cases} \qquad (3.1.2)$$

where α_i, $i = 1, 2,\ldots, n + 1$, are pertinent constants and ϵ is the constant gain. We expect that $\hat{x}_i(t)$ track $x_i(t)$ for all $i = 1, 2,\ldots, n + 1$ as $t \to \infty$.

The following Assumption 3.1.1 is about prior assumption on the unknown nonlinear function $f(t, \cdot)$ and the external disturbance $w(t)$.

Assumption 3.1.1 The possibly unknown functions $f(t, x)$ and $w(t)$ are continuously differentiable with respect to their variables and

$$|u(t)| + |f(t, x)| + |\dot{w}(t)| + \left|\frac{\partial f(t, x)}{\partial t}\right| + \left|\frac{\partial f(t, x)}{\partial x_i}\right| \le c_0 + \sum_{j=1}^{n} c_j |x_j|^k,$$

$$\forall\, t \ge 0, \quad x = (x_1, x_2,\ldots, x_n), \qquad (3.1.3)$$

for some positive constants c_j, $j = 0, 1, \ldots, n$ and positive integer k.

The Assumption 3.1.2 is a priori assumption about the solution.

Assumption 3.1.2 The solution of (3.0.2) and the external disturbance $w(t)$ satisfy $|w(t)| + |x_i(t)| \le B$ for some constant $B > 0$ and all $i = 1, 2 \ldots, n$, and $t \ge 0$.

Theorem 3.1.1 *If the matrix E defined by (3.1.4) below:*

$$E = \begin{pmatrix} -\alpha_1 & 1 & 0 & \cdots & 0 \\ -\alpha_2 & 0 & 1 & \cdots & 0 \\ \vdots & \vdots & \vdots & \ddots & \vdots \\ -\alpha_n & 0 & 0 & \cdots & 1 \\ -\alpha_{n+1} & 0 & 0 & \cdots & 0 \end{pmatrix}, \qquad (3.1.4)$$

is Hurwitz and Assumptions 3.1.1 and 3.1.2 are satisfied, then

(i) For every positive constant $a > 0$,

$$\lim_{\epsilon \to 0} |x_i(t) - \hat{x}_i(t)| = 0 \text{ uniformly in } t \in [a, \infty).$$

(ii) For any $\epsilon > 0$ there exist $t_\epsilon > 0$ and $\Gamma_i > 0$ such that

$$|x_i(t) - \hat{x}_i(t)| \leq \Gamma_i \epsilon^{n+2-i}, \quad \forall\, t \geq t_\epsilon,$$

where Γ_i is an ϵ-independent constant, $x_i(t)$ and $\hat{x}_i(t)$ are solutions of (3.1.1) and (3.1.2), respectively, $i = 1, 2, \ldots, n+1$, and $x_{n+1}(t) = f(t, \cdot) + w(t)$ is the extended state for system (3.1.1).

Proof. For the Hurwitz matrix E, by Theorem 1.3.6, there exists a positive definite matrix P satisfying the Lyapunov equation $PE + E^\top P = -I_{(n+1)\times(n+1)}$. Define the Lyapunov functions $V, W : \mathbb{R}^{n+1} \to \mathbb{R}$ by

$$V(\eta) = \langle P\eta, \eta \rangle, \quad W(\eta) = \langle \eta, \eta \rangle, \quad \forall\, \eta \in \mathbb{R}^{n+1}. \tag{3.1.5}$$

Then

$$\lambda_{\min}(P)\|\eta\|^2 \leq V(\eta) \leq \lambda_{\max}(P)\|\eta\|^2, \tag{3.1.6}$$

$$\sum_{i=1}^{n} \frac{\partial V(\eta)}{\partial \eta_i}(\eta_{i+1} - \alpha_i \eta_1) - \frac{\partial V(\eta)}{\partial \eta_{n+1}} \alpha_{n+1} \eta_1 = -\eta^\top \eta = -\|\eta\|^2 = -W(\eta), \tag{3.1.7}$$

and

$$\left| \frac{\partial V(\eta)}{\partial \eta_{n+1}} \right| \leq \left\| \frac{\partial V(\eta)}{\partial \eta} \right\| = \|2\eta^\top P\| \leq 2\|P\|\|\eta\| = 2\lambda_{\max}(P)\|\eta\|, \tag{3.1.8}$$

where $\lambda_{\max}(P)$ and $\lambda_{\min}(P)$ are the maximal and minimal eigenvalues of P, respectively.

By the extended state $x_{n+1}(t) = f(t, \cdot) + w(t)$, system (3.0.2) can be written as

$$\begin{cases} \dot{x}_1(t) = x_2(t), x_1(0) = x_{10}, \\ \dot{x}_2(t) = x_3(t), x_2(0) = x_{20}, \\ \quad \vdots \\ \dot{x}_n(t) = x_{n+1}(t) + u(t), x_n(0) = x_{n0}, \\ \dot{x}_{n+1}(t) = \dot{L}(t), x_{n+1}(0) = L(0), \\ y(t) = x_1(t), \end{cases} \tag{3.1.9}$$

where $L(t) = f(t, x_1(t), x_2(t), \ldots, x_n(t)) + w(t)$. We first notice that

$$\Delta(t) = \frac{\partial f(s, x_1(\epsilon t), \ldots, x_n(\epsilon t))}{\partial s} + \sum_{i=1}^{n-1} x_{i+1}(\epsilon t) \frac{\partial f(\epsilon t, x_1(\epsilon t), \ldots, x_n(\epsilon t))}{\partial x_i}$$

$$+ u(\epsilon t) \frac{\partial f(\epsilon t, x_1(\epsilon t), \ldots, x_n(\epsilon t))}{\partial x_n} + \dot{w}(\epsilon t), \quad s = \epsilon t. \tag{3.1.10}$$

From Assumptions 3.1.1 and 3.1.2, there is a positive constant $M > 0$ such that $|\Delta(t)| \leq M$ uniformly for $t \geq 0$. Set

$$e_i(t) = x_i(t) - \hat{x}_i(t), \quad \eta_i(t) = \frac{e_i(\epsilon t)}{\epsilon^{n+1-i}}, \quad i = 1, 2, \ldots, n+1. \tag{3.1.11}$$

A direct computation shows that $\eta(t) = (\eta_1(t), \eta_2(t), \ldots, \eta_{n+1}(t))^\top$ satisfies

$$\begin{cases} \dot{\eta}_1(t) = \eta_2(t) - g_1(\eta_1(t)), & \eta_1(0) = \dfrac{e_1(0)}{\epsilon^n}, \\[2mm] \dot{\eta}_2(t) = \eta_3(t) - g_2(\eta_1(t)), & \eta_2(0) = \dfrac{e_2(0)}{\epsilon^{n-1}}, \\ \quad\vdots \\ \dot{\eta}_n(t) = \eta_{n+1}(t) - g_n(\eta_1(t)), & \eta_n(0) = \dfrac{e_n(0)}{\epsilon}, \\[2mm] \dot{\eta}_{n+1}(t) = -g_{n+1}(\eta_1(t)) + \epsilon\Delta(t), & \eta_{n+1}(0) = e_{n+1}(0). \end{cases} \quad (3.1.12)$$

Finding the derivative of $V(\eta(t))$ with respect to t along the solution $\eta(t)$ of system (3.1.12) gives

$$\left. \frac{dV(\eta(t))}{dt} \right|_{\text{along (3.1.12)}} = \sum_{i=1}^n \frac{\partial V(\eta(t))}{\partial \eta_i}(\eta_{i+1}(t) - g_i(\eta_1(t))) - \frac{\partial V(\eta(t))}{\partial \eta_{n+1}} g_{n+1}(\eta_1(t))$$

$$+ \frac{\partial V(\eta(t))}{\partial \eta_{n+1}} \epsilon \Delta(t)$$

$$\leq -W(\eta(t)) + 2\epsilon M \lambda_{\max}(P)\|\eta(t)\| \leq -\frac{1}{\lambda_{\max}(P)} V(\eta(t))$$

$$+ 2\frac{\sqrt{\lambda_{\min}(P)}}{\lambda_{\min}(P)} \epsilon M \lambda_{\max}(P) \sqrt{V(\eta(t))}. \quad (3.1.13)$$

It follows that

$$\frac{d}{dt}\sqrt{V(\eta(t))} \leq -\frac{1}{2\lambda_{\max}(P)}\sqrt{V(\eta(t))} + \frac{2\sqrt{\lambda_{\min}(P)}\epsilon M \lambda_{\max}(P)}{2\lambda_{\min}(P)}. \quad (3.1.14)$$

By (3.1.6) to (3.1.8), we have

$$\|\eta(t)\| \leq \sqrt{\frac{V(\eta(t))}{\lambda_{\min}(P)}} \leq \frac{\sqrt{\lambda_{\min}(P)V(\eta(0))}}{\lambda_{\min}(P)} e^{-\frac{1}{2\lambda_{\max}(P)}t} + \frac{\epsilon M \lambda_{\max}(P)}{\lambda_{\min}(P)} \int_0^t e^{-\frac{1}{2\lambda_{\max}(P)}(t-s)} ds. \quad (3.1.15)$$

This together with (3.1.11) yields

$$|e_i(t)| = \epsilon^{n+1-i}|\eta_i(t/\epsilon)| \leq \epsilon^{n+1-i}\|\eta(t/\epsilon)\|$$

$$\leq \epsilon^{n+1-i}\left[\frac{\sqrt{\lambda_{\min}(P)V(\eta(0))}}{\lambda_{\min}(P)} e^{-\frac{t}{2\lambda_{\max}(P)\epsilon}} + \frac{\epsilon M \lambda_{\max}(P)}{\lambda_{\min}(P)} \int_0^{t/\epsilon} e^{-\frac{1}{2\lambda_{\max}(P)}(t/\epsilon-s)} ds \right]$$

$$\rightarrow 0 \text{ uniformly in } t \in [a, \infty) \text{ as } \epsilon \rightarrow 0. \quad (3.1.16)$$

Both (i) and (ii) of Theorem 3.1.1 then follow from (3.1.16). This completes the proof of the theorem. $\quad\square$

Now, we give a numerical simulation to illustrate Theorem 3.1.1.

Example 3.1.1 *For the system*

$$
\begin{cases}
\dot{x}_1(t) = x_2(t), \\
\dot{x}_2(t) = -x_1(t) - x_2(t) + w(t) + u(t), \\
y(t) = x_1(t),
\end{cases}
\tag{3.1.17}
$$

we design an LESO according to Theorem 3.1.1:

$$
\begin{cases}
\dot{\hat{x}}_1(t) = \hat{x}_2(t) + \dfrac{3}{\epsilon}(y(t) - \hat{x}_1(t)), \\
\dot{\hat{x}}_2(t) = \hat{x}_3(t) + \dfrac{3}{\epsilon^2}(y(t) - \hat{x}_1(t)) + u(t), \\
\dot{\hat{x}}_3(t) = \dfrac{1}{\epsilon^3}(y(t) - \hat{x}_1(t)).
\end{cases}
\tag{3.1.18}
$$

For this example, the corresponding matrix

$$
E = \begin{pmatrix} -3 & 1 & 0 \\ -3 & 0 & 1 \\ -1 & 0 & 0 \end{pmatrix},
\tag{3.1.19}
$$

for which all eigenvalues are equal to -1, is Hurwitz. For any bounded control $u(t)$ and bounded disturbance $w(t)$ and $\dot{w}(t)$ (for instance the finite superposition of sinusoidal disturbance $w(t) = \sum_{i=1}^{p} a_i \sin b_i t$), the solution of (3.1.17) is bounded. Figure 3.1.1 below gives a numerical simulation for Example 3.1.1 where we take

$$
x_1(0) = x_2(0) = 1, \ \hat{x}_1(0) = \hat{x}_2(0) = \hat{x}_3(0) = 0, \ u(t) = \sin t, \ w(t) = \cos t, \ \epsilon = 0.01.
\tag{3.1.20}
$$

It is seen from Figure 3.1.1 that the LESO (3.1.18) is very effective in tracking the system (3.1.17), not only for the state $(x_1(t), x_2(t))$ but also for the extended state (total disturbance) $x_3(t)$.

One of the problems in the high-gain observer is the robustness to the time delay. In the following we illustrate this point by numerical simulation for the system (3.1.17) if the output is $y(t) = x_1(t + \tau)$ with time delay τ. By the linear ESO (3.1.18), the numerical results are also very satisfactory. Here we take $\tau = 0.03$ and other parameters are the same as those in Figure 3.1.1. The result is plotted in Figure 3.1.2. It is seen that the extended state observer can tolerate a small output time delay.

In what follows, we explain briefly the filtering function of the ESO for high-frequency noise. Consider the following nonlinear system (3.1.1) where the real output $y(t) = x_1(t) + \sin \omega t$, that is, the output is contaminated by the high-frequency noise $\sin \omega t$. The linear ESO is designed as (3.1.2) for which the matrix E defined by (3.1.4) is Hurwitz. Let $\eta_i(t) = (x_i(t) - \hat{x}_i(t))/\epsilon^{n+1-i}$, $\eta(t) = (\eta_1(t), \ldots, \eta_{n+1}(t))^\top$. Then a straightforward computation shows that

$$
\dot{\eta}(t) = \frac{1}{\epsilon} E \eta(t) + \frac{\sin \omega t}{\epsilon^{n+1}} (\alpha_1, \ldots, \alpha_{n+1})^\top + (0, \ldots, \dot{x}_{n+1}(t))^\top,
\tag{3.1.21}
$$

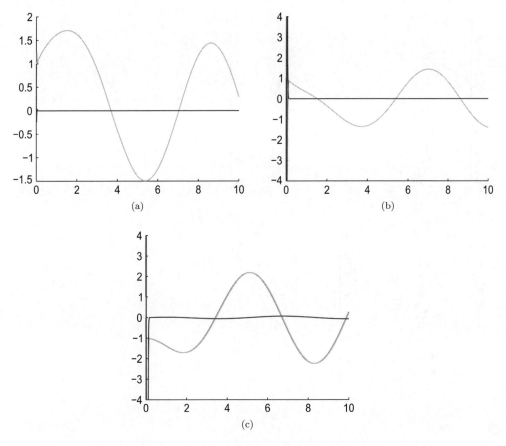

Figure 3.1.1   Linear ESO (3.1.18) for system (3.1.17).

where $x_{n+1}(t)$ is the total disturbance. It gives

$$
\begin{pmatrix} \eta_1(t) \\ \vdots \\ \eta_n(t) \\ \eta_{n+1}(t) \end{pmatrix} = e^{\frac{Et}{\epsilon}} \begin{pmatrix} \dfrac{\hat{x}_1(t_0) - x_1(t_0)}{\epsilon^n} \\ \vdots \\ \dfrac{\hat{x}_n(t_0) - \hat{x}_n(t_0)}{\epsilon} \\ \hat{x}_{n+1}(t_0) - x_{n+1}(t_0) \end{pmatrix}
$$

$$
+ \frac{1}{\epsilon^{n+1}} \int_0^t e^{\frac{E(t-s)}{\epsilon}} \begin{pmatrix} \alpha_1 \sin \omega s \\ \vdots \\ \alpha_n \sin \omega s \\ \alpha_{n+1} \sin \omega s + \dot{x}_{n+1}(s) \end{pmatrix} ds. \qquad (3.1.22)
$$

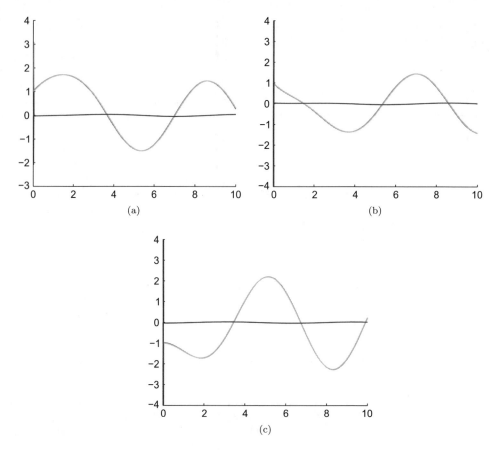

Figure 3.1.2 Linear ESO (3.1.18) for system (3.1.17) with time delay.

Further computation shows that

$$\overline{\lim_{t\to\infty}}|\hat{x}_i(t) - x_i(t)| \le \overline{M}\epsilon^{n+2-i} + \frac{1}{\epsilon^i}(\|(\alpha_1,\ldots,\alpha_{n+1})\|)\frac{1}{\omega}, \tag{3.1.23}$$

for some $\overline{M} > 0$. In particular,

$$\overline{\lim_{t\to\infty}}|\hat{x}_1(t) - x(t)| \le \overline{M}\epsilon^{n+1} + \frac{1}{\epsilon}(\|(\alpha_1,\ldots,\alpha_{n+1})\|)\frac{1}{\omega}. \tag{3.1.24}$$

Therefore, for the given $\epsilon > 0$, when the noise frequency ω is large, $\hat{x}_1(t)$ tracks $x_1(t)$ and is little affected by the noise $\sin \omega t$.

3.2 Nonlinear Extended State Observer for SISO Systems

In this section, we present general nonlinear ESO for SISO systems and its convergence. Some special ESOs, such as ESO for external disturbance only, are also discussed.

3.2.1 Nonlinear ESO for SISO Systems

The nonlinear ESO for system (3.1.1) is constructed as follows:

$$\begin{cases} \dot{\hat{x}}_1(t) = \hat{x}_2(t) + \epsilon^{n-1} g_1\left(\dfrac{y(t) - \hat{x}_1(t)}{\epsilon^n}\right), \\[2mm] \dot{\hat{x}}_2(t) = \hat{x}_3(t) + \epsilon^{n-2} g_2\left(\dfrac{y(t) - \hat{x}_1(t)}{\epsilon^n}\right), \\[2mm] \qquad \vdots \\[2mm] \dot{\hat{x}}_n(t) = \hat{x}_{n+1}(t) + g_n\left(\dfrac{y(t) - \hat{x}_1(t)}{\epsilon^n}\right) + u(t), \\[2mm] \dot{\hat{x}}_{n+1}(t) = \dfrac{1}{\epsilon} g_{n+1}\left(\dfrac{y - \hat{x}_1(t)}{\epsilon^n}\right). \end{cases} \tag{3.2.1}$$

The nonlinear ESO (3.2.1) is a special case of (3.0.1) and a nonlinear generalization of the LESO (3.1.2) for gain ϵ and pertinent chosen functions $g_i(\cdot), i = 1, 2, \ldots, n+1$.

The nonlinear functions $g_i(\cdot)(i = 1, 2, \ldots, n+1)$ are chosen such that the following condition holds true:

Assumption 3.2.1 There exist constants $\lambda_i (i = 1, 2, 3, 4), \beta$, and positive definite, continuous differentiable functions $V, W : \mathbb{R}^{n+1} \to \mathbb{R}$ such that

- $\lambda_1 \|y\|^2 \le V(y) \le \lambda_2 \|y\|^2, \quad \lambda_3 \|y\|^2 \le W(y) \le \lambda_4 \|y\|^2,$
- $\displaystyle\sum_{i=1}^n \frac{\partial V(y)}{\partial y_i}(y_{i+1} - g_i(y_1)) - \frac{\partial V(y)}{\partial y_{n+1}} g_{n+1}(y_1) \le -W(y),$
- $\left|\dfrac{\partial V(y)}{\partial y_{n+1}}\right| \le \beta \|y\|,$

where $y = (y_1, y_2, \ldots, y_{n+1})$.

Theorem 3.2.1 *Suppose that Assumptions 3.1.1 to 3.1.2, and 3.2.1 are satisfied. Then*

(i) *For every positive constant $a > 0$,*

$$\lim_{\epsilon \to 0} |x_i(t) - \hat{x}_i(t)| = 0 \text{ uniformly in } t \in [a, \infty);$$

(ii) $\displaystyle\varlimsup_{t \to \infty} |x_i(t) - \hat{x}_i(t)| \le O(\epsilon^{n+2-i}),$

where $x_i(t)$ and $\hat{x}_i(t)$ are the solutions of (3.1.1) and (3.2.1), respectively, $i = 1, 2, \ldots, n+1$, and $x_{n+1}(t) = f(t, \cdot) + w(t)$ is the extended state variable of system (3.1.1).

Proof. As before, we write system (3.1.1) as

$$
\begin{cases}
\dot{x}_1(t) = x_2(t),\, x_1(0) = x_{10}, \\
\dot{x}_2(t) = x_3(t),\, x_2(0) = x_{20}, \\
\quad\vdots \\
\dot{x}_n(t) = x_{n+1}(t) + u(t),\, x_n(0) = x_{n0}, \\
\dot{x}_{n+1}(t) = \dot{L}(t),\, x_{n+1}(0) = L(0), \\
y(t) = x_1(t),
\end{cases}
\tag{3.2.2}
$$

where $L(t) = f(t, x_1(t), x_2(t), \ldots, x_n(t)) + w(t)$. We first notice that

$$
\Delta(t) = \frac{\partial f(s, x_1(\epsilon t), \ldots, x_n(\epsilon t))}{\partial s} + \sum_{i=1}^{n} x_{i+1}(\epsilon t) \frac{\partial f(\epsilon t, x_1(\epsilon t), \ldots, x_n(\epsilon t))}{\partial x_i}
$$

$$
+ u(\epsilon t) \frac{\partial f(\epsilon t, x_1(\epsilon t), \ldots, x_n(\epsilon t))}{\partial x_n} + \dot{w}(\epsilon t),\ s = \epsilon t.
\tag{3.2.3}
$$

From Assumptions 3.1.1 and 3.1.2, there is a positive constant $M > 0$ such that $|\Delta(t)| \le M$ uniformly for $t \ge 0$. Set

$$
e_i(t) = x_i(t) - \hat{x}_i(t), \quad \eta_i(t) = \frac{e_i(\epsilon t)}{\epsilon^{n+1-i}}, \quad i = 1, 2, \ldots, n+1.
\tag{3.2.4}
$$

Then a direct computation shows that $\eta(t) = (\eta_1(t), \eta_2(t), \ldots, \eta_{n+1}(t))^\top$ satisfies

$$
\begin{cases}
\dot{\eta}_1(t) = \eta_2(t) - g_1(\eta_1(t)), \quad \eta_1(0) = \dfrac{e_1(0)}{\epsilon^n}, \\[2mm]
\dot{\eta}_2(t) = \eta_3(t) - g_2(\eta_1(t)), \quad \eta_2(0) = \dfrac{e_2(0)}{\epsilon^{n-1}}, \\
\quad\vdots \\
\dot{\eta}_n(t) = \eta_{n+1}(t) - g_n(\eta_1(t)), \quad \eta_n(0) = \dfrac{e_n(0)}{\epsilon}, \\[2mm]
\dot{\eta}_{n+1}(t) = -g_{n+1}(\eta_1(t)) + \epsilon\Delta(t), \quad \eta_{n+1}(0) = e_{n+1}(0).
\end{cases}
\tag{3.2.5}
$$

By Assumption 3.2.1, finding the derivative of $V(\eta(t))$ with respect to t along the solution $\eta(t)$ of system (3.2.5) gives

$$
\left. \frac{dV(\eta(t))}{dt} \right|_{\text{along (3.2.5)}}
$$

$$
= \sum_{i=1}^{n} \frac{\partial V(\eta(t))}{\partial \eta_i} (\eta_{i+1}(t) - g_i(\eta_1(t))) - \frac{\partial V(\eta(t))}{\partial \eta_{n+1}} g_{n+1}(\eta_1(t)) + \frac{\partial V(\eta(t))}{\partial \eta_{n+1}} \epsilon\Delta(t)
$$

$$
\le -W(\eta(t)) + \epsilon M \beta \|\eta(t)\| \le -\frac{\lambda_3}{\lambda_2} V(\eta(t)) + \frac{\sqrt{\lambda_1}}{\lambda_1} \epsilon M \beta \sqrt{V(\eta(t))}.
\tag{3.2.6}
$$

It follows that

$$
\frac{d}{dt} \sqrt{V(\eta(t))} \le -\frac{\lambda_3}{2\lambda_2} \sqrt{V(\eta(t))} + \frac{\sqrt{\lambda_1} \epsilon M \beta}{2\lambda_1}.
\tag{3.2.7}
$$

By Assumption 3.2.1 again, we have

$$\|\eta(t)\| \leq \sqrt{\frac{V(\eta(t))}{\lambda_1}} \leq \frac{\sqrt{\lambda_1 V(\eta(0))}}{\lambda_1} e^{-\frac{\lambda_3}{2\lambda_2}t} + \frac{\epsilon M\beta}{2\lambda_1} \int_0^t e^{-\frac{\lambda_3}{2\lambda_2}(t-s)} ds. \qquad (3.2.8)$$

This together with (3.2.4) yields

$$|e_i(t)| = \epsilon^{n+1-i} |\eta_i(t/\epsilon)| \leq \epsilon^{n+1-i} \|\eta(t/\epsilon)\|$$

$$\leq \epsilon^{n+1-i} \left[\frac{\sqrt{\lambda_1 V(\eta(0))}}{\lambda_1} e^{-\frac{\lambda_3 t}{2\lambda_2 \epsilon}} + \frac{\epsilon M\beta}{2\lambda_1} \int_0^{\frac{t}{\epsilon}} e^{-\frac{\lambda_3}{2\lambda_2}(t/\epsilon-s)} ds \right] \qquad (3.2.9)$$

$$\to 0 \text{ uniformly in } t \in [a, \infty) \text{ as } \epsilon \to 0.$$

Both (i) and (ii) of Theorem 3.2.1 follow from (3.2.9). This completes the proof of the theorem. □

Now, we give an example of a nonlinear ESO. For system

$$\begin{cases} \dot{x}_1(t) = x_2(t), \\ \dot{x}_2(t) = -x_1(t) - x_2(t) + w(t) + u(t), \\ y(t) = x_1(t), \end{cases} \qquad (3.2.10)$$

we design an NLESO as follows:

$$\begin{cases} \dot{\hat{x}}_1(t) = \hat{x}_2(t) + \frac{3}{\epsilon}(y(t) - \hat{x}_1(t)) + \epsilon\varphi\left(\frac{y(t) - \hat{x}_1(t)}{\epsilon^2}\right), \\ \dot{\hat{x}}_2(t) = \hat{x}_3(t) + \frac{3}{\epsilon^2}(y(t) - \hat{x}_1(t)) + u(t), \\ \dot{\hat{x}}_3(t) = \frac{1}{\epsilon^3}(y(t) - \hat{x}_1(t)), \end{cases} \qquad (3.2.11)$$

where $\varphi : \mathbb{R} \to \mathbb{R}$ is defined as

$$\varphi(r) = \begin{cases} -\dfrac{1}{4}, & r \in \left(-\infty, -\dfrac{\pi}{2}\right), \\ \dfrac{1}{4}\sin r, & r \in \left(-\dfrac{\pi}{2}, \dfrac{\pi}{2}\right), \\ \dfrac{1}{4}, & r \in \left(\dfrac{\pi}{2}, -\infty\right). \end{cases} \qquad (3.2.12)$$

In this case, $g_i(\cdot)$ in (3.2.1) can be specified as

$$g_1(y_1) = 3y_1 + \varphi(y_1), \quad g_2(y_1) = 3y_1, \quad g_3(y_1) = y_1. \qquad (3.2.13)$$

The Lyapunov function $V : \mathbb{R}^3 \to \mathbb{R}$ for this case is given by

$$V(y) = \langle Py, y \rangle + \int_0^{y_1} \varphi(s) ds, \qquad (3.2.14)$$

where

$$P = \begin{pmatrix} 1 & -\dfrac{1}{2} & -1 \\ -\dfrac{1}{2} & 1 & -\dfrac{1}{2} \\ -1 & -\dfrac{1}{2} & 4 \end{pmatrix}$$

is the positive definite solution of the Lyapunov equation $PE + E^{\top}P = -I_{3\times3}$ for

$$E = \begin{pmatrix} -3 & 1 & 0 \\ -3 & 0 & 1 \\ -1 & 0 & 0 \end{pmatrix}. \tag{3.2.15}$$

A direct computation shows that

$$\sum_{i=1}^{2} \frac{\partial V(y)}{\partial y_i}(y_{i+1} - g_i(y_1)) - \frac{\partial V(y)}{\partial y_3}g_3(y_1)$$

$$= -y_1^2 - y_2^2 - y_3^2 - (2y_1 - y_2 - 2y_3 + \varphi(y_1))\varphi(y_1) + (y_2 - 3y_1)\varphi(y_1) \tag{3.2.16}$$

$$\leq -\left(\frac{y_1^2}{8} + \frac{7y_2^2}{8} + \frac{3y_3^2}{4}\right) \triangleq -W(y_1, y_2, y_3).$$

Therefore, all conditions of Assumption 3.2.1 are satisfied and (3.2.11) serves as a well-defined NLESO for (3.1.17) according to Theorem 3.2.1. Now take the same data as (3.1.20). The numerical results for NLESO (3.2.11) are plotted as Figure 3.2.1. It is seen from Figure 3.2.1 that the NLESO (3.2.11) is at least as good as LESO (3.1.18) in tracking the state and the extended state of the system (3.1.17).

In what follows, we relax the conditions of Assumption 3.2.1 by Assumption 3.2.2.

Assumption 3.2.2 There exist constants $R, \alpha > 0$, and positive definite, continuous differentiable functions $V, W : \mathbb{R}^{n+1} \to \mathbb{R}$ such that for $y = (y_1, y_2, \ldots, y_{n+1})$,

- $\{y | V(y) \leq d\}$ is bounded for any $d > 0$,

- $\displaystyle\sum_{i=1}^{n} \frac{\partial V(y)}{\partial y_i}(y_{i+1} - g_i(y_1)) - \frac{\partial V(y)}{\partial y_{n+1}}g_{n+1}(y_1) \leq -W(y)$,

- $\left|\dfrac{\partial V(y)}{\partial y_{n+1}}\right| \leq \alpha W(y)$ for $\|y\| > R$.

We then have a weak convergence.

Theorem 3.2.2 *Under Assumptions 3.1.1, 3.1.2, and 3.2.2, the nonlinear extended state observer (3.2.1) is convergent in the sense that, for any $\sigma \in (0, 1)$, there exists $\epsilon_\sigma \in (0, 1)$ such that for any $\epsilon \in (0, \epsilon_\sigma)$,*

$$|x_i(t) - \hat{x}_i(t)| < \sigma, \quad \forall\, t \in (T_\epsilon, \infty),$$

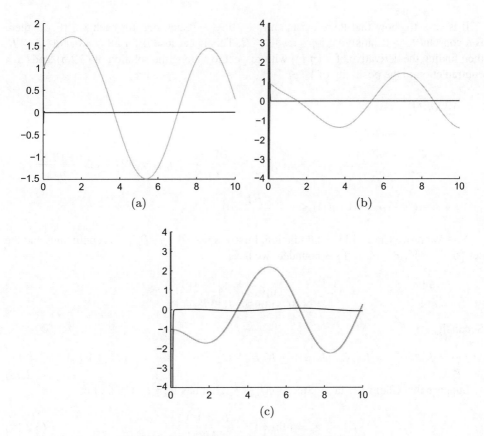

Figure 3.2.1   NLESO (3.2.11) for system (3.1.17).

where $T_\epsilon > 0$ depends on ϵ, $x_i(t)$ and $\hat{x}_i(t)$ are the solutions of (3.1.1) and (3.2.1), respectively, $i = 1, 2, \ldots, n + 1$, and $x_{n+1}(t) = f(t, \cdot) + w(t)$ is the extended state variable of system (3.1.1).

Proof. For positive definite functions $V(\cdot)$ and $W(\cdot)$, there exist class $\mathcal{K}_\infty$ functions $K_i : [0, \infty) \to [0, \infty)$, $i = 1, 2, 3, 4$, such that

- $K_1(\|(y_1, y_2, \ldots, y_{n+1})\|) \leq V(y_1, y_2, \ldots, y_{n+1}) \leq K_2(\|(y_1, y_2, \ldots, y_{n+1})\|)$,

- $K_3(\|(y_1, y_2, \ldots, y_{n+1})\|) \leq W(y_1, y_2, \ldots, y_{n+1}) \leq K_4(\|(y_1, y_2, \ldots, y_{n+1})\|)$.

Denote by $\eta(t)$ the solution of (3.2.5) starting from $\eta_0 = \left(\frac{e_1(0)}{\epsilon^n}, \frac{e_2(0)}{\epsilon^{n-1}}, \ldots, e_{n+1}(0) \right)^\top$. The proof will be split into several claims.

Claim 1. There exists an $\epsilon_1 \in (0, 1)$ such that for any $\epsilon \in (0, \epsilon_1)$ there exists a $t_\epsilon > 0$ such that

$$\{\eta(t) \mid t \in [t_\epsilon, \infty)\} \subset \{\eta \mid V(\eta) \leq C\}, \tag{3.2.17}$$

where $C = \max_{\|y\| \leq R} V(y) < \infty$.

It is easy to show that there exists an $\epsilon_0 \in (0, \frac{1}{2M\alpha})$ such that for each $\epsilon \in (0, \epsilon_0)$ there is a constant $t_\epsilon > 0$ satisfying $\|\eta(t_\epsilon; \eta_0)\| \leq R$. This is because if for all $t > 0, \|\eta(t)\| > R$, then finding the derivative of $V(\eta(t))$ with respect to t along the solution of (3.2.5) leads to a contradiction to the positivity of $V(y)$:

$$\left. \frac{dV(\eta(t))}{dt} \right|_{\text{along (3.2.5)}}$$

$$= \sum_{i=1}^{n} \frac{\partial V(\eta(t))}{\partial \eta_i}(\eta_{i+1}(t) - g_i(\eta_1(t))) - \frac{\partial V(\eta(t))}{\partial \eta_{n+1}} g_{n+1}(\eta_1(t)) + \epsilon \Delta \frac{\partial V(\eta(t))}{\partial \eta_{n+1}}$$

$$\leq -(1 - \alpha\epsilon M)W(\eta(t)) \leq -\frac{K_3(R)}{2} < 0.$$

Now we prove Claim 1 by contradiction. Firstly, since $\partial V(y)/\partial y_{n+1}$ is continuous and the set $\{y|\ C \leq V(y) \leq C+1\}$ is bounded, we have

$$A = \sup_{Y \in \{y|\ C \leq V(y) \leq C+1\}} \left| \frac{\partial V(y)}{\partial y_{n+1}} \right| < \infty.$$

Secondly,

$$W(\eta) \geq K_3(\|\eta\|) \geq K_3 K_2^{-1}(V(\eta)) \geq K_3 K_2^{-1}(C) > 0, \quad \forall\, \eta \in \{y|\ C \leq V(y) \leq C+1\}. \tag{3.2.18}$$

Suppose that Claim 1 is false. Since $\|\eta(t_\epsilon)\| \leq R$, it has $V(\eta(t_\epsilon)) \leq C$. Let

$$\epsilon_1 = \min\left\{1, \frac{K_3 K_2^{-1}(C)}{AM}\right\}. \tag{3.2.19}$$

Then there exist an $\epsilon < \epsilon_1$ and $t_1^\epsilon, t_2^\epsilon \in (t_\epsilon, \infty), t_1^\epsilon < t_2^\epsilon$, such that

$$\eta(t_1^\epsilon) \in \{\eta|V(\eta) = C\}, \ \eta(t_2^\epsilon) \in \{\eta|V(\eta) > C\} \tag{3.2.20}$$

and

$$\{\eta(t)|t \in [t_1^\epsilon, t_2^\epsilon]\} \subset \{y|\ C \leq V(y) \leq C+1\}. \tag{3.2.21}$$

Combining (3.2.18) and (3.2.21) yields

$$\inf_{t \in [t_1^\epsilon, t_2^\epsilon]} W(\eta(t)) \geq K_3 K_2^{-1}(C). \tag{3.2.22}$$

Therefore, for $t \in [t_1^\epsilon, t_2^\epsilon]$,

$$\frac{dV(\eta(t))}{dt} = \left. \frac{dV(\eta(t))}{dt} \right|_{\text{along (3.2.5)}}$$

$$\leq -W(\eta(t)) + AM\epsilon$$

$$\leq -K_3 K_2^{-1}(C) + AM \frac{K_3 K_2^{-1}(C)}{AM} = 0,$$

which shows that $V(\eta(t))$ is nonincreasing in $[t_1^\epsilon, t_2^\epsilon]$, and hence

$$V(\eta(t_2^\epsilon)) \le V(\eta(t_1^\epsilon)) = C.$$

This contradicts the second inequality of (3.2.20). Claim 1 follows.

Claim 2. There is an $\epsilon_\sigma \in (0, \epsilon_1)$ such that for any $\epsilon \in (0, \epsilon_\sigma)$ there exists a $T_\epsilon \in \left[t_\epsilon, t_\epsilon + \frac{2c}{K_3 K_2^{-1}(\delta)}\right]$ such that $\|\eta(T_\epsilon; \eta_0)\| < \delta$.

Actually, for any given $\sigma > 0$, since $V(\eta)$ is continuous, there exists a $\delta \in (0, \sigma)$ such that

$$0 \le V(\eta) \le K_1(\sigma), \quad \forall \|\eta\| \le \delta. \tag{3.2.23}$$

Now, for every $\eta \in \{\eta | V(\eta) \ge \delta\}$,

$$W(\eta) \ge K_3(\|\eta\|) \ge K_3 K_2^{-1}(V(\eta)) \ge K_3 K_2^{-1}(\delta) > 0. \tag{3.2.24}$$

By Claim 1, for any $\epsilon \in (0, \epsilon_1)$, $\{\eta(t) | t \in [t_\epsilon, \infty)\} \subset \{\eta | V(\eta) \le C\}$, and hence

$$H = \sup_{t \in [t_\epsilon, \infty)} \left| \frac{\partial V(\eta(t))}{\partial \eta_{n+1}} \right| \le \sup_{\eta \in \{\eta | V(\eta) \le C\}} \left| \frac{\partial V(\eta)}{\partial \eta_{n+1}} \right| < \infty.$$

Suppose that Claim 2 is false. Then for

$$\epsilon_\sigma = \min\left\{\epsilon_1, \frac{K_3 K_2^{-1}(\delta)}{2HM}\right\}, \tag{3.2.25}$$

there exists an $\epsilon < \epsilon_\sigma$ such that $\|\eta(t)\| \ge \delta$ for any $t \in \left[t_\epsilon, t_\epsilon + \frac{2C}{K_3 K_2^{-1}(\delta)}\right]$. This together with (3.2.24) concludes that for any $\epsilon \in (0, \epsilon_\sigma)$ and all $t \in \left[t_\epsilon, t_\epsilon + \frac{2C}{K_3 K_2^{-1}(\delta)}\right]$,

$$\frac{dV(\eta(t))}{dt} = \frac{dV(\eta(t))}{dt}\bigg|_{\text{along } (3.2.5)}$$

$$\le -W(\eta(t)) + \left| \frac{\partial V(\eta(t))}{\partial \eta_{n+1}} M\epsilon \right| - \frac{K_3 K_2^{-1}(\delta)}{2} < 0.$$

Use the integral above the inequality over $\left[t_\epsilon, t_\epsilon + \frac{2C}{K_3 K_2^{-1}(\delta)}\right]$ to give

$$V\left(\eta\left(\frac{2C}{K_3 K_2^{-1}(\delta)}\right)\right) = \int_{t_\epsilon}^{t_\epsilon + \frac{2C}{K_3 K_2^{-1}(\delta)}} \frac{dV(\eta(t))}{dt} dt + V(\eta(t_\epsilon))$$

$$\le -\frac{K_3 K_2^{-1}(\delta)}{2} \frac{2C}{K_3 K_2^{-1}(\delta)} + V(\eta(t_\epsilon)) \le 0.$$

This is a contradiction since for any $t \in \left[t_\epsilon, t_\epsilon + \frac{2C}{K_3 K_2^{-1}(\delta)}\right]$, $\|\eta(t)\| \ge \delta$. Claim 2 also follows.

Claim 3. For any $\epsilon \in (0, \epsilon_\sigma)$, if there exists a $T_\epsilon \in [t_\epsilon, \infty)$ such that

$$\eta(T_\epsilon) \in \{\eta | \|\eta\| \le \delta\},$$

then

$$\{\eta(t) | t \in (T_\epsilon, \infty]\} \subset \{\eta | \|\eta\| \le \sigma\}. \tag{3.2.26}$$

Suppose that Claim 3 is not valid. Then there exist $t_2^\epsilon > t_1^\epsilon \geq T_\epsilon$ such that

$$\|\eta(t_1^\epsilon)\| = \delta, \quad \|\eta(t_2^\epsilon)\| > \sigma, \quad \{\eta(t)|t \in [t_1^\epsilon, t_2^\epsilon]\} \subset \{\eta|\|\eta\| \geq \delta\}. \tag{3.2.27}$$

This together with (3.2.24) concludes that, for $t \in [t_1^\epsilon, t_2^\epsilon]$,

$$K_1(\|\eta(t_2^\epsilon)\|) \leq V(\eta(t_2^\epsilon)) = \int_{t_1^\epsilon}^{t_2^\epsilon} \frac{dV(\eta(t))}{dt} dt + V(\eta(t_1^\epsilon))$$

$$\leq \int_{t_1^\epsilon}^{t_2^\epsilon} -\frac{K_3 K_2^{-1}(\delta)}{2} dt + V(\eta(t_1^\epsilon)) \leq V(\eta(t_1^\epsilon)). \tag{3.2.28}$$

By (3.2.23) and the fact that $\|\eta(t_1^\epsilon)\| = \delta$, we have

$$V(\eta(t_1^\epsilon)) \leq K_1(\sigma).$$

This, together with (3.2.28), gives

$$K_1(\|\eta(t_2^\epsilon)\|) \leq K_1(\sigma). \tag{3.2.29}$$

Since the wedge function $K_1(\cdot)$ is increasing, (3.2.29) implies that $\|\eta(t_2^\epsilon)\| \leq \sigma$, which contradicts the middle inequality of (3.2.27). Claim 3 is verified.

Theorem 3.2.2 then follows by combining Claims 1 to 3. $\qquad\square$

It should be pointed out that Theorem 3.2.2 is obtained based on Assumption 3.2.2 rather than Assumption 3.2.1, which is less restrictive than Assumption 3.2.1. This is because in Assumption 3.2.1, positive definite functions $V(y)$ and $W(y)$ should satisfy conditions $\lambda_1\|y\|^2 \leq V(y) \leq \lambda_2\|y\|^2$, $\lambda_3\|y\|^2 \leq W(y) \leq \lambda_4\|y\|^2$, which are not required in Assumption 3.2.2. Therefore, under the assumptions of Theorem 3.2.2, it is more flexible to construct examples than under the assumptions of Theorem 3.2.1.

Now we construct an example that satisfies Assumption 3.2.2 but is hard to verify whether or not it satisfies Assumption 3.2.1.

By Definition 1.3.11, we can verify that if $V(x)$ is homogeneous with weights $\{r_1, r_2, \ldots, r_n\}$ and is differentiable with respect to x_n, then the partial derivative of $V(x)$ with respect to x_n satisfies

$$\lambda^{r_n} \frac{\partial V(\lambda^{r_1}x_1, \lambda^{r_2}x_1, \ldots, \lambda^{r_n}x_n)}{\partial \lambda^{r_n}x_n} = \lambda^d \frac{\partial V(x_1, x_2, \ldots, x_n)}{\partial x_n}. \tag{3.2.30}$$

The above equality is very convenient to be used for checking the homogeneity of $\partial V(x)/\partial x_n$ provided that we have known the homogeneity of $V(x)$.

Let $n = 2$, $g_1(y_1) = 3[y_1]^\alpha$, $g_2(y_1) = 3[y_1]^{2\alpha-1}$, and $g_3(y_1) = [y_1]^{3\alpha-2}$ in Theorem 3.2.2, where $[y_1] = \text{sign}(y_1)|y_1|$. Define the vector field:

$$F(y) = \begin{pmatrix} y_2 - g_1(y_1) \\ y_3 - g_2(y_1) \\ -g_3(y_1) \end{pmatrix}. \tag{3.2.31}$$

It is easy to verify that the vector field $F(y)$ in (3.2.31) is homogeneous of degree $\alpha - 1$ with respect to the weights $\{1, \alpha, 2\alpha - 1\}$.

Since the matrix E given in (3.1.19) is Hurwitz, it follows from Theorem 1.3.8 that for some $\alpha \in \left(\frac{2}{3}, 1\right)$, the system $\dot{y}(t) = F(y(t))$ is finite-time stable. From Theorem 1.3.9, we find that there exists a positive definite, radially unbounded function $V : \mathbb{R}^3 \to \mathbb{R}$ such that $V(y)$ is homogeneous of degree γ with respect to the weights $\{1, \alpha, 2\alpha - 1\}$, and $\frac{\partial V(y)}{\partial y_1}(y_2 - g_1(y_1)) + \frac{\partial V(y)}{\partial y_2}(y_3 - g_2(y_1)) - \frac{\partial V(y)}{\partial y_3}g_3(y_1)$ is negative definite and homogeneous of degree $\gamma + \alpha - 1$. From (3.2.30) and the homogeneity of $V(y)$, we find that $\left|\frac{\partial V(y)}{\partial y_3}\right|$ is homogeneous of degree $\gamma + 1 - 2\alpha$. By Property 1.3.3, there exist positive constants b_1, b_2, and $b_3 > 0$ such that

$$\left|\frac{\partial V(y)}{\partial y_3}\right| \le b_1 (V(y))^{\frac{\gamma - (2\alpha - 1)}{\gamma}} \tag{3.2.32}$$

and

$$-b_2 (V(y))^{\frac{\gamma - (1 - \alpha)}{\gamma}} \le \frac{\partial V(y)}{\partial y_1}(y_2 - g_1(y_1)) + \frac{\partial V(y)}{\partial y_2}(y_3 - g_2(y_1))$$
$$-\frac{\partial V(y)}{\partial y_3}g_3(y_1) \le -b_3 (V(y))^{\frac{\gamma - (1 - \alpha)}{\gamma}}. \tag{3.2.33}$$

Let $W(y) = c_2 (V(y))^{\frac{\gamma - (1 - \alpha)}{\gamma}}$. Since $V(y)$ is a radially unbounded positive definite function, we have, for any $d > 0$, that $\{y | V(y) \le d\}$ is bounded, and $\lim_{\|y\| \to \infty} V(y) = \infty$. This together with (3.2.32) yields that, for $\alpha \in \left(\frac{2}{3}, 1\right)$, $\lim_{\|y\| \to \infty} \frac{W(y)}{\left|\frac{\partial V(y)}{\partial y_3}\right|} = \infty$. Hence, there is a

$B > 0$ such that for $\|y\| \ge B$, $\left|\frac{\partial V(y)}{\partial y_3}\right| \le W(y)$. Therefore, Assumption 3.2.2 is satisfied.

By Theorem 3.2.2, we can then construct an NLESO:

$$\begin{cases} \dot{\hat{x}}_1(t) = \hat{x}_2(t) + 3\epsilon \left[\dfrac{y(t) - \hat{x}_1(t)}{\epsilon^2}\right]^\alpha, \\[2mm] \dot{\hat{x}}_2(t) = \hat{x}_3(t) + 3 \left[\dfrac{y(t) - \hat{x}_1(t)}{\epsilon^2}\right]^{2\alpha - 1} + u(t), \\[2mm] \dot{\hat{x}}_3(t) = \dfrac{1}{\epsilon} \left[\dfrac{y(t) - \hat{x}_1(t)}{\epsilon^2}\right]^{3\alpha - 2}. \end{cases} \tag{3.2.34}$$

Set $\alpha = 0.8, \epsilon = 0.05$, and other parameters as in (3.1.20). We plot the numerical results for LESO (3.1.18) in Figure 3.2.2 and the NLESO (3.2.34) in Figure 3.2.3, both for system (3.1.17).

The numerical results show that for the same tuning parameter ϵ, the NLESO (3.2.34) is more accurate with a small peaking value compared with the LESO (3.1.18). In Figure 3.2.2, the peaking value of $\hat{x}_3(t)$ almost reaches 100, while in Figure 3.2.3, the peaking value of $\hat{x}_3(t)$ is less than 15.

3.2.2 Some Special ESO

In this section, we state some special ESOs. First of all, we point out that if we only estimate the state rather than the extended state (total disturbance), Assumptions 3.1.1 and 3.1.2 can be replaced by the following Assumptions 3.2.3 or 3.2.4, where the boundedness of the derivative of disturbance is removed.

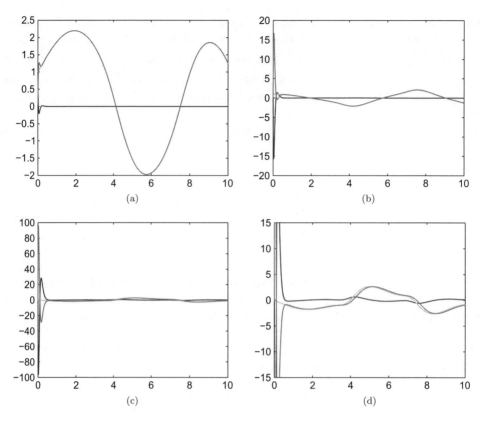

Figure 3.2.2   Linear ESO (3.1.18) for system (3.1.17). Magnification of (c).

Assumption 3.2.3 All $u, w \in C(\mathbb{R}, \mathbb{R})$ and $f \in C(\mathbb{R}^{n+1}, \mathbb{R})$ are bounded.

Assumption 3.2.4 The solution of (3.1.1) and $u, w \in C(\mathbb{R}, \mathbb{R})$ are bounded and $f \in C$ $(\mathbb{R}^{n+1}, \mathbb{R})$ satisfies

$$|f(t, x_1, x_2, \ldots, x_n)| \leq c_0 + \sum_{j=1}^{n} c_j |x_j|^{k_j}.$$

Under Assumption 3.2.3 or 3.2.4, the state observer can be designed as succeeding (3.2.35) to estimate the state of (3.1.1):

$$\begin{cases} \dot{\hat{x}}_1(t) = \hat{x}_2(t) + \epsilon^{n-2} g_1 \left(\dfrac{y(t) - \hat{x}_1(t)}{\epsilon^{n-1}} \right), \\[2mm] \dot{\hat{x}}_2(t) = \hat{x}_3(t) + \epsilon^{n-3} g_2 \left(\dfrac{y(t) - \hat{x}_1(t)}{\epsilon^{n-1}} \right), \\[2mm] \quad \vdots \\[2mm] \dot{\hat{x}}_n(t) = \frac{1}{\epsilon} g_n \left(\dfrac{y(t) - \hat{x}_1(t)}{\epsilon^{n-1}} \right) + u(t), \end{cases} \qquad (3.2.35)$$

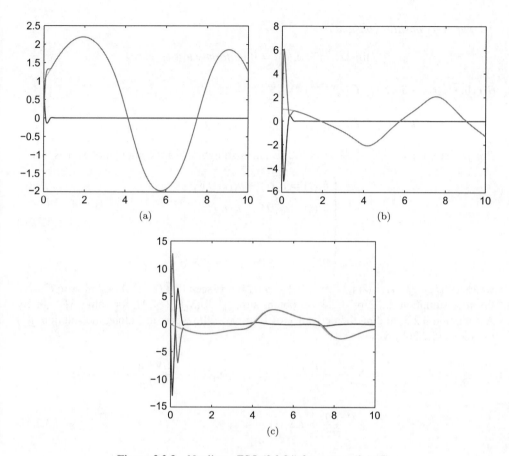

Figure 3.2.3   Nonlinear ESO (3.2.34) for system (3.1.17).

Also, we need Assumption 3.2.5, which is similar to Assumption 3.2.1.

Assumption 3.2.5 There exist constants $\lambda_i (i = 1, 2, 3, 4)$, α, β, and positive definite, continuous differentiable functions $V, W : \mathbb{R}^n \to \mathbb{R}$ such that

- $\lambda_1 \|y\|^2 \leq V(y) \leq \lambda_2 \|y\|^2, \lambda_3 \|y\|^2 \leq W(y) \leq \lambda_4 \|y\|^2,$

- $\displaystyle\sum_{i=1}^{n-1} \frac{\partial V(y)}{\partial y_i}(y_{i+1} - g_i(y_1)) - \frac{\partial V(y)}{\partial y_n} g_n(y_1) \leq -W(y),$

- $\left| \dfrac{\partial V}{\partial y_n} \right| \leq \beta \|y\|,$

where $y = (y_1, y_2, \ldots, y_n)$.

Proposition 3.2.1 *Suppose that Assumption 3.2.3 or 3.2.4 is satisfied. If Assumption 3.2.5 is satisfied, then*

(i) For every positive constant $a > 0$,

$$\lim_{\epsilon \to 0} |x_i(t) - \hat{x}_i(t)| = 0 \ \text{uniformly in } t \in [a, \infty).$$

(ii) $\overline{\lim}_{t \to \infty} |x_i(t) - \hat{x}_i(t)| \le O(\epsilon^{n+1-i})$

where $x_i(t)$ and $\hat{x}_i(t)$ are the solutions of (3.1.1), and (3.2.35) respectively, $i = 1, 2, \ldots, n$.

Proof. Let $\eta_i(t) = \frac{x_i(\epsilon t) - \hat{x}_i(\epsilon t)}{\epsilon^{n-i}}, i = 1, 2, \ldots, n$. A straightforward computation shows that

$$
\begin{cases}
\dot{\eta}_1(t) = \eta_2(t) - g_1(\eta_1(t)), \\
\dot{\eta}_2(t) = \eta_3(t) - g_2(\eta_1(t)), \\
\quad \vdots \\
\dot{\eta}_n(t) = -g_n(\eta_1(t)) + \epsilon \Delta_1(t),
\end{cases}
\tag{3.2.36}
$$

where $\Delta_1(t) = f(\epsilon t, x_1(\epsilon t), \ldots, x_n(\epsilon t)) + w(\epsilon t)$. It is seen that (3.2.36) is similar to (3.2.5). From Assumption 3.2.3 or 3.2.4 we obtain $\sup_{t \in [0,\infty)} |\Delta_1(t)| \le M$ for some $M > 0$. By Assumption 3.2.5, finding the derivative of $V(\eta(t))$ with respect to t along the solution $\eta(t)$ of system (3.2.36) gives

$$
\left. \frac{dV(\eta(t))}{dt} \right|_{\text{along (3.2.36)}} = \sum_{i=1}^{n-1} \frac{\partial V(\eta(t))}{\partial \eta_i} (\eta_{i+1}(t) - g_i(\eta_1(t))) - \frac{\partial V(\eta(t))}{\partial \eta_n} g_n(\eta_1(t))
$$

$$
+ \frac{\partial V(\eta(t))}{\partial \eta_n} \epsilon \Delta_1(t)
\tag{3.2.37}
$$

$$
\le -W(\eta(t)) + \epsilon M \beta \|\eta(t)\| \le -\frac{\lambda_3}{\lambda_2} V(\eta(t)) + \frac{\sqrt{\lambda_1}}{\lambda_1} \epsilon M \beta \sqrt{V(\eta(t))}.
$$

It follows that

$$
\frac{d}{dt} \sqrt{V(\eta(t))} \le -\frac{\lambda_3}{2\lambda_2} \sqrt{V(\eta(t))} + \frac{\sqrt{\lambda_1} \epsilon M \beta}{2\lambda_1}.
\tag{3.2.38}
$$

By Assumption 3.2.1 again, we have

$$
\|\eta(t)\| \le \sqrt{\frac{V(\eta(t))}{\lambda_1}} \le \frac{\sqrt{\lambda_1 V(\eta(0))}}{\lambda_1} e^{-\frac{\lambda_3}{2\lambda_2} t} + \frac{\epsilon M \beta}{2\lambda_1} \int_0^t e^{-\frac{\lambda_3}{2\lambda_2}(t-s)} ds.
\tag{3.2.39}
$$

This leads to

$$
|x_i(t) - \hat{x}_i(t)| = \epsilon^{n-i} |\eta_i(t/\epsilon)| \le \epsilon^{n-i} \|\eta(t/\epsilon)\|
$$

$$
\le \epsilon^{n-i} \left[\frac{\sqrt{\lambda_1 V(\eta(0))}}{\lambda_1} e^{-\frac{\lambda_3 t}{2\lambda_2 \epsilon}} + \frac{\epsilon M \beta}{2\lambda_1} \int_0^{\frac{t}{\epsilon}} e^{-\frac{\lambda_3}{2\lambda_2}(t/\epsilon-s)} ds \right]
$$

$$\to 0 \text{ uniformly in } t \in [a, \infty) \text{ as } \epsilon \to 0. \tag{3.2.40}$$

Both (i) and (ii) of Proposition 3.2.1 then follow from (3.2.40). □

We also point out that the ESO also can be used as a tracking differentiator. Suppose that $v(t)$ is the tracked signal. Let $x_i(t) = v^{(i-1)}(t)$. Then $x_i(t), i = 1, 2, \ldots, n$ satisfy

$$\begin{cases} \dot{x}_1(t) = x_2(t), \\ \dot{x}_2(t) = x_3(t), \\ \quad \vdots \\ \dot{x}_n(t) = v^{(n-1)}(t), \\ y(t) = x_1(t) = v(t). \end{cases} \tag{3.2.41}$$

The corresponding NLESO (3.2.1) becomes

$$\begin{cases} \dot{\hat{x}}_1(t) = \hat{x}_2(t) + \epsilon^{n-1} g_1 \left(\dfrac{v(t) - \hat{x}_1(t)}{\epsilon^n} \right), \\[2mm] \dot{\hat{x}}_2(t) = \hat{x}_3(t) + \epsilon^{n-2} g_2 \left(\dfrac{v(t) - \hat{x}_1(t)}{\epsilon^n} \right), \\[2mm] \quad \vdots \\[2mm] \dot{\hat{x}}_n(t) = \hat{x}_{n+1}(t) + g_n \left(\dfrac{v(t) - \hat{x}_1(t)}{\epsilon^n} \right), \\[2mm] \dot{\hat{x}}_{n+1}(t) = \dfrac{1}{\epsilon} g_{n+1} \left(\dfrac{v(t) - \hat{x}_1(t)}{\epsilon^n} \right). \end{cases} \tag{3.2.42}$$

The following Proposition 3.2.2 is actually a consequence of Theorem 3.2.1.

Proposition 3.2.2 (Tracking differentiator) *Suppose that Assumption 3.2.1 holds and $v^{(n+1)}(t)$ is bounded. Then*

(i) For every positive constant $a > 0$,

$$\lim_{\epsilon \to 0} |v^{(i-1)}(t) - \hat{x}_i(t)| = 0 \text{ uniformly in } t \in [a, \infty).$$

(ii) $\overline{\lim}_{t \to \infty} |v^{(i-1)}(t) - \hat{x}_i(t)| \leq 0(\epsilon^{n+2-i})$,
where $\hat{x}_i(t)$ is the solution of (3.2.42), $i = 1, 2 \ldots, n + 1$.

Next, we consider a special case where the system function $f(t, \cdot)$ is known, that is, the unknown part is only the external disturbance $w(t)$. In this case, we try to utilize the known information as much as possible. Our NLESO in this case can be modified as

$$
\begin{cases}
\dot{\hat{x}}_1(t) = \hat{x}_2(t) + \epsilon^{n-1} g_1 \left(\dfrac{x_1(t) - \hat{x}_1(t)}{\epsilon^n} \right), \\[2mm]
\dot{\hat{x}}_2(t) = \hat{x}_3(t) + \epsilon^{n-2} g_2 \left(\dfrac{x_1(t) - \hat{x}_1(t)}{\epsilon^n} \right), \\[2mm]
\quad \vdots \\[2mm]
\dot{\hat{x}}_n(t) = \hat{x}_{n+1}(t) + g_n \left(\dfrac{x_1(t) - \hat{x}_1(t)}{\epsilon^n} \right) + f(t, \hat{x}_1(t), \hat{x}_2(t), \dots, \hat{x}_n(t)) + u(t), \\[2mm]
\dot{\hat{x}}_{n+1}(t) = \frac{1}{\epsilon} g_{n+1} \left(\dfrac{x_1(t) - \hat{x}_1(t)}{\epsilon^n} \right),
\end{cases}
\tag{3.2.43}
$$

which is used to estimate not only the state $(x_1(t), x_2(t), \dots, x_n(t))$ but also the extended state $w(t)$. Using the same notation as that in (3.2.4) and setting $x_{n+1}(t) = w(t)$ in this case, we obtain

$$
\begin{cases}
\dot{\eta}_1(t) = \eta_2(t) - g_1(\eta_1(t)), \quad \eta_1(0) = \dfrac{e_1(0)}{\epsilon^n}, \\[2mm]
\dot{\eta}_2(t) = \eta_3(t) - g_2(\eta_1(t)), \quad \eta_2(0) = \dfrac{e_2(0)}{\epsilon^{n-1}}, \\[2mm]
\quad \vdots \\[2mm]
\dot{\eta}_n(t) = \eta_{n+1}(t) - g_n(\eta_1(t)) + \delta_1(t), \quad \eta_n(0) = \dfrac{e_n(0)}{\epsilon}, \\[2mm]
\dot{\eta}_{n+1}(t) = -g_{n+1}(\eta_1(t)) + \epsilon \delta_2(t), \quad \eta_{n+1}(0) = e_{n+1}(0),
\end{cases}
\tag{3.2.44}
$$

where

$$
\delta_1(t) = f(t, x_1(\epsilon t), \dots, x_n(\epsilon t)) - f(t, \hat{x}_1(\epsilon t), \dots, \hat{x}_n(\epsilon t)), \quad \delta_2(t) = \dot{w}(\epsilon t). \tag{3.2.45}
$$

Proposition 3.2.3 (Modified extended state observer) *In addition to the conditions in Assumption 3.2.1, we assume that $|\partial V(y)/\partial y_n| \le \alpha \|y\|$, $\alpha \rho < \lambda_3$, where ρ is the Lipschitz constant of $f(t, \cdot)$:*

$$
|f(t, x_1, \dots, x_n) - f(t, y_1, \dots, y_n)| \le \rho \|x - y\|,
$$
$$
\forall\, t \ge 0,\, x = (x_1, x_2, \dots, x_n)^\top,\, y = (y_1, y_2, \dots, y_n)^\top \in \mathbb{R}^n. \tag{3.2.46}
$$

If $\dot{w}(t)$ is bounded in (3.1.1), then

(i) For every positive constant $a > 0$,

$$
\lim_{\epsilon \to 0} |x_i(t) - \hat{x}_i(t)| = 0 \text{ uniformly in } t \in [a, \infty).
$$

(ii) $\displaystyle \overline{\lim_{t \to \infty}} |x_i(t) - \hat{x}_i(t)| \le O(\epsilon^{n+2-i})$,

where $x_i(t)$ and $\hat{x}_i(t)$ are the solutions of (3.1.1) and (3.2.43), respectively, $i = 1, 2, \dots, n+1$, $x_{n+1}(t) = w(t)$.

Proof. Finding the derivative of $V(\eta(t))$ with respect to t along the solution $\eta(t) = (\eta_1(t), \ldots, \eta_{n+1}(t))^\top$ of system (3.2.44) yields

$$\left.\frac{dV(\eta(t))}{dt}\right|_{\text{along}(3.2.44)} = \sum_{i=1}^{n} \frac{\partial V(\eta(t))}{\partial \eta_i}(\eta_{i+1}(t) - g_i(\eta_1(t))) - \frac{\partial V(\eta(t))}{\partial \eta_{n+1}}g_{n+1}(\eta_1(t))$$

$$+ \frac{\partial V(\eta(t))}{\partial \eta_n}\delta_1(t) + \frac{\partial V(\eta(t))}{\partial \eta_{n+1}}\epsilon\delta_2(t)$$

$$\leq -W(\eta(t)) + \alpha\rho\|\eta(t)\|^2 + \epsilon M\beta\|\eta(t)\|$$

$$\leq -\frac{\lambda_3 - \alpha\rho}{\lambda_2}V(\eta(t)) + \frac{\sqrt{\lambda_1}}{\lambda_1}\epsilon M\beta\sqrt{V(\eta(t))}.$$

$$(3.2.47)$$

It follows that

$$\frac{d}{dt}\sqrt{V(\eta(t))} \leq -\frac{\lambda_3 - \alpha\rho}{2\lambda_2}\sqrt{V(\eta(t))} + \frac{\sqrt{\lambda_1}\epsilon M\beta}{2\lambda_1}. \qquad (3.2.48)$$

This together with Assumption 3.2.1 gives

$$\|\eta(t)\| \leq \sqrt{\frac{V(\eta(t))}{\lambda_1}} \leq \frac{\sqrt{\lambda_1 V(\eta(0))}}{\lambda_1}e^{-\frac{\lambda_3-\alpha\rho}{2\lambda_2}t} + \frac{\epsilon M\beta}{2\lambda_1}\int_0^t e^{-\frac{\lambda_3-\alpha\rho}{2\lambda_2}(t-s)}ds. \qquad (3.2.49)$$

By (3.2.4), it follows that

$$|e_i(t)| = \epsilon^{n+1-i}|\eta_i(t/\epsilon)| \leq \epsilon^{n+1-i}\|\eta(t/\epsilon)\|$$

$$\leq \epsilon^{n+1-i}\left[\frac{\sqrt{\lambda_1 V(\eta(0))}}{\lambda_1}e^{-\frac{(\lambda_3-\alpha\rho)t}{2\lambda_2\epsilon}} + \frac{\epsilon M\beta}{2\lambda_1}\int_0^{\frac{t}{\epsilon}} e^{-\frac{\lambda_3-\alpha\rho}{2\lambda_2}(t/\epsilon-s)}ds\right]$$

$$\to 0 \text{ uniformly in } t \in [a, \infty) \text{ as } \epsilon \to 0. \qquad (3.2.50)$$

The (ii) of Theorem 3.2.3 also follows from (3.2.50). This completes the proof of the proposition. $\qquad\square$

Proposition 3.2.3 is a special case of Theorem 3.2.1. The only difference is that in Proposition 3.2.3, $f(t, \cdot)$ is known while in Theorem 3.2.1, $f(t, \cdot)$ is not. This results in the difference in designing the observer: in (3.2.43), we are able to utilize the known information of $f(t, \cdot)$, while in (3.2.1) this information is lacking. Nevertheless, the proof of Proposition 3.2.3 is similar to Theorem 3.2.1; we put it here for the sake of completeness.

Example 3.2.1 *For the system*

$$\begin{cases} \dot{x}_1(t) = x_2(t), \\ \dot{x}_2(t) = \dfrac{\sin(x_1(t)) + \sin(x_2(t))}{4\pi} + w(t) + u(t), \\ y(t) = x_1(t), \end{cases} \qquad (3.2.51)$$

where $w(t)$ is the external disturbance, we design the corresponding modified linear extended state observer as

$$
\begin{cases}
\dot{\hat{x}}_1(t) = \hat{x}_2(t) + \dfrac{6}{\epsilon}(y(t) - \hat{x}_1(t)), \\[2mm]
\dot{\hat{x}}_2(t) = \hat{x}_3(t) + \dfrac{11}{\epsilon^2}(y(t) - \hat{x}_1(t)) + \dfrac{\sin(\hat{x}_1(t)) + \sin(\hat{x}_2(t))}{4\pi} + u(t), \\[2mm]
\dot{\hat{x}}_3(t) = \dfrac{6}{\epsilon^3}(y(t) - \hat{x}_1(t)).
\end{cases}
\tag{3.2.52}
$$

For this example, the eigenvalues of associated matrix

$$
E = \begin{pmatrix} -6 & 1 & 0 \\ -11 & 0 & 1 \\ -6 & 0 & 0 \end{pmatrix}
$$

are $\{-1, -2, -3\}$, so it is Hurwitz.
　　Use Matlab to solve the Lyapunov equation $PE + E^\top P = -I$, to find that the eigenvalues of P satisfying $\lambda_{\max}(P) \approx 2.3230 < \pi$.
　　Let $V, W : \mathbb{R}^{n+1} \to \mathbb{R}$ be defined as

$$
V(\eta) = \langle P\eta, \eta \rangle, \quad W(\eta) = -\langle \eta, \eta \rangle.
$$

Then

$$
\lambda_{\min}(P)\|\eta\|^2 \le \|V(\eta)\| \le \lambda_{\max}(P)\|\eta\|^2, \quad \left\| \frac{\partial V}{\partial \eta} \right\| \le 2\lambda_{\max}(P)\|\eta\|
$$

and

$$
\frac{\partial V(\eta)}{\partial \eta_1}(\eta_2 - 6\eta_1) + \frac{\partial V(\eta)}{\partial \eta_2}(\eta_3 - 11\eta_1) - 6\eta_1 \frac{\partial V(\eta)}{\partial \eta_1} = \eta^\top (PE + E^\top P)\eta = -W(\eta).
\tag{3.2.53}
$$

Now $f(x_1, x_2) = \frac{\sin x_1 + \sin x_2}{4\pi}$ and we find that $L = 1/2\pi$. Hence $L\lambda_{\max}(P) < \frac{1}{2}$. Therefore, for any bounded control $u(t)$ and bounded $w(t)$ and $\dot{w}(t)$ (for instance, the finite superposition of sinusoidal disturbance $w(t) = \sum_{i=1}^{p} a_i \sin b_i t$), by Theorem 3.2.3, for any $a > 0$,

$$
\lim_{\epsilon \to 0} |x_i(t) - \hat{x}_i(t)| = 0 \text{ uniformly in } t \in [a, \infty)
$$

and

$$
\overline{\lim_{t \to \infty}} |x_i(t) - \hat{x}_i(t)| \le O(\epsilon^{n+2-i}),
$$

where $x_3(t) = w(t)$ and $x_i(t)$ and $\hat{x}_i(t)$ are the solutions of (3.2.51) and (3.2.52), respectively.

　　We use the data $x_1(0) = x_2(0) = 1$, $\hat{x}_1(0) = \hat{x}_2(0) = \hat{x}_3(0) = 0$, $u(t) = \cos t$, and $w(t) = \sin t$ to simulate Example 3.2.1. The results are plotted in Figure 3.2.4.

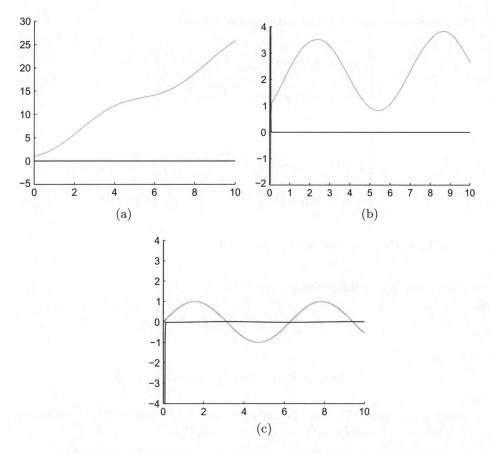

Figure 3.2.4 The modified extended state observer (3.2.52) for system (3.2.51).

It is seen from Figure 3.2.4 that the modified extended state observer (3.2.52) tracks very satisfactorily both the state and extended state of system (3.2.51), respectively.

The last case is more special as the system function $f(t, \cdot)$ is known and the external $w(t) = 0$. In this case, system (3.1.1) becomes a special deterministic nonlinear system. To design the observer, we make the following assumption.

Assumption 3.2.6 $f(t, \cdot)$ is locally Lipschitz continuous and

$$h(t, x) = \frac{\partial f(t, x)}{\partial t} + \sum_{i=1}^{n-1} x_{i+1} \frac{\partial f(t, x)}{\partial x_i} + x_{n+1}(t) \frac{\partial f(t, x)}{\partial x_n} + \frac{\partial f(t, x)}{\partial x_n} u(t) \qquad (3.2.54)$$

is globally Lipschitz continuous in $x = (x_1, x_2, \ldots, x_n)$ uniformly for $t \in (0, \infty)$, where $x_{n+1}(t) = f(t, x)$.

The state observer for (3.1.1) now is designed as follows:

$$
\begin{cases}
\dot{\hat{x}}_1(t) = \hat{x}_2(t) + \epsilon^{n-1} g_1\left(\dfrac{y(t) - \hat{x}_1(t)}{\epsilon^n} \right), \\[2mm]
\dot{\hat{x}}_2(t) = \hat{x}_3(t) + \epsilon^{n-2} g_2\left(\dfrac{y(t) - \hat{x}_1(t)}{\epsilon^n} \right), \\[2mm]
\quad\vdots \\[2mm]
\dot{\hat{x}}_n(t) = \hat{x}_{n+1}(t) + g_n\left(\dfrac{y(t) - \hat{x}_1(t)}{\epsilon^n} \right) + u(t), \\[2mm]
\dot{\hat{x}}_{n+1}(t) = h(t, \hat{x}(t)) + \dfrac{1}{\epsilon} g_{n+1}\left(\dfrac{y - \hat{x}_1(t)}{\epsilon^n} \right).
\end{cases}
\tag{3.2.55}
$$

Theorem 3.2.3 *Under assumptions 3.2.1 and 3.2.6:*

(i) There exists an $\epsilon_0 > 0$ such that for any $\epsilon \in (0, \epsilon_0)$,

$$
\lim_{t \to \infty} |x_i(t) - \hat{x}_i(t)| = 0.
$$

(ii) For any $a > 0$,

$$
\lim_{\epsilon \to 0} |x_i(t) - \hat{x}_i(t)| = 0 \ \ \text{uniformly in } t \in [a, \infty),
$$

where $x_i(t)$ and $\hat{x}_i(t)$ are the solutions of (3.1.1) and (3.2.55), respectively, $i = 1, 2, \ldots, n + 1$, and $x_{n+1}(t) = f(t, x_1(t), \ldots, x_n(t))$.

Proof. Using the same notation as in (3.2.4), since $x_{n+1}(t) = f(t, x_1(t), \ldots, x_n(t))$, we have the error system:

$$
\begin{cases}
\dot{\eta}_1(t) = \eta_2(t) - g_1(\eta_1(t)), \quad \eta_1(0) = \dfrac{e_1(0)}{\epsilon^n}, \\[2mm]
\dot{\eta}_2(t) = \eta_3(t) - g_2(\eta_1(t)), \quad \eta_2(0) = \dfrac{e_2(0)}{\epsilon^{n-1}}, \\[2mm]
\quad\vdots \\[2mm]
\dot{\eta}_n(t) = \eta_{n+1}(t) - g_n(\eta_1(t)), \quad \eta_n(0) = \dfrac{e_n(0)}{\epsilon}, \\[2mm]
\dot{\eta}_{n+1}(t) = -g_{n+1}(\eta_1(t)) + \epsilon \delta_4(t), \quad \eta_{n+1}(0) = e_{n+1}(0),
\end{cases}
\tag{3.2.56}
$$

where $\delta_4(t) = h(\epsilon t, x(\epsilon t)) - h(\epsilon t, \hat{x}(\epsilon t))$, $x(t) = (x_1(t), x_2(t), \ldots, x_n(t))^\top$, and $\hat{x}(t) = (\hat{x}_1(t), \hat{x}_2(t), \ldots, \hat{x}_n(t))^\top$.

A direct computation shows that the derivative of $V(\eta(t))$ with respect to t along the solution of system (3.2.56) satisfies, with $\eta(t) = (\eta_1(t), \eta_2(t), \cdots, \eta_{n+1}(t))^\top$,

$$
\begin{aligned}
\left. \frac{dV(\eta(t))}{dt} \right|_{\text{along (3.2.56)}} &= \sum_{i=1}^{n} \frac{\partial V(\eta(t))}{\partial \eta_i} (\eta_{i+1}(t) - g_i(\eta_1(t))) - \frac{\partial V(\eta(t))}{\partial \eta_{n+1}} g_{n+1}(\eta_1(t)) \\
&\quad + \frac{\partial V(\eta(t))}{\partial \eta_{n+1}} \epsilon \delta_4(t) \le -W(\eta(t)) + \epsilon \beta \rho_1 \|\eta(t)\|^2 \\
&\le -\frac{\lambda_3 - \epsilon \beta \rho_1}{\lambda_2} V(\eta(t)),
\end{aligned}
$$

(3.2.57)

for $\epsilon \in (0, \frac{\lambda_3}{\beta \rho_1})$, where ρ_1 is the Lipschitz constant of $h(t, x)$. It follows that

$$
\frac{d}{dt} \sqrt{V(\eta(t))} \le -\frac{\lambda_3 - \epsilon \beta \rho_1}{2\lambda_2} \sqrt{V(\eta(t))}.
$$

(3.2.58)

By Assumption 3.2.1, we have

$$
\|\eta(t)\| \le \sqrt{\frac{V(\eta(t))}{\lambda_1}} \le \frac{\sqrt{\lambda_1 V(\eta(0))}}{\lambda_1} e^{-\frac{\lambda_3 - \epsilon \beta \rho_1 t}{2\lambda_2}}.
$$

(3.2.59)

Return back to $e(t)$ by (3.2.4), to obtain finally that

$$
|e_i(t)| = \epsilon^{n+1-i} |\eta_i(t/\epsilon)| \le \epsilon^{n+1-i} \|\eta(t/\epsilon)\| \le \frac{\sqrt{\lambda_1 V(\eta(0))}}{\lambda_1} e^{-\frac{(\lambda_3 - \epsilon \beta \rho_1)t}{2\lambda_2 \epsilon}} \to 0, \quad (3.2.60)
$$

uniformly in $t \in [a, \infty)$ as $\epsilon \to 0$, or for any $\epsilon \in (0, \epsilon_0)$, $\epsilon_0 = \frac{\lambda_3}{\beta \rho_1}$, as $t \to \infty$. This completes the proof of the theorem. $\qquad \square$

3.3 The ESO for SISO Systems with Time-Varying Gain

In this section, we consider a more complicated nonlinear system of the following:

$$
\begin{cases}
\dot{x}(t) = A_n x(t) + B_n [f(t, x(t), \zeta(t), w(t)) + b(t)u(t)], \\
\dot{\zeta}(t) = f_0(t, x(t), \zeta(t), w(t)),
\end{cases}
$$

(3.3.1)

where $x(t) \in \mathbb{R}^n$ and $\zeta(t) \in \mathbb{R}^m$ are system states, A_n and B_n are defined as

$$
A_n = \begin{pmatrix} 0 & I_{n-1} \\ 0 & 0 \end{pmatrix}, \quad B_n^\top = C_n = (0, \ldots, 0, 1)
$$

(3.3.2)

and $f, f_0 \in C^1(\mathbb{R}^{n+m+2}, \mathbb{R})$ are possibly unknown nonlinear functions, $u(t) \in \mathbb{R}$ is the input (control), and $y(t) = C_n x(t) = x_1(t)$ is the output (measurement), while $b \in C^1(\mathbb{R}, \mathbb{R})$ contains some uncertainty around the constant nominal value b_0.

We design the following nonlinear time-varying gain ESO for system (3.3.1) as follows:

$$
\begin{cases}
\dot{\hat{x}}_1(t) = \hat{x}_2(t) + \dfrac{1}{r^{n-1}(t)} g_1(r^n(t)(y(t) - \hat{x}_1(t))), \\[2mm]
\dot{\hat{x}}_2(t) = \hat{x}_3(t) + \dfrac{1}{r^{n-2}(t)} g_2(r^n(t)(y(t) - \hat{x}_1(t))), \\[2mm]
\quad\vdots \\[2mm]
\dot{\hat{x}}_n(t) = \hat{x}_{n+1}(t) + g_n(r^n(t)(y(t) - \hat{x}_1(t))) + b_0 u(t), \\[2mm]
\dot{\hat{x}}_{n+1}(t) = r(t) g_{n+1}(r^n(t)(y(t) - \hat{x}_1(t))),
\end{cases}
\tag{3.3.3}
$$

to estimate the state $(x_1(t), x_2(t), \ldots, x_n(t))$ and the total disturbance

$$
x_{n+1}(t) \triangleq f(t, x(t), \zeta(t), w(t)) + [b(t) - b_0] u(t),
\tag{3.3.4}
$$

which is called, as before, the extended state, where $r(t)$ is the time-varying gain to be increased gradually. When $r(t) \equiv 1/\epsilon$, (3.3.3) is reduced to the constant gain ESO (3.2.1). The suitable time-varying gain can reduce dramatically the peaking value of both the ESO and the ADRC, given in the next chapter.

We first make the following assumptions. Assumption 3.3.1 is about the conditions on system (3.3.1) and the system functions.

Assumption 3.3.1 Any solution of (3.3.1) is uniformly bounded and there exist positive constant K and functions $\varpi_1 \in C(\mathbb{R}^n, [0, \infty))$ and $\varpi_2 \in C(\mathbb{R}, [0, \infty))$ such that $\|f_0(t, x, \zeta, w)\| \leq K + \varpi_1(x) + \varpi_2(w)$ for all $(t, x, \zeta, w) \in \mathbb{R}^{n+m+2}$.
Assumption 3.3.2 is on the external disturbance and the input.

Assumption 3.3.2 $w \in C^1(\mathbb{R}, \mathbb{R})$ and $u \in C^1(\mathbb{R}, \mathbb{R})$ satisfy

$$
\max \sup_{t \in [0, +\infty)} \{|w(t)|, |\dot{w}(t)|, |u(t)|, |\dot{u}(t)|, |b(t)|, |\dot{b}(t)|\} < +\infty.
$$

Assumption 3.3.3 is on the time-varying gain $r(t)$ in ESO (3.3.3).

Assumption 3.3.3 $r(t) \in C^1([0, \infty), [0, \infty))$, $r(t), \dot{r}(t) > 0$, $\lim\limits_{t \to +\infty} r(t) = +\infty$ and there exists a constant $M > 0$ such that $\lim\limits_{t \to +\infty} \dot{r}(t)/r(t) \leq M$.
Assumption 3.3.4 is on functions $g_i(\cdot)$ in the ESO (3.3.3).

Assumption 3.3.4 There exist positive constants R and $N > 0$ and radially unbounded, positive definite functions $\mathcal{V} \in C^1(\mathbb{R}^{n+1}, [0, \infty))$ and $\mathcal{W} \in C(\mathbb{R}^{n+1}, [0, \infty))$ such that

- $\displaystyle\sum_{i=1}^{n} (x_{i+1} - g_i(x_1)) \frac{\partial \mathcal{V}(x)}{\partial x_i} - g_{n+1}(x_1) \frac{\partial \mathcal{V}(x)}{\partial x_{n+1}} \leq -\mathcal{W}(x);$

- $\displaystyle\max_{\{i=1,..,n\}} \left\{ \left| x_i \frac{\partial \mathcal{V}(x)}{\partial x_i} \right| \right\} \leq N\mathcal{W}(x)$ and $\left| \dfrac{\partial \mathcal{V}(x)}{\partial x_{n+1}} \right| \leq N\mathcal{W}(x), \forall x = (x_1, x_2, \ldots, x_{n+1}),$
  $\|x\| \geq R.$

Theorem 3.3.1 *Assume Assumptions 3.3.1 to 3.3.4. Then the states of ESO (3.3.3) converge to the states and the extended state of (3.3.1) in the sense that*

$$\lim_{t\to+\infty} |\hat{x}_i(t) - x_i(t)| = 0, i = 1, 2, \ldots, n+1. \tag{3.3.5}$$

Proof. Let $\eta_i(t) = r^{n+1-i}(t)(x_i(t) - \hat{x}_i(t))$ be the error between the solutions of (3.3.3) and (3.3.1). Then $\eta_i(t)$ satisfies

$$\begin{cases} \dot{\eta}_1(t) = r(t)(\eta_2(t) - g_1(\eta_1(t))) + \dfrac{n\dot{r}(t)}{r(t)}\eta_1(t), \\ \quad\vdots \\ \dot{\eta}_n(t) = r(t)(\eta_{n+1}(t) - g_n(\eta_1(t))) + \dfrac{\dot{r}(t)}{r(r)}\eta_n(t), \\ \dot{\eta}_{n+1}(t) = -r(t)g_{n+1}(t)(\eta_1(t)) + \dot{x}_{n+1}(t). \end{cases} \tag{3.3.6}$$

We first consider the derivative of the extended sate $x_{n+1}(t)$ of (3.3.1) defined in (3.3.4):

$$\dot{x}_{n+1}(t) = \sum_{i=1}^{n-1} x_{i+1}(t)\frac{\partial f(t, x(t), \zeta(t), w(t))}{\partial x_i} + (f(t, x(t), \zeta(t), w(t))$$

$$+ b(t)u(t))\frac{\partial f(t, x(t), \zeta(t), w(t))}{\partial x_n} + \frac{\partial f(t, x(t), \zeta(t), w(t))}{\partial \zeta} \cdot f_0(t, x(t), \zeta(t), w(t))$$

$$+ \frac{\partial f(t, x(t), \zeta(t), w(t))}{\partial t} + \dot{w}(t)\frac{\partial f(t, x(t), \zeta(t), w(t))}{\partial w}$$

$$+ \dot{b}(t)u(t) + (b(t) - b_0)\dot{u}(t). \tag{3.3.7}$$

Considering $f(t, \cdot) \in C^1(\mathbb{R}^{n+m+2}, \mathbb{R})$, by Assumptions 3.3.1 and 3.2.2, it follows that there exists a positive constant $B > 0$ such that $|\dot{x}_{n+1}(t)| \le B$ for all $t \ge 0$.

Finding the derivatives of Lyapunov function $\mathcal{V}(\eta(t))$ along the solution of (3.3.6) gives

$$\frac{d\mathcal{V}(\eta(t))}{dt}\bigg|_{\text{along }(3.3.6)} = \dot{\eta}_1(t)\frac{\partial \mathcal{V}(\eta(t))}{\partial \eta_1} + \dot{\eta}_2(t)\frac{\partial \mathcal{V}(\eta(t))}{\partial \eta_2} + \cdots + \dot{\eta}_{n+1}(t)\frac{\partial \mathcal{V}(\eta(t))}{\partial \eta_{n+1}}$$

$$= r(t)\left(\sum_{i=1}^{n}(\eta_{i+1}(t) - g_i(\eta_1(t)))\frac{\partial \mathcal{V}(\eta(t))}{\partial \eta_i} - g_{n+1}(\eta_1(t))\frac{\partial \mathcal{V}(\eta(t))}{\partial \eta_{n+1}}\right)$$

$$+ \sum_{i=1}^{n}\frac{(n+1-i)\dot{r}(t)\eta_i(t)}{r(t)}\frac{\partial \mathcal{V}(\eta(t))}{\partial \eta_i} + \dot{x}_{n+1}(t)\frac{\partial \mathcal{V}(\eta(t))}{\partial \eta_{n+1}}$$

$$\le -r(t)\mathcal{W}(\eta(t)) + \sum_{i=1}^{n}\frac{(n+1-i)\dot{r}(t)\eta_i(t)}{r(t)}\left|\frac{\partial \mathcal{V}(\eta(t))}{\partial \eta_i}\right|$$

$$+ |\dot{x}_{n+1}(t)|\left|\frac{\partial \mathcal{V}(\eta(t))}{\partial \eta_{n+1}}\right|, \eta(t) = (\eta_1(t), \eta_2(t), \cdots, \eta_{n+1}(t)). \tag{3.3.8}$$

By Assumption 3.3.3, there exists a $t_1 > 0$ such that $r(t) > r_0 \triangleq n(n+1)MN + BN + 1$, $\dot{r}(t)/r(t) \leq 2M$ for all $t > t_1$, This together with (3.3.8) and Assumption 3.3.4 shows that (3.3.8) can be further written as follows. For any $t > t_1$ and $\|\eta(t)\| > R$,

$$\left.\frac{dV(\eta(t))}{dt}\right|_{\text{along }(3.3.6)} \leq -r(t)W(\eta(t)) + n(n+1)MNW(t) + BNW(\eta(t))$$

$$\leq -(r_0 - n(n+1)MN - BN)W(\eta(t)) \leq -\inf_{\|x\|\geq R} W(x) < 0. \tag{3.3.9}$$

Let

$$\mathcal{A} = \left\{ x \in \mathbb{R}^{n+1} \,\middle|\, V(x) \leq \max_{\|y\|\leq R} V(y) \right\}.$$

It follows from the radially unboundedness and continuity of $V(y)$ that $\mathcal{A}$ is a bounded subset of $\mathbb{R}^{n+1}$. It is easily shown that $\|x\| > R$ for all $x \in \mathcal{A}^C$, so there exists a $t_2 > t_1$ such that $\eta(t)$, the solution of (3.3.6), satisfies $\{\eta(t) \,|\, t \in [t_2, \infty)\} \subset \mathcal{A}$.

Now consider the derivative of the Lyapunov function $V(\eta(t))$ along the solution of (3.3.6) for $t > t_2$ to obtain

$$\left.\frac{dV(\eta(t))}{dt}\right|_{\text{along }(3.3.6)} \leq -r(t)W(\eta(t)) + C, \tag{3.3.10}$$

where $C = (n(n+1)M + B)N \sup_{x\in\mathcal{A}} \left\|\frac{\partial V(x)}{\partial x}\right\| < \infty$.

Since $W(\cdot)$ is radially unbounded and is continuous positive definite, it follows from Theorem 1.3.1 that there are continuous class $\mathcal{K}_\infty$ functions $\kappa_i(\cdot)$, $i = 1, 2, 3, 4$ such that

$$\kappa_1(\|\eta\|) \leq V(\eta) \leq \kappa_2(\|\eta\|), \quad \kappa_3(\|\eta\|) \leq W(\eta) \leq \kappa_4(\|\eta\|). \tag{3.3.11}$$

It follows that $W(\eta) \geq \kappa_3(\kappa_2^{-1}(V(\eta)))$.

For any $\sigma > 0$, since $\lim_{t\to\infty} r(t) = \infty$, we may suppose a $t_3 > t_2$ so that $r(t) > \frac{2C}{\kappa_3(\kappa_2^{-1}(\sigma))}$ for all $t > t_3$.

It then follows from (3.3.10) that for any $t > t_3$ with $V(\eta(t)) > \sigma$,

$$\left.\frac{dV(\eta(t))}{dt}\right|_{\text{along }(3.3.6)} \leq -r(t)\kappa_3(\kappa_2^{-1}(\sigma)) + C < -C < 0. \tag{3.3.12}$$

Therefore, one can find a $t_0 > t_3$ such that $V(\eta(t)) < \sigma$ for all $t > t_0$. Since σ is arbitrarily chosen, we obtain immediately that $\lim_{t\to\infty} V(\eta(t)) = 0$. This together with the facts that $\kappa_1(\|\eta(t)\|) \leq V(\eta(t))$ and $\|\eta(t)\| \leq \kappa_1^{-1}(V(\eta(t)))$ shows that $\lim_{t\to\infty} \|\eta(t)\| = 0$. The proof is then complete by the relation $\eta_i(t) = r^{n+1-i}(t)(x_i(t) - \hat{x}_i(t))$. $\qquad\square$

The simplest ESO satisfying the conditions of Theorem 3.3.1 is the linear time-varying ESO of the following:

$$\begin{cases} \dot{\hat{x}}_1(t) = \hat{x}_2(t) + \alpha_1 r(t)(y(t) - \hat{x}_1(t)), \\ \quad\vdots \\ \dot{\hat{x}}_n(t) = \hat{x}_{n+1}(t) + \alpha_n r^n(t)(y(t) - \hat{x}_1(t)) + b_0 u(t), \\ \dot{\hat{x}}_{n+1}(t) = \alpha_{n+1} r^{n+1}(t)(y(t) - \hat{x}_1(t)), \end{cases} \tag{3.3.13}$$

Corollary 3.3.1 *Assume Assumptions 3.3.1 to 3.3.3, and that the following matrix is Hurwitz:*

$$A = \begin{pmatrix} -\alpha_1 & 1 & 0 & \cdots & 0 \\ -\alpha_2 & 0 & 1 & \cdots & 0 \\ \vdots & \vdots & \vdots & \ddots & \vdots \\ -\alpha_{n+1} & 0 & 0 & \cdots & 0 \end{pmatrix}. \tag{3.3.14}$$

Then the states of ESO (3.3.13) converge to the states and extended state of (3.3.1) in the sense that

$$\lim_{t \to +\infty} |\hat{x}_i(t) - x_i(t)| = 0, \ i = 1, 2, \ldots, n+1. \tag{3.3.15}$$

Proof. By Theorem 3.3.1, we need only construct positive definite functions $\mathcal{V}(\cdot)$ and $\mathcal{W}(\cdot)$ that satisfy Assumption 3.3.4.

Let P be the positive definite matrix solution of the Lyapunov equation $PA + A^\top P = -I_{(n+1)\times(n+1)}$. Define the Lyapunov functions $\mathcal{V}, \mathcal{W} : \mathbb{R}^{n+1} \to \mathbb{R}$ by

$$\mathcal{V}(\eta) = \langle P\eta, \eta \rangle, \mathcal{W}(\eta) = \langle \eta, \eta \rangle, \ \forall \, \eta \in \mathbb{R}^{n+1}. \tag{3.3.16}$$

Then

$$\lambda_{\min}(P)\|\eta\|^2 \leq V(\eta) \leq \lambda_{\max}(P)\|\eta\|^2, \tag{3.3.17}$$

$$\sum_{i=1}^{n} (\eta_{i+1} - \alpha_i \eta_1) \frac{\partial \mathcal{V}(\eta)}{\partial \eta_i} - \alpha_{n+1} \eta_1 \frac{\partial \mathcal{V}(\eta)}{\partial \eta_{n+1}} = -\eta^\top \eta = -\|\eta\|^2 = -\mathcal{W}(\eta).$$

Thus the first condition of Assumption 3.3.4 is satisfied. Next since

$$\left| \frac{\partial \mathcal{V}(\eta)}{\partial \eta_i} \right| \leq \left\| \frac{\partial \mathcal{V}(\eta)}{\partial \eta} \right\| = \|2\eta^\top P\| \leq 2\|P\|\|\eta\| = 2\lambda_{\max}(P)\|\eta\|, i = 1, 2, \ldots, n+1,$$

where $\lambda_{\max}(P)$ and $\lambda_{\min}(P)$ are the maximal and minimal eigenvalues of P, respectively, the second condition of Assumption 3.3.4 is also satisfied. The result then follows from Theorem 3.3.1. □

It is noticed that in Theorem 3.3.1, although the exact approximation is achieved by increasing the gain to infinity, it is not realistic in practice because the large gain requires the small integration step. The real situation in numerical experiment and engineering application is that the time-varying gain $r(t)$ is small in the initial time to avoid the peaking value, and then increases rapidly to a certain acceptable large constant r_0 such that the approximation error in the given area is in a relatively short time.

From this idea, we modify the gain $r(t)$ in (3.3.3) as follows:

$$r(t) = \begin{cases} e^{at}, & 0 \leq t < \frac{1}{a} \ln r_0, \\ r_0, & t \geq \frac{1}{a} \ln r_0, \end{cases} \tag{3.3.18}$$

where r_0 is a large number so that the errors between the solutions of (3.3.3) and (3.3.1) are in the prescribed scale. Under this gain parameter, Assumption 3.3.4 is replaced by Assumption 3.3.5.

Assumption 3.3.5 There exist positive constants R and $N > 0$ and radially unbounded, positive definite functions $\mathcal{V} \in C^1(\mathbb{R}^{n+1}, [0, \infty))$ and $\mathcal{W}(\cdot) \in C(\mathbb{R}^{n+1}, [0, \infty))$ such that, for any $x = (x_1, x_2, \ldots, x_{n+1})$,

- $\displaystyle\sum_{i=1}^{n}(x_{i+1} - g_i(x_1))\frac{\partial \mathcal{V}(x)}{\partial x_i} - g_{n+1}(x_1)\frac{\partial \mathcal{V}(x)}{\partial x_{n+1}} \leq -\mathcal{W}(x);$

- $\left|\dfrac{\partial \mathcal{V}(x)}{\partial x_{n+1}}\right| \leq N\mathcal{W}(x), \forall\ \|x\| \geq R.$

Theorem 3.3.2 *Assume Assumptions 3.3.1, 3.3.2 and 3.3.5. Let the time-varying gain in the ESO (3.3.3) be chosen as that of (3.3.18). Then the states of ESO (3.3.3) converge to the states and extended state of (3.3.1) in the sense that for any given $\sigma > 0$, there exists an $r^* > 0$ such that for any $r_0 > r^*$,*

$$|\hat{x}_i(t) - x_i(t)| < \sigma, \ \forall\ t > t_{r_0}, i = 1, 2, \ldots, n+1,$$

where t_{r_0} is an r_0-dependent constant.

Proof. Again let $\eta_i(t) = r^{n+1-i}(x_i(t) - \hat{x}_i(t))$, $i = 1, 2, \ldots, n+1$, be the errors between the solutions of (3.3.3) and (3.3.1). In the time interval $t \in (0, \ln r_0/a)$, $\eta_i(t)$ satisfies

$$\begin{cases} \dot{\eta}_1(t) = r(t)(\eta_2(t) - g_1(\eta_1(t))) + \frac{n\dot{r}(t)}{r(t)}\eta_1(t), \\ \quad\vdots \\ \dot{\eta}_n(t) = r(t)(\eta_{n+1}(t) - g_n(\eta_1(t))) + \frac{\dot{r}(t)}{r(t)}\eta_n(t), \\ \dot{\eta}_{n+1}(t) = -r(t)g_{n+1}(t)(\eta_1(t)) + \dot{x}_{n+1}(t), \end{cases} \tag{3.3.19}$$

while in $(\ln r_0/a, \infty)$, $\eta_i(t)$ satisfies

$$\begin{cases} \dot{\eta}_1(t) = r(t)(\eta_2(t) - g_1(\eta_1(t))), \\ \quad\vdots \\ \dot{\eta}_n(t) = r(t)(\eta_{n+1}(t) - g_n(\eta_1(t))), \\ \dot{\eta}_{n+1}(t) = -r(t)g_{n+1}(t)(\eta_1(t)) + \dot{x}_{n+1}(t). \end{cases} \tag{3.3.20}$$

Find the derivative of $\mathcal{V}(\eta(t))$ for $t > (1/a)\ln r_0$ along the solution $\eta(t) = (\eta_1(t), \eta_2(t), \ldots, \eta_{n+1}(t))$ of (3.3.20) to obtain

$$\left.\frac{d\mathcal{V}(\eta(t))}{dt}\right|_{\text{along (3.3.20)}} \leq -r_0\mathcal{W}(\eta(t)) + \dot{x}_{n+1}(t)\frac{\partial\mathcal{V}(\eta(t))}{\partial\eta_{n+1}}, \ \forall\ t > \ln r_0/a. \tag{3.3.21}$$

Similar to the proof of Theorem 3.3.1, it follows from Assumptions 3.3.1 and 3.3.2 that $|\dot{x}_{n+1}(t)| < B$ for all $t \geq 0$ with some positive constant B.

If $\|\eta(t)\| \geq R$ with $t > \ln r_0/a$, then it follows from the first condition of Assumption 3.3.5 and (3.3.21) that

$$\left.\frac{dV(\eta(t))}{dt}\right|_{\text{along } (3.3.20)} \leq -(r_0 - NB)W(\eta(t)) < 0, \quad \forall \ r_0 > NB. \tag{3.3.22}$$

Same as the proof of Theorem 3.3.1, one can find a $t_1 > \ln r_0/a$ such that $\eta(t)$ is uniformly bounded on $[t_1, \infty)$. This shows, with the continuity of $V(\eta(t))$, that $|\dot{x}_{n+1}(t)\partial V(\eta(t))/\partial x_{n+1}| < C$ for all $t > t_1$ with some positive constant $C > 0$. Using the inequalities (3.3.11) and (3.3.21), we then have

$$\left.\frac{dV(\eta(t))}{dt}\right|_{\text{along } (3.3.20)} \leq -r_0\kappa_3(\kappa_2^{-1}(V(\eta(t)))) + C < 0, \quad \forall \ t > t_1. \tag{3.3.23}$$

For the given $\sigma > 0$, let $r^* = \max\left\{1, NB, \frac{C}{\kappa_3(\kappa_2^{-1}\kappa_1(\sigma))}\right\}$. Then for any $r_0 > r^*, t > r_1$, and $V(\eta(t)) > \kappa_1(\sigma)$, we have

$$\left.\frac{dV(\eta(t))}{dt}\right|_{\text{along } (3.3.20)} < -(r_0\kappa_3(\kappa_2^{-1}(\kappa_1(\sigma))) - C) < 0. \tag{3.3.24}$$

Therefore, there exists a $t_{r_0} > 0$ such that $V(\eta(t)) \leq \kappa_1(\sigma)$ for all $r_0 > r^*, t > t_{r_0}$. This completes the proof of the theorem by an obvious fact that $\|\eta(t)\| \leq \kappa_1^{-1}(\sigma) \leq \sigma$. $\qquad\square$

Here we present two types of ESOs that are the direct consequences of Theorem 3.3.2. The first is the linear ESO.

Corollary 3.3.2 *Assume Assumptions 3.3.1 and 3.3.2, and that the matrix A in (3.3.14) is Hurwitz. Let the time-varying gain in ESO (3.3.13) be chosen as (3.3.18). Then the states of linear ESO (3.3.13) converge to the states and extended state of (3.3.1) in the sense that for every $\sigma > 0$ there exists an $r^* > 0$ such that, for all $r_0 > r^*$,*

$$|\hat{x}_i(t) - x_i(t)| < \sigma, \forall \ t > t_{r_0}, i = 1, 2, \ldots, n+1,$$

where t_{r_0} is an r_0-dependent positive constant.

The second is the nonlinear ESO given by

$$\begin{cases} \dot{\hat{x}}_1(t) = \hat{x}_2(t) + \alpha_1 r^{n\theta-(n-1)}(t)[y(t) - \hat{x}_1(t)]^\theta, \\ \quad\vdots \\ \dot{\hat{x}}_n(t) = \hat{x}_{n+1}(t) + \alpha_n r^{n(n\theta-(n-1))}(t)[y(t) - \hat{x}_1(t)]^{n\theta-(n-1)}, \\ \dot{\hat{x}}_{n+1}(t) = \alpha_{n+1} r^{(n+1)(n\theta-(n-1))}(t)[y(t) - \hat{x}_1(t)]^{(n+1)\theta-n}, \end{cases} \tag{3.3.25}$$

where $[x]^\theta \triangleq \text{sign}(x)|x|^\theta$ for $\theta \in (0,1)$, $x \in \mathbb{R}$. Actually, let the vector field be defined as

$$F(y) = \begin{pmatrix} y_2 - \alpha_1[y_1]^\theta \\ \vdots \\ y_n - \alpha_n[y_1]^{n\theta - (n-1)} \\ -\alpha_{n+1}[y_1]^{(n+1)\theta - n} \end{pmatrix}. \tag{3.3.26}$$

It is easy to verify that the vector field $F(y)$ in (3.3.26) is homogeneous of degree $\theta - 1$ with the weights $\{1, \theta, 2\theta - 1, \ldots, n\theta - (n-1)\}$.

Since the matrix A given in (3.3.14) is Hurwitz, it follows from Theorem 1.3.8 that for all $\theta \in (n/(n+1), 1)$ that is sufficiently close to one, the system $\dot{y}(t) = F(y(t))$ must be finite-time stable. From Theorem 1.3.9, we get that there exists a positive definite, radially unbounded function $V : \mathbb{R}^{n+1} \to \mathbb{R}$ such that $V(y)$ is homogeneous of degree γ with the weights $\{1, \alpha, 2\alpha - 1, \ldots, n\alpha - (n-1)\}$, and the Lie derivative of $V(y)$ along the vector field $F(y)$ given by

$$L_F V(y) \triangleq \sum_{i=1}^{n} \frac{\partial V(y)}{\partial y_i}(y_{i+1} - \alpha_i[y_1]^{i\theta - (i-1)}) - \frac{\partial V(y)}{\partial y_{n+1}}\alpha_{n+1}[y_1]^{(n+1)\alpha - n}$$

is negative definite and is homogeneous of degree $\gamma + \theta - 1$. From the homogeneity of $V(y)$, we see that $|\partial V(y)/\partial y_{n+1}|$ is homogeneous of degree $\gamma - (n\theta - (n-1))$. By Property 1.3.3 there exist positive constants b_1, b_2, and $b_3 > 0$ such that

$$\left| \frac{\partial V(y)}{\partial y_{n+1}} \right| \le b_1 (V(y))^{(\gamma - (n\theta - (n-1)))/\gamma} \tag{3.3.27}$$

and

$$-b_2(V(y))^{\frac{\gamma - (1-\theta)}{\gamma}} \le L_F V(y) \le -b_3(V(y))^{\frac{\gamma - (1-\theta)}{\gamma}}. \tag{3.3.28}$$

Let $W(y) = b_2(V(y))^{\frac{\gamma - (1-\theta)}{\gamma}}$. Since $V(y)$ is radially unbounded positive definite, $\{y|V(y) \le d\}$ is bounded for any $d > 0$ and $\lim_{\|y\|\to\infty} V(y) = \infty$. This fact together with (3.3.27) shows that $\lim_{\|y\|\to\infty} W(y)/\left|\frac{\partial V}{\partial y_{n+1}}(y)\right| = \infty$ for any $\theta \in (n/(n+1), 1)$. So there is an $R > 0$ such that $|\partial V(y)/\partial y_3| \le W(y)$ for all $\|y\| \ge B$. Hence, Assumption 3.3.4 is satisfied. We have thus proved the following Corollary 3.3.3.

Corollary 3.3.3 *Assume Assumptions 3.3.1 and 3.3.2. Suppose that A in (3.3.14) is a Hurwitz matrix. Let the time-varying gain in ESO (3.3.25) be chosen as (3.3.18). Then for any $\theta \in (n/(n+1), 1)$ that makes the system $\dot{y}(t) = F(y(t))$ finite-time stable, the states of ESO (3.3.25) converge to the states and extended state of (3.3.1) in the sense that for every $\sigma > 0$, there exists a constant $r^* > 0$ such that for any $r_0 > r^*$*

$$|\hat{x}_i(t) - x_i(t)| < \sigma, t > t_{r_0}, i = 1, 2, \ldots, n+1,$$

where t_{r_0} is an r_0-dependent positive constant.

To illustrate the effectiveness of the time-varying gain ESO in dealing with the peaking phenomenon near the initial time and the convergence, we give some numerical simulations. Consider the nonlinear system

$$\begin{cases} \dot{x}_1(t) = x_2(t), \\ \dot{x}_2(t) = -x_1(t) - x_2(t) + \sin(x_1(t) + x_2(t)) + w(t) + u(t), \\ y(t) = x_1(t), \end{cases} \quad (3.3.29)$$

where $f(x_1(t), x_2(t)) = -x_1(t) - x_2(t) + \sin(x_1(t) + x_2(t))$ is the unmodeled part of the dynamics, $w(t)$ is the external disturbance, and $x_3(t) \triangleq f(x_1(t), x_2(t)) + \sin(x_1(t) + x_2(t)) + w(t)$ is the total disturbance.

The linear ESO is designed as follows:

$$\begin{cases} \dot{\hat{x}}_1(t) = \hat{x}_2(t) + 3r(t)(y(t) - \hat{x}_1(t)), \\ \dot{\hat{x}}_2(t) = \hat{x}_3(t) + 3r^2(t)(y(t) - \hat{x}_1(t)), \\ \dot{\hat{x}}_3(t) = r^3(t)(y(t) - \hat{x}_1(t)). \end{cases} \quad (3.3.30)$$

We expect that $\hat{x}_i(t)$ can approximate $x_i(t)$, $i = 1, 2, 3$. In (3.3.29), the matrix of the linear principle part is

$$\begin{pmatrix} 0 & 1 \\ -1 & -1 \end{pmatrix}. \quad (3.3.31)$$

The eigenvalues of this matrix are equal to -0.5, so for any bounded $w(t)$ and $u(t)$, the solution of system (3.3.29) must be bounded. Let $u(t) = \sin t, w(t) = 1 + \cos t + \sin 2t$. Then all conditions of Assumptions 3.3.1 and 3.3.2 are satisfied.

For the linear ESO (3.3.30), the matrix A is

$$A = \begin{pmatrix} -3 & 1 & 0 \\ -3 & 0 & 1 \\ -1 & 0 & 0 \end{pmatrix}. \quad (3.3.32)$$

The eigenvalues of A are -1, so all conditions of Corollary 3.3.2 are satisfied.

We apply the constant gain parameter $r(t) \equiv 100$ and the time-varying gain parameter as (3.3.18), with $a = 5$ and $r_0 = 100$ in the numerical simulations, respectively. The initial values are chosen as $x(0) = (1, 1)$ and $\hat{x}(0) = (0, 0, 0)$. The numerical results using the constant gain parameter are plotted in Figure 3.3.1 and that for the time-varying gain are plotted in Figure 3.3.2.

From these figures, we see that by the time-varying gain parameter, the states $\hat{x}_i(t)$ of ESO (3.3.30) approximate the state $x_i(t)(i = 1, 2)$ and the total disturbance $x_3(t)$ of system (3.3.29) satisfactorily. At the same time, the peaking value is dramatically reduced compared with the constant gain parameter.

Next, we design a nonlinear ESO for system (3.3.29) following Corollary 3.3.3, which is given as follows:

$$\begin{cases} \dot{\hat{x}}_1(t) = \hat{x}_2(t) + 3r^{2\theta-1}(t)[y(t) - \hat{x}_1(t)]^\theta, \\ \dot{\hat{x}}_2(t) = \hat{x}_3(t) + 3r^{2(2\theta-1)}(t)[y(t) - \hat{x}_1(t)]^{2\theta-1}, \\ \dot{\hat{x}}_3(t) = r^{3(2\theta-1)}(t)[y(t) - \hat{x}_1(t)]^{3\theta-2}. \end{cases} \quad (3.3.33)$$

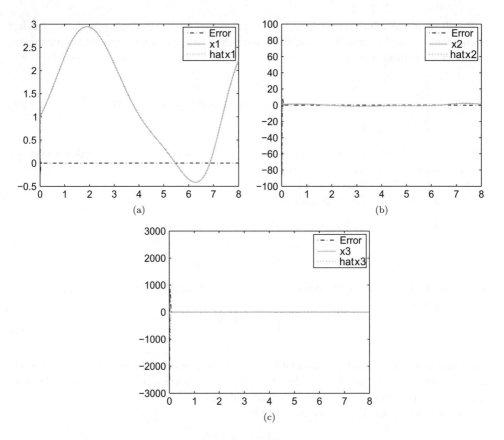

Figure 3.3.1   ESO (3.3.30) for system (3.3.29) using the constant gain parameter.

Again we apply the constant gain parameter $r(t) \equiv 100$ and the time-varying gain as in Figure 3.3.2 in ESO (3.3.33) in numerical simulations, respectively. The initial values are also chosen as $x(0) = (1,1)$ and $\hat{x}(0) = (0,0,0)$. Taking $\theta = 0.8$, the numerical results by the constant gain parameter are plotted in Figure 3.3.3 and that by the time-varying gain are plotted in Figure 3.3.4.

From Figures 3.3.3 and 3.3.4, we also see that this nonlinear ESO with the time-varying gain reduces the peaking value significantly.

In addition, comparing Figure 3.3.4 with Figure 3.3.2, we see that the error curve in Figure 3.3.4 is almost a straight line after a short time, which indicates that under the same gain parameters, this nonlinear ESO (3.3.33) is much more accurate than the linear ESO (3.3.30) for the estimation of the total disturbance.

To end this section, we discuss a special ESO and its convergence. If in the problems the "total disturbance" contains some known information, the ESO should make use of this information as much as possible. Here is the case:

$$f(t, x(t), \zeta(t), w(t)) = \tilde{f}(x(t)) + \bar{f}(t, \zeta(t), w(t)), \qquad (3.3.34)$$

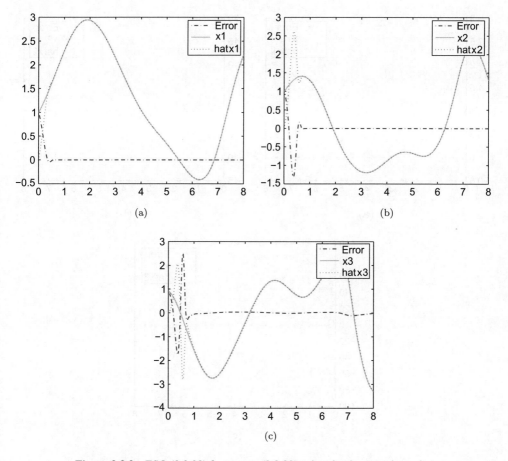

Figure 3.3.2 ESO (3.3.30) for system (3.3.29) using the time-varying gain.

where $\tilde{f}(\cdot)$ is known. The ESO for (3.3.1) is changed to be the following:

$$\begin{cases} \dot{\hat{x}}_1(t) = \hat{x}_2(t) + \dfrac{1}{r^{n-1}(t)} g_1(r^n(t)(y(t) - \hat{x}_1(t))), \\ \quad\vdots \\ \dot{\hat{x}}_n(t) = \hat{x}_{n+1}(t) + g_n(r^n(t)(y(t) - \hat{x}_1(t))) + \tilde{f}(\hat{x}_1(t),\dots,\hat{x}_n(t)) + b_0 u(t), \\ \dot{\hat{x}}_{n+1}(t) = r(t)g_{n+1}(r^n(t)(y(t) - \hat{x}_1(t))), \end{cases} \tag{3.3.35}$$

where the total disturbance in this case becomes

$$x_{n+1}(t) \triangleq \bar{f}(t, \zeta(t), w(t)) + (b(t) - b_0)u(t). \tag{3.3.36}$$

To deal with this problem, we need another assumption for $\tilde{f}(\cdot), \bar{f}(t, \cdot), f_0(t, \cdot)$.

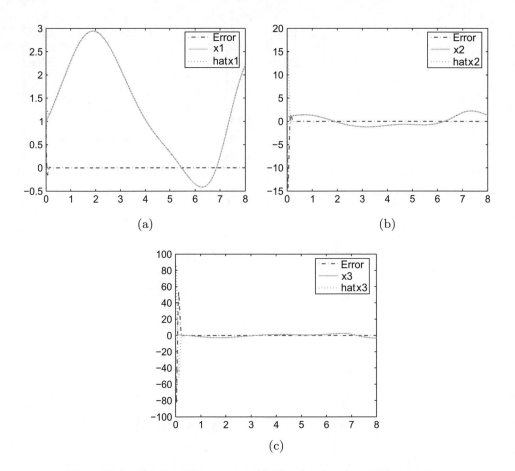

(a)   (b)

(c)

Figure 3.3.3   ESO (3.3.33) for system (3.3.29) using the constant gain parameter.

Assumption 3.3.6

- $\tilde{f}(\cdot)$ is Hölder continuous, that is,

$$|\tilde{f}(x) - \tilde{f}(\hat{x})| \le L\|x - \hat{x}\|^{\beta}, \ \forall \, x, \hat{x} \in \mathbb{R}^n, \beta > 0. \tag{3.3.37}$$

- There exists constant $K > 0$ and function $\varpi \in (\mathbb{R}, \bar{\mathbb{R}}^+)$ such that

$$\sum_{i=1}^{m} \left| \frac{\partial \bar{f}(t, \zeta, w)}{\partial \zeta_i} \right| + \left| \frac{\partial \bar{f}(t, \zeta, w)}{\partial w} \right| + \left| \frac{\partial \bar{f}(t, \zeta, w)}{\partial t} \right| + \|f_0(t, x, \zeta, w)\| \le K + \varpi(w),$$

for all $(t, x, \zeta, w) \in \mathbb{R}^{n+m+2}$.

The following Assumption 3.3.7 is the supplement of Assumption 3.3.4 or 3.3.6.

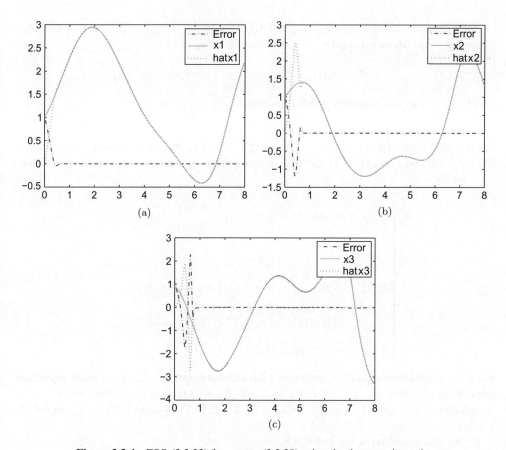

Figure 3.3.4   ESO (3.3.33) for system (3.3.29) using the time-varying gain.

Assumption 3.3.7   The functions $\mathcal{V}(x)$ and $\mathcal{W}(x)$ in Assumption 3.3.4 or 3.3.6 satisfy

$$\|x\|^{\beta} \left| \frac{\partial \mathcal{V}(x)}{\partial x_{n+1}}(x) \right| \leq N\mathcal{W}(x), \quad \forall \|x\| \geq R.$$

Theorem 3.3.3   *Assume Assumptions 3.3.6 and 3.3.2. Then the following assertions hold true.*

1. If the time-varying gain parameter $r(t)$ satisfies Assumption 3.3.3 and there are functions $\mathcal{V}(x)$ and $\mathcal{W}(x)$ that satisfy Assumptions 3.3.4 and 3.3.7, then the states of ESO (3.3.35) converge to the states and extended state $x_{n+1}(t) = \bar{f}(t, \zeta(t), w(t)) + (b(t) - b_0)u(t)$ of (3.3.1) in the sense that

$$\lim_{t \to +\infty} |\hat{x}_i(t) - x_i(t)| = 0, \quad i = 1, 2, \ldots, n + 1. \tag{3.3.38}$$

2. If the gain parameter is chosen as (3.3.18), and there are functions $\mathcal{V}(x)$ and $\mathcal{W}(x)$ that satisfy Assumptions 3.3.6 and 3.3.7, then the states of ESO (3.3.35) converge to the states *and extended state of (3.3.1)* $x_{n+1}(t) = \bar{f}(t, \zeta(t), w(t)) + (b(t) - b_0)u(t)$ *in the sense that for any* $\sigma > 0$ *there exists an* $r^* > 0$ *such that for all* $r_0 > r^*$

$$|\hat{x}_i(t) - x_i(t)| < \sigma, \quad t > t_{r_0}, \quad i = 1, 2, \dots, n+1,$$

where t_{r_0} *is an* r_0-*dependent positive constant.*

Proof. Once again, let $\eta_i(t) = r^{n+1-i}(t)(x_i(t) - \hat{x}_i(t))$ be the errors between the states of ESO (3.3.35) and system (3.3.1). Let the gain parameter $r(t)$ satisfy Assumption 3.3.3. Then $\eta_i(t)$ satisfies

$$
\begin{cases}
\dot{\eta}_1(t) = r(t)(\eta_2(t) - g_1(\eta_1(t))) + \dfrac{n\dot{r}(t)}{r(t)}\eta_1(t), \\
\quad\vdots \\
\dot{\eta}_n(t) = r(t)(\eta_{n+1}(t) - g_n(\eta_1(t))) + \dfrac{\dot{r}(t)}{r(r)}\eta_n(t) \\
\qquad\qquad + [\tilde{f}(x_1(t), \dots, x_n(t)) - \tilde{f}(\hat{x}_1(t), \dots, \hat{x}_n(t))], \\
\dot{\eta}_{n+1}(t) = -r(t)g_{n+1}(\eta_1(t)) + \dot{x}_{n+1}(t).
\end{cases}
\tag{3.3.39}
$$

From the second condition of Assumption 3.3.6 and Assumption 3.3.2, there exists a positive constant B such that $|\dot{x}_{n+1}(t)| < B$ for all $t > 0$. By Assumption 3.3.3, there exists a $t_1 > 0$ such that $r(t) > \max\{1, 2n(n+1)NM + 2BN, (2LN)^{1/\beta}\}$ and $\dot{r}(t)/r(t) < 2M$ for all $t > t_1$.

By the first condition of Assumption 3.3.6,

$$|\tilde{f}(x_1(t), \dots, x_n(t)) - \tilde{f}(\hat{x}_1(t), \dots, \hat{x}_n(t))|$$

$$\le L\|(x_1(t) - \hat{x}_1(t), x_2(t) - \hat{x}_2(t), \dots, x_n(t) - \hat{x}_n(t))\|^\beta$$

$$= L\left\|\left(\frac{\eta_1(t)}{r^n(t)}, \frac{\eta_2(t)}{r^{n-1}(t)}, \dots, \frac{\eta_n(t)}{r(t)}\right)\right\|^\beta \le \frac{L}{r^\beta(t)}\|(\eta_1(t), \eta_2(t), \dots, \eta_n(t))\|^\beta.$$
$$\tag{3.3.40}$$

Now find the derivative of $\mathcal{V}(\eta(t))$ along the solution $\eta(t) = (\eta_1(t), \eta_2(t), \dots, \eta_{n+1}(t))$ of the error equation (3.3.39), together with the Assumptions 3.3.4 and 3.3.7, to obtain, for every $t > t_1$, if $\|\eta(t)\| > R$,

$$\left.\frac{d\mathcal{V}(\eta(t))}{dt}\right|_{\text{along (3.3.39)}} = r(t)\left(\sum_{i=1}^n (\eta_{i+1}(t) - g_i(\eta_1(t)))\frac{\partial \mathcal{V}(\eta(t))}{\partial \eta_i} - g_{n+1}(\eta_1(t))\frac{\partial \mathcal{V}(\eta(t))}{\partial \eta_{n+1}}\right)$$

$$+ \sum_{i=1}^n \frac{(n+1-i)\dot{r}(t)\eta_i(t)}{r(t)}\frac{\partial \mathcal{V}(\eta(t))}{\partial \eta_i} + \dot{x}_{n+1}(t)\frac{\partial \mathcal{V}(\eta(t))}{\partial \eta_{n+1}}$$

$$+ r(t)(\tilde{f}(x_1(t), \dots, x_n(t)) - \tilde{f}(\hat{x}_1(t), \dots, \hat{x}_n(t)))\frac{\partial \mathcal{V}(\eta(t))}{\partial \eta_n}$$

$$\leq - r(t)\mathcal{W}(\eta(t)) + LN\ r^{1-\beta}(t)\mathcal{W}(t) + n(n+1)MN\mathcal{W}(t) + BN\mathcal{W}(\eta(t))$$

$$\leq - \frac{1}{2}[r(t) - 2(n(n+1)MN - 2BN)]\mathcal{W}(\eta(t)) < 0, \tag{3.3.41}$$

which shows that the solution $\eta(t)$ of (3.3.39) is uniformly bounded for all $t \geq t_2$ for some $t_2 > t_1$. Hence one can find a constant $C > 0$ such that

$$\left.\frac{d\mathcal{V}(\eta(t))}{dt}\right|_{\text{along (3.3.6)}} \leq - \frac{1}{2}r(t)\mathcal{W}(\eta(t)) + C, \quad \forall\, t > t_2. \tag{3.3.42}$$

The remainder of the proof is similar to the corresponding part of the proof of Theorem 3.3.1. The details are omitted.

As for the case of time-varying gain $r(t)$ satisfying (3.3.18), the proof is similar to the proof of Theorem 3.3.2. The details are omitted as well. $\qquad\square$

3.4   The ESO for MIMO Systems with Uncertainty

In this section, we consider ESO for a class of multi-input and multi-output (MIMO) non-linear systems with large uncertainty. In Subsection 3.4.1, we construct a linear ESO for the MIMO system with "total disturbance" coming both from external and internal uncertainty. The convergence is presented. A special ESO for a system with external disturbance only is considered in Subsection 3.4.2. The difference of Subsection 3.4.2 with Subsection 3.4.1 lies in the way that we try to make use of the known information of system function as much as possible in the design of the ESO. In Subsection 3.4.3, the numerical simulations are carried out for some examples for illustration of the effectiveness of the ESO.

3.4.1   ESO for Systems with Total Disturbance

We consider the following MIMO system with large uncertainty:

$$\begin{cases} x_1^{(n_1)}(t) = f_1\left(x_1(t),\ldots,x_1^{(n_1-1)}(t),\ldots,x_m^{(n_m-1)}(t), w_1(t)\right) + g_1(u_1(t),\ldots,u_k(t)), \\ x_2^{(n_2)}(t) = f_2(x_1(t),\ldots,x_1^{(n_1-1)}(t),\ldots,x_m^{(n_m-1)}(t), w_2(t)) + g_2(u_1(t),\ldots,u_k(t)), \\ \quad\vdots \\ x_m^{(n_m)}(t) = f_m(x_1(t),\ldots,x_1^{(n_1-1)}(t),\ldots,x_m^{(n_m-1)}(t), w_m(t)) + g_m(u_1(t),\ldots,u_k(t)), \\ y_i(t) = x_i(t), i = 1,2,\ldots,m, \end{cases} \tag{3.4.1}$$

where $n_i \in \mathbb{Z}$, $f_i \in C(\mathbb{R}^{n_1+n_2+\cdots+n_m+1})$ represents the system function, $w_i \in C([0,\infty),\mathbb{R})$ the external disturbance, $u_i \in C([0,\infty),\mathbb{R})$ the control (input), $y_i(t)$ the observations (output), and $g_i \in C(\mathbb{R}^k,\mathbb{R})$.

System (3.4.1) can be rewritten as a first-order system described by m subsystems of the first-order differential equations:

$$\begin{cases} \dot{x}_{i,1}(t) = x_{i,2}(t), \\ \dot{x}_{i,2}(t) = x_{i,3}(t), \\ \quad \vdots \\ \dot{x}_{i,n_i}(t) = f_i\left(x_{1,1}(t),\ldots,x_{1,n_1}(t),\ldots,x_{m,n_m}(t),w_i(t)\right) + g_i(u_1(t),u_2(t),\ldots,u_k(t)), \\ y_i(t) = x_{i,1}(t), i = 1,2,\ldots,m, \end{cases}$$

(3.4.2)

where $x_{i,j}(t) = x_i^{(j-1)}(t)$, $j = 1,2\ldots,n_i$.

The nonlinear ESO for system (3.4.2) is also composed of m subsystems:

$$\begin{cases} \dot{\hat{x}}_{i,1}(t) = \hat{x}_{i,2}(t) + \epsilon^{n_i-1}\varphi_{i,1}\left(\dfrac{x_{i,1}(t) - \hat{x}_{i,1}(t)}{\epsilon^{n_i}}\right), \\ \dot{\hat{x}}_{i,2}(t) = \hat{x}_{i,3}(t) + \epsilon^{n_i-2}\varphi_{i,2}\left(\dfrac{x_{i,1}(t) - \hat{x}_{i,1}(t)}{\epsilon^{n_i}}\right), \\ \quad \vdots \\ \dot{\hat{x}}_{i,n_i}(t) = \hat{x}_{i,n_i+1}(t) + \varphi_{i,n_i}\left(\dfrac{x_{i,1}(t) - \hat{x}_{i,1}(t)}{\epsilon^{n_i}}\right) + g_i(u_1(t),u_2(t),\ldots,u_k(t)), \\ \dot{\hat{x}}_{i,n_i+1}(t) = \frac{1}{\epsilon}\varphi_{i,n_i+1}\left(\dfrac{x_{i,1}(t) - \hat{x}_{i,1}(t)}{\epsilon^{n_i}}\right), i = 1,2,\ldots,m, \end{cases}$$

(3.4.3)

where $\hat{x}_{i,n_i+1}(t)$ is used to estimate the total disturbance $f_i(\cdot)$, ϵ is the high-gain tuning parameter, and $\varphi_{i,j}(r)$ is the pertinent function to be specified. When $\varphi_{i,j}(r) = k_{i,j}r$ for all $r \in \mathbb{R}$ and some $k_{i,j} \in \mathbb{R}$, the corresponding ESO (3.4.3) is reduced to the linear ESO considered in Section 3.1.

To establish convergence for ESO, we propose the following assumptions. Assumption 3.4.1 is on the system (3.4.2) itself, while Assumption 3.4.2 is on the functions in (3.4.3), which will be shown trivially satisfied by linear functions.

Assumption 3.4.1 For every $i \in \{1,2,\ldots,m\}$, all $u_i(t)$, $w_i(t)$, $\dot{w}_i(t)$, and the solution of (3.4.2) are bounded; $g_i \in C(\mathbb{R}^k,\mathbb{R})$, $f_i \in C^1(\mathbb{R}^{n_1+\ldots+n_m+1},\mathbb{R})$.

Assumption 3.4.2 For every $i \in \{1,2,\ldots,m\}$, there exist positive constants $\lambda_{i,j}$ $(j = 1,2,3,4)$, β_i, and positive definite functions $V_i, W_i : \mathbb{R}^{n_i+1} \to \mathbb{R}$ that satisfy the following conditions:

- $\lambda_{i,1}\|y\|^2 \le V_i(y) \le \lambda_{2,i}\|y\|^2$, $\quad \lambda_{i,3}\|y\|^2 \le W_i(y) \le \lambda_{i,4}\|y\|^2$,

- $\displaystyle\sum_{l=1}^{n_i} \dfrac{\partial V_i(y)}{\partial y_l}(y_{l+1} - \varphi_{i,l}(y_1)) - \dfrac{\partial V(y)}{\partial y_{n_i+1}}\varphi_{i,n_i+1}(y_1) \le -W_i(y)$,

- $\left|\dfrac{\partial V_i(y)}{\partial y_{n_i+1}}\right| \le \beta_i\|y\|, \forall\, y = (y_1,y_2,\ldots,y_{n_i+1})$.

Theorem 3.4.1 *Under Assumptions 3.4.1 and 3.4.2, for any given initial values of (3.4.2) and (3.4.3),*

(i) for every positive constant $a > 0$,

$$\lim_{\epsilon \to 0} |x_{i,j}(t) - \hat{x}_{i,j}(t)| = 0 \text{ uniformly for } t \in [a, \infty);$$

(ii) there exists an $\epsilon_0 > 0$ such that for every $\epsilon \in (0, \epsilon_0)$ there exists a $t_\epsilon > 0$ such that

$$|x_{i,j}(t) - \hat{x}_{i,j}(t)| \leq K_{ij} \epsilon^{n_i + 2 - j}, \ t \in (t_\epsilon, \infty),$$

where $\hat{x}_{i,j}(t)$, $j = 1, 2 \ldots, n_i + 1, i = 1, 2 \ldots, m$, are the solutions of (3.4.3), $x_{i,j}(t)$, $j = 1, 2, \ldots, n_i, i = 1, 2, \ldots, m$, are the solutions of (3.4.2), $x_{i,n_i+1}(t) = f_i(x_{1,1}(t), \ldots, x_{1,n_1}(t), \ldots, x_{m,n_m}(t), w_i(t))$ is the extended state, and K_{ij} is positive constant independent of ϵ but depending on initial values.

Proof. We first notice that

$$\Delta_i(t) = \sum_{l=1}^{m} \sum_{j=1}^{n_l} x_{l,j+1}(\epsilon t) \frac{\partial f_i(x_{1,1}(\epsilon t), \ldots, x_{1,n_1}(\epsilon t), \ldots, x_{m,n_m}(\epsilon t), w_i(\epsilon t))}{\partial x_{l,j}}$$

$$+ \sum_{l=1}^{m} g_l(u_1(\epsilon t), \ldots, u_k(\epsilon t)) \frac{\partial f_i(x_{1,1}(\epsilon t), \ldots, x_{1,n_1}(\epsilon t), \ldots, x_{m,n_m}(\epsilon t), w_i(\epsilon t))}{\partial x_{l,n_l}}$$

$$+ \dot{w}_i(\epsilon t) \frac{\partial f_i(x_{1,1}(\epsilon t), \ldots, x_{1,n_1}(\epsilon t), \ldots, x_{m,n_m}(\epsilon t), w_i(\epsilon t))}{\partial w_i}. \tag{3.4.4}$$

From Assumption 3.4.1, $|\Delta_i(t)| \leq M$ is uniformly bounded for some $M > 0$ and all $t \geq 0, i \in \{1, 2, \ldots, m\}$.

For each $j = 1, 2, \ldots, n_i + 1, i = 1, 2, \ldots, m$, set

$$e_{i,j}(t) = x_{i,j}(t) - \hat{x}_{i,j}(t), \eta_{i,j}(t) = \frac{e_{i,j}(\epsilon t)}{\epsilon^{n_i + 1 - j}}, \quad \eta_i(t) = (\eta_{i,1}(t), \ldots, \eta_{i,n_i+1}(t))^\top. \tag{3.4.5}$$

It follows from (3.4.2) and (3.4.3) that for each $i \in \{1, 2, \ldots, m\}, j \in \{1, 2, \ldots, n_i\}$, $\eta_{i,j}(t)$ satisfies

$$\frac{d\eta_{i,j}(t)}{dt} = \frac{d}{dt} \left(\frac{x_{i,j}(\epsilon t) - \hat{x}_{i,j}(\epsilon t)}{\epsilon^{n_i + 1 - j}} \right) = \frac{x_{i,j+1}(\epsilon t) - \hat{x}_{i,j+1}(\epsilon t)}{\epsilon^{n_i - j}} = \eta_{i,j+1}(t) - \varphi_{i,j}(\eta_{i,1}(t))$$

and for each $i \in \{1, 2, \ldots, m\}$,

$$\frac{d\eta_{i,n_i+1}(t)}{dt} = \epsilon(\dot{x}_{i,n_i+1}(\epsilon t) - \dot{\hat{x}}_{i,n_i+1}(\epsilon t)) = -\varphi_{i,n_i+1}(\eta_{i,1}(t)) + \epsilon \Delta_i(t).$$

We then put all these equations together into the following differential equations satisfied by $\eta_{i,j}(t)$:

$$\begin{cases} \dot{\eta}_{i,1}(t) = \eta_{i,2}(t) - \varphi_{i,1}(\eta_{i,1}(t)), \quad \eta_{i,1}(0) = \dfrac{e_{i,1}(0)}{\epsilon^{n_i}}, \\[2mm] \dot{\eta}_{i,2}(t) = \eta_{i,3}(t) - \varphi_{i,2}(\eta_{i,1}(t)), \quad \eta_{i,2}(0) = \dfrac{e_{i,2}(0)}{\epsilon^{n_i-1}}, \\[2mm] \quad\vdots \\[2mm] \dot{\eta}_{i,n_i}(t) = \eta_{i,n_i+1}(t) - \varphi_{i,n_i}(\eta_{i,1}(t)), \quad \eta_{i,n_i}(0) = \dfrac{e_{i,n_i}(0)}{\epsilon}, \\[2mm] \dot{\eta}_{i,n_i+1}(t) = -\varphi_{i,n_i+1}(\eta_{i,1}(t)) + \epsilon\Delta_i(t), \quad \eta_{i,n_i+1}(0) = e_{i,n_i+1}(0), \end{cases} \tag{3.4.6}$$

where $e_{i,j}(0) = x_{i,j}(0) - \hat{x}_{i,j}(0)$ is the ϵ-independent initial value.

By Assumption 3.4.2, we can find the derivative of $V(\eta_i(t))$ with respect to t along the solution $\eta_i(t) = (\eta_{i,1}(t), \eta_{i,2}(t), \ldots, \eta_{i,n_i+1}(t))$ of system (3.4.6) to be

$$\begin{aligned} \left.\frac{dV_i(\eta_i(t))}{dt}\right|_{\text{along (3.4.6)}} &= \sum_{j=1}^{n_i} \frac{\partial V_i(\eta_i(t))}{\partial \eta_{i,j}}(\eta_{i,j+1}(t) - \varphi_{i,j}(\eta_{i,1}(t))) \\ &\quad - \frac{\partial V_i(\eta_i(t))}{\partial \eta_{i,n_i+1}}\varphi_{i,n_i+1}(\eta_{i,1}(t)) \\ &\quad + \frac{\partial V_i(\eta_i(t))}{\partial \eta_{i,n_i+1}}\epsilon\Delta_i(t) \le -W_i(\eta_i(t)) + \epsilon M\beta_i\|\eta_i(t)\| \\ &\le -\frac{\lambda_{i,3}}{\lambda_{i,2}}V(\eta_i(t)) + \frac{\sqrt{\lambda_{i,1}}}{\lambda_{i,1}}\epsilon M\beta_i\sqrt{V(\eta_i(t))}. \end{aligned} \tag{3.4.7}$$

It then follows that

$$\frac{d}{dt}\sqrt{V(\eta_i(t))} \le -\frac{\lambda_{i,3}}{2\lambda_{i,2}}\sqrt{V(\eta_i(t))} + \frac{\sqrt{\lambda_{i,1}}\epsilon M\beta_i}{2\lambda_{i,1}}. \tag{3.4.8}$$

By virtue of Assumption 3.4.2 again, we have

$$\|\eta_i(t)\| \le \sqrt{\frac{V(\eta_i(t))}{\lambda_{i,1}}} \le \frac{\sqrt{\lambda_{i,1}V(\eta_i(0))}}{\lambda_{i,1}}e^{-\lambda_{i,3}/(2\lambda_{i,2})t} + \frac{\epsilon M\beta_i}{2\lambda_{i,1}}\int_0^t e^{-\lambda_{i,3}/(2\lambda_{i,2})(t-s)}ds. \tag{3.4.9}$$

This together with (3.4.5) yields

$$\begin{aligned} |e_{i,j}(t)| = \epsilon^{n_i+1-i}\left|\eta_{i,j}\left(\frac{t}{\epsilon}\right)\right| &\le \epsilon^{n+1-i}\left\|\eta_i\left(\frac{t}{\epsilon}\right)\right\| \\ &\le \epsilon^{n_i+1-i}\left(\frac{\sqrt{\lambda_{i,1}V(\eta(0))}}{\lambda_{i,1}}e^{-\lambda_{i,3}/(2\lambda_{i,2}\epsilon)t} + \frac{\epsilon M\beta_i}{2\lambda_{i,1}}\int_0^{\frac{t}{\epsilon}} e^{-\lambda_{i,3}/(2\lambda_{i,2})(t/\epsilon-s)}ds\right) \\ &\to 0 \text{ uniformly in } t \in [a,\infty) \text{ as } \epsilon \to 0. \end{aligned} \tag{3.4.10}$$

Both (i) and (ii) of Theorem 3.4.1 then follow from (3.4.10). This completes the proof of the theorem. □

A typical example of ESO satisfying the conditions of Theorem 3.4.1 is the linear ESO. This is the case where all $\varphi_{i,j}$ are linear functions: $\varphi_{i,j}(r) = k_{i,j}r$ for all $r \in \mathbb{R}$, and $k_{i,j}$ is real number such that the following matrix E_i is Hurwitz:

$$E_i = \begin{pmatrix} -k_{i,1} & 1 & 0 & \cdots & 0 \\ -k_{i,2} & 0 & 1 & \cdots & 0 \\ \vdots & \vdots & \vdots & \ddots & \vdots \\ -k_{i,n_i} & 0 & 0 & \cdots & 1 \\ -k_{i,n_i+1} & 0 & 0 & \cdots & 0 \end{pmatrix}. \tag{3.4.11}$$

In this case, the ESO (3.4.3) is of the linear form:

$$\begin{cases} \dot{\hat{x}}_{i,1}(t) = \hat{x}_{i,2}(t) + \dfrac{1}{\epsilon}k_{i,1}\left(x_{i,1}(t) - \hat{x}_{i,1}(t)\right), \\[2mm] \dot{\hat{x}}_{i,2}(t) = \hat{x}_{i,3}(t) + \dfrac{1}{\epsilon^2}k_{i,2}(x_{i,1}(t) - \hat{x}_{i,1}(t)), \\[2mm] \quad\vdots \\[2mm] \dot{\hat{x}}_{i,n_i}(t) = \hat{x}_{i,n_i+1}(t) + \dfrac{1}{\epsilon^{n_i}}k_{i,n_i}(x_{i,1}(t) - \hat{x}_{i,1}(t)) + g_i(u_1(t), u_2(t),\ldots, u_k(t)), \\[2mm] \dot{\hat{x}}_{i,n_i+1}(t) = \dfrac{1}{\epsilon^{n_i+1}}k_{i,n_i+1}(x_{i,1}(t) - \hat{x}_{i,1}(t)), i = 1, 2,\ldots, m. \end{cases} \tag{3.4.12}$$

Corollary 3.4.1 *Suppose that all matrices E_i in (3.4.11) are Hurwitz and Assumption 3.4.1 is satisfied. Then for any given initial values of (3.4.2) and (3.4.12), the following conclusions hold.*

(i) For every positive constant $a > 0$,

$$\lim_{\epsilon \to 0} |x_{i,j}(t) - \hat{x}_{i,j}(t)| = 0 \text{ uniformly for } t \in [a, \infty).$$

(ii) There exists an $\epsilon_0 > 0$ such that for every $\epsilon \in (0, \epsilon_0)$, there exists a $t_\epsilon > 0$ such that

$$|x_{i,j}(t) - \hat{x}_{i,j}(t)| \leq K_{ij}\epsilon^{n_i+2-j}, \quad t \in (t_\epsilon, \infty),$$

where $\hat{x}_{i,j}(t)$, $j = 1, 2,\ldots, n_i + 1, i = 1, 2,\ldots, m$, are the solutions of (3.4.12), $x_{i,j}(t)$, $j = 1, 2,\ldots, n_i, i = 1, 2, \ldots, m$, are the solutions of (3.4.1), $x_{i,n_i+1}(t) = f_i(x_{1,1}(t),\ldots,x_{1,n_1}(t),\ldots,x_{m,n_m}(t), w_i(t))$ is the extended state, and K_{ij} is a positive number independent of ϵ but depending on initial values.

Proof. By Theorem 3.4.1, we need only verify Assumption 3.4.2. To this end, let

$$V_i(y) = \langle P_i y, y \rangle, \quad W_i(y) = \langle y, y \rangle, \quad \forall\, y \in \mathbb{R}^{n_i+1}, \tag{3.4.13}$$

where P_i is the positive definite solution of the Lyapunov equation $P_i E_i + E_i^\top P_i = I_{(n_i+1)\times(n_i+1)}$. By basic linear algebra, it is easy to verify that

$$\lambda_{\min}(P_i)\|y\|^2 \le V_i(y) \le \lambda_{\max}(P_i)\|y\|^2, \tag{3.4.14}$$

$$\sum_{j=1}^{n_i} \frac{\partial V_i(y)}{\partial y_j}(y_{j+1} - k_{i,j}y_1) - \frac{\partial V_i(y)}{\partial y_{n_i+1}}k_{i,n_i+1}y_1 = -W_i(y), \forall\, y = (y_1, y_2, \cdots, y_{n_i+1}),$$

$$\tag{3.4.15}$$

and

$$\left|\frac{\partial V_i(y)}{\partial y_{n_i+1}}\right| \le 2\lambda_{\max}(P_i)\|y\|, \tag{3.4.16}$$

where $\lambda_{\max}(P_i)$ and $\lambda_{\min}(P_i)$ denote the maximal and minimal eigenvalues of P_i, respectively. So Assumption 3.4.2 is satisfied. The results then follow from Theorem 3.4.1. □

Now we construct a special class of nonlinear ESO and discuss its convergence.

Set $\varphi_{i,j}(r) = k_{i,j}[r]^{ja_i-(j-1)}$ in (3.4.3), where $a_i \in (1-\frac{1}{n_i}, 1)$, $[r]^\alpha = \text{sign}(r)|r|^\alpha$, and $k_{i,j}, j = 1, 2, \ldots, n_i + 1$ are constants such that every matrix E_i in (3.4.11) is Hurwitz. This reduces (3.4.3) to the following form:

$$\begin{cases} \dot{\hat{x}}_{i,1}(t) = \hat{x}_{i,2}(t) + \epsilon^{n_i-1}k_{i,1}\left[\dfrac{x_{i,1}(t) - \hat{x}_{i,1}(t)}{\epsilon^{n_i}}\right]^{a_i}, \\[2mm] \dot{\hat{x}}_{i,2}(t) = \hat{x}_{i,3}(t) + \epsilon^{n_i-2}k_{i,2}\left[\dfrac{x_{i,1}(t) - \hat{x}_{i,1}(t)}{\epsilon^{n_i}}\right]^{2a_i-1}, \\[2mm] \quad\vdots \\[2mm] \dot{\hat{x}}_{i,n_i}(t) = \hat{x}_{i,n_i+1}(t) + k_{i,n}\left[\dfrac{x_{i,1}(t) - \hat{x}_{i,1}(t)}{\epsilon^{n_i}}\right]^{n_ia_i-(n_i-1)} + g_i(u_1(t), u_2(t), \ldots, u_k(t)), \\[2mm] \dot{\hat{x}}_{i,n_i+1}(t) = \dfrac{1}{\epsilon}k_{i,n_i+1}\left[\dfrac{x_{i,1}(t) - \hat{x}_{i,1}(t)}{\epsilon^{n_i}}\right]^{(n_i+1)a_i-n_i}, \quad i = 1, 2, \ldots, m. \end{cases}$$

$$\tag{3.4.17}$$

It is easy to verify that the vector field

$$F_i(y) = \begin{pmatrix} y_2 + k_{i,1}[y_1(t)]^{a_i} \\ y_3 + k_{i,2}[y_1]^{2a_i-1} \\ \vdots \\ y_{n_i+1} + k_{i,n_i}[y_1]^{n_ia_i-(n_i-1)} \\ k_{i,n_i+1}[y_1]^{(n_i+1)a_i-n_i} \end{pmatrix} \tag{3.4.18}$$

is homogeneous of degree $-d_i = a_i - 1$ with respect to weights $\{r_{i,j} = (j-1)a_i - (j-2)\}_{j=1}^{n_i+1}$, where $[r]^{a_i} = \text{sign}(r)|r|^{a_i}$, $\forall\, r \in \mathbb{R}$.

By Theorem 1.3.8, for some $a_i \in (1 - \frac{1}{n_i}, 1)$, if all matrices E_i in (3.4.11) are Hurwitz, then system $\dot{y}(t) = F_i(y(t))$ is globally finite-time stable.

Now we are in a position to show the convergence of nonlinear ESO (3.4.17).

Theorem 3.4.2 *Suppose that every matrix E_i in (3.4.11) is Hurwitz and Assumption 3.4.1 is satisfied. Then there exists a constant $\epsilon_0 > 0$ such that for any $\epsilon \in (0, \epsilon_0)$ there exists a constant $T_\epsilon > 0$ such that:*

(i) If $a_i \in (n_i/(n_i + 1), 1)$, then

$$|x_{i,j}(t) - \hat{x}_{i,j}(t)| \leq K_{i,j}\epsilon^{n_i+1-j+((j-1)a_i-(j-2))/((n_i+1)a_i-n_i)}, \quad \forall t \geq T_\epsilon.$$

(ii) If $a_i = n_i/(n_i + 1)$, then

$$|x_{i,j}(t) - \hat{x}_{i,j}(t)| = 0, \quad \forall t \geq T_\epsilon,$$

where $K_{i,j}$ is positive constant independent of ϵ but depending on initial values. $x_{i,j}(t)$ and $\hat{x}_{i,j}(t)$, $j = 1, 2, \ldots, n_i, i = 1, 2, \ldots, n_i + 1$, are solutions of (3.4.2) and (3.4.17), respectively, and $x_{i,n_i+1}(t) = f_i(x_{1,1}(t), \ldots, x_{m,n_m}(t))$ is the extended state.

Proof. For each $i \in \{1, 2, \ldots, m\}$, let

$$\eta_{i,j}(t) = \frac{x_{i,j}(\epsilon t) - \hat{x}_{i,j}(\epsilon t)}{\epsilon^{n_i+1-j}}, \quad j = 1, 2, \ldots, n_i + 1. \tag{3.4.19}$$

A direct computation shows that $\eta_i(t) = (\eta_{i,1}(t), \eta_{i,2}(t), \ldots, \eta_{i,n_i+1}(t))^\top$ satisfies

$$\begin{cases} \dot{\eta}_{i,1}(t) = \eta_{i,2}(t) + k_{i,1}[\eta_{i,1}(t)]^{a_i}, \\ \dot{\eta}_{i,2}(t) = \eta_{i,3}(t) + k_{i,2}[\eta_{i,1}(t)]^{2a_i-1}, \\ \quad \vdots \\ \dot{\eta}_{i,n_i}(t) = \eta_{i,n_i+1}(t) + k_{i,n_i}[\eta_{i,1}(t)]^{n_ia_i-(n_i-1)}, \\ \dot{\eta}_{i,n_i+1}(t) = k_{i,n_i+1}[\eta_{i,1}(t)]^{(n_i+1)a_i-n_i} + \epsilon\Delta_i(t), \end{cases} \tag{3.4.20}$$

with $\Delta_i(t)$ given by (3.4.4). Therefore the error equation (3.4.20) is a perturbed system of global finite-time stable system $\dot{y} = F_i(y), y \in \mathbb{R}^{n_i+1}$. For the homogeneous global finite-time stable system, it follows from Theorem 1.3.9 that there exists a positive definite, radial unbounded, differentiable function $V_i : \mathbb{R}^n \to \mathbb{R}$ such that $V_i(x)$ is homogeneous of degree γ_i with respect to weights $\{r_{i,j}\}_{j=1}^{n_i}$, and the Lie derivative of $V_i(x)$ along the vector fields $F_i(x)$,

$$L_{F_i}V_i(x) = \sum_{j=1}^{n_i} \frac{\partial V_i(x)}{\partial \eta_{i,j}}(\eta_{i,j+1} + k_{i,j}[\eta_{i,1}]^{ja_i-(j-1)}) + \frac{\partial V_i(x)}{\partial \eta_{i,n_i+1}}k_{i,n_i+1}[\eta_{i,1}]^{(n_i+1)a_i-n_i},$$

$$\forall x = (\eta_{i,1}, \eta_{i,2}, \ldots, \eta_{i,n_i+1})^\top,$$

is negative definite, where $\gamma_i \geq \max\{d_i, r_{i,j}\}$. We note here that by radial unbounded for $V_i(x)$ we mean $\lim_{\|x\| \to +\infty} V_i(x) = +\infty$, where $F_i(x)$ is defined in (3.4.18).

From homogeneous of $V_i(x)$, for any positive constant λ,

$$V_i(\lambda^{r_{i,1}}x_1, \lambda^{r_{i,2}}x_2, \ldots, \lambda^{r_{i,n_i}}x_{n_i+1}) = \lambda^{\gamma_i}V_i(x_1, x_2, \ldots, x_{n_i+1}). \tag{3.4.21}$$

Finding the derivatives of both sides of the above equation with respect to the arguments x_j yields

$$\lambda^{r_{i,j}} \frac{\partial V_i(\lambda^{r_{i,1}} x_1, \lambda^{r_{i,2}} x_2, \ldots, \lambda^{r_{i,n_i+1}} x_{n_i+1})}{\partial \lambda^{r_{i,j}} x_j} = \lambda^{\gamma_i} \frac{\partial V_i(x_1, x_2, \ldots, x_{n_i+1})}{\partial x_i}. \tag{3.4.22}$$

This shows that $\partial V_i(x)/\partial x_j$ is homogeneous of degree $\gamma_i - r_{i,j}$ with respect to weights $\{r_{i,j}\}_{j=1}^{n_i+1}$.

Furthermore, the Lie derivative of $V_i(x)$ along the vector field $F_i(y)$ satisfies

$$L_{F_i} V_i(\lambda^{r_{i,1}} x_1, \lambda^{r_{i,2}} x_2, \ldots, \lambda^{r_{i,n_i+1}} x_{n_i+1})$$

$$= \sum_{j=1}^{n_i+1} \frac{\partial V_i(\lambda^{r_{i,1}} x_1, \lambda^{r_{i,2}} x_2, \ldots, \lambda^{r_{i,n_i}} x_{n_i+1})}{\partial \lambda^{r_{i,j}} x_j} F_{i,j}(\lambda^{r_{i,1}} x_1, \lambda^{r_{i,2}} x_2, \ldots, \lambda^{r_{i,n_i+1}} x_{n_i+1})$$

$$= \lambda^{\gamma_i - d_i} \sum_{i=1}^{n_i+1} \frac{\partial V(x_1, x_2, \ldots, x_{n_i+1})}{\partial \lambda^{r_{i,j}} x_j} f_i(x_1, x_2, \ldots, x_{n_i+1})$$

$$= \lambda^{\gamma_i - d_i} L_{F_i} V_i(x_1, x_2, \ldots, x_{n_i+1}). \tag{3.4.23}$$

So $L_{F_i} V_i(x)$ is homogeneous of degree $\gamma_i - d_i$ with respect to weights $\{r_{i,j}\}_{j=1}^{n_i+1}$.

By Property 1.3.3, we have the following inequalities:

$$\left| \frac{\partial V_i(x)}{\partial x_{n_i+1}} \right| \leq b_i (V_i(x))^{(\gamma_i - r_{i,n_i+1})/\gamma_i}, \quad \forall\, x \in \mathbb{R}^{n_i+1} \tag{3.4.24}$$

and

$$L_{F_i} V_i(x) \leq -c_i (V_i(x))^{(\gamma_i - d_i)/\gamma_i}, \quad \forall\, x \in \mathbb{R}^{n_i+1}, \tag{3.4.25}$$

where b_i and c_i are positive constants.

From Assumption 3.4.1, there exist constants $M_i > 0$ such that $|\Delta_i(t)| \leq M_i$ for all $t > 0$ and $i \in \{1, 2, \ldots, m\}$. Now finding the derivative of $V_i(\eta_i(t))$ along the solution of (3.4.20) gives

$$\left. \frac{dV(\eta_i(t))}{dt} \right|_{\text{along } (3.4.20)} = L_{F_i} V_i(\eta_i(t)) + \epsilon \Delta_i(t) \frac{\partial V_i(\eta_i(t))}{\partial x_{n_i+1}} F_i(\eta_i(t))$$

$$\leq -c_i (V_i(\eta_i(t)))^{(\gamma_i - d_i)/\gamma_i} + \epsilon M_i b_i (V_i(\eta_i(t)))^{(\gamma_i - r_{i,n_i+1})/\gamma_i}. \tag{3.4.26}$$

Let $\epsilon_i = c_i/(2M_i b_i)$. If $a_i = 1 - 1/(1 + n_i)$, then for any $\epsilon \in (0, \epsilon_i)$,

$$\frac{dV_i(\eta_i(t))}{dt} \leq -\frac{c_i}{2} (V_i(\eta_i(t)))^{(\gamma_i - r_{i,n_i+1})/\gamma_i}. \tag{3.4.27}$$

By Theorem 1.3.7, there exists a $T_i > 0$ such that $\eta_i(t) = 0$ for all $t \geq T_i$. This is (ii) for $T_\epsilon = \max\{T_1, T_2, \ldots, T_m\}$.

When $a_i > n_i/(1 + n_i)$, let

$$\mathcal{A} = \left\{ x \in \mathbb{R}^{n+1} \middle| V_i(x) \geq \left(\frac{2M_i b_i}{c_i} \epsilon \right)^{\gamma_i/(r_{i,n_i+1}-d_i)} \right\}.$$

For any $\epsilon < \epsilon_i$ and $\eta_i(t) \in \mathcal{A}$, $V_i(\eta_i(t)) \geq \left(\frac{2M_i b_i}{c_i} \epsilon \right)^{\frac{\gamma_i}{r_{i,n_i+1}-d_i}}$, which yields $M_i b_i \epsilon \leq \frac{c_i}{2} (V_i(\eta_i(t)))^{\frac{r_{i,n_i+1}-d_i}{\gamma_i}}$. This together with (3.4.26) leads to

$$
\begin{aligned}
\frac{dV_i(\eta_i(t))}{dt} &\leq -c_i (V_i(\eta_i(t)))^{(r_{i,n_i+1}-d_i)/\gamma_i} \\
&\quad + \frac{c_i}{2} (V_i(\eta_i(t)))^{(r_{i,n_i+1}-d_i)/\gamma_i} (V_i(\eta_i(t)))^{(\gamma_i - r_{i,n_i+1})/\gamma_i} \qquad (3.4.28) \\
&= -\frac{c_i}{2} (v_i(\eta_i(t)))^{(r_{i,n_i+1}-d_i)/\gamma_i} < 0,
\end{aligned}
$$

which shows that there exists a constant $T_i > 0$ such that $\eta_i(t) \in \mathcal{A}^c$ for all $t > T_i$.

Considering $|x_j|$ as the function of $(x_1, x_2, \dots, x_{n_i+1})$, it is easy to verify that $|x_j|$ is an homogeneous function of degree $r_{i,j}$ with respect to weights $\{r_{i,j}\}_{j=1}^{n_i+1}$. By Property 1.3.3, there exists an $L_{i,j} > 0$ such that

$$|x_i| \leq L_{i,j} (V_i(x))^{r_{i,j}/\gamma_i}, \quad \forall x \in \mathbb{R}^{n_i+1}. \qquad (3.4.29)$$

This together with the fact that $\eta_i(t) \in \mathcal{A}^c$ for $t > T_i$ gives

$$|\eta_{i,j}(t)| \leq K_{i,j} |\epsilon|^{r_{i,j}/(r_{i,n_i+1}-d_i)}, \quad t > T_i, \qquad (3.4.30)$$

where $K_{i,j}$, $j = 1, 2, \dots, n_i$, $i = 1, 2, \dots, m$ are positive constants. Claim (ii) then follows from (3.4.19) with $\epsilon_0 = \min\{\epsilon_1, \epsilon_2, \dots, \epsilon_m\}$, $T_\epsilon = \max\{T_1, T_2, \dots, T_m\}$. This completes the proof of the theorem. $\qquad \square$

3.4.2 ESO for Systems with External Disturbance Only

In this section, we construct ESO for MIMO system (3.4.2), in which

$$
\begin{aligned}
f_i(x_{1,1}(t), &\dots, x_{1,n_1}(t), \dots, x_{m,n_m}(t), w_i(t)) \\
&= \tilde{f}_i(x_{1,1}(t), \dots, x_{1,n_1}(t), \dots, x_{m,n_m}(t)) + w_i(t), \qquad (3.4.31)
\end{aligned}
$$

and $\tilde{f}_i(\cdot)$ is known. For this situation, we try to use, as before, the information of $\tilde{f}_i(\cdot)$ as much as possible in designing the ESO, which is composed of following m subsystems to estimate

$x_{i,j}(t)$ and $w_i(t)$:

$$\begin{cases} \dot{\hat{x}}_{i,1}(t) = \hat{x}_{i,2}(t) + \epsilon^{n_i-1}\varphi_{i,1}\left(\dfrac{x_i(t) - \hat{x}_{i,1}(t)}{\epsilon^{n_i}}\right), \\[2mm] \dot{\hat{x}}_{i,2}(t) = \hat{x}_{i,3}(t) + \epsilon^{n_i-2}\varphi_{i,2}\left(\dfrac{x_i(t) - \hat{x}_{i,1}(t)}{\epsilon^{n_i}}\right), \\[1mm] \quad\vdots \\[1mm] \dot{\hat{x}}_{n_i}(t) = \hat{x}_{n_i+1}(t) + \varphi_{i,n_i}\left(\dfrac{x_i(t) - \hat{x}_{i,1}(t)}{\epsilon^{n_i}}\right) + \tilde{f}_i(\hat{x}_{1,1}(t) \ldots, \hat{x}_{1,n_1}(t), \ldots, \hat{x}_{m,n_m}(t)) \\[1mm] \qquad\qquad\qquad\qquad\qquad\qquad\qquad\qquad + g_i(u_1(t), u_2(t), \ldots, u_k(t)), \\[2mm] \dot{\hat{x}}_{n_i+1}(t) = \dfrac{1}{\epsilon}\varphi_{i,n_i+1}\left(\dfrac{x_i(t) - \hat{x}_{i,1}(t)}{\epsilon^{n_i}}\right). \end{cases}$$

(3.4.32)

For the convergence of (3.4.32), we use the following Assumption 3.4.3 instead of Assumption 3.4.1.

Assumption 3.4.3 For each $i \in \{1, 2, \ldots, m\}$, $w_i(t)$ and $\dot{w}_i(t)$ are uniformly bounded in $\mathbb{R}$, and $\tilde{f}_i(\cdot)$ is Lipschitz continuous with Lipschitz constant L_i, that is,

$$|\tilde{f}_i(x_{1,1}, \ldots, x_{i,n_i}, \ldots, x_{m,n_m}) - \tilde{f}_i(y_{1,1} \ldots, y_{1,n_1}, \ldots, y_{m,n_m})| \le L_i\|x - y\|$$

for all $x = (x_{1,1}, \ldots, x_{i,n_i}, \ldots, x_{m,n_m})$ and $y = (y_{1,1} \ldots, y_{1,n_1}, \ldots, y_{m,n_m})$ in $\mathbb{R}^{n_1 + \cdots + n_m}$. Moreover,

$$\left|\frac{\partial V_i(x)}{\partial x_{i,n_i}}\right| \le \rho_i\|x_i\|, \forall\, x_i = (x_{i,1}, x_{i,2}, \ldots, x_{i,n_i}). \tag{3.4.33}$$

where L_i and ρ_i are constants satisfying $\lambda_{i,3} > L_1\rho_1 + \cdots + L_{i-1}\rho_{i-1} + 2L_i\rho_i + L_{i+1}\rho_{i+1} + \cdots + L_m\rho_m$, and $V_i, W_i, \lambda_{i,1}, \lambda_{i,2}$, and $\lambda_{i,3}$ are the constants in Assumption 3.4.2.

Theorem 3.4.3 *Under Assumptions 3.4.2 and 3.4.3, for any given initial values of (3.4.2) and (3.4.32), there exists a constant $\epsilon_0 > 0$, such that for any $\epsilon \in (0, \epsilon_0)$, it holds:*

(i) For any $a > 0$,

$$\lim_{\epsilon \to 0} |x_{i,j}(t) - \hat{x}_{i,j}(t)| = 0 \text{ uniformly for } t \in [a, \infty).$$

(ii) There exists an $\epsilon_0 > 0$ such that for any $\epsilon \in (0, \epsilon_0)$, there exists a $t_\epsilon > 0$ such that

$$|x_{i,j}(t) - \hat{x}_{i,j}(t)| \le K_{ij}\epsilon^{n_i+2-i}, \quad t \in (t_\epsilon, \infty),$$

where $i \in \{1, 2, \ldots, m\}$, $x_{i,n_i+1}(t) = w_i(t)$ is the extended state of (3.4.2), $x_{i,j}(t), j = 1, 2, \ldots, n_i$, are the states of (3.4.2), $\hat{x}_{i,j}(t), j = 1, 2, \ldots, n_i + 1$, are the states of (3.4.32), and K_{ij} is a positive constant independent of ϵ but depending on initial values.

Proof. For any $i \in \{1, 2, \ldots, m\}$, let $x_{i,n_i+1}(t) = w_i(t)$ and

$$\eta_{i,j}(t) = \frac{x_{i,j}(\epsilon t) - \hat{x}_{i,j}(\epsilon t)}{\epsilon^{n_i+1-j}}, \quad j = 1, 2, \ldots, n_i + 1. \tag{3.4.34}$$

A direct computation shows that $\eta_{i,j}(t)$ satisfies the following differential equation:

$$
\begin{cases}
\begin{cases}
\dot{\eta}_{1,1}(t) = \eta_{1,2}(t) - \varphi_{1,1}(\eta_{1,1}(t)), \\
\dot{\eta}_{1,2}(t) = \eta_{1,3}(t) - \varphi_{1,2}(\eta_{1,1}(t)), \\
\quad \vdots \\
\dot{\eta}_{1,n_i}(t) = \eta_{1,n_i+1}(t) - \varphi_{1,n_i}(\eta_{1,1}(t)) + \delta_{1,1}(t), \\
\dot{\eta}_{1,n_i+1}(t) = -\varphi_{1,n_i+1}(\eta_{1,1}(t)) + \epsilon\delta_{1,2}(t),
\end{cases} \\
\quad \vdots \\
\begin{cases}
\dot{\eta}_{m,1}(t) = \eta_{m,2}(t) - \varphi_{m,1}(\eta_{m,1}(t)), \\
\dot{\eta}_{m,2}(t) = \eta_{m,3}(t) - \varphi_{m,2}(\eta_{m,1}(t)), \\
\quad \vdots \\
\dot{\eta}_{m,n_m}(t) = \eta_{m,n_m+1}(t) - \varphi_{m,n_m}(\eta_{m,1}(t)) + \delta_{m,1}(t), \\
\dot{\eta}_{m,n_m+1}(t) = -\varphi_{m,n_m+1}(\eta_{m,1}(t)) + \epsilon\delta_{m,2}(t),
\end{cases}
\end{cases} \tag{3.4.35}
$$

where

$$
\begin{aligned}
\delta_{i,1}(t) &= f_i(x_{1,1}(\epsilon t), \ldots, x_{i,n_i}(\epsilon t), \ldots, x_{m,n_m}(\epsilon t)) \\
&\quad - \tilde{f}_i(\hat{x}_{1,1}(\epsilon t) \ldots, \hat{x}_{1,n_i}(\epsilon t), \ldots, \hat{x}_{m,n_m}(\epsilon t)), \tag{3.4.36} \\
\delta_{i,2}(t) &= \dot{w}_i(t).
\end{aligned}
$$

Set $\eta_i(t) = (\eta_{i,1}(t), \eta_{i,2}(t), \ldots, \eta_{i,n_i+1}(t))^\top, \eta(t) = (\eta_{1,1}(t), \ldots, \eta_{1,n_1+1}(t), \ldots, \eta_{m,n_m+1}(t))^\top,$

$$V(\eta(t)) = V_1(\eta_1(t)) + V_2(\eta_2(t)) + \cdots + V(\eta_m(t)).$$

Finding the derivative of $V(\eta(t))$ along the solution $\eta(t)$ of (3.4.35) with respect to t gives

$$
\begin{aligned}
\left. \frac{dV(\eta(t))}{dt} \right|_{\text{along (3.4.35)}} &= \sum_{i=1}^m \frac{dV_i(\eta_i(t))}{dt} = \sum_{i=1}^m \left(\sum_{j=1}^{n_i} (\eta_{i,j+1}(t) - \varphi_{i,j}(\eta_{i,1}(t)) \frac{\partial V_i(\eta_i(t))}{\partial \eta_{i,j}} \right. \\
&\quad \left. -\varphi_{i,n_i+1}(\eta_{i,1}(t)) \frac{\partial V_i(\eta_i(t))}{\partial \eta_{i,n_i+1}} + \delta_{i,1}(t) \frac{\partial V_i(\eta_i(t))}{\partial \eta_{i,n_i}} + \epsilon\delta_{i,2}(t) \frac{\partial V_i(\eta_i(t))}{\partial \eta_{i,n_i+1}} \right) \\
&\leq \sum_{i=1}^m (-W_i(\eta_i(t)) + L_i\rho_i\|\eta_i(t)\| \|\eta(t)\| + \epsilon M \beta_i \|\eta_i(t)\|) \\
&\leq -\Lambda V(\eta(t)) + \epsilon\Gamma\sqrt{V(\eta(t))},
\end{aligned} \tag{3.4.37}
$$

$\Lambda = \min_{i\in\{1,2,..,m\}}(\lambda_{i,3} - [L_1\rho_1 + \ldots + 2L_i\rho_i + \ldots + L_m\rho_m])/\lambda_{i,2}$, $\Gamma = (M(\beta_1 + \beta_2 + \ldots + \beta_m)\sqrt{\lambda_{i,1}})/\lambda_{i,1}$. It then follows that

$$\frac{d}{dt}\sqrt{V(\eta(t))} = \frac{1}{2V(\eta(t))}\frac{dV(\eta(t))}{dt} \leq -\frac{\Lambda}{2}\sqrt{V(\eta(t))} + \frac{\epsilon\Gamma}{2}. \tag{3.4.38}$$

This together with Assumption 3.4.3 gives, for each $i \in \{1, 2, \ldots, m\}$, that

$$\|\eta_i(t)\| \leq \sqrt{\frac{V_i(\eta_i(t))}{\lambda_{i,1}}} \leq \sqrt{\frac{V(\eta(t))}{\lambda_{i,1}}} \tag{3.4.39}$$

$$\leq \frac{\sqrt{\lambda_{i,1}}}{\lambda_{i,1}}\left(\sqrt{V(\eta(0))}e^{-\Lambda/2t} + \epsilon\Gamma\int_0^t e^{-\Lambda/2(t-s)}\right).$$

By (3.4.34), we obtain, for each $i \in \{1, 2, \ldots, m\}$, $j \in \{1, 2, \ldots, n_i + 1\}$, that

$$|x_{i,j}(t) - \hat{x}_{i,j}(t)| = \epsilon^{n_i+1-j}\left|\eta_{i,j}\left(\frac{t}{\epsilon}\right)\right|$$

$$\leq \epsilon^{n_i+1-j}\frac{\sqrt{\lambda_{i,1}}}{\lambda_{i,1}}\left(\sqrt{V(\eta(0))}e^{-\Lambda/(2\epsilon)t} + \epsilon\Gamma\int_0^{\frac{t}{\epsilon}}e^{-\Lambda/2(\frac{t}{\epsilon}-s)}\right) \tag{3.4.40}$$

$$\to 0 \text{ uniformly for } t \in [a, \infty) \text{ as } \epsilon \to 0.$$

This is Claim (i). Claim (ii) also follows from (3.4.40). This completes the proof of the theorem. $\qquad\square$

3.4.3   Examples and Numerical Simulations

Example 3.4.1   *Consider the following MIMO system:*

$$\begin{cases} \dot{x}_{11}(t) = x_{12}(t), \\ \dot{x}_{12}(t) = -\dfrac{1}{48}(x_{11}(t) + x_{12}(t)) - \dfrac{1}{100}(x_{21}(t) + \sin(x_{22}(t))) + w_1(t) + u_1(t), \\ \dot{x}_{21}(t) = x_{22}(t), \\ \dot{x}_{22}(t) = -\dfrac{1}{100}(x_{11}(t) + \cos(x_{12}(t))) - \dfrac{1}{48}(x_{21}(t) + x_{22}(t)) + w_2(t) + u_2(t), \\ y_1(t) = x_1(t), y_2(t) = x_2(t), \end{cases} \tag{3.4.41}$$

where $y_1(t)$ and $y_2(t)$ are the the the outputs, $u_1(t)$ and $u_2(t)$ are the inputs, and $w_1(t)$ and $w_2(t)$ are the external disturbances. Compute the eigenvalues of the matrix below that is the linear part matrix of system (3.4.41):

$$A = \begin{pmatrix} 0 & 1 & 0 & 0 \\ -\dfrac{1}{48} & -\dfrac{1}{48} & -\dfrac{1}{100} & 0 \\ 0 & 0 & 0 & 1 \\ -\dfrac{1}{100} & 0 & -\dfrac{1}{48} & -\dfrac{1}{48} \end{pmatrix}$$

to obtain $\{-0.0104 + 0.1753i, -0.0104 - 0.1753i, -0.0104 + 0.1036i, -0.0104 - 0.1036i\}$. *Therefore, the real parts of all eigenvalues are negative, and hence the solution of (3.4.41) is bounded for bounded* $w_i(t)$ *and* $u_i(t)$, $i = 1, 2$.

Design linear ESO (3.4.42) and nonlinear ESO (3.4.43) according to Corollary 3.4.1 and Theorem 3.4.2 as follows, respectively:

$$
\begin{cases}
\dot{\hat{x}}_{11}(t) = \hat{x}_{12}(t) + \dfrac{6}{\epsilon}(x_{11}(t) - \hat{x}_{11}(t)), \\[2mm]
\dot{\hat{x}}_{12}(t) = \hat{x}_{13}(t) + \dfrac{11}{\epsilon^2}(x_{11}(t) - \hat{x}_{11}(t)) + u_1(t), \\[2mm]
\dot{\hat{x}}_{13}(t) = \dfrac{6}{\epsilon^3}(x_{11}(t) - \hat{x}_{11}(t)), \\[2mm]
\dot{\hat{x}}_{21}(t) = \hat{x}_{22}(t) + \dfrac{6}{\epsilon}(x_{21}(t) - \hat{x}_{21}(t)), \\[2mm]
\dot{\hat{x}}_{22}(t) = \hat{x}_{13}(t) + \dfrac{11}{\epsilon^2}(x_{21}(t) - \hat{x}_{21}(t)) + u_2(t), \\[2mm]
\dot{\hat{x}}_{23}(t) = \dfrac{6}{\epsilon^3}(x_{21}(t) - \hat{x}_{21}(t)),
\end{cases}
\tag{3.4.42}
$$

$$
\begin{cases}
\dot{\hat{x}}_{11}(t) = \hat{x}_{12}(t) + 6\epsilon\left[\dfrac{x_{11}(t) - \hat{x}_{11}(t)}{\epsilon^2}\right]^{a_1}, \\[3mm]
\dot{\hat{x}}_{12}(t) = \hat{x}_{13}(t) + 11\left[\dfrac{x_{11}(t) - \hat{x}_{11}(t)}{\epsilon^2}\right]^{2a_1-1} + u_1(t), \\[3mm]
\dot{\hat{x}}_{13}(t) = \dfrac{6}{\epsilon}\left[\dfrac{x_{11}(t) - \hat{x}_{11}(t)}{\epsilon^2}\right]^{3a_1-2}, \\[3mm]
\dot{\hat{x}}_{21}(t) = \hat{x}_{22}(t) + 6\epsilon\left[\dfrac{x_{21}(t) - \hat{x}_{21}(t)}{\epsilon^2}\right]^{a_2}, \\[3mm]
\dot{\hat{x}}_{22}(t) = \hat{x}_{13}(t) + 11\left[\dfrac{x_{21}(t) - \hat{x}_{21}(t)}{\epsilon^2}\right]^{2a_2-1} + u_2(t), \\[3mm]
\dot{\hat{x}}_{23}(t) = \dfrac{6}{\epsilon}\left[\dfrac{x_{21}(t) - \hat{x}_{21}(t)}{\epsilon^2}\right]^{3a_2-2}.
\end{cases}
\tag{3.4.43}
$$

The corresponding matrices E_1 and E_2 in Corollary 3.4.1 and Theorem 3.4.2 for linear ESO (3.4.42) and nonlinear ESO (3.4.43) are given as

$$
E_1 = E_2 = \begin{pmatrix} -6 & 1 & 0 \\ -11 & 0 & 1 \\ -6 & 0 & 0 \end{pmatrix},
\tag{3.4.44}
$$

which have eigenvalues $\{-1, -2, -3\}$, so E_1 and E_2 are Hurwitz. Therefore, all conditions of Corollary 3.4.1 and Theorem 3.4.2 are satisfied.

Set

$$
\begin{cases}
w_1(t) = \cos t, \ u_1(t) = \sin t, \ w_2(t) = 2\cos t, \ u_2(t) = 2\sin t, \ \epsilon = 0.05, \ a_1 = a_2 = 0.8, \\
h = 0.001, \ x_{11}(0) = x_{12}(0) = x_{21}(0) = x_{22}(0) = 1, \\
\hat{x}_{11}(0) = \hat{x}_{12}(0) = \hat{x}_{13}(0) = \hat{x}_{21}(0) = \hat{x}_{22}(0) = \hat{x}_{23}(0) = 0.
\end{cases}
\tag{3.4.45}
$$

The numerical results of (3.4.42) for (3.4.41) are plotted in Figure 3.4.1. The numerical results of (3.4.43) for (3.4.41) are plotted in Figure 3.4.2.

In both figures, states $x_{ij}(t)(i = 1, 2, j = 1, 2)$ from (3.4.41) and their estimated values $\hat{x}_{ij}(t)$ from (3.4.42) or (3.4.43), together with their errors in a sub-figure are plotted, where we use $x_{13}(t)$ and $x_{23}(t)$ to denote the total disturbance $-[x_{11}(t) + x_{12}(t)]/48 - [x_{21}(t) + \sin(x_{22}(t))]/100 + w_1(t)$ and $-[x_{11}(t) + \cos(x_{12}(t))]/100 - [x_{21}(t) + x_{22}(t)]/48 + w_2(t)$ in (3.4.41), $\hat{x}_{13}(t)$ and $\hat{x}_{23}(t)$ from (3.4.42) or (3.4.43) are their estimated values, respectively.

It is seen from Figures 3.4.1 and 3.4.2 that both linear ESO (3.4.42) and nonlinear ESO (3.4.43) are convergent quite satisfactorily. From sub-figures 3.4.1(b), 3.4.1(c), 3.4.1(e), 3.4.1(f) and 3.4.2(b), 3.4.2(c), 3.4.2(e), 3.4.2(f), the nonlinear ESO (3.4.43) takes advantage of much smaller peaking values compared with the linear ESO (3.4.42) .

In addition, we use (3.4.41) to illustrate the results of Theorem 3.4.3 where the unknown parts are the external disturbances only. According to Theorem 3.4.3, the ESO now takes the form:

$$
\begin{cases}
\dot{\hat{x}}_{11}(t) = \hat{x}_{12}(t) + \dfrac{6}{\epsilon}(x_{11}(t) - \hat{x}_{11}(t)), \\[2mm]
\dot{\hat{x}}_{12}(t) = \hat{x}_{13}(t) + \dfrac{11}{\epsilon^2}(x_{11}(t) - \hat{x}_{11}(t)) - \dfrac{1}{48}[\hat{x}_{11}(t) + \hat{x}_{12}(t)] - \dfrac{1}{100}[\hat{x}_{21}(t) + \sin(\hat{x}_{22}(t))] \\[2mm]
\qquad + u_1(t), \\[2mm]
\dot{\hat{x}}_{13}(t) = \dfrac{6}{\epsilon^3}(x_{11}(t) - \hat{x}_{11}(t)), \\[2mm]
\dot{\hat{x}}_{21}(t) = \hat{x}_{22}(t) + \dfrac{6}{\epsilon}(x_{21}(t) - \hat{x}_{21}(t)), \\[2mm]
\dot{\hat{x}}_{22}(t) = \hat{x}_{13}(t) + \dfrac{11}{\epsilon^2}(x_{21}(t) - \hat{x}_{21}(t)) - \dfrac{1}{100}[\hat{x}_{11}(t) + \cos(\hat{x}_{12}(t))] - \dfrac{1}{48}[\hat{x}_{21}(t) + \hat{x}_{22}(t)] \\[2mm]
\qquad + u_2(t), \\[2mm]
\dot{\hat{x}}_{23}(t) = \dfrac{6}{\epsilon^3}(x_{21}(t) - \hat{x}_{21}(t)),
\end{cases}
$$

$$(3.4.46)$$

which is used to estimate external disturbances $w_1(t)$ and $w_2(t)$. Since $f_1(x(t)) = -[x_{11}(t) + x_{12}(t)]/48 - [x_{21}(t) + \cos(x_{22}(t))]/100$ and $f_2(x(t)) = -[x_{11}(t) + \cos(x_{12}(t))]/100 - [x_{21}(t) + x_{22}(t)]/48$ are Lipschitz continuous with the Lipschitz constant $1/48$. Let Lyapunov functions $V_1, V_2 : \mathbb{R}^3 \to \mathbb{R}$ be defined by $V_i(y) = \langle P_i y, y \rangle$. Let $W_i : \mathbb{R}^3 \to \mathbb{R}$ be defined by $W_i(y) = \langle y, y \rangle$, where P_i is the solution of the Lyapunov equation $P_i E_i + E_i^\top P_i = -I_{3 \times 3}$,

$$
P_1 = P_2 = \begin{pmatrix} 1.7000 & -0.5000 & -0.7000 \\ -0.5000 & 0.7000 & -0.5000 \\ -0.7000 & -0.5000 & 1.5333 \end{pmatrix}, \tag{3.4.47}
$$

and $\lambda_{\max}(P_i) = 2.3230, \lambda_{\min}(P_i) = 0.0917$. It is easy to verify that Assumptions 3.4.2 and 3.4.3 are satisfied with parameters $\lambda_{i1} = 0.0917, \lambda_{i2} = 2.3230, \lambda_{i3} = \lambda_{i4} = 1, \rho_i = 4.646, L_i = 1/48$. We use the same values of (3.4.33) to plot the numerical results of (3.4.46) in Figure 3.4.3.

Now in Figure 3.4.3, the states $x_{ij}(t)(i, j = 1, 2)$ from (3.4.41) and their estimated values $\hat{x}_{ij}(t)$ from (3.4.46) together with their errors are plotted in a sub-figure in Figure 3.4.3, where

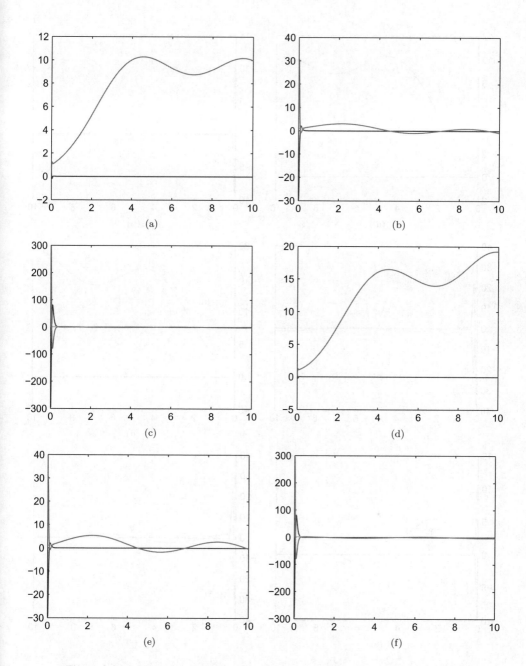

Figure 3.4.1 ESO (3.4.42) for (3.4.41) to estimate the states and total disturbances.

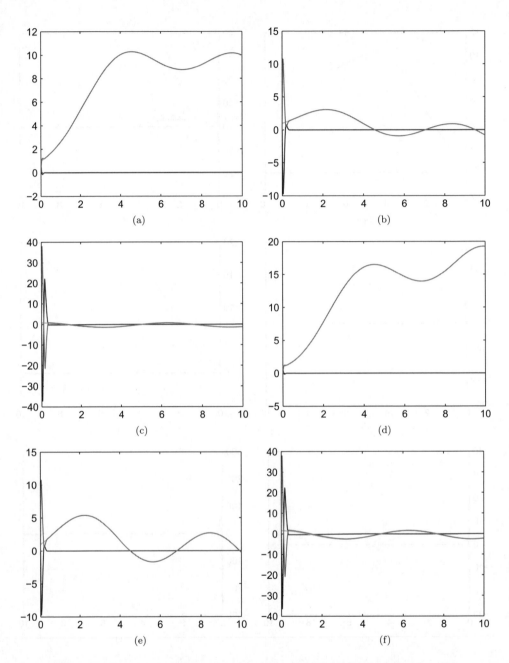

Figure 3.4.2   ESO (3.4.43) for (3.4.41) to estimate the states and total disturbances.

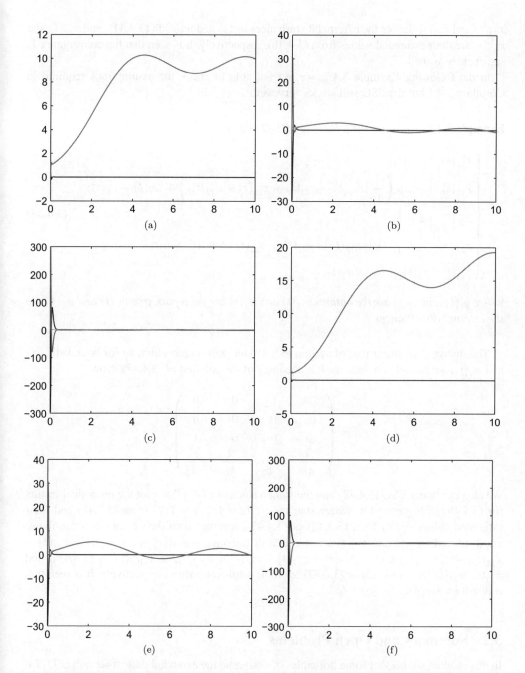

Figure 3.4.3 Linear ESO (3.4.46) for (3.4.41) to estimate the states and external disturbance.

$x_{13}(t)$ and $x_{23}(t)$ denote the external disturbances $w_1(t)$ and $w_2(t)$ in (3.4.41), and $\hat{x}_{13}(t)$ and $\hat{x}_{23}(t)$ are their estimated values from (3.4.46), respectively. It is seen that the convergence is satisfactory as well.

In the following Example 3.4.2, we are not able to check the assumptions required in Corollary 3.4.1 but the ESO still works very well.

Example 3.4.2 *Consider the following MIMO systems:*

$$
\begin{cases}
\dot{x}_{11}(t) = x_{12}(t), \\
\dot{x}_{12}(t) = -\sin\left(\dfrac{1}{48}[x_{11}(t) + x_{12}(t) + x_{21}(t) + x_{22}(t)]\right) + w_1(t) + u_1(t), \\
\dot{x}_{21}(t) = x_{22}(t), \\
\dot{x}_{22} = -\dfrac{1}{48}[x_{11}(t) + x_{12}(t) + x_{21}(t) + x_{22}(t)] + w_2(t) + u_2(t), \\
y_1(t) = x_1(t), y_2(t) = x_2(t),
\end{cases}
\tag{3.4.48}
$$

where $y_1(t)$ and $y_2(t)$ are the outputs, $u_1(t)$ and $u_2(t)$ are the inputs, and $w_1(t)$ and $w_2(t)$ are the external disturbances.

The matrix A of linear part of system (3.4.48) has zero eigenvalues, so for bounded $u_i(t)$ and $w_i(t)$, we cannot conclude the boundedness of the solution of (3.4.48). Now,

$$
A = \begin{pmatrix}
0 & 1 & 0 & 0 \\
0 & 0 & 0 & 0 \\
0 & 0 & 0 & 1 \\
-\dfrac{1}{48} & -\dfrac{1}{48} & -\dfrac{1}{48} & -\dfrac{1}{48}
\end{pmatrix}.
$$

We also use linear ESO (3.4.42) and the same values of (3.4.33) to plot the numerical results for (3.4.48) in Figure 3.4.4, where states $x_{ij}(t)(i = 1, 2, j = 1, 2)$ from (3.4.48) and their estimated values $\hat{x}_{ij}(t)$ from (3.4.42) or (3.4.43), together with their errors in a sub-figure are displayed, $x_{13}(t)$ and $x_{23}(t)$ are the total disturbance $-\sin([x_{11}(t) + x_{12}(t) + x_{21}(t) + x_{22}(t)]/48) + w_1(t)$ and $-[x_{11}(t) + x_{12}(t) + x_{21}(t) + x_{22}(t)]/48 + w_2(t)$ in (3.4.41), and $\hat{x}_{13}(t), \hat{x}_{23}(t)$ from (3.4.42) or (3.4.43) are their estimated values, respectively. It is seen that it is still convergent.

3.5 Summary and Open Problems

In this chapter, we present some principles of designing the extended state observer (ESO) for a class of nonlinear systems with vast uncertainty, that is, the total disturbance that comes from the unmodeled system dynamics, external disturbance, and deviation of the control coefficient away from its nominal value. The ESO is designed not only for estimating the system state but also the total disturbance by the output. In Section 3.1 we analyze the linear ESO for an SISO system with vast uncertainty. The design principle is given and a convergence proof is

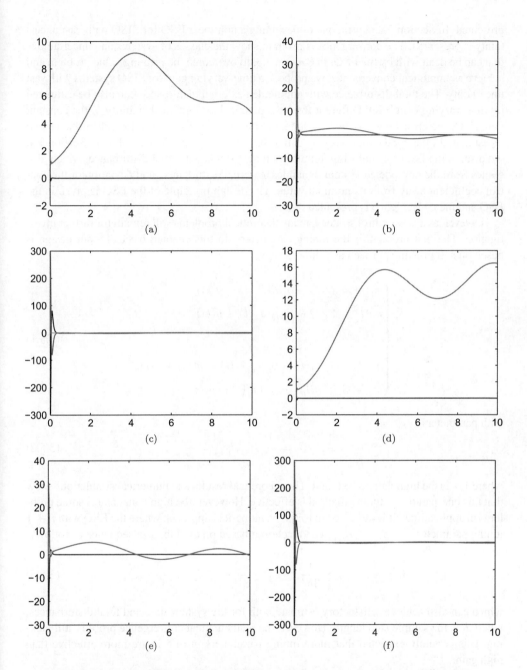

Figure 3.4.4   Linear ESO (3.4.42) for (3.4.48) to estimate the states and external disturbance.

presented. In Section 3.2, a principle of designing a nonlinear ESO for SISO with vast uncertainty is presented. The convergence is given to show the classes of systems and uncertainties that can be dealt with by the ESO. In Section 3.3, to overcome the peaking value problem and achieve asymptotical convergence, we propose a time-varying gain for SISO systems with vast uncertainty. The total disturbance with unbounded external disturbance can also be estimated by time-varying gain ESO. Different from the practical asymptotical stability of the constant gain ESO, the error between the real state and the estimated state, the total disturbance and its estimated value by the time-varying gain ESO can be asymptotically stable. In Section 3.4, we discuss the ESO for multi-input–multi-output systems with total disturbance, which also comes from the unmodeled system dynamics, external disturbance, and deviation of the control coefficient away from its nominal value. The design principle of the ESO is given in the ESO and the convergence is presented.

However, as a new control strategy, many theoretical problems still remain for further investigation. The first problem is the tuning parameters. In this chapter, basically, our design is based on a high-gain approach as follows:

$$
\begin{cases}
\dot{\hat{x}}_1(t) = \hat{x}_2(t) - \beta_1 g_1(\hat{x}_1(t) - y(t)), \\
\dot{\hat{x}}_2(t) = \hat{x}_3(t) - \beta_2 g_2(\hat{x}_1(t) - y(t)), \\
\quad\vdots \\
\dot{\hat{x}}_n(t) = \hat{x}_{n+1}(t) - \beta_n g_n(\hat{x}_1(t) - y(t)) + u(t), \\
\dot{\hat{x}}_{n+1}(t) = -\beta_{n+1} g_{n+1}(\hat{x}_1(t) - y(t)),
\end{cases}
\tag{3.5.1}
$$

with parameters

$$
\beta_1 = \frac{1}{\epsilon}, \quad \beta_2 = \frac{1}{\epsilon^2}, \ldots, \beta_{n+1} = \frac{1}{\epsilon^{n+1}},
\tag{3.5.2}
$$

where $1/\epsilon$ is the high gain to be tuned. The theoretical results and numerical simulations show that this one-parameter tuning method is effective. However, the high gain creates some problems in applications. It is indicated in [62] by a numerical approach where the ESO with $n + 1$ tuning parameters $\beta_1, \beta_2, \ldots, \beta_{n+1}$ can be chosen based on the Fibnacci sequence as follows:

$$
\beta_1 = \frac{1}{h}, \quad \beta_2 = \frac{1}{3h^2}, \quad \beta_3 = \frac{2}{8^2 h^3}, \quad \beta_4 = \frac{5}{13^3 h^4}, \ldots,
\tag{3.5.3}
$$

which can also achieve satisfactory estimate both for the system state and total disturbances, where h is the sample step length. However, the mathematical convergence proof is still lacking. It is certainly expected that more tuning parameters in (3.5.1) are more effective than high gain.

In [62], the numerical results show that the nonlinear ESO

$$
\begin{cases}
\dot{\hat{x}}_1(t) = \hat{x}_2(t) - \beta_1(\hat{x}_1(t) - x_1(t)), \\
\dot{\hat{x}}_2(t) = \hat{x}_3(t) - \beta_2 \, \mathrm{fal}(\hat{x}_1(t) - x_1(t), 1/2, \delta), \\
\dot{\hat{x}}_3(t) = -\beta_3 \, \mathrm{fal}(\hat{x}_1(t) - x_1(t), 1/4, \delta),
\end{cases}
\tag{3.5.4}
$$

can also recover the state and total disturbance for the second-order control systems with vast uncertainty, where

$$\mathrm{fal}(e, \alpha, \delta) = \begin{cases} \dfrac{e}{\delta^{\alpha-1}}, & |e| \le \delta, \\ |e|^{\alpha} \, \mathrm{sign}(e), & |e| > \delta, \end{cases} \qquad \delta > 0, \ , \alpha > 0, \qquad (3.5.5)$$

with $\beta_{1,2,3}$ being chosen as (3.5.3). However, the convergence of this special nonlinear ESO is still open.

3.6 Remarks and Bibliographical Notes

For many other designs of the nonlinear state observer, we refer to [22, 23, 40, 85, 90, 118, 132, 133], and references therein and the recent book [14]. Particular attention is paid to [71] where a comparison of the approach discussed in this work was made with other existing approaches and presented in detail.

The extended state observer (3.0.1) was first proposed in [60].

The idea of the ESO can also be found in the high-gain observer with extended state variable used in [35] (see also an expository survey in [83]). The numerical studies (e.g., [60]) and many other studies over the following years have shown that for some nonlinear functions $g_i(\cdot)$ and parameters α_i, the observer (3.0.1) performs very satisfactorily with regard to adaptability, robustness, and anti-chattering. For other perspectives on this remarkable type of research, we refer the reader to [63].

For engineering applications, we refer to [106, 68, 161, 162, 163, 164, 165]. The high-gain ESO (3.1.2) was first proposed in [37].

The LESO (3.1.2) is essentially similar to the "extended high-gain observer" used in [35] to stabilize a slightly more complex system than (3.0.2). The convergence of the linear high-gain observer for determined systems without an extended variable state is studied in [82] and Theorem 6 of [14] on page 18 (see also Chapter 14 of [84] on page 610). A good survey of a high-gain observer with or without an extended state variable can be found in [83].

For LESO (3.1.2), [160] presents some convergence results where either $f(t, \ldots) + w(t)$ is completely known or $f(t, \cdot) + w(t)$ is completely unknown. The convergence results for some special linear ESO are also available in [150].

Section 3.2 This section is taken from paper [54] and is reproduced by permission of the Elsevier license. If no extended state variable $x_{n+1}(t)$ is taken into account and the NLESO observer is taken as LESO, Theorem 3.2.3 is just a special case of Theorem 6 of [14] on page 18.

Section 3.3 System (3.3.1) under (3.3.34) is a special case of a system considered in [8]. In addition, if $\overline{f} \equiv 0$, it is also a special case considered in [116]. In [8] the nonlinear function $f(t, \cdot)$ in control channel satisfies a super-linear like condition for $x_i(t), i = 2, 3, \ldots, n$, and in [116], it satisfies global Lipschitz continuous. In our Theorem 3.3.3, it can be sub-linear when $0 < \beta < 1$, which does not satisfy the corresponding conditions of [8] or [116]. Moreover, in [8, 116] only the state is estimated, and there is no disturbance estimation.

We remark that the time-varying gain (updated gain or dynamic gain) can also be used to deal with the nonlinearity of some known functions of the system. Examples can be found in [8, 89, 110]. The disturbance attenuation is considered in [8] where a sufficiently large positive constant C is given, for which the initial value of gain satisfies $r(0) \geq C$, and the error from the state and the corresponding observer are estimated by this constant C. The larger C is, the smaller is the error. However, in these works, no estimation is made for the disturbance.

The first condition of Assumptions 3.3.5 and 3.3.4 are the same, which is used to guarantee the global asymptotic stability of the following system:

$$\begin{cases} \dot{x}_1(t) = x_2(t) - g_1(x_1(t)), \\ \quad \vdots \\ \dot{x}_n(t) = x_{n+1}(t) - g_n(x_1(t)), \\ \dot{x}_{n+1}(t) = -g_{n+1}(x_1(t)). \end{cases} \tag{3.6.1}$$

The second condition of Assumption 3.3.5 is less restrictive than that in Assumption 3.3.4.
Section 3.4 This part is taken from [56] and is reproduced by permission of the Institution of Engineering and Technology.
Section 3.4.1 The requirement for the boundedness of $\dot{w}_i(t)$ only arises from the fact that we want to estimate the extended state. Otherwise, this requirement can be removed. A typical external disturbance of finite sum of sinusoidal $w_i(t) = \sum a_{ij} \sin \omega_{ij} t$ satisfies this assumption.

4

The Active Disturbance Rejection Control

Let us first briefly recall the main idea of the active disturbance rejection control (ADRC) by investigating SISO systems with vast uncertainty. For an n-dimensional SISO nonlinear system

$$
\begin{cases}
x^{(n)}(t) = f(x(t), \dot{x}(t), \ldots, x^{(n-1)}(t), w(t)) + bu(t), \\
y(t) = x(t),
\end{cases}
$$

which can be written as

$$
\begin{cases}
\dot{x}_1(t) = x_2(t), \\
\dot{x}_2(t) = x_3(t), \\
\quad \vdots \\
\dot{x}_n(t) = f(x_1(t), \ldots, x_n(t), w(t)) + bu(t), \\
y(t) = x_1(t),
\end{cases}
\tag{4.0.1}
$$

where $y(t)$ is the output (observation), $u(t)$ is the input (control), $w(t)$ is the external disturbance, $f(\cdot)$ represents the nonlinear dynamic function of the plant, which is possibly unknown, and $b > 0$ is a parameter that is also unknown in some circumstances. The objective of the ADRC is to design an extended state observer (ESO)-based output feedback control so that the output $y(t)$ tracks a given reference signal $v(t)$, and at the same time $x_i(t)$ tracks $v^{(i-1)}(t)$ for every $i = 2, 3, \ldots, n$, provided that the latter exist in some sense.

We shall see later that this general formulation covers not only the special output regulation problem but also the output feedback stabilization by setting $v(t) \equiv 0$.

Active Disturbance Rejection Control for Nonlinear Systems: An Introduction, First Edition.
Bao-Zhu Guo and Zhi-Liang Zhao.
© 2016 John Wiley & Sons Singapore Pte. Ltd. Published 2016 by John Wiley & Sons, Ltd.

To begin with, suppose we are given a reference system:

$$
\begin{cases}
\dot{x}_1^*(t) = x_2^*(t), \\
\dot{x}_2^*(t) = x_3^*(t), \\
\quad \vdots \\
\dot{x}_n^*(t) = \varphi(x_1^*(t), \ldots, x_n^*(t)), \varphi(0, 0, \ldots, 0) = 0.
\end{cases}
\tag{4.0.2}
$$

Assume throughout this chapter that the system (4.0.2) is globally asymptotically stable. Specifically, it satisfies two conditions: (a) the state $x_i^*(t)$ is stable in the sense of Definition 1.3.4; (b) the zero state of (4.0.2) is globally attractive in the sense of Definition 1.3.5.

The ADRC is composed of three components. The first part is to recover all $v^{(i-1)}(t), i = 2, \ldots, n+1$, through the reference $v(t)$ itself. This is realized by tracking differentiator (TD) (2.2.21), which is rewritten as

$$
\text{TD:}
\begin{cases}
\dot{z}_1(t) = z_2(t), \\
\dot{z}_2(t) = z_3(t), \\
\quad \vdots \\
\dot{z}_n(t) = z_{n+1}(t), \\
\dot{z}_{n+1}(t) = R^n \psi \left(z_1(t) - v(t), \dfrac{z_2(t)}{R}, \ldots, \dfrac{z_{n+1}(t)}{R^n} \right), \psi(0, 0, \ldots, 0) = 0.
\end{cases}
\tag{4.0.3}
$$

The control objective of ADRC is to make $x_i(t) - z_i(t) \approx x_i^*(t)$. That is, $x_i(t)$ converges to $z_i(t)$ or $v^{(i-1)}(t)$ in the desirable way of reference system $x_i^*(t)$ converging to zero.

The second component of ADRC is the extended state observer (ESO) (3.2.42) for system (4.0.1), which is written here as

$$
\text{ESO:}
\begin{cases}
\dot{\hat{x}}_1(t) = \hat{x}_2(t) + \varepsilon^{n-1} g_1(\theta(t)), \\
\dot{\hat{x}}_2(t) = \hat{x}_3(t) + \varepsilon^{n-2} g_2(\theta(t)), \\
\quad \vdots \\
\dot{\hat{x}}_n(t) = \hat{x}_{n+1}(t) + g_n(\theta(t)) + b_0 u(t), \\
\dot{\hat{x}}_{n+1}(t) = \dfrac{1}{\varepsilon} g_{n+1}(\theta(t)),
\end{cases}
\tag{4.0.4}
$$

where $\theta(t) = (y(t) - \hat{x}_1(t))/\varepsilon^n$. The ESO is used to estimate, in real time, both the states $x_i(t) (i = 1, 2, \ldots, n)$ of (4.0.1) and the extended state $x_{n+1}(t) = f(\cdot) + (b - b_0)u(t)$, where $b_0 > 0$ is a nominal parameter of b and $\varepsilon > 0$ is the regulatable high gain. It is pointed out that it is this remarkable component in which the ADRC is rooted.

The convergence of ESO is presented in Theorem 3.2.1 where it is shown that (4.0.4) is convergent in the sense that $\hat{x}_i(t) \to x_i(t)$ as $\varepsilon \to 0$ and $t \to \infty$ for all $i = 1, 2, \ldots, n+1$.

The third and the last link of ADRC is to design an ESO-based output feedback control:

$$
\text{ADRC:} \quad u(t) = \frac{1}{b_0}[\varphi(\hat{x}(t) - z(t)) + z_{n+1}(t) - \hat{x}_{n+1}(t)],
\tag{4.0.5}
$$

where $(\hat{x}(t) = (\hat{x}_1(t), \hat{x}_2(t), \ldots, \hat{x}_n(t)), \hat{x}_{n+1}(t))$ is the solution of (4.0.4) and $(z(t) = (z_1(t), z_2(t), \ldots, z_n(t)), z_{n+1}(t))$ is the solution of (4.0.3).

It is observed from the expression of feedback control (4.0.5) that the third term $\hat{x}_{n+1}(t)$ is used to compensate (cancel), in real time, the effect of the total disturbance $x_{n+1}(t) = f(\cdot) + (b - b_0)u(t)$, and the first two terms of $\varphi(\hat{x}(t) - z(t)) + z_{n+1}(t)$ are used, after recovering the differentials $v^{(i-1)}(t)$ by $z(t)$, to achieve the reference stable system (4.0.2) with difference $x_i(t) - z_i(t) \approx x_i^*(t)$. From these observations, we see that the ADRC, unlike the traditional design, adopts an entirely new strategy in dealing with the uncertainty.

Definition 4.0.1 *Let $x_i(t)(1 \leq i \leq n)$ and $\hat{x}_i(t)(1 \leq i \leq n + 1)$ be the solutions of the closed-loop system (4.0.1) under the feedback (4.0.5) with ESO (4.0.4), coupling TD (4.0.3), and reference system (4.0.2). Let $x_{n+1}(t) = f(\cdot) + (b - b_0)u(t)$ be the extended state variable. We say that the ADRC is convergent if, for any given initial values of (4.0.1), (4.0.3), and (4.0.4), there exists a constant $R_0 > 0$ such that for any $R > R_0$,*

$$\lim_{\substack{t \to \infty \\ \varepsilon \to 0}} [x_i(t) - \hat{x}_i(t)] = 0, 1 \leq i \leq n+1, \lim_{\substack{t \to \infty \\ \varepsilon \to 0}} [x_i(t) - z_i(t)] = 0, 1 \leq i \leq n. \qquad (4.0.6)$$

Moreover, for any given $a > 0$, $\lim_{R \to \infty} |z_1(t) - v(t)| = 0$ uniformly for $t \in [a, \infty)$.

The remainder of this chapter is organized as follows. In Section 4.1, we discuss the linear ADRC, which is simple to design for control practices. In Section 4.2 we present a nonlinear ADRC for SISO systems. Section 4.3 focuses on the time-varying gain ADRC for SISO systems, and Section 4.4 is the ADRC for MIMO systems. In Sections 4.5 and 4.6, we compare the ADRC with the internal model principle and high-gain control, respectively. Finally, as an example of distributed control systems with disturbance, we apply ADRC to boundary stabilization for wave equations with in-domain anti-damping and boundary disturbances.

4.1   Linear ADRC for SISO Systems

In this section, we demonstrate a simple ADRC for SISO systems with vast uncertainty, where the ESOs are linear ones and the functions in the feedback control design systems are linear functions. Two design methods, namely the global design and semi-global design in feedback control, are discussed, respectively.

4.1.1   Global Convergence of Linear ADRC for SISO Systems

In this section, we consider output regulation for system (4.0.1). Firstly, the nonlinear function $f(\cdot)$ and the external disturbance $w(t)$ are assumed to satisfy Assumption 4.1.1.

Assumption 4.1.1 The $f \in C^1(\mathbb{R}^{n+1}, \mathbb{R})$, $w \in C^1([0, \infty), \mathbb{R})$; both $w(t)$ and $\dot{w}(t)$ are bounded on $[0, \infty)$; and all partial derivatives of $f(\cdot)$ with respect to its independent variables are bounded over $\mathbb{R}^{n+1}$.

As indicated in the beginning of this chapter, the linear ADRC, which is based on linear TD and the linear ESO feedback control, can be designed as follows:

$$u(t) = \frac{1}{b_0}[\alpha_1(\hat{x}_1(t) - z_1(t)) + \cdots + \alpha_n(\hat{x}_n(t) - z_n(t)) + z_{n+1}(t) - \hat{x}_{n+1}(t)], \quad (4.1.1)$$

where $\hat{x}_i(t), i = 1, 2, \ldots, n$ are used to estimate the system states $x_i(t), i = 1, 2, \ldots, n$, and $\hat{x}_{n+1}(t)$ is used to estimate the total disturbance $x_{n+1}(t)$. All $\hat{x}_i(t), i = 1, 2, \ldots, n+1$, are coming from the following LESO:

$$\text{LESO:} \begin{cases} \dot{\hat{x}}_1(t) = \hat{x}_2(t) + \dfrac{k_1}{\varepsilon}(y(t) - \hat{x}_1(t)), \\[2mm] \dot{\hat{x}}_2(t) = \hat{x}_3(t) + \dfrac{k_2}{\varepsilon^2}(y(t) - \hat{x}_1(t)), \\[2mm] \quad\vdots \\[2mm] \dot{\hat{x}}_n(t) = \hat{x}_{n+1}(t) + \dfrac{k_n}{\varepsilon^n}(y(t) - \hat{x}_1(t)) + b_0 u(t), \\[2mm] \dot{\hat{x}}_{n+1}(t) = \dfrac{k_{n+1}}{\varepsilon^{n+1}}(y(t) - \hat{x}_1(t)). \end{cases} \tag{4.1.2}$$

Also the $z_i(t), i = 1, 2, \ldots, n+1$, are coming from the linear tracking differentiator as follows:

$$\text{LTD:} \begin{cases} \dot{z}_1(t) = z_2(t), \\[2mm] \dot{z}_2(t) = z_3(t), \\[2mm] \quad\vdots \\[2mm] \dot{z}_n(t) = z_n(t), \\[2mm] \dot{z}_{(n+1)}(t) = R^{n+1}\left(a_1(z_1(t) - v(t)) + \dfrac{a_2 z_2(t)}{R} + \cdots + \dfrac{a_{n+1} z_n(t)}{R^n}\right), \end{cases} \tag{4.1.3}$$

to estimate the derivatives $\dot{v}(t), \ldots, v^{(n)}(t)$, respectively. If we replace $x_i(t)$ by $\hat{x}_i(t)$ and $v^{(i-1)}(t)$ by $z_i(t), i = 1, 2, \ldots, n+1$, then, under the control $u(t)$, system (3.1.1) becomes

$$\begin{cases} \dot{x}_1(t) = x_2(t), \\[2mm] \dot{x}_2(t) = x_3(t), \\[2mm] \quad\vdots \\[2mm] \dot{x}_n(t) = \alpha_1(x_1(t) - v(t)) + \cdots + \alpha_n(x_n(t) - v^{(n-1)}(t)) + v^{(n)}(t). \end{cases} \tag{4.1.4}$$

Let $\nu_i(t) = x_i(t) - v^{(i-1)}(t)$. Then (4.1.4) can be transformed into

$$\begin{cases} \dot{\nu}_1(t) = \nu_2(t), \\[2mm] \dot{\nu}_2(t) = \nu_3(t), \\[2mm] \quad\vdots \\[2mm] \dot{\nu}_n(t) = \alpha_1 \nu_1(t) + \alpha_2 \nu_2(t) + \cdots + \alpha_n \nu_n(t). \end{cases} \tag{4.1.5}$$

If the following matrix

$$A = \begin{pmatrix} 0 & 1 & 0 & \cdots & 0 & 0 \\ 0 & 0 & 1 & \cdots & 0 & 0 \\ \vdots & \vdots & \vdots & \ddots & \vdots & \vdots \\ 0 & 0 & 0 & \cdots & 0 & 1 \\ \alpha_1 & \alpha_2 & \alpha_3 & \cdots & \alpha_{n-1} & \alpha_n \end{pmatrix}_{n \times n} \tag{4.1.6}$$

is Hurwitz; then $\nu_i(t)(i = 1, 2, \ldots, n+1)$ converges to zero as time goes to infinity. In Section 2.1, we have proved that if the following matrix

$$\tilde{A} = \begin{pmatrix} 0 & 1 & 0 & \cdots & 0 & 0 \\ 0 & 0 & 1 & \cdots & 0 & 0 \\ \vdots & \vdots & \vdots & \ddots & \vdots & \vdots \\ 0 & 0 & 0 & \cdots & 0 & 1 \\ a_1 & a_2 & a_3 & \cdots & a_n & a_{n+1} \end{pmatrix}_{(n+1) \times (n+1)} \tag{4.1.7}$$

is Hurwitz and $\sup_{t \in [0,\infty), i=1,2,\ldots,n+1} |v^{(i-1)}(t)| < \infty$, then $z_i(t), i = 1, 2, \ldots, n+1$, can approximate $v^{(i-1)}(t)$. The main problem here is therefore to show that the state $\hat{x}_i(t)$ of LESO converges to the state and total disturbance of system (4.0.1). Although in Section 3.1 we present the convergence of LESO, what we need, however, is the bounded assumption for the system state. As a basic condition on LESO, we assume that the following matrix:

$$K = \begin{pmatrix} k_1 & 1 & 0 & \cdots & 0 \\ k_2 & 0 & 1 & \cdots & 0 \\ \vdots & \vdots & \vdots & \ddots & \vdots \\ k_n & 0 & 0 & \cdots & 1 \\ k_{n+1} & 0 & 0 & \cdots & 0 \end{pmatrix}_{(n+1) \times (n+1)}. \tag{4.1.8}$$

is Hurwitz. The closed-loop is written compactly as follows:

$$\begin{cases} \dot{x}_1(t) = x_2(t), \\ \dot{x}_2(t) = x_3(t), \\ \quad \vdots \\ \dot{x}_n(t) = f(x_1(t), \ldots, x_n(t), w(t)) + (b - b_0)u(t) + b_0 u(t), \\ \dot{\hat{x}}_1(t) = \hat{x}_2(t) + \dfrac{k_1}{\varepsilon}(x_1(t) - \hat{x}_1(t)), \\ \dot{\hat{x}}_2(t) = \hat{x}_3(t) + \dfrac{k_2}{\varepsilon^2}(x_1(t) - \hat{x}_1(t)), \\ \quad \vdots \\ \dot{\hat{x}}_n(t) = \hat{x}_{n+1}(t) + \dfrac{k_n}{\varepsilon^n}(x_1(t) - \hat{x}_1(t)) + b_0 u(t), \\ \dot{\hat{x}}_{n+1}(t) = \dfrac{k_{n+1}}{\varepsilon^{n+1}}(x_1(t) - \hat{x}_1(t)), \\ u(t) = \dfrac{1}{b_0}[\alpha_1(\hat{x}_1(t) - z_1(t)) + \cdots + \alpha_n(\hat{x}_n(t) - z_n(t)) + z_{(n+1)}(t) - \hat{x}_{n+1}(t)], \end{cases} \tag{4.1.9}$$

where $z_i(t)(1 \leq i \leq n)$ is the solution of (4.1.3). Since the TD loop is relatively independent (the convergence of the TD loop does not depend on the state of system (4.0.1)), we drop the TD loop in Equation (4.1.9).

Theorem 4.1.1 *Suppose that Assumption 4.1.1 holds. Assume that the matrices A, $\tilde{A}$, and K given by (4.1.6), (4.1.7), and (4.1.8) are Hurwitz, and that $(|b - b_0|/b)k_{n+1} < 1/(2\lambda_{\max}(P_K))$, where b_0 is the nominal parameter of b in (4.0.1). Suppose that $\lambda_{\max}(P_K)$ is the maximal eigenvalue of the positive definite matrix P_K, which is the solution of the Lyapunov equation $P_K K + K^\top P_K = -I_{(n+1)\times(n+1)}$. Let $z_1(t)$ be the solution of (4.0.3), $x_i(t)(1 \leq i \leq n)$. Let $\hat{x}_i(t)(1 \leq i \leq n + 1)$ be the solution of the closed-loop system (4.1.9) and let $x_{n+1}(t) = f(x(t), w(t)) + (b(t) - b_0)u$ be the extended state. Then*

(i) *For any $\sigma > 0$ and $\tau > 0$, there exists a constant $R_0 > 0$ such that $|z_1(t) - v(t)| < \sigma$ uniformly in $t \in [\tau, \infty)$ for all $R > R_0$.*

(ii) *For every $R > R_0$, there is an R-dependent constant $\varepsilon_0 > 0$ such that for any $\varepsilon \in (0, \varepsilon_0)$, there exists a $t_\varepsilon > 0$ such that for all $R > R_0$, $\varepsilon \in (0, \varepsilon_0)$, $t > t_\varepsilon$,*

$$|x_i(t) - \hat{x}_i(t)| \leq \Gamma_1 \varepsilon^{n+2-i}, i = 1, 2, \ldots, n + 1, \tag{4.1.10}$$

and

$$|x_i(t) - z_i(t)| \leq \Gamma_2 \varepsilon, i = 1, 2, \ldots, n, \tag{4.1.11}$$

where Γ_1 and Γ_2 are ε-independent positive constants.

(iii) *For any $\sigma > 0$, there exist $R_1 > R_0$ and $\varepsilon_1 \in (0, \varepsilon_0)$ such that for any $R > R_1$ and $\varepsilon \in (0, \varepsilon_1)$, there exists a $t_{R\varepsilon} > 0$ such that $|x_1(t) - v(t)| < \sigma$ for all $R > R_1$, $\varepsilon \in (0, \varepsilon_1)$, and $t > t_{R\varepsilon}$.*

Proof. Statement (i) follows directly from Theorem 2.1.2 in Section 2.1. In addition, from the proof of Theorem 2.1.2, we know that for any $R > R_0$, there exists an $M_R > 0$ such that

$$\|(z_1(t), z_2(t), \ldots, z_{(n+1)}(t), \dot{z}_{(n+1)}(t))\| \leq M_R \tag{4.1.12}$$

for all $t \geq 0$. In addition, note that statement (iii) is a direct consequence of (i) and (ii). We list it here to indicate the output tracking function of ADRC only. Therefore, it suffices to prove (ii). Let

$$e_i(t) = \frac{1}{\varepsilon^{n+1-i}}[x_i(t) - \hat{x}_i(t)], i = 1, 2, \ldots, n + 1. \tag{4.1.13}$$

Then we have

$$\begin{cases} \varepsilon \dot{e}_i(t) = \dfrac{1}{\varepsilon^{n+1-(i+1)}}[\dot{x}_i(t) - \dot{\hat{x}}_i(t)] = \dfrac{1}{\varepsilon^{n+1-(i+1)}}[x_{i+1}(t) - \hat{x}_{i+1}(t) - \varepsilon^{n-i}g_i(e_1(t))] \\ \qquad = e_{i+1}(t) - g_i(e_1(t)), 1 \leq i \leq n, \\ \varepsilon \dot{e}_{n+1}(t) = \varepsilon(\dot{x}_{n+1}(t) - \dot{\hat{x}}_{n+1}(t)) = \varepsilon h(t) - g_{n+1}(e_1(t)), \end{cases}$$

$$\tag{4.1.14}$$

where

$$x_{n+1}(t) = f(x(t), w(t)) + (b - b_0)u(t)$$

$$= f(x(t), w(t)) + \frac{b - b_0}{b_0}[\alpha_1(\hat{x}_1(t) - z_1(t)) + \cdots + \alpha_n(\hat{x}_n(t) - z_n(t)) \quad (4.1.15)$$

$$+ z_{(n+1)}(t) - \hat{x}_{n+1}(t)]$$

is the extended state of system (4.0.1) and $h(t)$ is defined by

$$h(t) = \frac{d[f(x(t), w(t)) + (b - b_0)u(t)]}{dt}\Big|_{\text{along }(4.1.9)}$$

$$= \sum_{i=1}^{n-1} x_{i+1}(t)\frac{\partial f(x(t), w(t))}{\partial x_i} + [f(x(t), w(t)) + bu(t)]\frac{\partial f(x(t), w(t))}{\partial x_n} \quad (4.1.16)$$

$$+ \dot{w}(t)\frac{\partial f(x(t), w(t))}{\partial w} + (b - b_0)\dot{u}(t).$$

Under feedback control (4.1.1), since $x_{n+1}(t) = f(x(t), w(t)) + (b - b_0)u(t)$, and

$$(f(x(t), w(t)) + bu(t))\frac{\partial f(x(t), w(t))}{\partial x_n} = x_{n+1}(t)\frac{\partial f(x(t), w(t))}{\partial x_n} + b_0 u(t)\frac{\partial f(x(t), w(t))}{\partial x_n},$$

we can compute $h(t)$ as

$$h(t) = \frac{d[f(x(t), w(t)) + (b - b_0)u(t)]}{dt}\Big|_{\text{along }(4.1.9)}$$

$$= \sum_{i=1}^{n} x_{i+1}(t)\frac{\partial f(x(t), w(t))}{\partial x_i} + \dot{w}(t)\frac{\partial f(x(t), w(t))}{\partial w}$$

$$+ [\alpha_1(\hat{x}_1(t) - z_1(t)) + \cdots + \alpha_n(\hat{x}_n(t) - z_n(t))] + z_{(n+1)}(t) - \hat{x}_{n+1}(t)]\frac{\partial f(x(t), w(t))}{\partial x_n}$$

$$(4.1.17)$$

$$+ \frac{b - b_0}{b_0}\left\{\sum_{i=1}^{n} \alpha_i\left[\hat{x}_{i+1}(t) + \varepsilon^{n-i}k_i e_1(t) - z_{(i+1)}(t)\right] + \dot{z}_{(n+1)}(t) - \frac{k_{n+1}}{\varepsilon}e_1(t)\right\},$$

where $\partial f(x(t), w(t))/\partial x_i$ denotes the value of the ith partial derivative of $f \in C^1(\mathbb{R}^{n+1}, \mathbb{R})$ at $(x(t), w(t)) \in \mathbb{R}^{n+1}$. Set

$$\eta_i(t) = x_i(t) - z_i(t), i = 1, 2, \ldots, n, \eta(t) = (\eta_1(t), \eta_2(t), \ldots, \eta_n(t)).$$

It follows from (4.1.9) and (4.1.14) that

$$
\begin{cases}
\dot{\eta}_1(t) = \eta_2(t), \\
\quad \vdots \\
\dot{\eta}_n(t) = \alpha_1 \eta_1(t) + \alpha_2 \eta_2(t) + \cdots + \alpha_n \eta_n(t)) + e_{n+1}(t) \\
\qquad + [\alpha_1(\hat{x}_1(t) - x_1(t)) + \alpha_2(\hat{x}_2(t) - x_2(t)) + \cdots + \alpha_n(\hat{x}_n(t) - x_n(t))], \\
\varepsilon \dot{e}_1(t) = e_2(t) - k_1 e_1(t), \\
\quad \vdots \\
\varepsilon \dot{e}_n(t) = e_{n+1}(t) - k_n e_1(t), \\
\varepsilon \dot{e}_{n+1}(t) = \varepsilon h(t) - k_{n+1} e_1(t).
\end{cases}
$$

$$(4.1.18)$$

Furthermore, set

$$
\tilde{e}(t) = (e_1(t), e_2(t), \ldots, e_n(t)), e(t) = (e_1(t), e_2(t), \ldots, e_{n+1}(t)).
$$

From the boundedness of the partial derivatives of $f(\cdot)$, $z(t)$ (see (4.1.12) and $w(t)$, it follows that for some R-dependent positive numbers $M, N_0, N_1, N > 0$ (by the mean value theorem with $f(0,0)$),

$$
|f(x(t), w(t))| \le M(\|x(t)\| + |w(t)|) + N_0
$$

$$
\le M(\|\eta(t)\| + \|z(t)\| + |w(t)|) + N_1 \le M\|\eta(t)\| + N. \qquad (4.1.19)
$$

Notice that by (4.1.15), we have

$$
\hat{x}_{n+1}(t) = \frac{b_0}{b}[-e_{n+1}(t) + f(x(t), w(t)) + \frac{b - b_0}{b_0}(\alpha_1(\hat{x}_1(t) - z_1(t)) + \cdots + \alpha_n(\hat{x}_n(t)
$$

$$
- z_n(t)) + z_{(n+1)}(t)]. \qquad (4.1.20)
$$

By (4.1.12), (4.1.19), and (4.1.20), and from Assumption 4.1.1, the function $h(t)$ given by (4.2.10) satisfies

$$
|h(t)| \le B_0 + B_1 \|e(t)\| + B_2 \|\eta(t)\| + \frac{B}{\varepsilon} \|e(t)\|, \quad B = \frac{|b - b_0|}{b_0} k_{n+1}, \qquad (4.1.21)
$$

where B_0, B_1, and B_2 are R-dependent positive numbers.

We proceed the proof in three steps as follows.

Step 1. *For every $R > R_0$, there exists an R-dependent constant $\varepsilon_0 > 0$ such that for any $\varepsilon \in (0, \varepsilon_0)$, there exist $t_{1\varepsilon}$ and $r > 0$ such that the solution of (4.1.18) satisfies $\|(e(t), \eta(t))\| \le r$ for all $t > t_{1\varepsilon}$, where r is an R-dependent constant.*

Let

$$
V_1(e) = \langle P_K e, e \rangle, W_1(e) = \|e\|, \quad \forall\, e = (e_1, e_2, \ldots, e_{n+1})^{\top} \in \mathbb{R}^{n+1}. \qquad (4.1.22)
$$

Then

$$
\lambda_{\min}(P_K) \|e\|^2 \le V_1(e) \le \lambda_{\max}(P_K) \|e\|^2, \qquad (4.1.23)
$$

$$\sum_{i=1}^{n-1}(e_{i+1} - k_i e_1)\frac{\partial V_1(e)}{\partial e_i} - k_{n+1}e_1\frac{\partial V_1(e)}{\partial e_{n+1}} = -\langle e, e\rangle = -W_1(e), \qquad (4.1.24)$$

and

$$\left|\frac{\partial V_1(e)}{\partial e_{n+1}}\right| \leq 2\lambda_{\max}(P_K)\|e\|, \qquad (4.1.25)$$

where $\lambda_{\min}(P_K)$ and $\lambda_{\max}(P_K)$ are the minimal and maximal eigenvalues of P_K respectively.

Next let

$$V_2(x) = \langle P_A x, x\rangle, \ W_2(x) = \|x\|, \ \forall \, x = (x_1, x_2, \ldots, x_n)^\top \in \mathbb{R}^n, \qquad (4.1.26)$$

where P_A is the positive definite matrix that is the (unique) solution of the Lyapunov equation $P_A A + A^\top P_A = -I_{n\times n}$. Then

$$\lambda_{\min}(P_A)\|x\|^2 \leq V_2(x) \leq \lambda_{\max}(P_A)\|x\|^2, \qquad (4.1.27)$$

$$\sum_{i=1}^{n} x_{i+1}\frac{\partial V_1(x)}{\partial x_i} - (\alpha_1 x_1 + \alpha_2 x_2 + \cdots + \alpha_n x_n)\frac{\partial V_2(x)}{\partial x_n} = -\langle x, x\rangle = -W_2(x),$$
$$(4.1.28)$$

and

$$\left|\frac{\partial V_2(x)}{\partial x_n}\right| \leq 2\lambda_{\max}(P_A)\|x\|, \qquad (4.1.29)$$

where $\lambda_{\min}(P_A)$ and $\lambda_{\max}(P_A)$ are the minimal and maximal eigenvalues of P_A respectively.

Let

$$V(e_1, \ldots, e_{n+1}, \eta_1, \ldots, \eta_n) = V_1(e_1, \ldots, e_{n+1}) + V_2(\eta_1, \ldots, \eta_n), \qquad (4.1.30)$$

where $V_1(e)$ and $V_2(\eta)$ are positive definite functions given in (4.1.22) and (4.1.26) respectively. Computing the derivative of $V(e(t), \eta(t))$ along the solution of (4.1.18), we obtain

$$\frac{dV(e(t), \eta(t))}{dt}\bigg|_{\text{along (4.1.18)}}$$

$$= \frac{1}{\varepsilon}\left[\sum_{i=1}^{n}(e_{i+1}(t) - k_i e_1(t))\frac{\partial V_1(e(t))}{\partial e_i} - k_{n+1}e_1(t)\frac{\partial V_1(e(t))}{\partial e_{n+1}} + \varepsilon h(t)\frac{\partial V_1(e(t))}{\partial e_{n+1}}\right]$$

$$+ \sum_{i=1}^{n-1}\eta_{i+1}(t)\frac{\partial V_2(\eta(t))}{\partial \eta_i}$$

$$+ \{\alpha_1\eta_1(t) + \cdots + \alpha_n\eta_n(t) + e_{n+1}(t) + [\alpha_1(\hat{x}_1(t) - x_1(t))) + \cdots + \alpha_n(\hat{x}_n(t) - x_n(t)))]\}\frac{\partial V_2(\eta(t))}{\partial x_n}$$

$$\leq \frac{1}{\varepsilon}\left\{-W_1(e(t)) + 2\varepsilon\left(B_0 + B_1\|e(t)\| + B_2\|\eta(t)\| + \frac{B}{\varepsilon}\|e(t)\|\right)\lambda_{\max}(P_K)\|e(t)\|\right\}$$

$$- W_2(\eta(t)) + 2\left(\sum_{i=1}^{n}|\alpha_i| + 1\right)\lambda_{\max}(P_A)\|e(t)\|\|\eta(t)\|$$

$$\leq -\frac{W_1(e(t))}{\varepsilon} + 2\lambda_{\max}(P_K)B_0\|e(t)\| + 2\lambda_{\max}(P_K)B_1\|e(t)\|^2 - W_2(\eta(t))$$

$$+ \left(2\lambda_{\max}(P_K)B_2 + 2\lambda_{\max}(P_A)\left(\sum_{i=1}^{n} |\alpha_i| + 1 \right) \right) \|e(t)\| \|\eta(t)\| + \frac{2\lambda_{\max}(P_K)B}{\varepsilon} \|e(t)\|^2$$

$$\leq - \left[\frac{1 - 2\lambda_{\max}(P_K)B}{\varepsilon} - 2\lambda_{\max}(P_K)B_1 \right] \|e(t)\|^2 + 2\lambda_{\max}(P_K)B_0 \|e(t)\| - \|\eta(t)\|^2$$

$$+ \sqrt{\frac{1 - 2\lambda_{\max}(P_K)B}{\varepsilon}} \|e(t)\| \frac{2\lambda_{\max}(P_K)B_2 + 2\lambda_{\max}(P_A)\left(\sum_{i=1}^{n} |\alpha_i| + 1 \right)}{\sqrt{1 - 2\lambda_{\max}(P_K)B}} \sqrt{\varepsilon} \|\eta(t)\|$$

$$\leq - \left[\frac{1 - 2\lambda_{\max}(P_K)B}{\varepsilon} - 2\lambda_{\max}(P_K)B_1 \right] \|e(t)\|^2 + 2\lambda_{\max}(P_K)B_0 \|e(t)\| - \|\eta(t)\|^2$$

$$+ \frac{1 - 2\lambda_{\max}(P_K)B}{2\varepsilon} \|e(t)\|^2 + \frac{(2\lambda_{\max}(P_K)B_2 + 2(\sum_{i=1}^{n}|\alpha_i| + 1)\lambda_{\max}(P_A))^2}{2(1 - 2\lambda_{\max}(P_K)B)} \varepsilon \|\eta(t)\|^2$$

$$\leq - \left[\frac{1 - 2\lambda_{\max}(P_K)B}{2\varepsilon} - 2\lambda_{\max}(P_K)B_1 \right] \|e(t)\|^2 + 2\lambda_{\max}(P_K)B_0 \|e(t)\|$$

$$- \left[1 - \frac{(2\lambda_{\max}(P_K)B_2 + \lambda_{\max}(P_A)(\sum_{i=1}^{n}|\alpha_i| + 1))^2}{2(1 - 2\lambda_{\max}(P_K)B)} \varepsilon \right] \|\eta(t)\|^2. \tag{4.1.31}$$

Set

$$r = \max\{2, 4(1 + 2\lambda_{\max}(P_K)B_0)\},$$

$$\varepsilon_0 = \min \left\{ 1, \frac{1 - 2\lambda_{\max}(P_K)B}{4\lambda_{\max}(P_K)(B_0 + B_1)}, \frac{1 - 2\lambda_{\max}(P_K)B}{(2\lambda_{\max}(P_K)B_2 + 2\lambda_{\max}(P_A)(\sum_{i=1}^{n}|\alpha_i| + 1))^2} \right\}. \tag{4.1.32}$$

For any $\varepsilon \in (0, \varepsilon_0)$ and $\|(e(t), \eta(t))\| \geq r$, we consider the derivative of $V(e(t), \eta(t))$ along the solution of (4.1.18) in two cases.

Case 1: $\|e(t)\| \geq r/2$. In this case, $\|e(t)\| \geq 1$ and hence $\|e(t)\|^2 \geq \|e(t)\|$. By the definition ε_0 of (4.1.32), $1 - (2\lambda_{\max}(P_K)B_2 + 2\lambda_{\max}(P_A)(\sum_{i=1}^{n}|\alpha_i| + 1)^2)/(2(1 - 2\lambda_{\max}(P_K)B)\varepsilon > 0$. It then follows from (4.1.31) that

$$\frac{dV(e(t), \eta(t))}{dt}\Big|_{\text{along (4.1.18)}} \leq - \left(\frac{1 - 2\lambda_{\max}(P_K)B}{2\varepsilon} - 2\lambda_{\max}(P_K)B_1 \right) \|e(t)\|^2$$

$$+ \lambda_{\max}(P_K)B_0 \|e(t)\|^2$$

$$\leq - \left(\frac{1 - \beta_1 B}{2\varepsilon} - 2\lambda_{\max}(P_K)B_1 - 2\lambda_{\max}(P_K)B_0 \right) \|e(t)\|^2$$

$$= - \frac{1 - \lambda_{\max}(P_K)B - 4\varepsilon\lambda_{\max}(P_K)(B_1 + B_0)}{2\varepsilon} \|e(t)\|^2 < 0,$$

where we used again the definition ε_0 of (4.1.32), which gives $1 - 2\lambda_{\max}(P_K)B - 4\varepsilon\lambda_{\max}(P_K)(B_1 + B_0) > 0$.

Case 2: $\|e(t)\| < r/2$. In this case, since $\|\eta(t)\| + \|e(t)\| \geq \|(e(t), \eta(t))\|$, it has $\|\eta(t)\| \geq r/2$. By the definition ε_0 of (4.1.32), $1 - \lambda_{\max}(P_K)B - 4\varepsilon\lambda_{\max}(P_K)B_1 > 0$. Thus it follows from (4.1.31) that

$$\frac{dV(e(t), \eta(t))}{dt}\Big|_{\text{along (4.1.18)}} \leq 2\lambda_{\max}(P_K)B_0 \|e(t)\|$$

$$- \left(1 - \frac{(2\lambda_{\max}(P_K)B_2 + 2\lambda_{\max}(P_A)(\sum_{i=1}^{n}|\alpha_i| + 1))^2}{2(1 - 2\lambda_{\max}(P_K)B)} \varepsilon \right) \|\eta(t)\|^2$$

$$\leq -\frac{1}{2}\|\eta(t)\|^2 + 2\lambda_{\max}(P_K)B_0\|e(t)\|$$

$$\leq -\frac{1}{2}\left(\frac{r}{2}\right)^2 + 2\lambda_{\max}(P_K)B_0\left(\frac{r}{2}\right) = -\frac{r}{2}\left(\frac{r - 8\lambda_{\max}(P_K)B_0}{4}\right) < 0,$$

where we used the definition r of (4.1.32), which gives $r - 8\lambda_{\max}(P_K)B_0 > 0$.

Combining the above two cases yields that for any $\varepsilon \in (0, \varepsilon_0)$, if $\|(e(t), \eta(t))\| \geq r$ then

$$\frac{dV(e(t), \eta(t))}{dt}\bigg|_{\text{along (4.1.18)}} < 0.$$

Therefore, there exists a $t_{1\varepsilon}$ such that $\|(e(t), \eta(t))\| \leq r$ for all $t > t_{1\varepsilon}$.

Step 2. *The convergence of $\|x_i(t) - \hat{x}_i(t)\|$.*

Consider the following subsystem, which is composed of the last $n + 1$ equations in system (4.1.18):

$$\begin{cases} \varepsilon\dot{e}_1(t) = e_2(t) + k_1 e_1(t), \\ \quad\vdots \\ \varepsilon\dot{e}_n(t) = e_{n+1}(t) - k_n e_1(t), \\ \varepsilon\dot{e}_{n+1}(t) = \varepsilon h(t) - k_{n+1}e_1(t). \end{cases} \tag{4.1.33}$$

Since $\|(e(t), \eta(t))\| \leq r$ for all $t > t_{1\varepsilon}$, we obtain, together with (4.1.21), that $|h(t)| \leq M_0 + B\|e(t)\|/\varepsilon$ for all $t > t_{1\varepsilon}$ and some R-dependent constant $M_0 > 0$. We can compute the derivative of $V_1(e(t))$ from the solution of (4.1.33) as follows:

$$\frac{dV_1(e(t))}{dt}\bigg|_{\text{along (4.1.33)}}$$

$$= \frac{1}{\varepsilon}\left(\sum_{i=1}^{n}(e_{i+1}(t) - k_i e_1(t))\frac{\partial V_1(e(t))}{\partial e_i} - k_{n+1}e_1(t)\frac{\partial V_1(e(t))}{\partial e_{n+1}} + \varepsilon h(t)\frac{\partial V_1(e(t))}{\partial e_{n+1}}\right)$$

$$\tag{4.1.34}$$

$$\leq -\frac{1 - 2\lambda_{\max}(P_K)B}{\varepsilon}\|e(t)\|^2 + 2\lambda_{\max}(P_K)M_0\|e(t)\|$$

$$\leq -\frac{1 - 2\lambda_{\max}(P_K)B}{\varepsilon\lambda_{\max}(P_K)}V_1(e(t)) + \frac{2M_0\sqrt{\lambda_{\min}(P_K)}}{\lambda_{\min}(P_K)}\sqrt{V_1(e(t))}, \forall\, t > t_{1\varepsilon}.$$

It follows that

$$\frac{d\sqrt{V_1(e(t))}}{dt} \leq -\frac{1 - 2B\lambda_{\max}(P_K)}{2\varepsilon\lambda_{\max}(P_K)}\sqrt{V_1(e(t))} + \frac{M_0\beta_1\sqrt{\lambda_{\min}(P_K)}}{2\lambda_{\min}(P_K)}, \forall\, t > t_{1\varepsilon},$$

$$\tag{4.1.35}$$

and hence

$$\sqrt{V_1(e(t))} \leq \sqrt{V_1(e(t_{1\varepsilon}))}e^{\frac{-1 - 2B\lambda_{\max}(P_K)}{2\varepsilon\lambda_{\max}(P_K)}(t - t_{1\varepsilon})}$$

$$+ \frac{2M_0\lambda_{\max}(P_K)\sqrt{\lambda_{\min}(P_K)}}{\lambda_{\min}(P_K)}\int_{t_{1\varepsilon}}^{t} e^{\frac{-1 - 2B\lambda_{\max}(P_K)}{2\varepsilon\lambda_{\max}(P_K)}(t - s)}ds, \forall\, t > t_{1\varepsilon}. \tag{4.1.36}$$

This together with (4.1.13) implies that there exist $t_{2\varepsilon} > t_{1\varepsilon}$ and R-dependent constant $\Gamma_1 > 0$ such that

$$|x_i(t) - \hat{x}_i(t)| = \varepsilon^{n+1-i}|e_i(t)|$$

$$\leq \varepsilon^{n+1-i}\|e(t)\| \leq \varepsilon^{n+1-i}\sqrt{\frac{V_1(e(t))}{\lambda_{\min}(P_K)}} \leq \Gamma_1\varepsilon^{n+2-i}, \forall\, t > t_{2\varepsilon},$$

$$(4.1.37)$$

where we used the facts $xe^{-x} < 1$ for all $x > 0$ and $\|(e(t), \eta(t))\| \leq r$ for all $t > t_{1\varepsilon}$, which are proved in Step 1.

Step 3. *The convergence of $\|x(t) - z(t)\|$.*

Consider the following system, which is composed of the first n equations in system (4.1.18):

$$\begin{cases} \dot{\eta}_1(t) = \eta_2(t), \\ \dot{\eta}_2(t) = \eta_3(t), \\ \quad\vdots \\ \dot{\eta}_n(t) = \alpha_1\eta_1(t) + \cdots + \alpha_n\eta_n(t) + e_{n+1}(t) + \alpha_1(\hat{x}_1(t) - x_1(t)) + \cdots + \alpha_n(\hat{x}_n(t) - x_n(t)). \end{cases}$$

$$(4.1.38)$$

We can compute the derivative of $V_2(\eta(t))$ along the solution of (4.1.38) as follows:

$$\frac{dV_2(\eta(t))}{dt}\Big|_{\text{along (4.1.38)}} = \sum_{i=1}^{n-1} \eta_{i+1}(t)\frac{\partial V_2(\eta(t))}{\partial\eta_i} + [\alpha_1\eta_1(t) + \cdots + \alpha_n\eta_n(t)$$

$$+ e_{n+1}(t) + \alpha_1(\hat{x}_1(t) - x_1(t)) + \cdots + \alpha_n(\hat{x}_n(t) - x_n(t))]\frac{\partial V_2(\eta(t))}{\partial\eta_n}$$

$$(4.1.39)$$

$$\leq -W_2(\eta(t)) + \left(\sum_{i=1}^n |\alpha_i| + 1\right)\beta_2\|e(t)\|\|\eta(t)\|$$

$$\leq -\frac{1}{\lambda_{\max}(P_A)}V_2(\eta(t)) + N_0\varepsilon\sqrt{V_2(\eta(t))}, \forall\, t > t_{2\varepsilon},$$

where N_0 is some R-dependent positive constant and we used the fact $\|e(t)\| \leq (n+1)B_1\varepsilon$ proved in (4.1.37). It then follows that

$$\frac{d\sqrt{V_2(\eta(t))}}{dt} \leq -\frac{1}{\lambda_{\max}(P_A)}\sqrt{V_2(\eta(t))} + N_0\varepsilon, \forall\, t > t_{2\varepsilon}. \tag{4.1.40}$$

This implies that for all $t > t_{2\varepsilon}$,

$$\|\eta(t)\| \leq \frac{\sqrt{\lambda_{\min}(P_A)}}{\lambda_{\min}(P_A)}\sqrt{V_2(\eta(t))} \tag{4.1.41}$$

$$\leq \frac{\sqrt{\lambda_{\min}(P_A)}}{\lambda_{\min}(P_A)}\left(e^{-\frac{1}{\lambda_{\max}(P_A)}(t-t_{2\varepsilon})}\sqrt{V_2(\eta(t_{2\varepsilon}))} + N_0\varepsilon\int_{t_{2\varepsilon}}^t e^{-\frac{1}{\lambda_{\max}(P_A)}(t-s)}ds\right).$$

Since the first term of the right-hand side of (4.1.41) tends to zero as t goes to infinity and the second term is bounded by ε multiplied by an ε-independent constant, it follows that there exist $t_\varepsilon > t_{2\varepsilon}$ and $\Gamma_2 > 0$ such that $\|x(t) - z(t)\| \le \Gamma_2\varepsilon$ for all $t > t_\varepsilon$. Thus (4.1.11) follows. This completes the proof of the theorem. $\qquad\square$

4.1.2 Global Convergence for Systems with External Disturbance Only

In this section, we consider the following system:

$$\begin{cases} \dot{x}_1(t) = x_2(t), \\ \dot{x}_2(t) = x_3(t), \\ \quad\vdots \\ \dot{x}_n(t) = \tilde{f}(x_1(t), \ldots, x_n(t)) + w(t) + bu(t), \\ y(t) = x_1(t), \end{cases} \tag{4.1.42}$$

here $\tilde{f}(\cdot)$ is known. This is the case when the uncertainty comes from external disturbance only. We assume that the system function satisfies the following Assumption 4.1.2.

Assumption 4.1.2 Both $w(t)$ and $\dot{w}(t)$ are bounded and all partial derivatives of $\tilde{f}(x)$ are bounded for all $x \in \mathbb{R}^n$.

The assumption on $\tilde{f}(x)$ implies that there is a constant $L_1 > 0$ such that $\tilde{f}(x)$ is Lipschitz continuous with the Lipschitz constant L_1:

$$|\tilde{f}(x) - \tilde{f}(y)| \le L_1\|x - y\|, \forall\, x, y \in \mathbb{R}^n. \tag{4.1.43}$$

Since $\tilde{f}(x)$ is available, we design the following ESO(f) to estimate the system states $x_i(t), i = 1, 2, \ldots, n$, and the extended state $x_{n+1}(t) = w(t) + (b - b_0)u(t)$.

The LESO for this special case is given as follows:

$$\text{LESO(f):} \begin{cases} \dot{\hat{x}}_1(t) = \hat{x}_2(t) + \dfrac{k_1}{\varepsilon}(y(t) - \hat{x}_1(t)), \\ \dot{\hat{x}}_2(t) = \hat{x}_3(t) + \dfrac{k_2}{\varepsilon^2}(y(t) - \hat{x}_1(t)), \\ \quad\vdots \\ \dot{\hat{x}}_n(t) = \hat{x}_{n+1}(t) + \dfrac{k_n}{\varepsilon^n}(y(t) - \hat{x}_1(t)) + \tilde{f}(\hat{x}(t)) + b_0 u(t), \\ \dot{\hat{x}}_{n+1}(t) = \dfrac{k_{n+1}}{\varepsilon^{n+1}}(y(t) - \hat{x}_1(t)), \end{cases} \tag{4.1.44}$$

and the linear feedback control in this case can be designed as

$$u(t) = \frac{1}{b_0}[\alpha_1(\hat{x}_1(t) - z_1(t)) + \cdots + \alpha_n(\hat{x}_1(t) - z_1(t)) - \tilde{f}(\hat{x}(t)) + z_{(n+1)}(t) - \hat{x}_{n+1}(t)], \tag{4.1.45}$$

where $\hat{x}_i(t)$ comes from the modified LESO(f) (4.1.44) and $z_i(t)$ comes from linear tracking differentiator (LTD)

$$
\text{LTD:} \quad
\begin{cases}
\dot{z}_1(t) = z_2(t), \\
\dot{z}_2(t) = z_3(t), \\
\quad \vdots \\
\dot{z}_n(t) = z_n(t), \\
\dot{z}_{(n+1)}(t) = R^{n+1}\left(a_1(z_1(t) - v(t)) + \dfrac{a_2 z_2(t)}{R} + \cdots + \dfrac{a_{n+1} z_n(t)}{R^n}\right).
\end{cases}
\tag{4.1.46}
$$

With the linear LESO(f) (4.1.44) and the linear feedback control (4.1.45), the closed-loop of linear ADRC for this case becomes

$$
\begin{cases}
\dot{x}_1(t) = x_2(t), \\
\dot{x}_2(t) = x_3(t), \\
\quad \vdots \\
\dot{x}_n(t) = \tilde{f}(x(t)) + w(t) + bu(t), \\
\dot{\hat{x}}_1(t) = \hat{x}_2(t) + \dfrac{k_1}{\varepsilon}(x_1(t) - \hat{x}_1(t)), \\
\dot{\hat{x}}_2(t) = \hat{x}_3(t) + \dfrac{k_2}{\varepsilon^2}(x_1(t) - \hat{x}_1(t)), \\
\quad \vdots \\
\dot{\hat{x}}_n(t) = \hat{x}_{n+1}(t) + \dfrac{k_n}{\varepsilon^n}(x_1(t) - \hat{x}_1(t)) + \tilde{f}(\hat{x}(t)) + b_0 u(t), \\
\dot{\hat{x}}_{n+1}(t) = \dfrac{k_{n+1}}{\varepsilon^{n+1}}(x_1(t) - \hat{x}_1(t)), \\
u(t) = \dfrac{1}{b_0}[\alpha_1(\hat{x}_1(t) - z_1(t)) + \cdots + \alpha_n(\hat{x}_n(t) - z_n(t)) - \tilde{f}(\hat{x}(t)) \\
\qquad\qquad + z_{n+1}(t) - \hat{x}_{n+1}(t)].
\end{cases}
\tag{4.1.47}
$$

The parameters $\alpha_i, i = 1, 2, \ldots, n$ in feedback control, $a_i, i = 1, 2, \ldots, n + 1$ in LTD, and $k_i, i = 1, 2, \ldots, n + 1$ in LESO satisfy Assumption 4.1.3.

Assumption 4.1.3 The following matrices A, $\tilde{A}$, and K are Hurwitz:

$$
A = \begin{pmatrix}
0 & 1 & 0 & \cdots & 0 & 0 \\
0 & 0 & 1 & \cdots & 0 & 0 \\
\vdots & \vdots & \vdots & \ddots & \vdots & \vdots \\
0 & 0 & 0 & \cdots & 0 & 1 \\
\alpha_1 & \alpha_2 & \alpha_3 & \cdots & \alpha_{n-1} & \alpha_n
\end{pmatrix}_{n \times n},
$$

$$\tilde{A} = \begin{pmatrix} 0 & 1 & 0 & \cdots & 0 & 0 \\ 0 & 0 & 1 & \cdots & 0 & 0 \\ \vdots & \vdots & \vdots & \ddots & \vdots & \vdots \\ 0 & 0 & 0 & \cdots & 0 & 1 \\ a_1 & a_2 & a_3 & \cdots & a_n & a_{n+1} \end{pmatrix}_{(n+1)\times(n+1)}, \tag{4.1.48}$$

$$K = \begin{pmatrix} k_1 & 1 & 0 & \cdots & 0 \\ k_2 & 0 & 1 & \cdots & 0 \\ \vdots & \vdots & \vdots & \ddots & \vdots \\ k_n & 0 & 0 & \cdots & 1 \\ k_{n+1} & 0 & 0 & \cdots & 0 \end{pmatrix}_{(n+1)\times(n+1)}.$$

Theorem 4.1.2 *Suppose that both $w(t)$ and $\dot{w}(t)$ are bounded over $\mathbb{R}$, and Assumption 4.1.2 and Assumption 4.1.3 are satisfied. Assume that*

$$\left(L_1 + \frac{|b - b_0|}{b_0} k_{n+1} \right) \lambda_{\max}(P_K) < \frac{1}{2},$$

where L_1 is the Lipschitz constant in Assumption 4.1.2, b, b_0, and k_{n+1} are parameters in (4.1.42) and (4.1.44), $\lambda_{\max}(P_K)$ and $\lambda_{\max}(P_A)$ are the maximal eigenvalues of P_K and P_A, which are the positive definite solutions of the Lyapunov equations $P_K K + K^\top P_K = -I_{(n+1)\times(n+1)}$ and $P_A A + A^\top P_A = -I_{n\times n}$ respectively.

Let $z_1(t)$ be the solution of (4.1.46), and let $x_i(t)(1 \le i \le n)$ and $\hat{x}_i(t)(1 \le i \le n+1)$ be the solutions of the closed-loop (4.1.47). Suppose that $x_{n+1}(t) = w(t) + (b - b_0)u(t)$ is the extended state. Then, for any given initial values of (4.1.46) and (4.1.47), the statements of (i) to (iii) of Theorem 4.2.2 hold true.

Proof. Statement (i) comes from Theorem 2.1.2. Also for any $R > R_0$, there exists an $M_R > 0$ such that

$$\|(z_1(t), z_2(t), \ldots, z_n(t), \dot{z}_{(n+1)}(t))\| \le M_R \tag{4.1.49}$$

for all $t \ge 0$.

Statement (iii) is the consequence of (i) and (ii), which is considered as an output regulation result in this case. It therefore suffices to prove statement (ii). Set

$$\eta_i(t) = x_i(t) - z_i(t), e_i(t) = \frac{1}{\varepsilon^{n+1-i}}[x_i(t) - \hat{x}_i(t)], e_{n+1}(t) = x_{n+1}(t) - \hat{x}_{n+1}(t),$$

$$i = 1, 2, \ldots, n.$$

By (4.1.47), $e_i(t)(1 \leq i \leq n+1)$ and $\eta_i(t)(1 \leq i \leq n)$ satisfy the following system of differential equations:

$$
\begin{cases}
\dot{\eta}_1(t) = \eta_2(t), \\
\dot{\eta}_2(t) = \eta_3(t), \\
\quad \vdots \\
\dot{\eta}_n(t) = \alpha_1(\hat{x}_1(t) - z_1(t)) + \cdots + \alpha_n(\hat{x}_n(t) - z_n(t)) + \tilde{f}(x(t)) - \tilde{f}(\hat{x}(t)) + e_{n+1}(t), \\
\varepsilon \dot{e}_1(t) = e_2(t) - k_1 e_1(t), \\
\quad \vdots \\
\varepsilon \dot{e}_n(t) = e_{n+1}(t) - k_n e_1(t) + \tilde{f}(x(t)) - \tilde{f}(\hat{x}(t)), \\
\varepsilon \dot{e}_{n+1}(t) = -k_{n+1} e_1(t) + \varepsilon \dfrac{d}{dt}\left(w(t) + (b - b_0)u(t)\right).
\end{cases}
\tag{4.1.50}
$$

Define

$$
\hbar(t) = \frac{d}{dt}\left(w(t) + (b - b_0)u(t)\right)
$$

$$
= \dot{w}(t) + \frac{b - b_0}{b_0}\left[\sum_{i=1}^{n} a_i\left[\hat{x}_{i+1}(t) - \varepsilon^{n-i}k_i e_1(t) - z_{i+1}(t)\right] \right. \tag{4.1.51}
$$

$$
\left. - \sum_{i=1}^{n}\left[\hat{x}_{n+1}(t) - \varepsilon^{n-i}k_i e_1(t)\right]\frac{\partial}{\partial x_i}\tilde{f}(\hat{x}(t)) + \dot{z}_{n+1}(t) - \frac{k_{n+1}e_1(t)}{\varepsilon}\right].
$$

Similar to (4.1.21), we also have

$$
|\hbar(t)| \leq B_0 + B_1\|\eta(t)\| + B_2\|e(t)\| + \frac{B}{\varepsilon}\|e(t)\|, \quad B = \frac{|b - b_0|}{b_0}k_{n+1}, \tag{4.1.52}
$$

where B_0, B_1, and B_2 are some R-dependent positive constants. Let

$$
\begin{cases}
V_1(e) = \langle P_K e, e\rangle, W_1(e) = \|e\|, \ \forall\, e = (e_1, e_2, \ldots, e_{n+1})^\top \in \mathbb{R}^{n+1}, \\
V_2(x) = \langle P_A x, x\rangle, \ W_2(x) = \|x\|, \forall\, x = (x_1, x_2, \ldots, x_n)^\top \in \mathbb{R}^n.
\end{cases}
\tag{4.1.53}
$$

It is easy to verify that

$$
\lambda_{\min}(P_K)\|e\|^2 \leq V_1(e) \leq \lambda_{\max}(P_K)\|e\|^2,
$$

$$
\sum_{i=1}^{n-1}(e_{i+1} - k_i e_1)\frac{\partial V_1(e)}{\partial e_i} - k_{n+1}e_1\frac{\partial V_1(e)}{\partial e_{n+1}} = -\langle e, e\rangle = -W_1(e), \tag{4.1.54}
$$

$$
\left|\frac{\partial V_1(e)}{\partial e_{n+1}}\right| \leq 2\lambda_{\max}(P_K)\|e\|,
$$

and

$$\lambda_{\min}(P_A)\|x\|^2 \le V_2(x) \le \lambda_{\max}(P_A)\|x\|^2,$$

$$\sum_{i=1}^{n} x_{i+1}\frac{\partial V_2(x)}{\partial x_i} - (\alpha_1 x_1 + \alpha_2 x_2 + \cdots + \alpha_n x_n)\frac{\partial V_2(x)}{\partial x_n} = -\langle x, x\rangle = -W_2(x), \quad (4.1.55)$$

$$\left|\frac{\partial V_2(x)}{\partial x_n}\right| \le 2\lambda_{\max}(P_A)\|x\|.$$

Now construct a Lyapunov function for system (4.1.50) as

$$V(e,\eta) = V_1(e) + V_2(\eta), \forall\, e \in \mathbb{R}^{n+1},\ \eta \in \mathbb{R}^n. \quad (4.1.56)$$

Computing the derivative of $V(e(t), \eta(t))$ along the solution of (4.1.50) yields

$$\frac{dV(e(t),\eta(t))}{dt}\Big|_{\text{along (4.1.50)}} = \frac{1}{\varepsilon}\left[\sum_{i=1}^{n}(e_{i+1}(t) - k_i e_1(t))\frac{\partial V_1(e(t))}{\partial e_i}\right.$$

$$\left. -k_{n+1}e_1(t)\frac{\partial V_1(e(t))}{\partial e_{n+1}} + (\tilde{f}(x(t)) - \tilde{f}(\hat{x}(t)))\frac{\partial V_1(e(t))}{\partial e_n} + \varepsilon \hbar(t)\frac{\partial V_1(e(t))}{\partial e_{n+1}}\right] \quad (4.1.57)$$

$$+\left[\sum_{i=1}^{n-1}\eta_{i+1}(t)\frac{\partial V_2(\eta(t))}{\partial \eta_i} + [\alpha_1(\hat{x}_1(t) - z_1(t)) + \cdots + \alpha_n(\hat{x}_n(t) - z_n(t))\right.$$

$$\left. +\tilde{f}(x(t)) - \tilde{f}(\hat{x}(t)) + e_{n+1}(t)]\frac{\partial V_2(\eta(t))}{\partial \eta_n}\right].$$

This, together with (4.1.54) and (4.1.55), gives

$$\frac{dV(e(t),\eta(t))}{dt}\Big|_{\text{along (4.1.50)}} \le -\frac{W_1(e(t))}{\varepsilon} + 2\lambda_{\max}(P_K)|\hbar(t)|\|e(t)\| - W_2(\eta(t))$$

$$+\left(1 + \sum_{i=1}^{n}|\alpha_i| + L_1\right)2\lambda_{\max}(P_A)\|e(t)\|\|\eta(t)\| + \frac{2\lambda_{\max}(P_K)L_1}{\varepsilon}\|e(t)\|^2$$

$$\le -\frac{W_1(e(t))}{\varepsilon} + \frac{2\lambda_{\max}(P_K)L_1 + 2\lambda_{\max}(P_K)B}{\varepsilon}\|e(t)\|^2$$

$$+ 2\lambda_{\max}(P_K)B_0\|e(t)\| + 2\lambda_{\max}(P_K)B_2\|e(t)\|^2 + [2\lambda_{\max}(P_K)B_1$$

$$+\left(1 + \sum_{i=1}^{n}|\alpha_i| + L_1\right)2\lambda_{\max}(P_A)]\|e(t)\|\|\eta(t)\| - W_2(\eta(t))$$

$$\le -\left[\frac{1 - (2\lambda_{\max}(P_K)L_1 + 2\lambda_{\max}(P_K)B)}{\varepsilon} - 2\lambda_{\max}(P_K)B_2\right]\|e(t)\|^2$$

$$+ 2\lambda_{\max}(P_K)B_0\|e(t)\| - \|\eta(t)\|^2$$

$$+\left[\left(1 + \sum_{i=1}^{n}|\alpha_i| + L_1\right)2\lambda_{\max}(P_A) + 2\lambda_{\max}(P_K)B_1\right]\|e(t)\|\|\eta(t)\| \quad (4.1.58)$$

$$
= - \left[\frac{1 - (2\lambda_{\max}(P_K)L_1 + 2\lambda_{\max}(P_K)B)}{\varepsilon} - 2\lambda_{\max}(P_K)B_2 \right] \|e(t)\|^2
$$

$$
+ 2\lambda_{\max}(P_K)B_0\|e(t)\| - \|\eta(t)\|^2
$$

$$
+ \sqrt{\frac{1 - (2\lambda_{\max}(P_K)L_1 + 2\lambda_{\max}(P_K)B)}{2\varepsilon}} \|e(t)\| \sqrt{\frac{2\varepsilon}{1 - (2\lambda_{P_K}L_1 + 2\lambda_{\max}(P_K)B)}}
$$

$$
\times \left[\left(1 + \sum_{i=1}^{n} |\alpha_i| + L_1 \right) 2\lambda_{\max}(P_A) + 2\lambda_{\max}(P_K)B_1 \right] \|\eta(t)\|
$$

$$
\leq - \left[\frac{1 - (2\lambda_{\max}(P_K)L_1 + 2\lambda_{\max}(P_K)B)}{2\varepsilon} - 2\lambda_{\max}(P_K)B_2 \right] \|e(t)\|^2
$$

$$
+ 2\lambda_{\max}(P_K)B_0\|e(t)\|
$$

$$
- \left(1 - \frac{[(1 + \sum_{i=1}^{n} |\alpha_i| + L_1)2\lambda_{\max}(P_A) + 2\lambda_{\max}(P_K)B_1]^2}{2(1 - (\alpha L_1 + 2\lambda_{\max}(P_K)B)} \varepsilon \right) \|\eta(t)\|^2,
$$

where we used $2\lambda_{\max}(P_K)L_1 + 2B\lambda_{\max}(P_K) < 1$. We accomplish the proof in three steps.

Step 1. *Show that for every $R > R_0$, there exist $r > 0$ and $\varepsilon_0 > 0$ such that for every $\varepsilon \in (0, \varepsilon_0)$, there exists a $t_{1\varepsilon} > 0$ such that $\|(e(t), \eta(t))\| \leq r$ for all $t > t_{1\varepsilon}$, where $(e(t), \eta(t))$ is the solution of (4.1.50).*

Set

$$
r = \max\{2, 8\lambda_{\max}(P_K)B_0 + 4\},
$$

$$
\varepsilon_0 = \min \left\{ 1, \frac{1 - (2\lambda_{\max}(P_K)L_1 + 2\lambda_{\max}(P_K)B)}{4\lambda_{\max}(P_K)(B_0 + B_2)}, \right. \tag{4.1.59}
$$

$$
\left. \frac{1 - 4(\lambda_{\max}(P_K))^2 L_1 B}{[(1 + L + L_1)2\lambda_{\max}(P_A) + 2\lambda_{\max}(P_K)B_2]^2} \right\}.
$$

For any $\varepsilon \in (0, \varepsilon_0)$ and $\|(e(t), \eta(t))\| \geq r$, we compute the derivative of $V(e(t), \eta(t))$ along the solution of (4.1.50) in the two cases.

Case 1: $\|e(t)\| \geq r/2$. In this case, $\|e(t)\| \geq 1$ and hence $\|e(t)\|^2 \geq \|e(t)\|$. By the definition ε_0 of (4.1.59),

$$
1 - \frac{[(1 + L + L_1)2\lambda_{\max}(P_A) + 2\lambda_{\max}(P_K)B_1]^2}{2(1 - (2\lambda_{\max}(P_K)L_1 + 2\lambda_{\max}(P_K)B)} \varepsilon > 0.
$$

It follows from (4.1.58) that

$$
\frac{dV(e(t), \eta(t))}{dt}\Big|_{\text{along (4.1.50)}} \leq - \left[\frac{1 - (2\lambda_{\max}(P_K)L_1 + 2\lambda_{\max}(P_K)B)}{2\varepsilon} \right.
$$

$$
\left. - 2\lambda_{\max}(P_K)B_2 \right] \|e(t)\|^2 + 2\lambda_{\max}(P_K)B_0\|e(t)\|
$$

$$
\leq - \left[\frac{1 - (2\lambda_{\max}(P_K)L_1 + 2\lambda_{\max}(P_K)B)}{2\varepsilon} - 2\lambda_{\max}(P_K)B_2 \right] \|e(t)\|^2
$$

$$+ 2\lambda_{\max}(P_K)B_0\|e(t)\|^2$$

$$= -\frac{1 - (2\lambda_{\max}(P_K)L_1 + 2\lambda_{\max}(P_K)B) - 4\varepsilon\lambda_{\max}(P_K)(B_0 + B)}{2\varepsilon}\|e(t)\|^2 < 0,$$

$$(4.1.60)$$

where we used the fact $1 - (2\lambda_{\max}(P_K)L_1 + 2\lambda_{\max}(P_K)B) - 2\varepsilon 2\lambda_{\max}$ $(P_K)(B_0 + B) > 0$ owing to (4.1.59).

Case 2: $\|e(t)\| < r/2$. In this case, it follows from $\|(e(t), \eta(t))\| \ge r$ and $\|e(t)\| + \|\eta(t)\| \ge \|(e(t), \eta(t))\|$ that $\|\eta(t)\| \ge r/2$. In view of (4.1.59), $1 - (2\lambda_{\max}(P_K)L_1 + 2\lambda_{\max}(P_K)B) - 4\varepsilon\lambda_{\max}(P_K)B_2 > 0$. By (4.1.58), it follows that

$$\frac{dV(e(t), \eta(t))}{dt}\bigg|_{\text{along } (4.1.50)}$$

$$\le -\left(1 - \frac{[(1 + L + L_1)2\lambda_{\max}(P_A) + 2\lambda_{\max}(P_K)B_1]^2}{2(1 - (2\lambda_{\max}(P_K)L_1 + 2\lambda_{\max}(P_K)B)}\varepsilon\right)\|\eta(t)\|^2$$

$$+ 2\lambda_{\max}(P_K)B_0\|e(t)\|$$

$$\le 2\lambda_{\max}(P_K)B_0\frac{r}{2} - \frac{1}{2}\left(\frac{r}{2}\right)^2 = -\frac{r}{8}[r - 42\lambda_{\max}(P_K)B_0] < 0,$$

$$(4.1.61)$$

where owing to (4.1.59), $r - 8\lambda_{\max}(P_K)B_0 > 0$ and $[(1 + L + L_1) 2\lambda_{\max}(P_A) + 2\lambda_{\max}(P_K)B_1]^2/(2(1 - (2\lambda_{\max}(P_K)L_1 + 2\lambda_{\max}(P_K)B))\varepsilon \le \frac{1}{2}$.

Combining the above two cases, we conclude that for every $\varepsilon \in (0, \varepsilon_0)$ and $\|(e(t), \eta(t))\| \ge r$, the derivative of $V(e(t), \eta(t))$ along the solution of system (4.1.50) satisfies

$$\frac{d(e(t), \eta(t))}{dt}\bigg|_{\text{along } (4.1.50)} < 0. \qquad (4.1.62)$$

This shows that there exists a $t_{1\varepsilon} > 0$ such that $\|(e(t), \eta(t))\| \le r$ for all $t > t_{1\varepsilon}$.

Step 2. *Establish the convergence of* $|x_i(t) - \hat{x}_i(t)|$.

Consider the subsystem that is composed of the last $n + 1$ equations in system (4.1.50):

$$\begin{cases} \varepsilon\dot{e}_1(t) = e_2(t) - k_1e_1(t), \\ \vdots \\ \varepsilon\dot{e}_n(t) = e_{n+1}(t) - k_ne_1(t) + \tilde{f}(x(t)) - \tilde{f}(\hat{x}(t)), \\ \varepsilon\dot{e}_{n+1}(t) = -k_{n+1}e_1(t) + \varepsilon\hbar(t), \end{cases} \qquad (4.1.63)$$

where $\hbar(t)$ is given by (4.1.51). Since $\|(e(t), x(t))\| \le r$ for all $t > t_{1\varepsilon}$, we have, in view of (4.1.52), that $|\hbar(t)| \le M_0 + \frac{B}{\varepsilon}\|e(t)\|$ for all $t > t_{1\varepsilon}$ and some $M_0 > 0$. This together with Assumption 4.1.3 shows that the derivative of $V_1(e(t))$ along the solution of (4.1.63) satisfies

$$
\frac{dV_1(e(t))}{dt}\Big|_{\text{along (4.1.63)}}
$$

$$
= \frac{1}{\varepsilon}\left(\sum_{i=1}^{n}(e_{i+1}(t) - k_i e_1(t))\frac{\partial V_1(e(t))}{\partial e_i} - k_{n+1}e_1(t)\frac{\partial V_1(e(t))}{\partial e_{n+1}}\right.
$$

$$
\left. + [\tilde{f}(x(t)) - \tilde{f}(\hat{x}(t))]\frac{\partial V_1(e(t))}{\partial e_n} + \varepsilon \hbar(t)\frac{\partial V_1(e(t))}{\partial e_{n+1}}\right)
$$

$$
\leq -\frac{1 - (2\lambda_{\max}(P_K)L_1 + 2\lambda_{\max}(P_K)B)}{\varepsilon}\|e(t)\|^2
$$

$$
+ 2M_0\lambda_{\max}(P_K)\|e(t)\|
$$

$$
\leq -\frac{1 - (2\lambda_{\max}(P_K)L_1 + 2\lambda_{\max}(P_K)B)}{\varepsilon\lambda_{\max}(P_K)}V_1(e(t))
$$

$$
+ \frac{2M_0\lambda_{\max}(P_K)\sqrt{\lambda_{\min}(P_K)}}{\lambda_{\min}(P_K)}\sqrt{V_1(e(t))}, \quad \forall\, t > t_{1\varepsilon}. \tag{4.1.64}
$$

It follows that

$$
\frac{d\sqrt{V_1(e(t))}}{dt} \leq -\frac{1 - (2\lambda_{\max}(P_K)L_1 + 2\lambda_{\max}(P_K)B)}{\varepsilon\lambda_{\max}(P_K)}\sqrt{V_1(e(t))}
$$

$$
+ \frac{2\varepsilon M_0\lambda_{\max}(P_K)\sqrt{\lambda_{\min}(P_K)}}{\lambda_{\min}(P_K)}, \forall\, t > t_{1\varepsilon}, \tag{4.1.65}
$$

and hence

$$
\sqrt{V_1(e(t))} \leq \sqrt{V_1(e(t_{1\varepsilon}))}e^{-\frac{1-(2\lambda_{\max}(P_K)L_1 + 2\lambda_{\max}(P_K)B)}{\varepsilon\lambda_{\max}(P_K)}(t-t_{1\varepsilon})}
$$

$$
+ \frac{2M\lambda_{\max}(P_K)\sqrt{\lambda_{\min}(P_K)}}{\lambda_{\min}(P_K)}\int_{t_{1\varepsilon}}^{t}
$$

$$
e^{-1-(2\lambda_{\max}(P_K)L_1 + 2\lambda_{\max}(P_K)B)/(\varepsilon\lambda_{\max}(P_K))(t-s)}ds, \forall\, t > t_{1\varepsilon}, \tag{4.1.66}
$$

where once again we used the assumption that $1 > 2\lambda_{\max}(P_K)L_1 + 2\lambda_{\max}(P_K)B$. Since the first term of the right-hand side of (4.1.66) tends to zero as t goes to infinity and the second term is bounded by ε multiplied by an ε-independent constant, by the definition of $e(t)$, we conclude that there exist $t_{2\varepsilon} > t_{1\varepsilon}$ and $\Gamma_1 > 0$ such that

$$
|x_i(t) - \hat{x}_i(t)| = \varepsilon^{n+1-i}|e_i(t)| \leq \|e(t)\| \leq \sqrt{\frac{V_1(e(t))}{\lambda_{\min}(P_K)}} \leq \Gamma_1\varepsilon^{n+2-i}, \forall\, t > t_{2\varepsilon}. \tag{4.1.67}
$$

Step 3. *Establish the convergence of $\|x(t) - z(t)\|$.*

Consider the following subsystem that is composed of the first n number equations in (4.1.50):

$$
\begin{cases}
\dot{\eta}_1(t) = \eta_2(t), \\
\dot{\eta}_2(t) = \eta_3(t), \\
\quad\vdots \\
\dot{\eta}_n(t) = \alpha_1(\hat{x}_1(t) - z_1(t)) + \cdots + \alpha_n(\hat{x}_n(t) - z_n(t)) \\
\qquad\quad + \tilde{f}(x(t)) - \tilde{f}(\hat{x}(t)) + e_{n+1}(t).
\end{cases}
\tag{4.1.68}
$$

Computing the derivative of $V_2(\eta(t))$ along the solution of system (4.1.68) gives

$$
\frac{dV_2(\eta(t))}{dt}\Big|_{\text{along } (4.1.68)}
$$

$$
= \sum_{i=1}^{n-1} \eta_{i+1}(t)\frac{\partial V_2(\eta(t))}{\partial \eta_i} + \{e_{n+1}(t) + \alpha_1\eta_1(t) + \cdots + \alpha_n\eta_n(t) + e_{n+1}(t)
$$

$$
+ [\alpha_1(\hat{x}_1(t) - x_1(t)) + \cdots + \alpha_n(\hat{x}_n(t) - x_n(t))]
$$

$$
+ \tilde{f}(x(t)) - \tilde{f}(\hat{x}(t))\}\frac{\partial V_2(\eta(t))}{\partial \eta_n}
$$

$$
\leq -W_2(\eta(t)) + \left(1 + \sum_{i=1}^{n}|\alpha_i| + L_1\right)2\lambda_{\max}(P_A)\|e(t)\|\|\eta(t)\|
$$

$$
\leq -\frac{1}{\lambda_{\max}(P_A)}V_2(\eta(t)) + N_0\varepsilon\sqrt{V_2(\eta(t))}, \forall\, t > t_{2\varepsilon},
\tag{4.1.69}
$$

where N_0 is an R-dependent positive constant. By (4.1.69), we have, for all $t > t_{2\varepsilon}$, that

$$
\frac{d\sqrt{V_2(\eta(t))}}{dt}\Big|_{\text{along } (4.1.68)} \leq -\frac{1}{\lambda_{\max}(P_A)}\sqrt{V_2(\eta(t))} + N_0\varepsilon,
\tag{4.1.70}
$$

and hence

$$
\sqrt{V_2(\eta(t))} \leq \sqrt{V_2(\eta(t_{2\varepsilon}))}e^{-\frac{1}{\lambda_{\max}(P_A)}(t-t_{2\varepsilon})}
$$

$$
+ N_0\varepsilon\int_{t_{2\varepsilon}}^{t} e^{-\frac{1}{\lambda_{\max}(P_A)}(t-s)}ds, \forall\, t > t_{2\varepsilon}.
\tag{4.1.71}
$$

Once again, the first term of the right-hand side of (4.1.71) tends to zero as t goes to infinity and the second term is bounded by ε multiplied by an ε-independent constant. Thus there exist $t_\varepsilon > t_{2\varepsilon}$ and $\Gamma_2 > 0$ such that $\|x(t) - z(t)\| \leq \Gamma_2\varepsilon$ for all $t > t_\varepsilon$. This completes the proof of the theorem. $\square$

4.1.3   Semi-Global Convergence of LADRC

In this section, by using the a priori estimate for the bounds of the initial state and disturbance, we design a semi-global convergence linear ADRC based on LESO. The system that we consider in this section is as follows:

$$
\begin{cases}
\dot{x}_1(t) = x_2(t), \\
\dot{x}_2(t) = x_3(t), \\
\quad \vdots \\
\dot{x}_n(t) = f(t, x(t), \zeta(t), w(t)) + b(x(t), \zeta(t), w(t))u(t), \\
\dot{\zeta}(t) = F_0(x(t), \zeta(t), w(t)), \\
y(t) = x_1(t),
\end{cases}
\tag{4.1.72}
$$

where $x(t) = (x_1(t), x_2(t), \ldots, x_n(t))^\top \in \mathbb{R}^n$, and $\zeta(t) \in \mathbb{R}^m$ are states, $y(t) \in \mathbb{R}$ is the output, $u(t) \in \mathbb{R}$ is the input, $f \in C^1(\mathbb{R}^{n+m+2}, \mathbb{R}), b \in C^1(\mathbb{R}^{n+m+1}, \mathbb{R})$, $F_0 \in C^1(\mathbb{R}^{n+m+1}, \mathbb{R}^m)$ and $w \in C^1(\mathbb{R}, \mathbb{R})$ is the external disturbance.

Let $\mathrm{sat}_M : \mathbb{R} \to \mathbb{R}$ be an odd continuous differentiable saturation function defined as follows:

$$
\mathrm{sat}_M(\nu) =
\begin{cases}
\nu, & 0 \leq \nu \leq M, \\
-\dfrac{1}{2}\nu^2 + (M+1)\nu - \dfrac{1}{2}M^2, & M < \nu \leq M+1, \\
M + \dfrac{1}{2}, & \nu > M+1,
\end{cases}
\tag{4.1.73}
$$

and the $\mathrm{sat}_M(\nu)$ for $\nu \leq 0$ is defined by symmetry. The observer-based feedback control is designed as follows:

$$
\text{ADRC(S):} \quad u(t) = \frac{1}{b_0(\hat{x}(t))}[\mathrm{sat}_{M_2}(\alpha_1(\hat{x}_1(t) - z_1(t)) + \alpha_2(\hat{x}_2(t) - z_2(t))
$$

$$
+ \cdots + \alpha_n(\hat{x}_n(t) - z_n(t))) - \mathrm{sat}_{M_1}(\hat{x}_{n+1}(t)) + \mathrm{sat}_{C_1+1}(z_{(n+1)}(t))],
\tag{4.1.74}
$$

where $z(t) = (z_1(t), z_2(t), \ldots, z_{n+1}(t))$ comes from the LTD:

$$
\text{LTD:} \quad
\begin{cases}
\dot{z}_1(t) = z_2(t), \\
\dot{z}_2(t) = z_3(t), \\
\quad \vdots \\
\dot{z}_n(t) = z_{n+1}(t), \\
\dot{z}_{n+1}(t) = R^{n+1}\left(a_1(z_1(t) - v(t)) + \dfrac{a_2 z_2(t)}{R} + \cdots + \dfrac{a_{n+1} z_n(t)}{R^n} \right),
\end{cases}
\tag{4.1.75}
$$

and $\hat{x}(t) = (\hat{x}_1(t), \hat{x}_2(t), \ldots, \hat{x}_{n+1}(t))$comes from the LESO:

$$
\begin{cases}
\dot{\hat{x}}_1(t) = \hat{x}_2(t) + \dfrac{k_1}{\varepsilon}(y(t) - \hat{x}_1(t)), \\[2mm]
\dot{\hat{x}}_2(t) = \hat{x}_3(t) + \dfrac{k_2}{\varepsilon^2}(y(t) - \hat{x}_1(t)), \\[2mm]
\qquad \vdots \\[2mm]
\dot{\hat{x}}_n(t) = \hat{x}_{n+1}(t) + \dfrac{k_n}{\varepsilon^n}(y(t) - \hat{x}_1(t)), \\[2mm]
\dot{\hat{x}}_{n+1}(t) = \dfrac{k_{n+1}}{\varepsilon^{n+1}}(y(t) - \hat{x}_1(t)).
\end{cases}
\tag{4.1.76}
$$

The constants M_1 and M_2 depend on a priori bounds of the initial state and external disturbance to be specified later.

For system function $f(\cdot)$, we need the following assumption.

Assumption 4.1.4 The $f \in C^1(\mathbb{R}^{n+m+2}, \mathbb{R})$ and there exist positive constant N and functions $\varpi_1 \in C(\mathbb{R}^n, [0,\infty))$, $\varpi_2 \in C(\mathbb{R}^m, [0,\infty))$, and $\varpi_3 \in C(\mathbb{R}, [0,\infty))$ such that

$$
\sum_{i=1}^{n} \left| \frac{\partial f(t, x, \zeta, w)}{\partial x_i} \right| + \sum_{i=1}^{m} \left| \frac{\partial f(t, x, \zeta, w)}{\partial \zeta_i} \right| + \left| \frac{\partial f(t, x, \zeta, w)}{\partial t} \right| + |f(t, x, \zeta, w)|
$$

$$
\leq N + \varpi_1(x) + \varpi_2(\zeta) + \varpi_3(w), \forall\, x \in \mathbb{R}^n, \zeta \in \mathbb{R}^m, w \in \mathbb{R}.
$$

The following Assumption 4.1.5 is on a priori bounds of the initial state and external disturbance.

Assumption 4.1.5 There exist positive constants C_1, C_2, and C_3 such that $\sup_{t \in [0,\infty)}$ $\|(v(t), \ldots, v^{(n)}(t))\| < C_1$, $\|x(0)\| < C_2$, $\|(w(t), \dot{w}(t))\| < C_3$ for all $t \in [0, \infty)$. Let $C_1^* = (C_1 + C_2 + 1)^2 \lambda_{\max}(P_A)$, let P_A be given in (4.1.80), and

$$
C_4 \geq \max \left\{ \sup_{\|x\| \leq C_1 + C_1^* + 2, \|w\| \leq C_3,} |\chi(x, w)|, \|\zeta(0)\| \right\}.
\tag{4.1.77}
$$

The constants M_1 and M_2 in feedback control (4.1.74) are chosen such that

$$
\begin{cases}
M_1 \geq 2 \left(1 + M_2 + C_1 + N + \sup_{\|x\| \leq C_1 + C_1^* + 2} \varpi_1(x) + \sup_{\|w\| \leq C_3} \varpi_3(w) + \sup_{\|\zeta\| \leq C_4} \varpi_2(\zeta) \right), \\[2mm]
M_2 \geq (C_1 + C_1^* + 2)(|\alpha_1| + |\alpha_2| + \cdots + |\alpha_n|).
\end{cases}
\tag{4.1.78}
$$

The following assumption is to guarantee the input-to-state stable for zero dynamics.

Assumption 4.1.6 There exist positive definite functions $V_0, W_0 \mathbb{R}^s \to \mathbb{R}$ such that $L_{F_0} V_0(\zeta) \leq -W_0(\zeta)$ for all $\zeta : \|\zeta\| > \chi(x, w)$, where $\chi : \mathbb{R}^{n+1} \to \mathbb{R}$ is a wedge function and $L_{F_0} V_0(\zeta)$ denotes the Lie derivative along the zero dynamics of system (4.1.72).

For the parameters $a_i, i = 1, 2, \ldots, n+1$ in the LTD loop, $k_i, i = 1, 2, \ldots, n+1$ in the LESO loop, and $\alpha_i, i = 1, 2, \ldots, n+1$ in the feedback loop, we assume that the matrices composed by those parameters are Hurwitz:

$$A = \begin{pmatrix} 0 & 1 & 0 & \cdots & 0 & 0 \\ 0 & 0 & 1 & \cdots & 0 & 0 \\ \vdots & \vdots & \vdots & \ddots & \vdots & \vdots \\ 0 & 0 & 0 & \cdots & 0 & 1 \\ \alpha_1 & \alpha_2 & \alpha_3 & \cdots & \alpha_{n-1} & \alpha_n \end{pmatrix}_{n \times n},$$

$$\tilde{A} = \begin{pmatrix} 0 & 1 & 0 & \cdots & 0 & 0 \\ 0 & 0 & 1 & \cdots & 0 & 0 \\ \vdots & \vdots & \vdots & \ddots & \vdots & \vdots \\ 0 & 0 & 0 & \cdots & 0 & 1 \\ a_1 & a_2 & a_3 & \cdots & a_n & a_{n+1} \end{pmatrix}_{(n+1) \times (n+1)}, \qquad (4.1.79)$$

$$K = \begin{pmatrix} -k_1 & 1 & 0 & \cdots & 0 \\ -k_2 & 0 & 1 & \cdots & 0 \\ \vdots & & \vdots & \vdots & \ddots & \vdots \\ -k_n & 0 & 0 & \cdots & 1 \\ -k_{n+1} & 0 & 0 & \cdots & 0 \end{pmatrix}_{(n+1) \times (n+1)}.$$

For Hurwitz matrices A and K, there exist, respectively, positive definite matrices P_A and P_K satisfying the following Lyapunov equations:

$$P_K K + K^\top P_K = -I_{(n+1) \times (n+1)}, \quad P_A A + A^\top P_A = -I_{n \times n}. \qquad (4.1.80)$$

For the nominal control parameter $b_0(x)$ of $b(x)$, we assume that it is close to $b(x)$.

Assumption 4.1.7 There exists a nominal parameter function $b_0 \in C^1(\mathbb{R}^n, \mathbb{R})$ such that

(i) $b_0(x) \neq 0$ for every $x \in \mathbb{R}^n$.
(ii) The $b_0(x)$, $1/b_0(x)$, and all their partial derivatives with respect to their arguments are globally bounded.
(iii)

$$\vartheta = \sup_{\|x\| \leq C_1 + C_1^* + 2, \|\zeta\| \leq C_4, \|w\| \leq C_3, \nu \in \mathbb{R}^n} \left| \frac{b(x, \zeta, w) - b_0(x)}{b_0(\nu)} \right|$$

$$< \min \left\{ \frac{1}{2}, \left(2\lambda_{\max}(P_K) |k_{n+1}| \left(M_1 + \frac{1}{2} \right) \right)^{-1} \right\}. \qquad (4.1.81)$$

Under feedback (4.1.74), the closed-loop of system (4.1.72) and ESO (4.1.76) are rewritten as

$$
\begin{cases}
\dot{x}_1(t) = x_2(t), \\[4pt]
\dot{x}_2(t) = x_3(t), \\[4pt]
\quad\vdots \\[4pt]
\dot{x}_n(t) = f(t, x(t), \zeta(t), w(t)) + b(x(t), \zeta(t), w(t))u(t), \\[4pt]
\dot{\zeta}(t) = F_0(x(t), \zeta(t), w(t)), \\[4pt]
\dot{\hat{x}}_1(t) = \hat{x}_2(t) + \dfrac{k_1}{\varepsilon}(y(t) - \hat{x}_1(t)), \\[10pt]
\dot{\hat{x}}_2(t) = \hat{x}_3(t) + \dfrac{k_2}{\varepsilon^2}(y(t) - \hat{x}_1(t)), \\[10pt]
\quad\vdots \\[4pt]
\dot{\hat{x}}_n(t) = \hat{x}_{n+1}(t) + \dfrac{k_n}{\varepsilon^n}(y(t) - \hat{x}_1(t)), \\[10pt]
\dot{\hat{x}}_{n+1}(t) = \dfrac{k_{n+1}}{\varepsilon^{n+1}}(y(t) - \hat{x}_1(t)), \\[10pt]
u(t) = \dfrac{1}{b_0(\hat{x}(t))}\left[\operatorname{sat}_{M_2}(\alpha_1(\hat{x}_1(t) - z_1(t)) + \alpha_2(\hat{x}_2(t) - z_2(t)) + \cdots \right. \\[10pt]
\qquad\qquad +\alpha_n(\hat{x}_n(t) - z_n(t))) \\[6pt]
\quad -\operatorname{sat}_{M_1}(\hat{x}_{n+1}(t)) + \operatorname{sat}_{C_1+1}(z_{(n+1)}(t))\big].
\end{cases}
\tag{4.1.82}
$$

Theorem 4.1.3 *Let $x_i(t)(1 \le i \le n)$ and $\hat{x}_i(t)(1 \le i \le n+1)$ be the solutions of system (4.1.72) and (4.1.76) coupled by (4.1.74), and let $x_{n+1}(t) = f(t, x(t), \zeta(t), w(t)) + (b(x(t), \zeta(t), w(t)) - b_0(x(t)))u(t)$ be the extended state. Suppose (4.1.75) and that Assumptions 4.1.4 to 4.1.7 are satisfied. Then for any given initial value of (4.1.72), the following assertions hold.*

(i) *For any $\sigma > 0$ and $\tau > 0$, there exists a constant $R_0 > 0$ such that $|z_1(t) - v(t)| < \sigma$ uniformly in $t \in [\tau, \infty)$ for all $R > R_0$, where $z_i(t)$ is the solution of (4.1.75).*

(ii) *For any $\sigma > 0$, there exist $\varepsilon_0 > 0$ and $R_1 > R_0$ such that for any $\varepsilon \in (0, \varepsilon_0)$, $R > R_1$, there exists an ε-independent constant $t_0 > 0$, such that*

$$|x_i(t) - \hat{x}_i(t)| < \sigma \text{ for all } t > t_0, i = 1, 2, \ldots, n+1. \tag{4.1.83}$$

(iii) *For any $\sigma > 0$, there exist $R_2 > R_1$ and $\varepsilon_1 \in (0, \varepsilon_0)$ such that for any $R > R_2$ and $\varepsilon \in (0, \varepsilon_1)$, there exists a $t_{R\varepsilon} > 0$ such that $|x_1(t) - v(t)| < \sigma$ for all $R > R_1$, $\varepsilon \in (0, \varepsilon_1)$, and $t > t_{R\varepsilon}$.*

The conclusion (i) follows directly from the convergence results of the linear tracking differentiator. Let

$$e_i(t) = \frac{x_i(t) - \hat{x}_i(t)}{\varepsilon^{n+1-i}}, \ i = 1, 2, \ldots, n+1, \ \eta_j(t) = x_j(t) - z_j(t), j = 1, 2, \ldots, n.$$

(4.1.84)

The proof of Theorem 4.1.3 is based on the boundedness of the solution stated in the following Lemma 4.1.1.

Lemma 4.1.1 *Assume that Assumptions 4.1.4 to 4.1.7 are satisfied. Let $\Omega_0 = \{y | V_2(y) \le C_1^*\}$ and $\Omega_1 = \{y | V_2(y) \le C_1^* + 1\}$, where the Lyapunov functions $V_1 : \mathbb{R}^{n+1} \to \mathbb{R}$ and $V_2 : \mathbb{R}^n \to \mathbb{R}$ are defined as $V_1(\nu) = \langle P_K \nu, \nu \rangle, \nu \in \mathbb{R}^{n+1}$ and $V_2(\iota) = \langle P_A \iota, \iota \rangle, \iota \in \mathbb{R}^n$ with P_K and P_A given in (4.1.80). Then there exists an $\varepsilon_1 > 0$ such that for any $\varepsilon \in (0, \varepsilon_1)$ and $t \in [0, \infty)$, $\eta(t) \in \Omega_1$.*

Proof. First, we see that for any $\varepsilon > 0$,

$$|\eta_j(t)| \le |\eta_j(0)| + |\eta_{j+1}(t)|t, 1 \le j \le n-1, 1 \le i \le m,$$
$$|\eta_n(t, \varepsilon)| \le |\eta_n(0)| + [C_1 + M_1 + mM_1^*(C_1 + M_1 + M_2)]t, \quad \eta(t) \in \Omega_1,$$

(4.1.85)

where

$$M_1^* = \sup_{\|x\| \le C_1 + (C_1^*+1)/2+1, \|w\| \le \bar{C}_3, \|\xi\| \le \bar{C}_4} |b(x, \xi, w)|.$$

Next, by the iteration process, we can show that all terms on the right-hand side of (4.1.85) are ε-independent. Since $\|\eta(0)\| < C_1 + C_2, \eta(0) \in \Omega_0$, there exists an ε-independent constant $t_0 > 0$ such that $\eta(t) \in \Omega_0$ for all $t \in [0, t_0]$.

Now we suppose Lemma 4.1.1 is false to obtain a contradiction. Then, for any $\varepsilon > 0$, there exists $t^* \in (0, \infty)$ such that

$$\eta(t^*) \in \mathbb{R}^n - \Omega_1.$$

(4.1.86)

Since for any $t \in [0, t_0], \eta(t, \varepsilon^*) \in \Omega_0$, and $\eta(t)$ is continuous in t, there exist $t_1 \in (t_0, t_2)$ such that

$$\eta(t_1) \in \partial\Omega_0 \text{ or } V_2(\eta(t_1)) = C_1^*,$$
$$\|\eta(t_2)\| \in \Omega_1 - \Omega_0 \text{ or } C_1^* < V_2(\eta(t_2)) \le C_1^* + 1,$$
$$\eta(t) \in \Omega_1 - \Omega_0^\circ, \forall t \in [t_1, t_2] \text{ or } C_1^* \le V_2(\eta(t)) \le C_1^* + 1,$$
$$\eta(t) \in \Omega_1, \forall t \in [0, t_2].$$

(4.1.87)

By (4.1.72) and (4.1.84), it follows that the error $e(t) = (e_1(t), e_2(t), \ldots, e_n(t), e_{n+1}(t))^\top$ in this case satisfies

$$\varepsilon \dot{e}(t) = Pe(t) + \Delta(t), \ \Delta(t) = (0, \ldots, 0, \Delta_1(t), \varepsilon \Delta_2(t))^\top,$$

(4.1.88)

where

$$\begin{cases} \Delta_1(t) = (b_0(x(t)) - b_0(\hat{x}(t)))u(t), \hat{x}(t) = (\hat{x}_1(t), \ldots, \hat{x}_n(t))^\top, \\ \Delta_2 = \frac{d}{dt}\left(f(t, x(t), \zeta(t), w(t)) + (b(x(t), \zeta(t), w(t)) - b_0(x(t))u(t))\right)\Big|_{\text{along } (4.1.82)}. \end{cases}$$

(4.1.89)

Since all the partial derivatives of $b_0(\cdot)$ are globally bounded, there exists a constant $N_0 > 0$ such that $|\Delta_1(t)| \leq \varepsilon N_0 \|e(t)\|$.

From Assumptions 4.1.4 to 4.1.6, $\|(w(t), \dot{w}(t))\|$, $\|x(t)\|$, and $\|\zeta(t)\|$ are bounded in $[t_1, t_2]$, and $\|z(t)\| = \|(z_1(t), z_2(t), \ldots, z_{n+1}(t))\|$ is bounded on $[0, \infty)$.

Considering the derivative of $x_{n+1}(t)$ with respect to t in the interval $[t_1, t_2]$, we have

$$\Delta_2(t) = \frac{d(f(x(t), \zeta(t), w_i(t)) + (b(x(t), \zeta(t), w(t)) - b_0(x(t)))u(t))}{dt}$$

$$= \left. \frac{df(x(t), \zeta(t), w(t))}{dt} \right|_{\text{along (4.1.82)}} + u(t) \left. \frac{d(b(x(t), \zeta(t), w(t)) - b_0(x(t)))}{dt} \right|_{\text{along (4.1.82)}}$$

$$+ \dot{u}(t)(b(x(t), \zeta(t), w(t)) - b_0(x(t))). \tag{4.1.90}$$

It is easy to verify that for any $t \in [0, t_2]$,

$$\left| \left. \frac{df(x(t), \zeta(t), w(t))}{dt} \right|_{\text{along (4.1.82)}} + u(t) \left. \frac{d(b(x(t), \zeta(t), w(t)) - b_0(x(t)))}{dt} \right|_{\text{along (4.1.82)}} \right|$$

$$\leq N_1 + N_2 \|e\|, \quad N_1, N_2 > 0. \tag{4.1.91}$$

Now we look at the derivative of $u(t)$. It follows from (4.1.74) that

$$\dot{u}(t) = \left. \frac{d}{dt} \left(\frac{1}{b_0(\hat{x}(t))} \right) \right|_{\text{along (4.1.82)}} \left[\text{sat}_{M_2}(\alpha_1(\hat{x}_1(t) - z_1(t)) + \alpha_2(\hat{x}_2(t) - z_2(t)) + \cdots \right.$$

$$\left. + \alpha_n(\hat{x}_n(t) - z_n(t))) - \text{sat}_{M_1}(\hat{x}_{n+1}(t)) + \text{sat}_{C_1+1}(z_{(n+1)}(t)) \right]$$

$$+ \left. \frac{1}{b_0(\hat{x}(t))} \frac{d}{dt} \left(\text{sat}_{M_2}(\alpha_1(\hat{x}_1(t) - z_1(t)) + \cdots + \alpha_n(\hat{x}_n(t) - z_n(t))) \right) \right|_{\text{along (4.1.82)}}$$

$$+ \frac{1}{b_0(\hat{x}(t))} \frac{d}{dt} \text{sat}_{C_1+1}(z_{(n+1)}(t)) + \frac{1}{b_0(\hat{x}(t))} \dot{\text{sat}}_{M_1}(\hat{x}_{n+1}(t)) \frac{k_{n+1} e_1(t)}{\varepsilon}. \tag{4.1.92}$$

It is easy to verify that

$$|\dot{u}(t)| \leq N_3 + N_4 \|e(t)\| + \frac{|k_{n+1}|}{\varepsilon b_0(\hat{x}(t))} \|e(t)\|, \quad N_3, N_4 > 0. \tag{4.1.93}$$

This, together with (4.1.91), gives

$$|\Delta_2(t)| \leq N_5 + N_6 \|e(t)\| + \frac{N}{\varepsilon} \|e(t)\|, \tag{4.1.94}$$

where $N_5, N_6 > 0$ and

$$N = |k_{n+1}| \left(M_1 + \frac{1}{2} \right) \sup_{\|x\| \leq C_1 + C_1^* + 2, \|\xi\| \leq C_4, \|w\| \leq C_3, \nu \in \mathbb{R}^n} \left| \frac{1}{b_0(\nu)} \right| |b(x, \xi, w) - b_0(x)|$$

$$= |k_{n+1}| \left(M_1 + \frac{1}{2} \right) \vartheta. \tag{4.1.95}$$

Finding the derivative of $V_1(e(t))$ along system (4.1.88) with respect to t shows that for any $\varepsilon \in (0, \ (1 - 2N\lambda_{\max}(P_K))/(2(N_0 + N_2 + N_4)\lambda_{\max}(P_K)))$ and $t \in [0, t_2]$,

$$\left. \frac{dV_1(e(t))}{dt} \right|_{\text{along (4.1.88)}}$$

$$= \left. \frac{d(e^\top(t)P_K e(t))}{dt} \right|_{\text{along (4.1.88)}}$$

$$= \frac{1}{\varepsilon}(Ke(t) + \Delta(t))^\top P_K e(t) + e^\top(t)P_K(Ke(t) + \Delta(t))$$

$$\leq -\frac{1 - 2\lambda_{\max}(P_K)N}{\varepsilon}\|e(t)\| + 2\lambda_{\max}(P_K)(N_0 + N_6)\|e(t)\|^2 + 2\lambda_{\max}(P_K)N_5\|e(t)\|$$

$$\leq -\frac{\Pi_1}{\varepsilon}V_1(e(t)) + \Pi_2 V_1(e(t)) + \Pi_3\sqrt{V_1(e(t))}, \tag{4.1.96}$$

where

$$\Pi_1 = \frac{1 - 2\lambda_{\max}(P_K)N}{\max\{\lambda_{\max}(P_K)\}}, \Pi_2 = \frac{2\lambda_{\max}(P_K)(N_0 + N_6)}{\lambda_{\min}(P_K)}, \Pi_3 = \frac{2\lambda_{\max}(P_K)N_5}{\lambda_{\min}(P_K)}. \tag{4.1.97}$$

Therefore, for every $0 < \varepsilon < 1 - 2\lambda_{\max}(P_K)N$ and $t \in [0, t_2]$,

$$\|e(t)\| \leq \frac{1}{\sqrt{\lambda_{\min}(P_K)}}\sqrt{V_1(e(t))} \leq \frac{1}{\sqrt{\lambda_{\min}(P_K)}}\left[e^{(-\Pi_1/\varepsilon + \Pi_2)t}\sqrt{V_1(e(0))} \right.$$

$$\left. + \Pi_3\int_0^t e^{(-\Pi_1/\varepsilon + \Pi_2)(t-s)}ds \right]. \tag{4.1.98}$$

Passing to the limit as $\varepsilon \to 0$ yields, for any $t \in [t_1, t_2]$, that

$$e^{(-\Pi_1/\varepsilon + \Pi_2)t}\sqrt{V_1(e(0))} \leq \frac{1}{\sqrt{\min\{\lambda_{11}^i\}}}e^{(-\Pi_1/\varepsilon + \Pi_2)t}$$

$$\times \sum_{i=1}^m \left\| \left(\frac{e_{i1}(t)}{\varepsilon^{n_i+1}}, \frac{e_{i2}(t)}{\varepsilon^{n_i}}, \ldots, e_{i(n_i+1)}(t) \right) \right\| \to 0. \tag{4.1.99}$$

Hence for any $\sigma \in (0, \min\{1/2, (C_1 + C_2)/N\})$, there exists an $\varepsilon_1 \in (0, 1)$ such that $\|e(t)\| \leq \sigma$ for all $\varepsilon \in (0, \varepsilon_1)$ and $t \in [t_1, t_2]$, where $N_7 = 2\lambda_{\max}(P_A)(1 + \sum_{i=1}^n |\alpha_i|)\}$. Notice that for any $0 < \varepsilon < \varepsilon_1$ and $t \in [t_1, t_2]$, $\eta \in \Omega_1$, $\|e(t)\| \leq \sigma$,

$$\|\hat{x}(t) - z(t)\| \leq \|x(t) - \hat{x}(t)\| + \|x(t) - z(t)\| \leq (C_1^* + 1)/\lambda_{\max}(P_A) + 1,$$

$$|\phi_i(\hat{x}(t) - z(t))| \leq M_2,$$

$$|\hat{x}_{n_i+1}(t)| \leq |e_{n+1}(t)| + |x_{n_i+1}(t)| \leq |e_{n+1}(t)| + \left| f(x(t), \xi(t), w(t)) \right.$$

$$\left. + (a(x(t), \xi(t), w(t)) - b(x(t)))u(t) \right|$$

$$\leq |e_{n+1}(t)| + |f(x(t), \xi(t), w(t))| + \vartheta(M_1 + M_2 + C_2)$$
$$\leq 1 + M_2 + C_1 + |f(x(t), \xi(t), w(t))| + \vartheta M_1 \leq M_1. \tag{4.1.100}$$

Therefore, $u_i(t)$ in (4.1.82) takes the form:

$$u_i(t) = u(t) = \frac{1}{b_0(\hat{x}(t))}[\alpha_1(\hat{x}_1(t) - z_1(t)) + \alpha_2(\hat{x}_2(t) - z_2(t)) + \cdots + \alpha_n(\hat{x}_n(t) - z_n(t)))$$
$$- \hat{x}_{n+1}(t) + z_{(n+1)}(t)]$$

for all $t \in [t_1, t_2]$. With this $u_i^*(t)$, the derivative of $V_2(\eta(t))$, along system (4.1.82) with respect to t in interval $[t_1, t_2]$, satisfies

$$\frac{dV_2(\eta(t))}{dt}\bigg|_{\text{along (4.1.82)}} = \sum_{i=1}^{m}(-W_2(\eta(t)) + N_6\sigma\|\eta(t)\|)$$

$$\leq -\|\eta(t)\|^2 + mN_6\|e(t)\|\|\eta(t)\| < 0, \tag{4.1.101}$$

which contradicts (4.1.87). This completes the proof of the lemma. $\square$

Proof of Theorem 4.1.3 From Lemma 4.1.1, $\eta(t) \in \Omega_1$ for all $\varepsilon \in (0, \varepsilon_1)$, and $t \in (0, \infty)$, it follows that (4.1.100) holds true for all $t \in [0, \infty)$. Therefore, (4.1.101) and (4.1.96) also hold true for any $\varepsilon \in (0, \varepsilon_1)$ and $t \in [0, \infty)$.

For any $\sigma > 0$, it follows from (4.1.101) that there exists a $\sigma_1 \in (0, \sigma/2)$ such that $\lim_{t\to\infty}\|\eta(t)\| \leq \sigma/2$ provided that $\|e(t)\| \leq \sigma_1$. From (4.1.96), for any $\tau > 0$ and determined $\sigma_1 > 0$, there exist an $\varepsilon_0 \in (1, \varepsilon_1)$ such that $\|x(t) - \hat{x}(t)\| \leq \sigma_1$ for any $\varepsilon \in (0, \varepsilon_0)$, $t > \tau$. This completes the proof of the theorem. $\square$

4.1.4 Numerical Simulations

Example 4.1.1 *Consider the following second-order system:*

$$\begin{cases} \dot{x}_1(t) = x_2(t), \\ \dot{x}_2(t) = \tilde{f}(x_1(t), x_2(t)) + 2\sin t + 10.5u(t), \\ y(t) = x_1(t), \end{cases} \tag{4.1.102}$$

where $y(t)$ is the output, $u(t)$ the input (control), and

$$\tilde{f}(x_1, x_2) = \frac{x_1 + x_2 + \sin x_1 + \sin x_2}{40}.$$

The control objective is to design an ESO-based output feedback control so that $x_1(t) \to v(t) = \sin t$ and $x_2(t) \to \dot{v}(t)$. Notice that in system (4.1.102), we have

$$f(x_1(t), x_2(t), w(t)) = \tilde{f}(x_1(t), x_2(t)) + w(t), w(t) = 2\sin t, b = 10.5,$$

where $w(t)$ is the external disturbance.

First of all, we construct the following reference system:

$$\begin{cases} \dot{x}_1^*(t) = x_2^*(t), \\ \dot{x}_2^*(t) = \varphi(x_1^*(t), x_2^*(t)) = -x_1^*(t) - 2x_2^*(t). \end{cases} \quad (4.1.103)$$

Next, we construct the following linear tracking differentiator to recover $\dot{v}(t)$ from $v(t)$:

$$\begin{cases} \dot{z}_1(t) = z_2(t), \\ \dot{z}_2(t) = z_3(t), \\ \dot{z}_3(t) = -R^3(z_1(t) - v(t)) - 3R^2 z_2(t) - 3R z_3(t). \end{cases} \quad (4.1.104)$$

We have two different cases for the design of ADRC.

Case 1. Linear ADRC with total disturbance.

In this case, we construct the following linear extended state observer to estimate the state of (4.1.102) and the extended state $f(\cdot) + (b - b_0)u(t)$ with $b = 10.5$ and its nominal parameter $b_0 = 10$:

$$\begin{cases} \dot{\hat{x}}_1(t) = \hat{x}_2(t) + \dfrac{6}{\varepsilon}(y(t) - \hat{x}_1(t)), \\ \dot{\hat{x}}_2(t) = \hat{x}_3(t) + \dfrac{11}{\varepsilon^2}(y(t) - \hat{x}_1(t)) + 10u(t), \\ \dot{\hat{x}}_3(t) = \dfrac{6}{\varepsilon^3}(y(t) - \hat{x}_1(t)). \end{cases} \quad (4.1.105)$$

The ESO-based linear output feedback control is designed as follows:

$$u(t) = \frac{1}{10}[-(\hat{x}_1(t) - z_1(t)) - 2(\hat{x}_2(t) - z_2(t)) + z_3(t) - \hat{x}_3(t)]. \quad (4.1.106)$$

The matrices K and A in (4.1.6) in this example are

$$K = \begin{pmatrix} -6 & 1 & 0 \\ -11 & 0 & 1 \\ -6 & 0 & 0 \end{pmatrix}, \quad A = \begin{pmatrix} 0 & 1 \\ -1 & -2 \end{pmatrix}. \quad (4.1.107)$$

Since the spectrum set $\sigma(K) = \{-1, -2, -3\}$, $\sigma(A) = \{-1\}$, K and A are Hurwitz. It is computed that $\lambda_{\max}(P_K) \approx 2.3$ and $\lambda_{\max}(P_A) \approx 1.7$. Since $\varphi(x_1, x_2) = -x_1 - 2x_2$ is Lipschitz continuous with Lipschitz constant $L = 2$, all conditions of Corollary 4.1.1 are satisfied.

Take

$$R = 50, \varepsilon = 0.005 \, , z(0) = (0.5, 0.5, 0.5), \; x(0) = (-0.5, -0.5), \; \hat{x}(0) = (0, 0, 0). \quad (4.1.108)$$

We obtain the numerical solutions of (4.1.102), (4.1.104), and (4.1.105) with (4.1.106) by the Euler method with integration step $h = 0.001$. The state variables are plotted in Figure 4.1.1.

From all sub-figures of Figure 4.1.1, we can see that the state $(\hat{x}_1(t), \hat{x}_2(t), \hat{x}_3(t))$ of the observer tracks the system state $(x_1(t), x_2(t), x_3(t) = f(\cdot) + (b - b_0)u(t))$

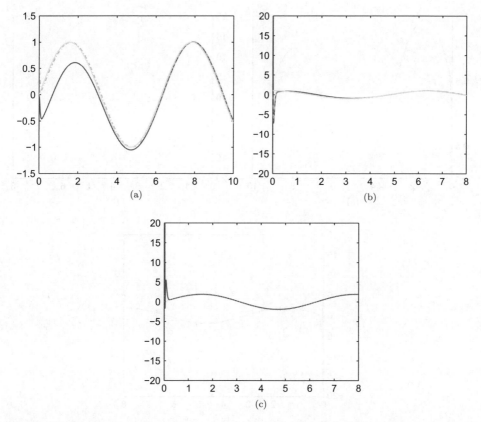

Figure 4.1.1 Numerical results of ADRC for systems (4.1.102), (4.1.104), (4.1.105), and (4.1.106) with total disturbance.

very quickly in spite of the uncertainties in system (4.1.102). From Figure 4.1.1(a), it is seen that $x_1(t)$ catches up with $v(t)$ in less than four seconds. The same observation is observed from Figure 4.1.1(b) for the tracking of $x_2(t)$ to $\dot{v}(t)$.

For the same system (4.1.102), we consider the following case where only the external disturbance is not known.

Case 2. Linear ADRC with the external disturbance only.

In this case, the system function $\tilde{f}(\cdot)$ is known and $\tilde{f}(\cdot)$ is Lipschitz continuous with Lipschitz constant $L_1 = \frac{1}{20}$.

The extended state observer is modified by using $\tilde{f}(\cdot)$ as the following form where the extended state becomes $x_3(t) = w(t) + (b - b_0)u(t)$:

$$\begin{cases} \dot{\hat{x}}_1(t) = \hat{x}_2(t) + \dfrac{6}{\varepsilon}(y(t) - \hat{x}_1(t)), \\[2mm] \dot{\hat{x}}_2(t) = \hat{x}_3(t) + \dfrac{11}{\varepsilon^2}(y(t) - \hat{x}_1(t)) + \tilde{f}(\hat{x}_1(t), \hat{x}_2(t)) + 10u(t), \\[2mm] \dot{\hat{x}}_3(t) = \dfrac{6}{\varepsilon^3}(y(t) - \hat{x}_1(t)). \end{cases} \qquad (4.1.109)$$

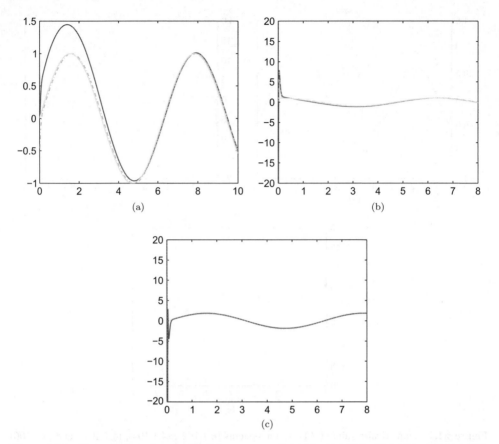

Figure 4.1.2   Numerical results of ADRC for systems (4.1.102), (4.1.104), (4.1.109), and (4.1.110) with external disturbance only.

The ESO-based output feedback control is designed to be

$$u(t) = \frac{1}{10}[-(\hat{x}_1(t) - z_1(t)) - 2(\hat{x}_2(t) - z_2(t)) + \tilde{f}(\hat{x}_1(t), \hat{x}_2(t)) + z_3(t) - \hat{x}_3(t)].$$

$$(4.1.110)$$

For the same data in (4.1.108), we obtain the numerical solutions shown in Figure 4.1.2.

From Figure 4.1.2, we can also see that the state $(\hat{x}_1(t), \hat{x}_2(t), \hat{x}_3(t))$ of the observer tracks the system state $(x_1(t), x_2(t), x_3(t) = \tilde{f}(x_1(t), x_2(t)) + w(t) + (b - b_0)u(t))$ very quickly.

Comparing Figure 4.1.1(a) with Figure 4.1.2(a), we can see that the catching time in Case 1 is slightly longer than that in Case 2 although the difference is quite small. This is natural in the sense that the observer in Case 1 takes more time to learn the uncertainties not only from external disturbance but also from the unknown system structure.

In the following, we use the semi-global ADRC for system (4.1.102). Under the design principle in Subsection 4.1.2, the semi-global feedback control can be designed as follows:

$$u(t) = \frac{1}{10}[\text{sat}_4(-(\hat{x}_1(t) - z_1(t)) - 2(\hat{x}_2(t) - z_2(t))) + \text{sat}_4(z_3(t) - \hat{x}_3(t))], \quad (4.1.111)$$

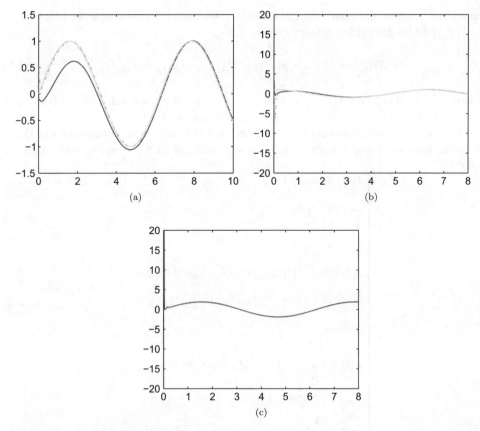

Figure 4.1.3 Numerical results of ADRC for systems (4.1.102), (4.1.104), (4.1.105), and (4.1.111) with total disturbance.

where the saturation function $\text{sat}_4(\cdot)$ is given in (4.1.73). By using the same other parameters as in Figure 4.1.1, the numerical results are plotted in Figure 4.1.3. From Figure 4.1.3 we can see that under the semi-global ADRC (4.1.111), the tracking results are also very satisfactory. Furthermore, the state $x_2(t)$ has a much smaller overshoot than that in Figure 4.1.1.

4.2 Nonlinear ADRC for SISO Systems

In this section, we discuss the nonlinear ADRC with nonlinear ESO for the SISO system (4.0.1). The tracking differentiator is designed as (4.0.3). The reference system is (4.0.2).

4.2.1 Global ADRC for SISO Systems with Total Disturbance

Firstly, we analyze the nonlinear ADRC with total disturbance, which comes from the unmodeled system dynamics, external disturbance, and the derivation of the nominal value of control parameters from its real value. The ESO used in this section is the nonlinear one (4.0.4).

The ADRC is designed as follows:

$$\text{ADRC:} \quad u(t) = \frac{1}{b_0}[\varphi(\hat{x}(t) - z(t)) + z_{n+1}(t) - \hat{x}_{n+1}(t)], \quad (4.2.1)$$

where $(\hat{x}(t) = (\hat{x}_1(t), \hat{x}_2(t), \ldots, \hat{x}_n(t)), \hat{x}_{n+1}(t))$ is the solution of (4.0.4) and $(z(t) = (z_1(t), z_2(t), \ldots, z_n(t)), z_{n+1}(t))$ is the solution of (4.0.3).

Due to the relative independency of (4.0.3) and (4.0.2) with the other components of ADRC, we write the closed-loop system of (4.0.1) under feedback (4.2.1) coupling with (4.0.4) as follows:

$$\begin{cases} \dot{x}_1(t) = x_2(t), \\[6pt] \dot{x}_2(t) = x_3(t), \\[4pt] \quad \vdots \\[4pt] \dot{x}_n(t) = f(x(t), w(t)) + (b - b_0)u(t) + b_0 u(t), \\[4pt] \dot{\hat{x}}_1(t) = \hat{x}_2(t) + \varepsilon^{n-1}g_1(\theta_1(t)), \\[4pt] \quad \vdots \\[4pt] \dot{\hat{x}}_n(t) = \hat{x}_{n+1}(t) + g_n(\theta_1(t)) + b_0 u(t), \\[4pt] \dot{\hat{x}}_{n+1}(t) = \frac{1}{\varepsilon}g_{n+1}(\theta_1(t)), \\[4pt] u(t) = \frac{1}{b_0}[\varphi(\hat{x}(t) - z(t)) + z_{n+1}(t) - \hat{x}_{n+1}(t)], \end{cases} \quad (4.2.2)$$

where $z(t) = (z_1(t), z_2(t), \ldots, z_n(t))$, $(z(t), z_{n+1}(t))$ is the solution of (4.0.3), $x(t) = (x_1(t), x_2(t), \ldots, x_n(t))$, $\hat{x}(t) = (\hat{x}_1(t), \hat{x}_2(t), \ldots, \hat{x}_n(t))$, and $\theta_1(t) = (x_1(t) - \hat{x}_1(t))/\varepsilon^n$.

The following assumptions about $f(\cdot)$, $w(t)$, and $\varphi(\cdot)$, b are needed in the establishment of the convergence. Assumption 4.2.1 is made for system (4.0.1) itself and the external disturbance, Assumption 4.2.2 is for ESO (4.0.4) and unknown parameter b, Assumption 4.2.3 is for reference system (4.0.2), and Assumption 4.2.4 is for TD (4.0.3).

Assumption 4.2.1 For $f \in C^1(\mathbb{R}^{n+m+2}, \mathbb{R})$, there exist positive constant N and functions $\varpi_1 \in C(\mathbb{R}^n, [0, +\infty))$, $\varpi_2 \in C(\mathbb{R}^m, [0, +\infty))$, and $\varpi_3 \in C(\mathbb{R}, [0, +\infty))$ such that

$$\sum_{i=1}^{n}\left|\frac{\partial f(t, x, \zeta, w)}{\partial x_i}\right| + \sum_{i=1}^{m}\left|\frac{\partial f(t, x, \zeta, w)}{\partial \zeta_i}\right| + \left|\frac{\partial f(t, x, \zeta, w)}{\partial t}\right| + |f(t, x, \zeta, w)|$$

$$\leq N + \varpi_1(x) + \varpi_2(\zeta) + \varpi_3(w), \forall\, x \in \mathbb{R}^n, \zeta \in \mathbb{R}^m, w \in \mathbb{R}.$$

Assumption 4.2.2 With $|g_i(r)| \leq k_i r$ for some positive constants k_i for all $i = 1, 2, \ldots, n + 1$, there exist constants $\lambda_{1i}(i = 1, 2, 3, 4)$, β_1, and positive definite continuous differentiable functions $V_1, W_1 : \mathbb{R}^{n+1} \to \mathbb{R}$ such that

1. $\lambda_{11}\|y\|^2 \leq V_1(y) \leq \lambda_{12}\|y\|^2, \quad \lambda_{13}\|y\|^2 \leq W_1(y) \leq \lambda_{14}\|y\|^2,$

2. $\displaystyle\sum_{i=1}^{n}(y_{i+1} - g_i(y_1))\frac{\partial V_1(y)}{\partial y_i} - g_{n+1}(y_1)\frac{\partial V_1(y)}{\partial y_{n+1}} \leq -W_1(y),$

3. $\left|\dfrac{\partial V_1(y)}{\partial y_{n+1}}\right| \leq \beta_1\|y\|, \forall\, y = (y_1, y_2, \ldots, y_n) \in \mathbb{R}^{n+1}.$

Moreover, the parameter b satisfies $\frac{|b-b_0|}{b_0}k_{n+1} < \lambda_{13}/\beta_1$, where b_0 is the nominal value of b. Here and throughout the paper, we always use $\|\cdot\|$ to denote the corresponding Euclidian norm.

Assumption 4.2.3 The $\varphi(y)$ is globally Lipschitz continuous with the Lipschitz constant L: $|\varphi(x) - \varphi(y)| \leq L\|x - y\|$ for all $x, y \in \mathbb{R}^n$. There exist constants λ_{2i} $(i = 1, 2, 3, 4)$, β_2, and positive definite continuous differentiable functions $V_2, W_2 : \mathbb{R}^n \to \mathbb{R}$ such that

1. $\lambda_{21}\|y\|^2 \leq V_2(y) \leq \lambda_{22}\|y\|^2, \quad \lambda_{23}\|y\|^2 \leq W_2(y) \leq \lambda_{24}\|y\|^2,$

2. $\displaystyle\sum_{i=1}^{n-1}y_{i+1}\frac{\partial V_2(y)}{\partial y_i} + \varphi(y_1, y_2, \ldots, y_n)\frac{\partial V_2(y)}{\partial y_n} \leq -W_2(y),$

3. $\left|\dfrac{\partial V_2(y)}{\partial y_n}\right| \leq \beta_2\|y\|, \forall\, y = (y_1, y_2, \ldots, y_n) \in \mathbb{R}^n.$

Assumption 4.2.4 Both $v(t)$ and $\dot{v}(t)$ are bounded over $[0, \infty)$, $\psi(\cdot)$ is locally Lipschitz continuous, and the system (4.0.3) with $v(t) = 0, R = 1$ is globally asymptotically stable.

Theorem 4.2.1 *Let $x_i(t)(1 \leq i \leq n)$ and $\hat{x}_i(t)(1 \leq i \leq n+1)$ be the solutions of the closed-loop system (4.2.2), $x_{n+1}(t) = f(x(t), w(t)) + (b - b_0)u(t)$ be the extended state, and let $z_1(t)$ be the solution of (4.0.3). Under Assumptions 4.2.1 to 4.2.4, the following statements hold true for any given initial values of (4.0.3) and the closed-loop system (4.2.2).*

(i) *For any $\sigma > 0$ and $\tau > 0$, there exists a constant $R_0 > 0$ such that $|z_1(t) - v(t)| < \sigma$ uniformly in $t \in [\tau, \infty)$ for all $R > R_0$.*

(ii) *For every $R > R_0$, there is an R-dependent constant $\varepsilon_0 > 0$ (specified by (4.2.18) later) such that for any $\varepsilon \in (0, \varepsilon_0)$, there exists a $t_\varepsilon > 0$ such that for all $R > R_0, \varepsilon \in (0, \varepsilon_0)$, $t > t_\varepsilon$,*

$$|x_i(t) - \hat{x}_i(t)| \leq \Gamma_1\varepsilon^{n+2-i}, \quad i = 1, 2, \ldots, n+1 \tag{4.2.3}$$

and

$$|x_i(t) - z_i(t)| \leq \Gamma_2\varepsilon, \quad i = 1, 2, \ldots, n. \tag{4.2.4}$$

where Γ_1 and Γ_2 are R-dependent positive constants only.

(iii) *For any $\sigma > 0$, there exist $R_1 > R_0, \varepsilon_1 \in (0, \varepsilon_0)$ such that for any $R > R_1$ and $\varepsilon \in (0, \varepsilon_1)$, there exists a $t_{R\varepsilon} > 0$ such that for all $R > R_1, \varepsilon \in (0, \varepsilon_1)$, and $t > t_{R\varepsilon}$, it holds that $|x_1(t) - v(t)| < \sigma$.*

Proof. Statement (i) follows directly from Theorem 2.2.2, which concludes the convergence of tracking differentiator (4.0.3) under Assumption 4.2.4. Moreover, for any $R > R_0$, there exists an $M_R > 0$ such that

$$\|(z_1(t), z_2(t), \ldots, z_{n+1}(t), \dot{z}_{n+1}(t))\| \leq M_R \tag{4.2.5}$$

for all $t \geq 0$.

Note that statement (iii) is a direct consequence of (i) and (ii). We list it here to indicate the output tracking function of ADRC only. Therefore, it suffices to prove (ii).

Let

$$e_i(t) = \frac{1}{\varepsilon^{n+1-i}}[x_i(t) - \hat{x}_i(t)], \quad i = 1, 2, \ldots, n+1. \tag{4.2.6}$$

Then

$$\begin{cases} \varepsilon \dot{e}_i(t) = \dfrac{1}{\varepsilon^{n+1-(i+1)}}[\dot{x}_i(t) - \dot{\hat{x}}_i(t)] = \dfrac{1}{\varepsilon^{n+1-(i+1)}}[x_{i+1}(t) - \hat{x}_{i+1}(t) - \varepsilon^{n-i}g_i(e_1(t))] \\[2mm] \qquad = e_{i+1}(t) - g_i(e_1(t)), \quad i \leq n, \\[2mm] \varepsilon \dot{e}_{n+1}(t) = \varepsilon(\dot{x}_{n+1}(t) - \dot{\hat{x}}_{n+1}(t)) = \varepsilon h(t) - g_{n+1}(e_1(t)), \end{cases} \tag{4.2.7}$$

where

$$x_{n+1}(t) = f(x(t), w(t)) + (b - b_0)u(t)$$
$$= f(x(t), w(t)) + \frac{b - b_0}{b_0}[\varphi(\hat{x}(t) - z(t)) + z_{n+1}(t) - \hat{x}_{n+1}(t)] \tag{4.2.8}$$

is the extended state of system (4.2.2) and $h(t)$ is defined by

$$h(t) = \frac{d[f(x(t), w(t)) + (b - b_0)u(t)]}{dt}\bigg|_{\text{along (4.0.1)}}$$

$$= \sum_{i=1}^{n-1} x_{i+1}(t)\frac{\partial f(x(t), w(t))}{\partial x_i} + [f(x(t), w(t)) + bu(t)]\frac{\partial f(x(t), w(t))}{\partial x_n}$$

$$+ \dot{w}(t)\frac{\partial f(x(t), w(t))}{\partial w} + (b - b_0)\dot{u}(t). \tag{4.2.9}$$

Under feedback control (4.2.1), since $x_{n+1}(t) = f(\cdot) + (b - b_0)u(t)$, $(f(\cdot) + bu(t))\partial f(\cdot)/\partial x_n = x_{n+1}(t)\frac{\partial f(\cdot)}{\partial x_n} + b_0 u\frac{\partial f(\cdot)}{\partial x_n}$, we can compute $h(t)$ as

$$h(t) = \frac{d[f(x(t), w(t)) + (b - b_0)u(t)]}{dt}\bigg|_{\text{along (4.2.2)}}$$

$$= \sum_{i=1}^{n} x_{i+1}(t)\frac{\partial f(x(t), w(t))}{\partial x_i} + \dot{w}(t)\frac{\partial f(x(t), w(t))}{\partial w}$$

$$+ [\varphi(\hat{x}(t) - z(t)) + z_{n+1}(t) - \hat{x}_{n+1}(t)]\frac{\partial f(x(t), w(t))}{\partial x_n}$$

$$+ \frac{b - b_0}{b_0} \left\{ \sum_{i=1}^{n} \left[\hat{x}_{i+1}(t) + \varepsilon^{n-i} g_i(e_1(t)) - z_{i+1}(t) \right] \frac{\partial \varphi(\hat{x}(t) - z(t))}{\partial y_i} + \dot{z}_{n+1}(t) \right.$$

$$\left. - \frac{1}{\varepsilon} g_{n+1}(e_1(t)) \right\}, \tag{4.2.10}$$

where $\frac{\partial f(x(t), w(t))}{\partial x_i}$ denotes the ith partial derivative of $f(\cdot)$ at $(x(t), w(t)) \in \mathbb{R}^{n+1}$ and $\frac{\partial \varphi(\hat{x}(t) - z(t))}{\partial y_i}$ denotes the ith partial derivative of $\varphi(\cdot)$ at $\hat{x}(t) - z(t)$. Set

$$\eta_i(t) = x_i(t) - z_i(t), i = 1, 2, \ldots, n, \eta(t) = (\eta_1(t), \eta_2(t), \ldots, \eta_n(t)).$$

It follows from (4.2.2) and (4.2.7) that

$$\begin{cases} \dot{\eta}_1(t) = \eta_2(t), \\ \quad \vdots \\ \dot{\eta}_n(t) = \varphi(\eta_1(t), \eta_2(t), \ldots, \eta_n(t)) + e_{n+1}(t) + [\varphi(\hat{x}(t) - z(t)) - \varphi(x(t) - z(t))], \\ \varepsilon \dot{e}_1(t) = e_2(t) - g_1(e_1(t)), \\ \quad \vdots \\ \varepsilon \dot{e}_n(t) = e_{n+1}(t) - g_n(e_1(t)), \\ \varepsilon \dot{e}_{n+1}(t) = \varepsilon h(t) - g_{n+1}(e_1(t)). \end{cases} \tag{4.2.11}$$

Furthermore, set

$$\tilde{e}(t) = (e_1(t), e_2(t), \ldots, e_n(t)), e(t) = (e_1(t), e_2(t), \ldots, e_{n+1}(t)).$$

By Assumption 4.2.3, we have

$$|\varphi(\hat{x}(t) - z(t)) - \varphi(x(t) - z(t))| \le L\|\hat{x}(t) - x(t)\| \le L\|\tilde{e}(t)\|, \forall \; \varepsilon \in (0, 1). \tag{4.2.12}$$

From the boundedness of the partial derivatives of $f(\cdot)$, $z(t)$ (see (4.2.5)) and $w(t)$, it follows that for some R-dependent positive numbers $M, N_0, N_1, N > 0$ (by the mean value theorem with $f(0, 0)$),

$$|f(x(t), w(t))| \le M(\|x(t)\| + |w(t)|) + N_0 \le M(\|\eta(t)\| + \|z(t)\| + |w(t)|) + N_1$$

$$\le M\|\eta(t)\| + N. \tag{4.2.13}$$

Notice that by (4.2.8),

$$\hat{x}_{n+1}(t) = \frac{b_0}{b} \left[-e_{n+1}(t) + f(x(t), w(t)) + \frac{b - b_0}{b_0}(\varphi(\hat{x}(t) - z(t)) + z_{n+1}(t) \right]. \tag{4.2.14}$$

By (4.2.5), (4.2.13), and (4.2.14), and Assumptions 4.2.1 and 4.2.3, the function $h(t)$ given by (4.2.10) satisfies

$$|h(t)| \le B_0 + B_1\|e(t)\| + B_2\|\eta(t)\| + \frac{B}{\varepsilon}\|e(t)\|, \quad B = \frac{|b - b_0|}{b_0} k_{n+1}, \tag{4.2.15}$$

where B_0, B_1, and B_2 are R-dependent positive numbers.

We split the proof into three steps.

Step 1. *Show that for every $R > R_0$, there exists an R-dependent $\varepsilon_0 > 0$ such that for any $\varepsilon \in (0, \varepsilon_0)$, there exist $t_{1\varepsilon}$ and $r > 0$ such that the solution of (4.2.11) satisfies $\|(e(t), \eta(t))\| \leq r$ for all $t > t_{1\varepsilon}$, where r is an R-dependent constant.*
Consider the positive definite function $V : \mathbb{R}^{2n+1} \to \mathbb{R}$ given by

$$V(e_1, \ldots, e_{n+1}, \eta_1, \ldots, \eta_n) = V_1(e_1, \ldots, e_{n+1}) + V_2(\eta_1, \ldots, \eta_n), \qquad (4.2.16)$$

where $V_1(\cdot)$ and $V_2(\cdot)$ are positive definite functions specified in Assumptions 4.2.2 and 4.2.3, respectively. Computing the derivative of $V(e(t), \eta(t))$ along the solution of (4.2.11), owing to Assumptions 4.2.2, 4.2.3, and the inequalities (4.2.12) and (4.2.15), we obtain

$$\left. \frac{dV(e(t), \eta(t))}{dt} \right|_{\text{along } (4.2.11)}$$

$$= \frac{1}{\varepsilon} \left[\sum_{i=1}^{n} (e_{i+1}(t) - g_i(e_1(t))) \frac{\partial V_1(e(t))}{\partial e_i} - g_{n+1}(e_1(t)) \frac{\partial V_1(e(t))}{\partial e_{n+1}} + \varepsilon h(t) \frac{\partial V_1(e(t))}{\partial e_{n+1}} \right]$$

$$+ \sum_{i=1}^{n-1} \eta_{i+1}(t) \frac{\partial V_2(\eta(t))}{\partial \eta_i}$$

$$+ \{ \varphi(\eta(t)) + e_{n+1}(t) + [\varphi(\hat{x}(t) - z(t)) - \varphi(x(t) - z(t))] \} \frac{\partial V_2(\eta(t))}{\partial \eta_n}$$

$$\leq \frac{1}{\varepsilon} \left\{ -W_1(e(t)) + \varepsilon \left(B_0 + B_1 \|e(t)\| + B_2 \|\eta(t)\| + \frac{B}{\varepsilon} \|e(t)\| \right) \beta_1 \|e(t)\| \right\}$$

$$- W_2(\eta(t)) + (L+1)\beta_2 \|e(t)\| \|\eta(t)\|$$

$$\leq -\frac{W_1(e(t))}{\varepsilon} + B_0 \beta_1 \|e(t)\| + B_1 \beta_1 \|e(t)\|^2 - W_2(\eta(t))$$

$$+ (B_2 \beta_1 + (L+1)\beta_2) \|e(t)\| \|\eta(t)\| + \frac{\beta_1 B}{\varepsilon} \|e(t)\|^2$$

$$\leq -\left[\frac{\lambda_{13} - \beta_1 B}{\varepsilon} - \beta_1 B_1 \right] \|e(t)\|^2 + \beta_1 B_0 \|e(t)\| - \lambda_{23} \|\eta(t)\|^2$$

$$+ \sqrt{\frac{\lambda_{13} - \beta_1 B}{\varepsilon}} \|e(t)\| \sqrt{\frac{\varepsilon}{\lambda_{13} - \beta_1 B}} (B_2 \beta_1 + (L+1)\beta_2) \|\eta(t)\|$$

$$\leq -\left[\frac{\lambda_{13} - \beta_1 B}{\varepsilon} - \beta_1 B_1 \right] \|e(t)\|^2 + \beta_1 B_0 \|e(t)\| - \lambda_{23} \|\eta(t)\|^2$$

$$+ \frac{\lambda_{13} - \beta_1 B}{2\varepsilon} \|e(t)\|^2 + \frac{(B_2 \beta_1 + (L+1)\beta_2)^2}{2(\lambda_{13} - \beta_1 B)} \varepsilon \|\eta(t)\|^2$$

$$\leq -\left[\frac{\lambda_{13} - \beta_1 B}{2\varepsilon} - \beta_1 B_1 \right] \|e(t)\|^2 + \beta_1 B_0 \|e(t)\|$$

$$- \left[\lambda_{23} - \frac{(B_2 \beta_1 + (L+1)\beta_2)^2}{2(\lambda_{13} - \beta_1 B)} \varepsilon \right] \|\eta(t)\|^2, \qquad (4.2.17)$$

where we used Assumption 4.2.2 and $B < \lambda_{13}/\beta_1$. Set

$$r = \max\left\{2, \frac{4(1 + B_0\beta_1)}{\lambda_{23}}\right\},$$

$$\varepsilon_0 = \min\left\{1, \frac{\lambda_{13} - \beta_1 B}{2\beta_1(B_0 + B_1)}, \frac{(\lambda_{13} - \beta_1 B)\lambda_{23}}{(B_2\beta_1 + (L+1)\beta_2)^2}\right\}. \tag{4.2.18}$$

For any $\varepsilon \in (0, \varepsilon_0)$ and $\|(e(t), \eta(t))\| \geq r$, we consider the derivative of $V(e(t), \eta(t))$ along the solution of (4.2.11) in two cases.

Case 1: $\|e(t)\| \geq r/2$. In this case, $\|e(t)\| \geq 1$ and hence $\|e(t)\|^2 \geq \|e(t)\|$. By the definition ε_0 of (4.2.18), it has $\lambda_{23} - (B_2\beta_1 + (L+1)\beta_2)^2/(2(\lambda_{13} - \beta_1 B))\varepsilon > 0$, and thus it follows from (4.2.17) that

$$\begin{aligned}
\frac{dV(e(t), \eta(t))}{dt}\bigg|_{\text{along (4.2.11)}} &\leq -\left(\frac{\lambda_{13} - \beta_1 B}{2\varepsilon} - \beta_1 B_1\right)\|e(t)\|^2 + \beta_1 B_0\|e(t)\|^2 \\
&\leq -\left(\frac{\lambda_{13} - \beta_1 B}{2\varepsilon} - \beta_1 B_1 - \beta_1 B_0\right)\|e(t)\|^2 \\
&= -\frac{\lambda_{13} - \beta_1 B - 2\varepsilon\beta_1(B_1 + B_0)}{2\varepsilon}\|e(t)\|^2 < 0,
\end{aligned}$$

where we used again the definition ε_0 of (4.2.18), which gives $\lambda_{13} - \beta_1 B - 2\varepsilon\beta_1(B_1 + B_0) > 0$.

Case 2: $\|e(t)\| < r/2$. In this case, from $\|\eta(t)\| + \|e(t)\| \geq \|(e(t), \eta(t))\|$, it has $\|\eta(t)\| \geq r/2$. By the definition ε_0 of (4.2.18), $\lambda_{13} - \beta_1 B - 2\varepsilon\beta_1 B_1 > 0$. Thus it follows from (4.2.17) that

$$\begin{aligned}
\frac{dV(e(t), \eta(t))}{dt}\bigg|_{\text{along (4.2.11)}} &\leq \beta_1 B_0\|e(t)\| - \left(\lambda_{23} - \frac{(B_2\beta_1 + (L+1)\beta_2)^2}{2(\lambda_{13} - \beta_1 B)}\varepsilon\right)\|\eta(t)\|^2 \\
&\leq -\frac{\lambda_{23}}{2}\|\eta(t)\|^2 + \beta_1 B_0\|e(t)\| \\
&\leq -\frac{\lambda_{23}}{2}\left(\frac{r}{2}\right)^2 + B_0\beta_1\left(\frac{r}{2}\right) = -\frac{r}{2}\left(\frac{r\lambda_{23} - 4B_0\beta_1}{4}\right) < 0,
\end{aligned}$$

where we used the definition r of (4.2.18), which gives $r\lambda_{23} - 4B_0\beta_1 > 0$.

Combining the above two cases yields that for any $\varepsilon \in (0, \varepsilon_0)$, if $\|(e(t), \eta(t))\| \geq r$ then

$$\frac{dV(e(t), \eta(t))}{dt}\bigg|_{\text{along (4.2.11)}} < 0.$$

Therefore, there exists a $t_{1\varepsilon}$ such that $\|(e(t), \eta(t))\| \leq r$ for all $t > t_{1\varepsilon}$.

Step 2. *Establish the convergence of* $\|x_i(t) - \hat{x}_i(t)\|$.

Consider the following subsystem, which is composed of the last $n + 1$ equations in system (4.2.11):

$$\begin{cases} \varepsilon \dot{e}_1(t) = e_2(t) - g_1(e_1(t)), \\ \qquad \vdots \\ \varepsilon \dot{e}_n(t) = e_{n+1}(t) - g_n(e_1(t)), \\ \varepsilon \dot{e}_{n+1}(t) = \varepsilon h(t) - g_{n+1}(e_1(t)). \end{cases} \qquad (4.2.19)$$

Since $\|(e(t), \eta(t))\| \leq r$ for all $t > t_{1\varepsilon}$, we obtain, together with (4.2.15), that $|h(t)| \leq M_0 + B\|e(t)\|/\varepsilon$ for all $t > t_{1\varepsilon}$ and some R-dependent constant $M_0 > 0$. Under Assumption 4.2.2, we can compute the derivative of $V_1(e(t))$ along the solution of (4.2.19) as follows:

$$\left. \frac{dV_1(e(t))}{dt} \right|_{\text{along (4.2.19)}}$$

$$= \frac{1}{\varepsilon} \left(\sum_{i=1}^n (e_{i+1}(t) - g_i(e_1(t))) \frac{\partial V_1(e(t))}{\partial e_i} - g_{n+1}(e_1(t)) \frac{\partial V_1(e(t))}{\partial e_{n+1}} + \varepsilon h(t) \frac{\partial V_1(e(t))}{\partial e_{n+1}} \right)$$

$$\leq -\frac{\lambda_{13} - B\beta_1}{\varepsilon} \|e(t)\|^2 + M_0 \beta_1 \|e(t)\|$$

$$\leq -\frac{\lambda_{13} - B\beta_1}{\varepsilon \lambda_{12}} V_1(e(t)) + \frac{M_0 \beta_1 \sqrt{\lambda_{11}}}{\lambda_{11}} \sqrt{V_1(e(t))}, \forall\, t > t_{1\varepsilon}. \qquad (4.2.20)$$

In the last step above, we used again Assumption 4.2.3 and $B \leq \lambda_{13}/\beta_1$. It follows that

$$\frac{d\sqrt{V_1(e(t))}}{dt} \leq -\frac{\lambda_{13} - B\beta_1}{2\varepsilon \lambda_{12}} \sqrt{V_1(e(t))} + \frac{M_0 \beta_1 \sqrt{\lambda_{11}}}{2\lambda_{11}}, \forall\, t > t_{1\varepsilon}, \qquad (4.2.21)$$

and hence

$$\sqrt{V_1(e(t))} \leq \sqrt{V_1(e(t_{1\varepsilon}))} e^{-\lambda_{13} - B\beta_1/2\varepsilon\lambda_{12}(t - t_{1\varepsilon})}$$

$$+ \frac{M_0 \beta_1 \sqrt{\lambda_{11}}}{\lambda_{11}} \int_{t_{1\varepsilon}}^t e^{-\lambda_{13} - B\beta_1/2\varepsilon\lambda_{12}(t-s)} ds, \forall\, t > t_{1\varepsilon}. \qquad (4.2.22)$$

This together with (4.2.6) implies that there exist $t_{2\varepsilon} > t_{1\varepsilon}$ and R-dependent constant $\Gamma_1 > 0$ such that

$$|x_i(t) - \hat{x}_i(t)| = \varepsilon^{n+1-i} |e_i(t)| \leq \varepsilon^{n+1-i} \|e(t)\|$$

$$\leq \varepsilon^{n+1-i} \sqrt{\frac{V_1(e(t))}{\lambda_{11}}} \leq \Gamma_1 \varepsilon^{n+2-i}, \forall\, t > t_{2\varepsilon}, \qquad (4.2.23)$$

where we used the facts $xe^{-x} < 1$ for all $x > 0$ and $\|(e(t), \eta(t))\| \leq r$ for all $t > t_{1\varepsilon}$, which is proved in Step 1.

Step 3. *Establish the convergence of $\|x(t) - z(t)\|$.*

Consider the following system, which is composed of the first n equations in system (4.2.11):

$$\begin{cases} \dot{\eta}_1(t) = \eta_2(t), \\ \dot{\eta}_2(t) = \eta_3(t), \\ \quad \vdots \\ \dot{\eta}_n(t) = \varphi(\eta(t)) + e_{n+1}(t) + [\varphi(\hat{x}(t) - z(t)) - \varphi(x(t) - z(t))]. \end{cases} \tag{4.2.24}$$

Under Assumption 4.2.3 and using (4.2.12), we can compute the derivative of $V_2(\eta(t))$ along the solution of (4.2.24) as follows:

$$\left. \frac{dV_2(\eta(t))}{dt} \right|_{\text{along (4.0.1)}}$$

$$= \sum_{i=1}^{n-1} \eta_{i+1}(t) \frac{\partial V_2(\eta(t))}{\partial \eta_i}$$

$$+ \{\varphi(\eta(t)) + e_{n+1}(t) + [\varphi(\hat{x}(t) - z(t)) - \varphi(x(t) - z(t))]\} \frac{\partial V_2(\eta(t))}{\partial \eta_n}$$

$$\leq -W_2(\eta(t)) + (L+1)\beta_2\|e(t)\|\|\eta(t)\| \leq -\frac{\lambda_{23}}{\lambda_{22}} V_2(\eta(t)) + N_0\varepsilon\sqrt{V_2(\eta(t))}, \quad \forall\, t > t_{2\varepsilon}, \tag{4.2.25}$$

where N_0 is some R-dependent positive constant and we used the fact that $\|e(t)\| \leq (n+1)B_1\varepsilon$ which was proved in Step 1. It then follows that

$$\frac{d\sqrt{V_2(\eta(t))}}{dt} \leq -\frac{\lambda_{23}}{\lambda_{22}} \sqrt{V_2(\eta(t))} + N_0\varepsilon, \forall\, t > t_{2\varepsilon}. \tag{4.2.26}$$

This together with Assumption 4.2.3 implies that for all $t > t_{2\varepsilon}$,

$$\|\eta(t)\| \leq \frac{\sqrt{\lambda_{21}}}{\lambda_{21}} \sqrt{V_2(\eta(t))} \leq \frac{\sqrt{\lambda_{21}}}{\lambda_{21}}$$

$$\times \left(e^{-\frac{\lambda_{23}}{\lambda_{22}}(t-t_{2\varepsilon})} \sqrt{V_2(\eta(t_{2\varepsilon}))} + N_0\varepsilon \int_{t_{2\varepsilon}}^{t} e^{-\frac{\lambda_{23}}{\lambda_{22}}(t-s)} ds \right). \tag{4.2.27}$$

Since the first term on the right-hand side of (4.2.27) tends to zero as t goes to infinity and the second term is bounded by ε multiplied by an ε-independent constant, it follows that there exist $t_\varepsilon > t_{2\varepsilon}$ and $\Gamma_2 > 0$ such that $\|x(t) - z_R(t)\| \leq \Gamma_2\varepsilon$ for all $t > t_\varepsilon$. Thus (4.2.4) follows. This completes the proof of the theorem.

4.2.2 Global ADRC for SISO System with External Disturbance Only

In this subsection, we consider a special class of the systems (4.0.1) where the dynamic function $f(x, w) = \tilde{f}(x) + w$, $x = (x_1, x_2, \ldots, x_n) \in \mathbb{R}^n$, with $\tilde{f}(\cdot)$ being known. This is the case when the uncertainty comes from external disturbance only.

Since $\tilde{f}(\cdot)$ is available, instead of using ESO (4.0.4) we design the following ESO(f) to estimate the system state $x_i(t), i = 1, 2, \ldots, n$ and the extended state $x_{n+1}(t) = w(t) + (b - b_0)u(t)$.

$$\text{ESO(f):} \begin{cases} \dot{\hat{x}}_1(t) = \hat{x}_2(t) + \varepsilon^{n-1} g_1(\theta(t)), \\ \dot{\hat{x}}_2(t) = \hat{x}_3(t) + \varepsilon^{n-2} g_2(\theta(t)), \\ \quad \vdots \\ \dot{\hat{x}}_n(t) = \hat{x}_{n+1}(t) + g_n(\theta(t)) + \tilde{f}(\hat{x}(t)) + b_0 u(t), \\ \dot{\hat{x}}_{n+1}(t) = \frac{1}{\varepsilon} g_{n+1}(\theta(t)), \end{cases} \tag{4.2.28}$$

where $\theta(t) = [y(t) - \hat{x}_1(t)]/\varepsilon^n$.

The ESO-based output feedback control is also modified to be the following form:

$$u(t) = \frac{1}{b_0}[\varphi(\hat{x}(t) - z(t)) - \tilde{f}(\hat{x}(t)) + z_{n+1}(t) - \hat{x}_{n+1}(t)], \tag{4.2.29}$$

where $(\hat{x}(t) = (\hat{x}_1(t), \ldots, \hat{x}_n(t)), \hat{x}_{n+1}(t))$ is the solution of (4.2.28) and $(z(t) = (z_1(t), \ldots, z_n(t)), z_{n+1}(t))$ is the solution of (4.0.3). The closed-loop system now becomes

$$\begin{cases} \dot{x}_1(t) = x_2(t), \\ \dot{x}_2(t) = x_3(t), \\ \quad \vdots \\ \dot{x}_n(t) = \tilde{f}(x(t)) + w(t) + (b - b_0)u(t) + b_0 u(t), \\ \dot{\hat{x}}_1(t) = \hat{x}_2(t) + \varepsilon^{n-1} g_1(\theta(t)), \\ \quad \vdots \\ \dot{\hat{x}}_n(t) = \hat{x}_{n+1}(t) + g_n(\theta(t)) + \tilde{f}(\hat{x}(t)) + b_0 u(t), \\ \dot{\hat{x}}_{n+1}(t) = \frac{1}{\varepsilon} g_{n+1}(\theta(t)), \\ u(t) = \frac{1}{b_0}[\phi(\hat{x}(t) - z(t)) - \tilde{f}(\hat{x}(t)) + z_{n+1}(t) - \hat{x}_{n+1}(t)], \end{cases} \tag{4.2.30}$$

where $x(t) = (x_1(t), x_2(t), \ldots, x_n(t))$.

The parallel convergence for this case is stated in Theorem 4.2.2 where Assumption 4.2.1 in Theorem 4.2.1 is replaced by Assumption 4.2.5.

Assumption 4.2.5 Both $w(t)$ and $\dot{w}(t)$ are bounded and all partial derivatives of $\tilde{f}(\cdot)$ are bounded.

The assumption on $\tilde{f}(\cdot)$ implies that there is a constant $L_1 > 0$ such that $\tilde{f}(\cdot)$ is Lipschitz continuous with the Lipschitz constant L_1:

$$|\tilde{f}(x) - \tilde{f}(y)| \leq L_1 \|x - y\|, \quad \forall x, y \in \mathbb{R}^n. \tag{4.2.31}$$

Theorem 4.2.2 *Suppose that Assumptions 4.2.2 to 4.2.5 hold, $|\partial V_1(e)/\partial e_n| \leq \alpha \|e\|$ for some $\alpha > 0$, and all $e \in \mathbb{R}^{n+1}$. And*

$$\alpha L_1 + \frac{|b - b_0|}{b_0} k_{n+1} \beta_1 < \lambda_{13}. \tag{4.2.32}$$

Let $z_1(t)$ be the solution of (4.0.3), $x_i(t)(1 \leq i \leq n)$, $\hat{x}_i(t)(1 \leq i \leq n+1)$ be the solutions of the closed-loop (4.2.30), and let $x_{n+1}(t) = w(t) + (b - b_0)u(t)$ be the extended state. Then, the following statements hold for any given initial values of (4.0.3) and (4.2.30).

(i) For any sufficiently small constant $\sigma > 0$ and $\tau > 0$, there exists a constant $R_0 > 0$ such that $|z_1(t) - v(t)| < \sigma$ uniformly in $t \in [\tau, \infty)$ for all $R > R_0$.

(ii) For every $R > R_0$, there is an $\varepsilon_0 > 0$ (specified by (4.2.43) later) such that for any $\varepsilon \in (0, \varepsilon_0)$, there is a $t_\varepsilon > 0$ such that for all $R > R_0$, $\varepsilon \in (0, \varepsilon_0)$ and $t > t_\varepsilon$,

$$|x_i(t) - \hat{x}_i(t)| \leq \Gamma_1 \varepsilon^{n+2-i}, \quad i = 1, 2, \ldots, n, n+1, \tag{4.2.33}$$

and

$$|x_i(t) - z_i(t)| \leq \Gamma_2 \varepsilon, \quad i = 1, 2, \ldots, n, \tag{4.2.34}$$

where Γ_1 and Γ_2 are R-dependent positive numbers.

(iii) For any $\sigma > 0$, there exist $R_1 > 0$, $\varepsilon_1 > 0$, for any $\varepsilon \in (0, \varepsilon_1)$, $R > R_1$, there is a $t_{R\varepsilon} > 0$, such that $|x_1(t) - z_1(t)| < \sigma$, for all $R > R_1$, $\varepsilon \in (0, \varepsilon_1)$, and $t > t_{R\varepsilon}$.

Proof. This is the same as the proof of Theorem 4.2.1, where statement (i) comes from Theorem 4.2.2 under Assumption 4.2.4. Also, it follows from the proof of Theorem 2.2.2 that for any $R > R_0$, there exists an $M_R > 0$ such that

$$\|(z_1(t), z_2(t), \ldots, z_n(t), z_{n+1}(t))\| \leq M_R \tag{4.2.35}$$

for all $t \geq 0$.

Statement (iii) is the consequence of (i) and (ii), which is considered as an output regulation result in this case. Thus it suffices to prove statement (ii).

Set

$$\eta_i(t) = x_i(t) - z_i(t), e_i(t) = \frac{1}{\varepsilon^{n+1-i}}[x_i(t) - \hat{x}_i(t)], e_{n+1}(t)$$

$$= x_{n+1}(t) - \hat{x}_{n+1}(t), i = 1, 2, \ldots, n.$$

By (4.2.30), $e_i(t), 1 \leq i \leq n+1$, and $\eta_i(t), 1 \leq i \leq n$ satisfy the system of following differential equations:

$$\begin{cases} \dot{\eta}_1(t) = \eta_2(t), \\ \dot{\eta}_2(t) = \eta_3(t), \\ \quad \vdots \\ \dot{\eta}_n(t) = \varphi(\hat{x}(t) - z(t)) + \tilde{f}(x(t)) - \tilde{f}(\hat{x}(t)) + e_{n+1}(t), \\ \varepsilon\dot{e}_1(t) = e_2(t) - g_1(e_1(t)), \\ \quad \vdots \\ \varepsilon\dot{e}_n(t) = e_{n+1}(t) - g_n(e_1(t)) + \tilde{f}(x(t)) - \tilde{f}(\hat{x}(t)), \\ \varepsilon\dot{e}_{n+1}(t) = -g_{n+1}(e_1(t)) + \varepsilon\frac{d}{dt}(w(t) + (b - b_0)u(t)) \Big|_{\text{along } (4.2.30)} \end{cases} \tag{4.2.36}$$

By Assumptions 4.2.3 and 4.2.5, and the fact that $\varphi(\hat{x} - z) = \varphi(x - z) - (\varphi(\hat{x} - z) - \varphi(x - z))$, we have

$$|\tilde{f}(x) - \tilde{f}(\hat{x})| \le L_1 \|x - \hat{x}\|, |\varphi(\hat{x} - z) - \varphi(x - z)| \le L\|\hat{x} - x\| \le L\|e\|. \qquad (4.2.37)$$

Define

$$\hbar(t) = \left. \frac{d(w(t) + (b - b_0)u(t))}{dt} \right|_{\text{along (4.2.30)}}$$

$$= \dot{w}(t) + \frac{b - b_0}{b_0} \left[\sum_{i=1}^{n} \left[\hat{x}_{i+1}(t) - \varepsilon^{n-i} g_i(e_1(t)) - z_{i+1}(t) \right] \frac{\partial \varphi(\hat{x}(t))}{\partial \hat{x}_i} - \dot{z}_{n+1}(t) \right)$$

$$- \sum_{i=1}^{n} \left[\hat{x}_{n+1}(t) - \varepsilon^{n-i} g_i(e_1(t)) \right] \frac{\partial \tilde{f}(\hat{x}(t))}{\partial \hat{x}_i} + \dot{z}_{n+1}(t) - \frac{g_{n+1}(e_1(t))}{\varepsilon} \right]. \qquad (4.2.38)$$

Similar to (4.2.15), we also have

$$|\hbar(t)| \le B_0 + B_1 \|\eta(t)\| + B_2 \|e(t)\| + \frac{B}{\varepsilon} \|e(t)\|, B = \frac{|b - b_0|}{b_0} |k_{n+1}|, \qquad (4.2.39)$$

where $B_0, B_1,$ and B_2 are some R-dependent positive constants.

Now construct a Lyapunov function for system (4.2.36) as

$$V(e, \eta) = V_1(e) + V_2(\eta), \quad \forall e \in \mathbb{R}^{n+1}, \quad \eta \in \mathbb{R}^n. \qquad (4.2.40)$$

Computing the derivative of $V(e(t), \eta(t))$ along the solution of (4.2.36) yields

$$\left. \frac{dV(e(t), \eta(t))}{dt} \right|_{\text{along (4.2.36)}}$$

$$= \frac{1}{\varepsilon} \left[\sum_{i=1}^{n} (e_{i+1}(t) - g_i(e_1(t))) \frac{\partial V_1(e(t))}{\partial e_i} \right.$$

$$\left. - g_{n+1}(e_1(t)) \frac{\partial V_1(e(t))}{\partial e_{n+1}} + (\tilde{f}(x(t)) - \tilde{f}(\hat{x}(t))) \frac{\partial V_1(e(t))}{\partial e_n} + \varepsilon \hbar(t) \frac{\partial V_1(e(t))}{\partial e_{n+1}} \right]$$

$$+ \left[\sum_{i=1}^{n-1} \eta_{i+1}(t) \frac{\partial V_2(\eta(t))}{\partial \eta_i} + [\varphi(\hat{x}(t) - z(t)) - \varphi(x(t) - z(t)) + \varphi(x(t) - z(t)) \right.$$

$$\left. + \tilde{f}(x(t)) - \tilde{f}(\hat{x}(t)) + e_{n+1}(t)] \frac{\partial V_2(\eta(t))}{\partial \eta_n} \right]. \qquad (4.2.41)$$

This, together with Assumptions 4.2.2, 4.2.3, 4.2.5, (4.2.37), and (4.2.39), gives

$$\left. \frac{dV(e(t), \eta(t))}{dt} \right|_{\text{along (4.2.36)}}$$

$$\le -\frac{W_1(e(t))}{\varepsilon} + \beta_1 |\hbar(t)| \|e(t)\| - W_2(\eta(t))$$

$$+ (1 + L + L_1)\beta_2 \|e(t)\| \|\eta(t)\| + \frac{\alpha L_1}{\varepsilon} \|e(t)\|^2$$

$$\leq -\frac{W_1(e(t))}{\varepsilon} + \frac{\alpha L_1 + \beta_1 B}{\varepsilon} \|e(t)\|^2 + \beta_1 B_0 \|e(t)\| + \beta_1 B_2 \|e(t)\|^2 + [\beta_1 B_1$$

$$+ (1 + L + L_1)\beta_2] \|e(t)\| \|\eta(t)\| - W_2(\eta(t))$$

$$\leq -\left[\frac{\lambda_{13} - (\alpha L_1 + \beta_1 B)}{\varepsilon} - \beta_1 B_2\right] \|e(t)\|^2 + \beta_1 B_0 \|e(t)\| - \lambda_{23} \|\eta(t)\|^2$$

$$+ [(1 + L + L_1)\beta_2 + \beta_1 B_1] \|e(t)\| \|\eta(t)\|$$

$$= -\left[\frac{\lambda_{13} - (\alpha L_1 + \beta_1 B)}{\varepsilon} - \beta_1 B_2\right] \|e(t)\|^2 + \beta_1 B_0 \|e(t)\| - \lambda_{23} \|\eta(t)\|^2$$

$$+ \sqrt{\frac{\lambda_{13} - (\alpha L_1 + \beta_1 B)}{2\varepsilon}} \|e(t)\| \sqrt{\frac{2\varepsilon}{\lambda_{13} - (\alpha L_1 + \beta_1 B)}} [(1 + L + L_1)\beta_2 + \beta_1 B_1] \|\eta(t)\|$$

$$\leq -\left[\frac{\lambda_{13} - (\alpha L_1 + \beta_1 B)}{2\varepsilon} - \beta_1 B_2\right] \|e(t)\|^2 + \beta_1 B_0 \|e(t)\|$$

$$- \left(\lambda_{23} - \frac{[(1 + L + L_1)\beta_2 + \beta_1 B_1]^2}{2(\lambda_{13} - (\alpha L_1 + \beta_1 B))} \varepsilon\right) \|\eta(t)\|^2, \tag{4.2.42}$$

where we used (4.2.32): $\alpha L_1 + B\beta_1 < \lambda_{13}$.

Similar to the proof of Theorem 4.2.1, we accomplish the proof in three steps.

Step 1. *Show that for every $R > R_0$, there exist $r > 0$ and $\varepsilon_0 > 0$ such that for every $\varepsilon \in (0, \varepsilon_0)$, there exists a $t_{1\varepsilon} > 0$ such that for all $t > t_{1\varepsilon}$, $\|(e(t), \eta(t))\| \leq r$, where $(e(t), \eta(t))$ is the solution of (4.2.36).*

Set

$$\begin{cases} r = \max\left\{2, \dfrac{4(\beta_1 B_0 + 1)}{\lambda_{23}}\right\}, \\ \varepsilon_0 = \min\left\{1, \dfrac{\lambda_{13} - (\alpha L_1 + \beta_1 B)}{2\beta_1(B_0 + B_2)}, \dfrac{\lambda_{23}(\lambda_{13} - \alpha L_1 \beta_1 B)}{[(1 + L + L_1)\beta_2 + \beta_1 B_2]^2}\right\}. \end{cases} \tag{4.2.43}$$

By (4.2.32), $\varepsilon_0 > 0$. For any $\varepsilon \in (0, \varepsilon_0)$ and $\|(e(t), \eta(t))\| \geq r$, we compute the derivative of $V(e(t), \eta(t))$ along the solution of (4.2.36) in the following two cases.

Case 1: $\|e(t)\| \geq r/2$. In this case, $\|e(t)\| \geq 1$ and hence $\|e(t)\|^2 \geq \|e(t)\|$. By the definition ε_0 of (4.2.43), $\lambda_{23} - [(1 + L + L_1)\beta_2 + \beta_1 B_1]^2/(2(\lambda_{13} - (\alpha L_1 + \beta_1 B))\varepsilon > 0$, it follows from (4.2.42) that

$$\left.\frac{dV(e(t), \eta(t))}{dt}\right|_{\text{along (4.2.36)}}$$

$$\leq -\left[\frac{\lambda_{13} - (\alpha L_1 + \beta_1 B)}{2\varepsilon} - \beta_1 B_2\right] \|e(t)\|^2 + \beta_1 B_0 \|e(t)\|$$

$$\leq -\left[\frac{\lambda_{13} - (\alpha L_1 + \beta_1 B)}{2\varepsilon} - \beta_1 B_2\right] \|e(t)\|^2 + \beta_1 B_0 \|e(t)\|^2$$

$$= -\frac{\lambda_{13} - (\alpha L_1 + \beta_1 B) - 2\varepsilon\beta_1(B_0 + B)}{2\varepsilon}\|e(t)\|^2 < 0, \qquad (4.2.44)$$

where we used the fact that $\lambda_{13} - (\alpha L_1 + \beta_1 B) - 2\varepsilon\beta_1(B_0 + B) > 0$ owing to (4.2.43).

Case 2: $\|e(t)\| < r/2$. In this case, it follows from $\|(e(t), \eta(t))\| \geq r$ and $\|e(t)\| + \|\eta(t)\| \geq \|(e(t), \eta(t))\|$ that $\|\eta(t)\| \geq r/2$. In view of (4.2.43), $\lambda_{13} - (\alpha L_1 + \beta_1 B) - 2\varepsilon\beta_1 B_2 > 0$. Hence, from (4.2.42), it follows that

$$\left.\frac{dV(e(t), \eta(t))}{dt}\right|_{\text{along (4.2.36)}}$$

$$\leq -\left(\lambda_{23} - \frac{[(1 + L + L_1)\beta_2 + \beta_1 B_1]^2}{2(\lambda_{13} - (\alpha L_1 + \beta_1 B))}\varepsilon\right)\|\eta(t)\|^2$$

$$+ \beta_1 B_0\|e(t)\| \leq \beta_1 B_0 \frac{r}{2} - \frac{\lambda_{23}}{2}\left(\frac{r}{2}\right)^2 = -\frac{r}{8}[r\lambda_{23} - 4\beta_1 B_0] < 0, \qquad (4.2.45)$$

where owing to (4.2.43), $r\lambda_{23} - 4\beta_1 B_0 > 0$ and $[(1 + L + L_1)\beta_2 + \beta_1 B_1]^2/(2(\lambda_{13} - (\alpha L_1 + \beta_1 B))\varepsilon \leq \lambda_{23}/2$.

Combining the above two cases, we conclude that, for every $\varepsilon \in (0, \varepsilon_0)$ and $\|(e(t), \eta(t))\| \geq r$, the derivative of $V(e(t), \eta(t))$ along the solution of system (4.2.36) satisfies

$$\left.\frac{dV(e(t), \eta(t))}{dt}\right|_{\text{along (4.2.36)}} < 0. \qquad (4.2.46)$$

This shows that there exists a $t_{1\varepsilon} > 0$ such that $\|(e(t), \eta(t))\| \leq r$ for all $t > t_{1\varepsilon}$.

Step 2. *Establish the convergence of* $|x_i(t) - \hat{x}_i(t)|$.

Consider the subsystem that is composed of the last $n + 1$ equations in system (4.2.36):

$$\begin{cases} \varepsilon\dot{e}_1(t) = e_2(t) - g_1(e_1(t)), \\ \quad\vdots \\ \varepsilon\dot{e}_n(t) = e_{n+1}(t) - g_n(e_1(t)) + \tilde{f}(x(t)) - \tilde{f}(\hat{x}(t)), \\ \varepsilon\dot{e}_{n+1}(t) = -g_{n+1}(e_1(t)) + \varepsilon\hbar(t), \end{cases} \qquad (4.2.47)$$

where $\hbar(t)$ is given by (4.2.38). Since $\|(e(t), x(t))\| \leq r$ for all $t > t_{1\varepsilon}$, we have, in view of (4.2.39), that $|\hbar(t)| \leq M_0 + B/\varepsilon\|e(t)\|$ for all $t > t_{1\varepsilon}$ and some $M_0 > 0$. This together with Assumption 4.2.2 gives the derivative of $V_1(e(t))$ along the solution of (4.2.47) as

$$\left.\frac{dV_1(e(t))}{dt}\right|_{\text{along (4.2.47)}}$$

$$= \frac{1}{\varepsilon}\left(\sum_{i=1}^{n}(e_{i+1}(t) - g_i(e_1(t)))\frac{\partial V_1(e(t))}{\partial e_i} - g_{n+1}(e_1(t))\frac{\partial V_1(e(t))}{\partial e_{n+1}}\right)$$

$$+[\tilde{f}(x(t)) - \tilde{f}(\hat{x}(t))]\frac{\partial V_1(e(t))}{\partial e_n} + \varepsilon\hbar(t)\frac{\partial V_1(e(t))}{\partial e_{n+1}}\Big)$$

$$\leq -\frac{\lambda_{13} - (\alpha L_1 + \beta_1 B)}{\varepsilon}\|e(t)\|^2 + M_0\beta_1\|e(t)\|$$

$$\leq -\frac{\lambda_{13} - (\alpha L_1 + \beta_1 B)}{\varepsilon\lambda_{12}}V_1(e(t)) + \frac{M_0\beta_1\sqrt{\lambda_{11}}}{\lambda_{11}}\sqrt{V_1(e(t))}, \quad \forall\, t > t_{1\varepsilon}.$$

$$(4.2.48)$$

It follows that

$$\frac{d\sqrt{V_1(e(t))}}{dt} \leq -\frac{\lambda_{13} - (\alpha L_1 + \beta_1 B)}{\varepsilon\lambda_{12}}\sqrt{V_1(e(t))} + \frac{\varepsilon M_0\beta_1\sqrt{\lambda_{11}}}{\lambda_{11}}, \quad \forall\, t > t_{1\varepsilon},$$

$$(4.2.49)$$

and hence

$$\sqrt{V_1(e(t))} \leq \sqrt{V_1(e(t_{1\varepsilon}))}e^{-\frac{\lambda_{13}-(\alpha L_1+\beta_1 B)}{\varepsilon\lambda_{12}}(t-t_{1\varepsilon})}$$

$$+ \frac{M\beta_1\sqrt{\lambda_{11}}}{\lambda_{11}}\int_{t_{1\varepsilon}}^t e^{-\frac{\lambda_{13}-(\alpha L_1+B\beta_1)}{\varepsilon\lambda_{12}}(t-s)}ds, \quad \forall\, t > t_{1\varepsilon}, \quad (4.2.50)$$

where once again we used (4.2.32): $\lambda_{13} > \alpha L_1 + \beta_1 B$. Since the first term on the right-hand side of (4.2.50) tends to zero as t goes to infinity and the second term is bounded by ε multiplied by an ε-independent constant, by the definition of $e(t)$, we conclude that there exist $t_{2\varepsilon} > t_{1\varepsilon}$ and $\Gamma_1 > 0$ such that

$$|x_i(t) - \hat{x}_i(t)| = \varepsilon^{n+1-i}|e_i(t)| \leq \|e(t)\| \leq \sqrt{\frac{V_1(e(t))}{\lambda_{11}}} \leq \Gamma_1\varepsilon^{n+2-i}, \quad \forall\, t > t_{2\varepsilon}.$$

$$(4.2.51)$$

Step 3. *Establish the convergence of* $\|x(t) - z(t)\|$.

Consider the following subsystem that is composed of the first n number equations in (4.2.36):

$$\begin{cases} \dot{\eta}_1(t) = \eta_2(t), \\ \dot{\eta}_2(t) = \eta_3(t), \\ \quad\vdots \\ \dot{\eta}_n(t) = \varphi(\hat{x}(t) - z(t)) + \tilde{f}(x(t)) - \tilde{f}(\hat{x}(t)) + e_{n+1}(t). \end{cases} \quad (4.2.52)$$

Computing the derivative of $V_2(\eta(t))$ along the solution of system (4.2.52) and applying Assumption 4.2.3 and (4.2.37), gives

$$\frac{dV_2(\eta(t))}{dt}\bigg|_{\text{along (4.2.52)}}$$

$$= \sum_{i=1}^{n-1}\eta_{i+1}(t)\frac{\partial V_2(\eta(t))}{\partial\eta_i} + \varphi(\eta(t))\frac{\partial V_2(\eta(t))}{\partial\eta_n}$$

$$+ \{e_{n+1}(t) + \varphi(\hat{x}(t) - z(t)) - \varphi(x(t) - z(t)) + \tilde{f}(x(t)) - \tilde{f}(\hat{x}(t))\} \frac{\partial V_2(\eta(t))}{\partial \eta_n}$$

$$\leq -W_2(\eta(t)) + (1 + L + L_1)\beta_2 \|e(t)\| \|\eta(t)\|$$

$$\leq -\frac{\lambda_{23}}{\lambda_{22}} V_2(\eta(t)) + N_0 \varepsilon \sqrt{V_2(\eta(t))}, \quad \forall \, t > t_{2\varepsilon}, \tag{4.2.53}$$

where N_0 is an R-dependent positive constant. By (4.2.53), we have, for all $t > t_{2\varepsilon}$, that

$$\frac{d\sqrt{V_2(\eta(t))}}{dt}\bigg|_{\text{along (4.2.52)}} \leq -\frac{\lambda_{23}}{\lambda_{22}} \sqrt{V_2(\eta(t))} + N_0 \varepsilon, \tag{4.2.54}$$

and hence

$$\sqrt{V_2(\eta(t))} \leq \sqrt{V_2(\eta(t_{2\varepsilon}))} e^{-\frac{\lambda_{23}}{\lambda_{22}}(t - t_{2\varepsilon})} + N_0 \varepsilon \int_{t_{2\varepsilon}}^{t} e^{-\frac{\lambda_{23}}{\lambda_{22}}(t - s)} ds, \, \forall \, t > t_{2\varepsilon}. \tag{4.2.55}$$

Once again, the first term on the right-hand side of (4.2.55) tends to zero as t goes to infinity and the second term is bounded by ε multiplied by an ε-independent constant; thus there exist $t_\varepsilon > t_{2\varepsilon}$ and $\Gamma_2 > 0$ such that $\|x(t) - z(t)\| \leq \Gamma_2 \varepsilon$ for all $t > t_\varepsilon$. This completes the proof of the theorem.

We consider system (4.1.72) once again. For system (4.1.72), if the control parameter depends only on the external disturbance, that is, $b(x, \zeta, w) \triangleq b(w)$ and satisfies Assumption 4.2.6 below, we can also design the same control (4.2.1), the tracking differentiator (4.0.3), and the extended state observer (4.0.4) to get the global convergence such as Theorem 4.2.1 for system (4.0.1).

Assumption 4.2.6 *(Assumption on control parameters)* There exists a constant $b_0 \neq 0$ such that $\sigma \triangleq \sup_{w \in \mathbb{R}} |b(w) - b_0| \leq \lambda_{13} |b_0| / (\beta_1 |k_{n+1}|)$.

Assumption 4.2.7 *(Assumption on zero dynamics)* There exist constants k_1 and k_2, and function $\varpi \in C(\mathbb{R}, \bar{\mathbb{R}}^+)$ such that

$$\|F_0(x, \zeta, w)\| \leq k_1 + k_2 \|x\| + \varpi(w), \quad \forall \, x \in \mathbb{R}^n, \quad \zeta \in \mathbb{R}^m, \quad w \in \mathbb{R}.$$

Theorem 4.2.3 *Let $x_i(t)(1 \leq i \leq n)$ and $\hat{x}_i(t)(1 \leq i \leq n+1)$ be the solutions of system (4.1.72) and (4.0.4) coupled by (4.2.1), and let $x_{n+1}(t) = f(t, x(t), \zeta(t), w(t)) + (b(w(t)) - b_0)u(t)$ be the extended state. Assume that $\|(w(t), \dot{w}(t))\| < \infty$ and Assumptions 4.2.1 to 4.2.4 and 4.2.6 and 4.2.7 hold true. Then for any given initial value of (4.1.72), the following conclusion hold.*

(i) For any $\sigma > 0$ and $\tau > 0$, there exists a constant $R_0 > 0$ such that $|z_1(t) - v(t)| < \sigma$ uniformly in $t \in [\tau, \infty)$ for all $R > R_0$, where $z_i(t)$ are solutions of (4.0.3).

(ii) For every $R > R_0$, there is an R-dependent constant $\varepsilon_0 > 0$ such that for any $\varepsilon \in (0, \varepsilon_0)$, there exists a $t_\varepsilon > 0$ such that for all $R > R_0$, $\varepsilon \in (0, \varepsilon_0)$, $t > t_\varepsilon$,

$$|x_i(t) - \hat{x}_i(t)| \leq \Gamma_1 \varepsilon^{n+2-i}, \quad i = 1, 2, \ldots, n+1, \tag{4.2.56}$$

and

$$|x_i(t) - z_i(t)| \leq \Gamma_2 \varepsilon, \quad i = 1, 2, \ldots, n, \tag{4.2.57}$$

where Γ_1 and Γ_2 are R-dependent positive constants only.

(iii) For any $\sigma > 0$, there exist $R_1 > R_0$ and $\varepsilon_1 \in (0, \varepsilon_0)$ such that for any $R > R_1$ and $\varepsilon \in (0, \varepsilon_1)$, there exists a $t_{R\varepsilon} > 0$ such that $|x_1(t) - v(t)| < \sigma$ for all $R > R_1$, $\varepsilon \in (0, \varepsilon_1)$, and $t > t_{R\varepsilon}$.

4.2.3   Semi-Global ADRC for SISO System with Vast Uncertainty

To establish the semi-global convergence, we need the following additional assumptions.

Assumption 4.2.8 For $f \in C^1(\mathbb{R}^{n+m+2}, \mathbb{R})$, there exists positive constant N and functions $\varpi_1 \in C(\mathbb{R}^n, [0, \infty))$, $\varpi_2 \in C(\mathbb{R}^m, [0, \infty))$, and $\varpi_3 \in C(\mathbb{R}, [0, \infty))$ such that

$$\sum_{i=1}^{n} \left| \frac{\partial f(t, x, \zeta, w)}{\partial x_i} \right| + \sum_{i=1}^{m} \left| \frac{\partial f(t, x, \zeta, w)}{\partial \zeta_i} \right| + \left| \frac{\partial f(t, x, \zeta, w)}{\partial t} \right|$$

$$+ |f(t, x, \zeta, w)| \leq N + \varpi_1(x) + \varpi_2(\zeta) + \varpi_3(w)$$

for all $x \in \mathbb{R}^n, \zeta \in \mathbb{R}^m$, and $w \in \mathbb{R}$.

Assumption 4.2.9 There exist positive constants C_1, C_2, and C_3 such that $\sup_{t \in [0, \infty)}$ $\|(v(t), \ldots, v^{(n)}(t))\| < C_1$, $\|x(0)\| < C_2$, $\|(w(t), \dot{w}(t))\| < C_3$ for all $t \in [0, \infty)$.

Let $C_1^* = \max_{\{y \in \mathbb{R}^n, \|y\| \leq C_1 + C_2 + 1\}} V_2(y)$. The following assumption is to guarantee the input-to-state stable for zero dynamics.

Assumption 4.2.10 There exist positive definite functions $V_0, W_0 : \mathbb{R}^s \to \mathbb{R}$ such that $L_{F_0} V_0(\zeta) \leq -W_0(\zeta)$ for all $\zeta : \|\zeta\| > \chi(x, w)$, where $\chi : \mathbb{R}^{n+1} \to \mathbb{R}$ is a class $\mathcal{K}_\infty$ function and $L_{F_0} V_0(\zeta)$ denotes the Lie derivative along the zero dynamics of system (4.1.72).

Let

$$C_4 \geq \max \left\{ \sup_{\|x\| \leq C_1 + (C_1^*+1)/\lambda_{23}+1, \|w\| \leq C_3,} |\chi(x, w)|, \|\zeta(0)\| \right\},$$

$$M_1 \geq 2 \left(1 + M_2 + C_1 + N + \sup_{\|x\| \leq C_1 + (C_1^*+1)/\lambda_{23}+1} \varpi_1(x) + \sup_{\|\zeta\| \leq C_4} \varpi_2(\zeta) + \sup_{\|w\| \leq C_3} \varpi_3(w) \right),$$

$$M_2 \geq \max_{\|x\| \leq C_1 + (C_1^*+1)/\lambda_{23}+1} |\varphi(x)|. \tag{4.2.58}$$

The following assumption is for the control parameters.

Assumption 4.2.11 There exists a nominal parameter function $b_0 \in C^1(\mathbb{R}^n, \mathbb{R})$ such that

(i) $b_0(x) \neq 0$ for every $x \in \mathbb{R}^n$.

(ii) $b_0(x)$, $\frac{1}{b_0}(x)$, and all their partial derivatives with respect to their arguments are globally bounded.

(iii)

$$\vartheta = \sup_{\|x\| \leq C_1 + (C_1^* + 1)/(\lambda_{23}^i) + 1, \|\zeta\| \leq C_4, \|w\| \leq C_3, \nu \in \mathbb{R}^n} \left| \frac{b(x, \zeta, w) - b_0(x)}{b_0(\nu)} \right|$$

$$< \min \left\{ \frac{1}{2}, \lambda_{13} \left(\beta_1 k_{n+1} \left(M_1 + \frac{1}{2} \right) \right)^{-1} \right\}.$$

(4.2.59)

Condition 3 of Assumption 4.2.2 is changed into

$$\left| \frac{\partial V_1(y)}{\partial y_n} \right| + \left| \frac{\partial V_1(y)}{\partial y_{n+1}} \right| \leq \beta_1 \|y\|, \quad \forall \, y = (y_1, y_2, \ldots, y_{n+1}) \in \mathbb{R}^{n+1}.$$

Let $\mathrm{sat}_M(\cdot) : \mathbb{R} \to \mathbb{R}$ be an odd continuous differentiable saturation function defined in (4.1.73). Using the same tracking differentiator (4.0.3), the ESO (4.0.4) where the parameter b_0 is replaced by $b_0(\hat{x}(t))$, the ESO-based feedback control is designed as follows:

ADRC(S): $u(t) = \dfrac{1}{b_0(\hat{x}(t))} [\mathrm{sat}_{M_2}(\varphi(\hat{x}(t) - z(t))) - \mathrm{sat}_{M_1}(\hat{x}_{n+1}(t)) + \mathrm{sat}_{C_1+1}(z_{n+1}(t))].$

(4.2.60)

Theorem 4.2.4 *Let $x_i(t)(1 \leq i \leq n)$ and $\hat{x}_i(t)(1 \leq i \leq n+1)$ be the solutions of systems (4.1.72) and (4.0.4) with $\varrho(t) \equiv \varepsilon$ coupled by (4.2.1), and let $x_{n+1}(t) = f(t, x(t), \zeta(t), w(t)) + (b(x(t), \zeta(t), w(t)) - b_0(x(t)))u(t)$ be the extended state. Let $\psi(x_1, \ldots, x_{n+1}) = \sum_{i=1}^{n+1} \lambda_i x_i, \lambda_i \in \mathbb{R}$ in (4.0.3), and assume that Assumptions 4.2.2 to 4.2.4 and 4.2.8 to 4.2.11 are satisfied. Then for any given initial value of (4.1.72), the following assertions hold.*

(i) *For any $\sigma > 0$ and $\tau > 0$, there exists a constant $R_0 > 0$ such that $|z_{1R}(t) - v(t)| < \sigma$ uniformly in $t \in [\tau, \infty)$ for all $R > R_0$, where $z_{iR}(t)$ are solutions of (4.0.3).*

(ii) *For any $\sigma > 0$, there exist $\varepsilon_0 > 0$ and $R_1 > R_0$ such that for any $\varepsilon \in (0, \varepsilon_0)$, $R > R_1$, there exists an ε-independent constant $t_0 > 0$, such that*

$$|x_i(t) - \hat{x}_i(t)| < \sigma \text{ for all } t > t_0, \quad i = 1, 2, \ldots, n+1.$$

(4.2.61)

(iii) *For any $\sigma > 0$, there exist $R_2 > R_1$ and $\varepsilon_1 \in (0, \varepsilon_0)$ such that for any $R > R_2$ and $\varepsilon \in (0, \varepsilon_1)$, there exists a $t_{R\varepsilon} > 0$ such that $|x_1(t) - v(t)| < \sigma$ for all $R > R_1$, $\varepsilon \in (0, \varepsilon_1)$, and $t > t_{R\varepsilon}$.*

Let $\Omega_0 = \{y| V_2(y) \leq C_1^*\}$, $\Omega_1 = \{y| V_2(y) \leq C_1^* + 1\}$. Theorem 4.2.4 can be proved by the fact that under the conditions of Theorem 4.2.4, there exists an $\varepsilon_1 > 0$ such that

$$(x_1(t), \ldots, x_n(t))^\top - (z_1(t), \ldots, z_n(t))^\top \in \Omega_1, \quad \forall \, \varepsilon \in (0, \varepsilon_1), t \in [0, \infty).$$

The above fact means that under the control (4.2.60), the solution of (4.1.72) is bounded. Then the proof can follow similarly from the proof of Theorem 4.1.3.

4.2.4 Examples and Numerical Simulations

In this subsection, we give an example of stabilization using nonlinear ESO.

Example 4.2.1 *Consider the following system:*

$$\begin{cases} \dot{x}_1(t) = x_2(t), \\ \dot{x}_2(t) = 2x_1(t) + 4x_2(t) + \sin\left(2t + \dfrac{\pi}{3}\right) + 10.1u(t), \\ y(t) = x_1(t), \end{cases} \tag{4.2.62}$$

Now $b = 10.1$ and $b_0 = 10$. Our aim is to design an ESO-based output feedback control to stabilize system (4.2.62). Thus it is assumed that $z(t) = v(t) \equiv 0$.

The ESO designed in the following is to estimate the total disturbance $x_3(t) = 2x_1(t) + 4x_2(t) + \sin(2t + \pi/3) + 0.1u(t)$:

$$\begin{cases} \dot{\hat{x}}_1(t) = \hat{x}_2(t) + \dfrac{3}{\varepsilon}(y(t) - \hat{x}_1(t)) + \varepsilon\Phi\left(\dfrac{y(t) - \hat{x}_1(t)}{\varepsilon^2}\right), \\ \dot{\hat{x}}_2(t) = \hat{x}_3(t) + \dfrac{3}{\varepsilon^2}(y(t) - \hat{x}_1(t)) + 10u(t), \\ \dot{\hat{x}}_3(t) = \dfrac{1}{\varepsilon^3}(y(t) - \hat{x}_1(t)), \end{cases} \tag{4.2.63}$$

where $\Phi : \mathbb{R} \to \mathbb{R}$ is defined as

$$\Phi(r) = \begin{cases} -\dfrac{1}{4}, & r \in \left(-\infty, -\dfrac{\pi}{2}\right), \\ \dfrac{1}{4}\sin r, & r \in \left(-\dfrac{\pi}{2}, \dfrac{\pi}{2}\right), \\ \dfrac{1}{4}, & r \in \left(\dfrac{\pi}{2}, -\infty\right). \end{cases} \tag{4.2.64}$$

In this case, $g_i(y)$ in (4.0.4) can be specified as

$$g_1(y_1) = 3y_1 + \Phi(y_1), \quad g_2(y_1) = 3y_1, \quad g_3(y_1) = y_1. \tag{4.2.65}$$

The Lyapunov function $V : \mathbb{R}^3 \to \mathbb{R}$ for this case is given by

$$V(y) = \langle Py, y \rangle + \int_0^{y_1} \Phi(s)ds, \tag{4.2.66}$$

where

$$P = \begin{pmatrix} 1 & -\dfrac{1}{2} & -1 \\ -\dfrac{1}{2} & 1 & -\dfrac{1}{2} \\ -1 & -\dfrac{1}{2} & 4 \end{pmatrix}$$

is the positive definite solution of the Lyapunov equation $PE + E^\top P = -I_{3\times 3}$ with

$$E = \begin{pmatrix} -3 & 1 & 0 \\ -3 & 0 & 1 \\ -1 & 0 & 0 \end{pmatrix}.$$

A direct computation shows that

$$\sum_{i=1}^{2} \frac{\partial V(y)}{\partial y_i}(y_i - g_i(y_1)) - \frac{\partial V(y)}{\partial y_3}g_3(y_1)$$

$$= -y_1^2 - y_2^2 - y_3^2 - (2y_1 - y_2 - 2y_3 + \Phi(y_1))\Phi(y_1) + (y_2 - 3y_1)\Phi(y_1) \qquad (4.2.67)$$

$$\leq -\left(\frac{y_1^2}{8} + \frac{7y_2^2}{8} + \frac{3y_3^2}{4}\right) \triangleq -W(y_1, y_2, y_3).$$

The parameters in Assumption 4.2.2 are specified here as $k_3 = 1, \lambda_{13} = 1/8, \beta_1 = 8$. Notice that $(b - b_0)k_3/b_0 = 1/100 < 1/64 = \lambda_{13}/\beta_1$, so Assumption 4.2.2 is satisfied.

The ESO-based output feedback control can be designed as follows:

$$u(t) = \frac{1}{10}[-\hat{x}_1(t) - 2\hat{x}_2(t) - \Phi(\hat{x}_1(t)) - \hat{x}_3(t)]. \qquad (4.2.68)$$

Take

$$x(0) = (-5, -5), \hat{x}(0) = (1, 2, 3), \quad \varepsilon = 0.005,$$

and the integration step $h = 0.001$. The numerical results of the closed loop that is composed of (4.2.62), (4.2.63), and (4.2.68) are plotted in Figure 4.2.1.

It is seen from Figures 4.2.1(a) and 4.2.1(b) that $(x_1(t), x_2(t)) \to 0$ and $(\hat{x}_1(t), \hat{x}_2(t)) \to (x_1(t), x_2(t))$ as $t \to \infty$.

In Figure 4.2.1, the peaking phenomenon is observed for both $x(t)$ and $\hat{x}(t)$. In particular, the large peaking values are observed for $(x_2(t), \hat{x}_2(t))$ and $(x_3(t), \hat{x}_3(t))$. This can be attributed to the difference in their initial values $(x_1(0), \hat{x}_1(0))$ and $(x_2(0), \hat{x}_2(0))$. In the following, we show that the peaking problem relies on the disturbance of the initial value. Here we take $x_1(0) = \hat{x}_1(0) = 0, x_2(0) = \hat{x}_2(0) = 1, \hat{x}_3(0) = 3$, and $\varepsilon = 0.001$ and $h = 0.0001$ and obtain the solutions for the closed-loop system composed of (4.2.62), (4.2.63), and (4.2.68) in Figure 4.2.2. Comparing this with Figure 4.2.1, we see that the peaking value does not appear in Figure 4.2.2.

4.3 ADRC with Time-Varying Tuning Parameter

In this section, we consider the the following nonlinear system:

$$\begin{cases} \dot{x}(t) = A_n x(t) + B_n[f(t, x(t), \zeta(t), w(t)) + b(t)u(t)], \\ \dot{\zeta}(t) = f_0(t, x(t), \zeta(t), w(t)), \\ y(t) = C_n x(t), \end{cases} \qquad (4.3.1)$$

where $x(t) \in \mathbb{R}^n$ and $\zeta(t) \in \mathbb{R}^m$ are the system states, A_n and B_n are defined as

$$A_n = \begin{pmatrix} 0 & I_{n-1} \\ 0 & 0 \end{pmatrix}, B_n^\top = C_n = (0, 0, \cdots, 1), \qquad (4.3.2)$$

$f \in C(\mathbb{R}^{n+m+2}, \mathbb{R}), f_0 \in C(\mathbb{R}^{n+m+2}, \mathbb{R}^m)$ are unknown nonlinear functions, $u(t) \in \mathbb{R}$ is the input (control), $y(t) = C_n x(t) = x_1(t)$ is the output (measurement), and $b \in C^1(\mathbb{R}, \mathbb{R})$ contains some uncertainty with nominal value $b_0 \neq 0$.

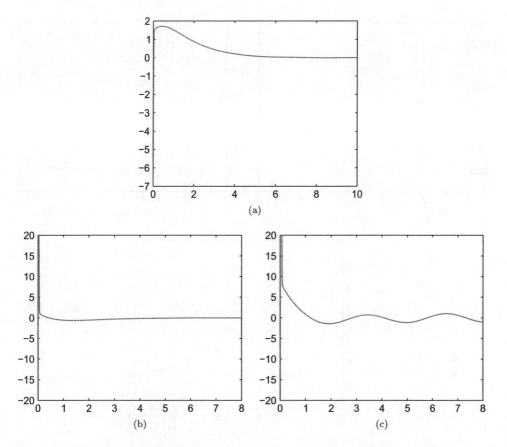

Figure 4.2.1 Stabilization of system (4.2.62), (4.2.63), and (4.2.68) by ADRC.

We design a nonlinear ESO with time-varying gain for system (4.3.1) as follows:

$$\begin{cases} \dot{\hat{x}}_1(t) = \hat{x}_2(t) + \dfrac{1}{r^{n-1}(t)} g_1(r^n(t)(\delta(t)), \\[2mm] \dot{\hat{x}}_2(t) = \hat{x}_3(t) + \dfrac{1}{r^{n-2}(t)} g_2(r^n(t)\delta(t)), \\[1mm] \qquad \vdots \\[1mm] \dot{\hat{x}}_n(t) = \hat{x}_{n+1}(t) + g_n(r^n(t)\delta(t)) + b_0 u(t), \\[2mm] \dot{\hat{x}}_{n+1}(t) = r(t) g_{n+1}(r^n(t)\delta(t)), \end{cases} \qquad (4.3.3)$$

where $\delta(t) = \hat{x}_1(t) - y(t)$, and $r(t)$ is the time-varying gain to be increasing gradually. When $r(t) \equiv 1/\varepsilon$, (4.3.3) is reduced to the constant high-gain nonlinear ESO in Section 4.2. The aim of ESO is to estimate the state $(x_1(t), x_2(t), \ldots x_n(t))$ and the total disturbance

$$x_{n+1}(t) \triangleq f(t, x(t), \zeta(t), w(t)) + [b(t) - b_0] u(t), \qquad (4.3.4)$$

which is also called the extended state.

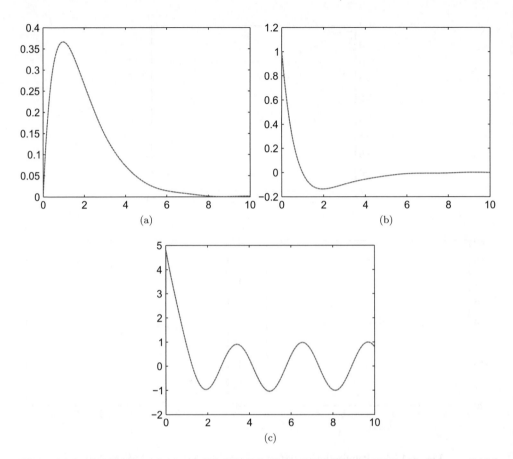

Figure 4.2.2   Stabilization of system (4.2.62), (4.2.63), and (4.2.68) by ADRC with the same initial value as in Figure 4.2.1.

Once again, the first part of ADRC is the tracking differentiator (TD). For the reference signal $v(t)$, we use the following TD to estimate its derivatives:

$$
\begin{cases}
\dot{z}_1(t) = z_2(t), \\
\quad \vdots \\
\dot{z}_n(t) = z_{n+1}(t), \\
\dot{z}_{n+1}(t) = -\rho^{n+1} k_1 (z_1(t) - v(t)) - \rho^n k_2 z_2(t) - \cdots - \rho k_{n+1} z_{n+1}(t).
\end{cases}
\tag{4.3.5}
$$

By Theorem 2.1.2, if $\sup_{t \in [0,\infty)} |v^{(i)}(t)| < \infty, i = 1, 2, \ldots, n$, and the following matrix is Hurwitz:

$$\begin{pmatrix} 0 & 1 & 0 & \cdots & 0 \\ 0 & 0 & 1 & \cdots & 0 \\ \vdots & \vdots & \vdots & \ddots & \vdots \\ 0 & 0 & 0 & \cdots & 1 \\ -k_1 & -k_2 & -k_3 & \cdots & -k_n \end{pmatrix},$$

then for any $a > 0$, $|z_i(t) - v^{(i-1)}(t)| \leq \bar{M}/\rho$, $i = 1, 2, \cdots, n+1$ uniformly in $t \in [a, \infty)$, where $\bar{M}$ is a ρ-independent constant. As before, if derivatives of $v(t)$ are available, we just let $z_i(t) = v^{(i-1)}(t)$.

Since the TD part is relatively independent of the other two parts of ADRC, we do not couple TD in the closed loop; instead, we use $z_i(t)$ directly in the feedback loop.

The second part of ADRC is again the ESO (4.3.3), which estimates both the state and the total disturbance of system (4.3.1).

Suppose that we have obtained estimates for both the state and total disturbance. We then use an estimation/cancelation strategy to design the ESO-based output feedback control as follows:

$$u(t) = \frac{1}{b_0}(u_0(\hat{x}_1(t) - z_1(t), \dots, \hat{x}_n(t) - z_n(t)) + z_{n+1}(t) - \hat{x}_{n+1}(t)), \qquad (4.3.6)$$

where $\hat{x}_{n+1}(t)$ is used to compensate the total disturbance $x_{n+1}(t)$ and $u_0(\cdot)$ is the nominal control to be specified later. The objective of the control is to make the error $(x_1(t) - z_1(t), x_2(t) - z_2(t), \dots, x_n(t) - z_n(t))$ convergent to zero as time goes to infinity in the prescribed way. Precisely,

$$x_i(t) - z_i(t) \approx y^{(i-1)}(t) \text{ for } i = 1, 2, \dots, n, t \to \infty,$$

where $y(t)$ satisfies

$$y^{(n)}(t) = u_0(y(t), y'(t), \dots, y^{(n-1)}(t)) \qquad (4.3.7)$$

and $u_0(\cdot)$ is chosen so that (4.3.7) is asymptotically stable. It is worth pointing out that, unlike the high-gain dominated approach in dealing with uncertainty, we do not need the high gain in the feedback loop.

To prove the convergence, we need some assumptions. The following assumption is about the unknown system functions $f(\cdot)$ and $f_0(\cdot)$, control coefficient $b(t)$, and the external disturbance $w(t)$.

Assumption 4.3.1 There exist positive constants M_1, K_0, K_1, K_2, and function $\varpi \in C(\mathbb{R}, [0, \infty))$ such that

- $\sup_{t \in [0, \infty)} \{|w(t)|, |\dot{w}(t)|, |b(t)|, |\dot{b}(t)|, \|(z_1(t), \dots, z_{n+1}(t), \dot{z}_{n+1}(t))\|\} \leq M_1$;

- $\displaystyle\sum_{i=1}^{m}\left|\frac{\partial f(t,x,\zeta,w)}{\partial \zeta_i}\right| + \sum_{i=1}^{n}\left|\frac{\partial f(t,x,\zeta,w)}{\partial x_i}\right| \leq K_0 + \varpi(w);$
- $\displaystyle\left|\frac{\partial f(t,x,\zeta,w)}{\partial t}\right| + \left|\frac{\partial f(t,x,\zeta,w)}{\partial w}\right| + |f(t,x,\zeta,w)| + \|f_0(t,x,\zeta,w)\| \leq K_1 + K_2\|x\| +$
$\varpi(w),$ for all $t \in [0,\infty), x \in \mathbb{R}^n, \zeta \in \mathbb{R}^m,$ and $w \in \mathbb{R}.$

For the nonlinear functions $g_i(\cdot)$ in ESO (4.3.3), we need the following Assumption.

Assumption 4.3.2 The nonlinear functions $g_i : \mathbb{R} \to \mathbb{R}$ are Hölder continuous and satisfy the following Lyapunov conditions: There exist radially unbounded, positive definite functions $\mathcal{V} \in C^1(\mathbb{R}^{n+1}, [0,\infty)), \mathcal{W} \in C(\mathbb{R}^{n+1}, [0,\infty))$ such that for any $\iota \in \mathbb{R}^{n+1},$

- $\displaystyle\sum_{i=1}^{n}(\iota_{i+1} - g_i(\iota_1))\frac{\partial \mathcal{V}(\iota)}{\partial \iota_i} - g_{n+1}(\iota_1)\frac{\partial \mathcal{V}(\iota)}{\partial \iota_{n+1}} \leq -\mathcal{W}(\iota);$
- $\displaystyle\left|g_{n+1}(\iota_1)\frac{\partial \mathcal{V}(\iota)}{\partial \iota_{n+1}}\right| \leq N_0\mathcal{W}(\iota), N_0 > 0;$
- $\displaystyle\|\iota\|^2 + \sum_{i=1}^{n}\left|\iota_i\frac{\partial \mathcal{V}(\iota)}{\partial \iota_i}\right| + \sum_{i=1}^{n}|g_i(\iota_1)|^2 + \|\iota\|\left|\frac{\partial \mathcal{V}(\iota)}{\partial \iota_{n+1}}\right| + \left|\frac{\partial \mathcal{V}(\iota)}{\partial \iota_{n+1}}\right|^2 \leq N_1\mathcal{W}(\iota), N_1 > 0.$

Essentially, the key step towards the existence of the Lyapunov functions $\mathcal{V}(\iota)$ and $\mathcal{W}(\iota)$ in Assumption 4.3.2 is that $g_i'(\cdot)$s should be chosen so that the following system is asymptotically stable:

$$\begin{cases} \dot{\xi}_1(t) = \xi_2(t) - g_1(\xi(t)), \\ \quad\vdots \\ \dot{\xi}_n(t) = \xi_{n+1}(t) - g_n(\xi(t)), \\ \dot{\xi}_{n+1}(t) = -g_{n+1}(\xi(t)). \end{cases} \qquad (4.3.8)$$

There are two ways to construct $g_i(\cdot)$. One is a linear function and other is a weighted homogeneous function. Concrete examples are given in Example 4.3.1.

Assumption 4.3.3 The gain function $r \in C^1([0,\infty), [0,\infty)), r(t) > 0, \dot{r}(t) > 0, r(0) = 1,$ $\lim_{t\to+\infty} r(t) = +\infty,$ and there exists a constant $M > 0$ such that $\lim_{t\to+\infty} \dot{r}(t)/r(t) \leq M.$

The next assumption is about the function $u_0(\cdot)$ in feedback control (4.3.6).

Assumption 4.3.4 All partial derivatives of $u_0(\cdot)$ are globally bounded by L and there exist continuous, radially unbounded positive definite functions $V \in C^1(\mathbb{R}^n, [0,\infty)), W \in C(\mathbb{R}^n, [0,\infty))$ such that for any $\tau \in \mathbb{R}^n,$

$$\sum_{i=1}^{n-1}\tau_{i+1}\frac{\partial V(\tau)}{\partial \tau_i} + u_0(\tau)\frac{\partial V(\tau)}{\partial \tau_n} \leq -W(\tau), \quad \|\tau\|^2 + \left|\frac{\partial V(\tau)}{\partial \tau_n}\right|^2 \leq N_2W(\tau), \quad N_2 > 0.$$

Since $\mathcal{V}(\iota), \mathcal{W}(\iota), V(\tau)$, and $W(\tau)$ are continuous, radially unbounded, and positive definite functions in Assumptions 4.3.2 and 4.3.4, it follows from Theorem 1.3.1 that there exist continuous class $\mathcal{K}_\infty$ functions $\kappa_i, \varkappa_i (i = 1, 2, 3, 4)$ such that for any $\iota \in \mathbb{R}^{n+1}$ and $\tau \in \mathbb{R}^n$,

$$\begin{cases} \kappa_1(\|\iota\|) \le \mathcal{V}(\iota) \le \kappa_2(\|\iota\|), & \kappa_3(\|\iota\|) \le \mathcal{W}(\iota) \le \kappa_4(\|\iota\|), \\ \varkappa_1(\|\tau\|) \le V(\tau) \le \varkappa_2(\|\tau\|), & \varkappa_3(\|\tau\|) \le W(\tau) \le \varkappa_4(\|\tau\|). \end{cases} \quad (4.3.9)$$

Assumption on these class $\mathcal{K}_\infty$ functions is given by Assumption 4.3.5.

Assumption 4.3.5 For any $s \ge 0$, $\kappa_4(s) \le N_3 \varkappa_3(s)$ for some $N_3 > 0$.

The numerical simulations at the end of this section show that the ESO with time-varying gain starting from a small value and then increasing with a given increasing rate can reduce significantly the peaking value observed by the constant high-gain ESO. Another advantage of the time-varying ESO is that by the time-varying gain, we can achieve asymptotic stability for ESO rather than practical stability by the constant high gain. However, the time-varying gain is sensitive to high-frequency noise. The recommended control strategy is to use the time-varying gain first to reduce the peaking value in the initial stage to a reasonable level and then apply the constant high gain. In this way, we can reduce the peaking value and meanwhile filter the high-frequency noise possessed by the constant high-gain approach. For this reason, we consider the following combined gain function:

$$r(t) = \begin{cases} e^{at}, & 0 \le t < \ln r_0/a, \\ r_0, & t \ge \ln r_0/a, \end{cases} \quad (4.3.10)$$

where r_0 is a large number so that the errors between the solutions of (4.3.3) and (4.3.1) are in the prescribed scale and $a > 0$ is a constant to control the increasing speed of $r(t)$.

Theorem 4.3.1 *Suppose that Assumptions 4.3.2–4.3.4, and 4.3.5 are satisfied, $b_0 > 0$, and $\sup_{t \in [0,\infty)} |(b(t) - b_0)/b_0| \le \min\{1/2, 1/N_0\}$. Then the closed-loop system composed of system (4.3.1), ESO (4.3.3), and output feedback (4.3.6) has the following convergence:*

(i) If the gain function satisfies Assumption 4.3.3, then for $i = 1, \dots, n+1, j = 1, \dots, n$,

$$\lim_{t \to \infty} |x_i(t) - \hat{x}_i(t)| = 0, \quad \lim_{t \to \infty} |x_j(t) - z_j(t)| = 0. \quad (4.3.11)$$

(ii) If the gain function satisfies (4.3.10), then for any given $\sigma > 0$, there exists a constant $r^* > 0$ (the relation with σ can be seen in (4.3.64)) such that for any $r_0 > r^*$,

$$|x_i(t) - \hat{x}_i(t)| < \sigma, \quad |x_j(t) - z_j(t)| < \sigma, \quad t > t_0,$$

where t_0 is an r_0-dependent constant, $i = 1, 2, \dots, n+1$.

Proof. Let

$$\begin{cases} \eta_i(t) = r^{n+1-i}(t) (x_i(t) - \hat{x}_i(t)), i = 1, \dots, n+1, \\ \mu_j(t) = x_j(t) - z_j(t), j = 1, \dots, n. \\ \eta(t) = (\eta_1(t), \dots, \eta_{n+1}(t))^\top, \mu(t) = (\mu_1(t), \dots, \mu_n(t))^\top. \end{cases} \quad (4.3.12)$$

According to the different choices of $r(t)$, the proof is divided into two parts.

Part I: the gain function $r(t)$ satisfying Assumption 4.3.3. In this case, the error equation can be written as

$$
\begin{cases}
\dot{\eta}_1(t) = r(t)(\eta_2(t) - g_1(\eta_1(t))) + \dfrac{n\dot{r}(t)}{r(t)}\eta_1(t), \\
\quad\vdots \\
\dot{\eta}_n(t) = r(t)(\eta_{n+1}(t) - g_n(\eta_1(t))) + \dfrac{\dot{r}(t)}{r(t)}\eta_n(t), \\
\dot{\eta}_{n+1}(t) = -r(t)g_{n+1}(\eta_1(t)) + \dot{x}_{n+1}(t), \\
\dot{\mu}(t) = A_n\mu(t) + B_n[u_0(\hat{x}_1(t) - z_1(t), \ldots, \hat{x}_n(t) - z_n(t)) + \eta_{n+1}(t)].
\end{cases}
\tag{4.3.13}
$$

For any $T > 0$, since $r(t)$ and $\dot{r}(t)/r(t)$ are bounded in $[0, T]$ and $u_0(\cdot)$ and $g_i(\cdot)$ are continuous, (4.3.13) admits a continuous solution on $[0, T]$. We first need to estimate the derivative of the total disturbance. By (4.3.1), (4.3.6), (4.3.4), and (4.3.12),

$$
\begin{aligned}
\dot{x}_n(t) &= f(t, x(t), \zeta(t), w(t)) + b(t)u(t) \\
&= x_{n+1}(t) + z_{n+1}(t) - \hat{x}_{n+1}(t) + u_0(\hat{x}_1(t) - z_1(t), \ldots, \hat{x}_n(t) - z_n(t)) \quad (4.3.14) \\
&= u_0(\hat{x}_1(t) - z_1(t), \ldots, \hat{x}_n(t) - z_n(t)) + \eta_{n+1}(t) + z_{n+1}(t).
\end{aligned}
$$

By (4.3.4), (4.3.1), (4.3.3), and (4.3.14),

$$
\begin{aligned}
\dot{x}_{n+1}(t) &= \frac{d[f(t, x(t), \zeta(t), w(t)) + (b(t) - b_0)u(t)]}{dt} \\
&= \frac{\partial f(t, x(t), \zeta(t), w(t))}{\partial t} + \sum_{i=1}^{n-1} x_{i+1}(t)\frac{\partial f(t, x(t), \zeta(t), w(t))}{\partial x_i} \\
&\quad + [u_0(\hat{x}_1(t) - z_1(t), \ldots, \hat{x}_n(t) - z_n(t)) + z_{n+1}(t) + \eta_{n+1}(t)]\frac{\partial f(t, x(t), \zeta(t), w(t))}{\partial x_n} \\
&\quad + f_0(t, x(t), \zeta(t), w(t))\frac{\partial f(t, x(t), \zeta(t), w(t))}{\partial \zeta} + \dot{w}(t)\frac{\partial f(t, x(t), \zeta(t), w(t))}{\partial w} \\
&\quad + \frac{\dot{b}(t)}{b_0}(u_0(\hat{x}_1(t) - z_1(t), \ldots, \hat{x}_n(t) - z_n(t)) + z_{n+1}(t) - \hat{x}_{n+1}(t)) \\
&\quad + \left(\frac{b(t) - b_0}{b_0}\right)\left[\sum_{i=1}^{n}\left(\hat{x}_{i+1}(t) + \frac{1}{r^{n-i}(t)}g_i(\eta_1(t)) - z_{i+1}(t)\right)\right. \\
&\quad \times \left.\frac{\partial u_0(\hat{x}_1(t) - z_1(t), \ldots, \hat{x}_n(t) - z_n(t))}{\partial \tau_i} + \dot{z}_{n+1}(t) - r(t)g_{n+1}(\eta_1(t))\right],
\end{aligned}
\tag{4.3.15}
$$

where $\partial u_0(\cdot)/\partial \tau_i$ is the partial derivative of $u_0(\cdot)$ with respect to its ith component.

From Assumption 4.3.1 and (4.3.12),

$$
\begin{aligned}
\left|\frac{\partial f(t, x(t), \zeta(t), w(t))}{\partial t}\right| &\leq K_1 + K_2(\|x(t) - z(t)\| + \|z(t)\|) + \varpi(w(t)) \\
&\leq (K_1 + K_2 M_1 + \varpi(w(t))) + K_2\|\mu(t)\|,
\end{aligned}
\tag{4.3.16}
$$

where $z(t) = (z_1(t), \ldots, z_n(t))^\top$. Similarly,

$$\left| \sum_{i=1}^{n-1} x_{i+1}(t) \frac{\partial f(t, x(t), \zeta(t), w(t))}{\partial x_i} \right| \leq (K_0 + \varpi(w(t))) \sum_{i=1}^{n-1} |x_{i+1}(t)| \tag{4.3.17}$$
$$\leq n(K_0 + \varpi(w(t)))\|\mu(t)\| + nM_1(K_0 + \varpi(w(t))),$$

$$\left| f_0(t, x(t), \zeta(t), w(t)) \frac{\partial f(t, x(t), \zeta(t), w(t))}{\partial w} \right|$$
$$\leq (K_1 + K_2\|x(t)\| + \varpi(w(t)))(K_0 + \varpi(w(t))) \tag{4.3.18}$$
$$\leq (K_0 + \varpi(w(t)))(K_1 + \varpi(w(t)) + K_2M_1) + (K_0 + \varpi(w(t)))K_2\|\mu(t)\|,$$

and

$$\left| \dot{w}(t) \frac{\partial f(t, x(t), \zeta(t), w(t))}{\partial w} \right| \leq M_1(K_1 + \varpi(w(t)) + K_2\|x(t)\|) \tag{4.3.19}$$
$$\leq M_1(K_1 + K_2M_1 + \varpi(w(t))) + M_1K_2\|\mu(t)\|.$$

Since by Assumption 4.3.3, $r(t) \geq 1$, it follows from (4.3.12) that

$$\|(\hat{x}_1(t) - x_1(t), \ldots, \hat{x}_n(t) - x_n(t))\| = \left\| \left(\frac{\eta_1(t)}{r^n(t)}, \ldots, \frac{\eta_n(t)}{r(t)} \right) \right\| \leq \|\eta(t)\|. \tag{4.3.20}$$

By Assumption 4.3.1 and (4.3.20),

$$\sum_{i=1}^{n} \left| \left(\hat{x}_{i+1}(t) + \frac{1}{r^{n-i}(t)} g_i(\eta_1(t)) - z_{i+1}(t) \right) \right|$$
$$\leq \sum_{i=2}^{n+1} |\hat{x}_i(t) - x_i(t)| + \sum_{i=2}^{n} |x_i(t) - z_i(t)| + |x_{n+1}(t)| + |z_{n+1}(t)|$$
$$+ \sum_{i=1}^{n} \frac{1}{r^{n-i}(t)} |g_i(\eta_1(t))| \tag{4.3.21}$$
$$\leq n(\|\mu(t)\| + \|\eta(t)\|) + \sum_{i=1}^{n} |g_i(\eta_1(t))| + M_1 + |x_{n+1}(t)|.$$

Furthermore, by (4.3.4) and (4.3.6),

$$x_{n+1}(t) = f(t, x(t), \zeta(t), w(t)) + (b(t) - b_0)u(t)$$
$$= f(t, x(t), \zeta(t), w(t)) + \frac{b(t) - b_0}{b_0} u_0(\hat{x}_1(t) - z_1(t), \ldots, \hat{x}_n(t) - z_n(t)) \tag{4.3.22}$$
$$+ \frac{b(t) - b_0}{b_0}((x_{n+1}(t) - \hat{x}_{n+1}(t)) + z_{n+1}(t)) - \frac{b(t) - b_0}{b_0} x_{n+1}(t).$$

It follows that

$$x_{n+1}(t) = \frac{b_0}{b(t)} f(t, x(t), \zeta(t), w(t)) + \frac{b(t) - b_0}{b(t)} (u_0(\hat{x}_1(t) - z_1(t), \ldots, \hat{x}_n(t) - z_n(t))$$
$$+ (x_{n+1}(t) - \hat{x}_{n+1}(t)) + z_{n+1}(t)). \tag{4.3.23}$$

Since $|(b(t) - b_0)/b_0| \leq 1/2$ by assumption, $b_0/2 \leq b(t) \leq 3b_0/2$, $|(b(t) - b_0)/b(t)| \leq 1$. This together with (4.3.23) yields

$$|x_{n+1}(t)| \leq 2(K_1 + K_2\|x(t)\| + \varpi(w(t)))$$
$$+ |u_0(\hat{x}_1(t) - z_1(t), \ldots, \hat{x}_n(t) - z_n(t))| + \|\eta(t)\| + M_1$$
$$\leq M_1 + 2(K_1 + K_2 M_1 + \varpi(w(t))) + 2K_2\|\mu(t)\| + \|\eta(t)\|$$
$$+ |u_0(\hat{x}_1(t) - z_1(t), \ldots, \hat{x}_n(t) - z_n(t))|, \tag{4.3.24}$$

where

$$u_0(\hat{x}_1(t) - z_1(t), \ldots, \hat{x}_n(t) - z_n(t)) = u_0(\mu_1(t), \ldots, \mu_n(t))$$
$$+ u_0(\hat{x}_1(t) - z_1(t), \ldots, \hat{x}_n(t) - z_n(t)) - u_0(x_1(t) - z_1(t), \ldots, x_n(t) - z_n(t)). \tag{4.3.25}$$

Again by assumption, since all partial derivatives of $u_0(\cdot)$ are bounded by L, we have

$$\left| u_0(\hat{x}_1(t) - z_1(t), \ldots, \hat{x}_n(t) - z_n(t)) - u_0(x_1(t) - z_1(t), \ldots, x_n(t) - z_n(t)) \right|$$

$$\leq \left| u_0(\hat{x}_1(t) - z_1(t), \hat{x}_2(t) - z_2(t), \ldots, \hat{x}_n(t) - z_n(t)) \right.$$

$$\left. - u_0(x_1(t) - z_1(t), \hat{x}_2(t) - z_2(t), \ldots, \hat{x}_n(t) - z_n(t)) \right|$$

$$+ \left| u_0(x_1(t) - z_1(t), \hat{x}_2(t) - z_2(t), \ldots, \hat{x}_n(t) - z_n(t)) \right.$$

$$\left. - u_0(x_1(t) - z_1(t), x_2(t) - z_2(t), \ldots, \hat{x}_n(t) - z_n(t)) \right|$$

$$+ \cdots$$

$$+ \left| u_0(x_1(t) - z_1(t), \ldots, \hat{x}_n(t) - z_n(t)) - u_0(x_1(t) - z_1(t), \ldots, x_n(t) - z_n(t)) \right|$$

$$\leq L(|\eta_1(t)| + |\eta_2(t)| + \cdots + |\eta_n(t)|) \leq Ln \, \|\eta(t)\|. \tag{4.3.26}$$

Similarly,

$$|u_0(\mu_1(t), \ldots, \mu_n(t))| \leq Ln \, \|\mu(t)\| + |u_0(0, \ldots, 0)|. \tag{4.3.27}$$

By (4.3.15) to (4.3.27) and Assumption 4.3.1, there exists $B > 0$ such that for all $t \in [0, \infty)$,

$$|\dot{x}_{n+1}(t)| \le B \left(1 + \|\eta(t)\| + \|\mu(t)\| + \sum_{i=1}^{n} |g_i(\eta_1(t))| \right) + \left| \frac{b(t) - b_0}{b_0} \right| r(t) |g_{n+1}(\eta_1(t))|.$$

(4.3.28)

Let $\mathcal{V}(\cdot), \mathcal{W}(\cdot), V(\cdot)$ and $W(\cdot)$ be the Lyapunov functions satisfying Assumptions 4.3.2 and 4.3.4. Define the Lyapunov functions $\mathfrak{V}, \mathfrak{W} \colon \mathbb{R}^{2n+1} \to [0, \infty)$:

$$\mathfrak{V}(x, y) = \mathcal{V}(x) + V(y), \quad \mathfrak{W}(x, y) = \mathcal{W}(x) + W(y). \tag{4.3.29}$$

Finding the derivative of $\mathfrak{V}(\eta(t), \mu(t))$ along the solution of (4.3.13) yields

$$\left. \frac{d\mathfrak{V}(\eta(t), \mu(t))}{dt} \right|_{\text{along (4.3.13)}} = \sum_{i=1}^{n+1} \dot{\eta}_i(t) \frac{\partial \mathcal{V}(\eta(t))}{\partial \eta_i} + \sum_{i=1}^{n} \dot{\mu}_i(t) \frac{\partial V(\mu(t))}{\partial \mu_i}$$

$$= r(t) \left(\sum_{i=1}^{n} (\eta_{i+1}(t) - g_i(\eta_1(t))) \frac{\partial \mathcal{V}(\eta(t))}{\partial \eta_i} - g_{n+1}(\eta_1(t)) \frac{\partial \mathcal{V}(\eta(t))}{\partial \eta_{n+1}} \right)$$

$$+ \sum_{i=1}^{n} \frac{(n+1-i)\dot{r}(t)\eta_i(t)}{r(t)} \frac{\partial \mathcal{V}(\eta(t))}{\partial \eta_i} + \dot{x}_{n+1}(t) \frac{\partial \mathcal{V}(\eta(t))}{\partial \eta_{n+1}} + \sum_{i=1}^{n-1} \mu_{i+1}(t) \frac{\partial V(\mu(t))}{\partial \mu_i}$$

$$+ [u_0(\hat{x}_1(t) - z_1(t), \dots, \hat{x}_n(t) - z_n(t)) + \eta_{n+1}(t)] \frac{\partial V(\mu(t))}{\partial \mu_n}. \tag{4.3.30}$$

By Assumptions 4.3.2 and 4.3.4, we have

$$\sum_{i=1}^{n} (\eta_{i+1}(t) - g_i(\eta_1(t))) \frac{\partial \mathcal{V}(\eta(t))}{\partial \eta_i} - g_{n+1}(\eta_1(t)) \frac{\partial \mathcal{V}(\eta(t))}{\partial \eta_{n+1}} \le -\mathcal{W}(\eta(t)) \tag{4.3.31}$$

and

$$\sum_{i=1}^{n-1} \mu_{i+1}(t) \frac{\partial V(\mu(t))}{\partial \mu_i} + (u_0(\mu_1(t), \dots, \mu_n(t))) \frac{\partial V(\mu(t))}{\partial \mu_n} \le -W(\mu(t)). \tag{4.3.32}$$

Applying (4.3.28), (4.3.30), (4.3.31), (4.3.25), (4.3.26), and (4.3.32) gives

$$\left. \frac{d\mathfrak{V}(\eta(t), \mu(t))}{dt} \right|_{\text{along (4.3.13)}} \le -r(t)\mathcal{W}(\eta(t)) + \sum_{i=1}^{n} \frac{(n+1-i)\dot{r}(t)|\eta_i(t)|}{r(t)} \left| \frac{\partial \mathcal{V}(\eta(t))}{\partial \eta_i} \right|$$

$$+ \left[B \left(1 + \|\eta(t)\| + \|\mu(t)\| + \sum_{i=1}^{n} |g_i(\eta_1(t))| \right) + \left| \frac{b(t) - b_0}{b_0} \right| r(t) |g_{n+1}(\eta_1(t))| \right] \left| \frac{\partial \mathcal{V}(\eta(t))}{\partial \eta_{n+1}} \right|$$

$$- W(\mu(t)) + (Ln + 1)\|\eta(t)\| \left| \frac{\partial V(\mu(t))}{\partial \mu_n} \right|. \tag{4.3.33}$$

By Assumption 4.3.2,

$$\sum_{i=1}^{n} \frac{(n+1-i)\dot{r}(t)|\eta_i(t)|}{r(t)} \left| \frac{\partial \mathcal{V}(\eta(t))}{\partial \eta_i} \right| \le \frac{n(n+1)MN_1}{2} \mathcal{W}(\eta(t)), \tag{4.3.34}$$

$$B\left(1 + \|\eta(t)\| + \sum_{i=1}^{n} |g_i(\eta_1(t))|\right)\left|\frac{\partial \mathcal{V}(\eta(t))}{\partial \eta_{n+1}}\right| \leq B\sqrt{N_1}\sqrt{\mathcal{W}(\eta(t))} + B(n+1)N_1\mathcal{W}(\eta(t)),$$

$$(4.3.35)$$

and

$$\left|\frac{b(t) - b_0}{b_0}\right| r(t)|g_{n+1}(\eta_1(t))| \left|\frac{\partial \mathcal{V}(\eta(t))}{\partial \eta_{n+1}}\right| \leq N_0 \left|\frac{b(t) - b_0}{b_0}\right| r(t)\mathcal{W}(\eta(t)). \qquad (4.3.36)$$

Furthermore, by Assumptions 4.3.2 and 4.3.4,

$$B\|\mu(t)\|\left|\frac{\partial \mathcal{V}(\eta(t))}{\partial \eta_{n+1}}\right| + (Ln+1)\|\eta(t)\|\left|\frac{\partial V(\mu(t))}{\partial \mu_n}\right|$$

$$\leq (Ln+1+B)\sqrt{N_1 N_2}\sqrt{\mathcal{W}(\eta(t))W(\mu(t))}. \qquad (4.3.37)$$

Combining (4.3.33) to (4.3.37) gives

$$\left.\frac{d\mathfrak{V}(\eta(t), \mu(t))}{dt}\right|_{\text{along (4.3.13)}} \leq -\left(1 - N_0\left|\frac{b(t) - b_0}{b_0}\right|\right) r(t)\mathcal{W}(\eta(t))$$

$$+ \frac{n(n+1)MN_1 + 2B(n+1)}{2}\mathcal{W}(\eta(t)) + (B+Ln+1)\sqrt{N_1 N_2}\sqrt{\mathcal{W}(\eta(t))W(\mu(t))}$$

$$- W(\mu(t)) + B\sqrt{N_1}\sqrt{\mathcal{W}(\eta(t))}. \qquad (4.3.38)$$

By Assumption 4.3.3, we may assume that $\dot{r}(t)/r(t) < 2M$ and

$$r(t) > \max\left\{\frac{2(B+Ln+1)^2 N_1 N_2}{\Delta}, \frac{2n(n+1)MN_1 + 2B(n+1)}{\Delta}\right\}, \forall\, t \geq t_1 \quad (4.3.39)$$

for some $t_1 > 0$, where

$$\Delta = 1 - N_0 \sup_{t \in [0, \infty)} |(b(t) - b_0))/b_0|. \qquad (4.3.40)$$

By (4.3.39) and (4.3.40),

$$-\left(1 - N_0\left|\frac{b(t) - b_0}{b_0}\right|\right) r(t)\mathcal{W}(\eta(t)) + \frac{n(n+1)MN_1 + 2B(n+1)}{2}\mathcal{W}(\eta(t))$$

$$(4.3.41)$$

$$\leq -\Delta r(t)\mathcal{W}(\eta(t)) + \frac{\Delta}{4}r(t)\mathcal{W}(\eta(t)) = -\frac{3\Delta}{4}r(t)\mathcal{W}(\eta(t)), \forall\, t > t_1,$$

and

$$(B+Ln+1)\sqrt{N_1 N_2}\sqrt{\mathcal{W}(\eta(t))W(\mu(t))} \leq \frac{(B+Ln+1)^2 N_1 N_2}{2}\mathcal{W}(\eta(t)) + \frac{W(\mu(t))}{2}$$

$$\leq \frac{\Delta}{4}r(t)\mathcal{W}(\eta(t)) + \frac{1}{2}W(\mu(t)), \forall\, t > t_1. \qquad (4.3.42)$$

In addition, by (4.3.38), (4.3.41), and (4.3.42), we have, for all $t > t_1$, that

$$\left. \frac{d\mathfrak{V}(\eta(t), \mu(t))}{dt} \right|_{\text{along (4.3.13)}} \leq -\frac{\Delta r(t)}{2} \mathcal{W}(\eta(t)) + B\sqrt{N_1}\sqrt{\mathcal{W}(\eta(t))} - \frac{W(\mu(t))}{2}.$$

$$(4.3.43)$$

The remainder of the proof is split into the following three steps.

Step 1. *The boundedness of* $\|(\eta(t), \mu(t))\|$. Let

$$R = \max\left\{ \kappa_3^{-1}\left(\left(\frac{4B\sqrt{N_1}}{\Delta r(t_1)}\right)^2 \right), \; \varkappa_3^{-1}((4B)^2(N_1 N_3)) \right\}, \quad (4.3.44)$$

where $\kappa_3(\cdot)$ and $\varkappa_3(\cdot)$ are given in (4.3.9). When $\|(\eta(t), \mu(t))\| > 2R$, there are two cases.

Case 1: $\|\eta(t)\| \geq R$. In this case, first, by Assumption 4.3.3, $r(t) > r(t_1)$ for all $t > t_1$. Next, it follows from (4.3.9) that

$$\mathcal{W}(\eta(t)) \geq \kappa_3(\|\eta(t)\|) \geq \kappa_3(R). \quad (4.3.45)$$

This together with (4.3.44) yields

$$-\frac{\Delta r(t)}{2}\sqrt{\mathcal{W}(\eta(t))} \leq -\frac{\Delta r(t_1)}{2}\sqrt{\kappa_3(R)} \leq -2B\sqrt{N_1}, \forall\, t > t_1. \quad (4.3.46)$$

By (4.3.43), (4.3.45), and (4.3.46),

$$\left. \frac{d\mathfrak{V}(\eta(t), \mu(t))}{dt} \right|_{\text{along (4.3.13)}} \leq -\frac{\Delta}{2} r(t)\mathcal{W}(\eta(t)) + B\sqrt{N_1}\sqrt{\mathcal{W}(\eta(t))}$$

$$= \sqrt{\mathcal{W}(\eta(t))}\left(-\frac{\Delta}{2} r(t_1)\sqrt{\mathcal{W}(\eta(t))} + B\sqrt{N_1} \right) \quad (4.3.47)$$

$$\leq -B\sqrt{N_1}\sqrt{\kappa_3(R)} < 0, \forall\, t > t_1.$$

Case 2: $\|\eta(t)\| < R$. In this case, by $\|\mu(t)\| + \|\eta(t)\| \geq \|(\eta(t), \mu(t))\| > R$, $\|\mu(t)\| > R$. This together with Assumption 4.3.5, (4.3.44), and (4.3.43) gives

$$\left. \frac{d\mathfrak{V}(\eta(t), \mu(t))}{dt} \right|_{\text{along (4.3.13)}}$$

$$\leq B\sqrt{N_1}\sqrt{\mathcal{W}(\eta(t))} - \frac{1}{2}W(\mu(t)) \leq B\sqrt{N_1}\sqrt{\kappa_4(\|\eta(t)\|)} - \frac{1}{2}\varkappa_3(\|\mu(t)\|)$$

$$\leq B\sqrt{N_1 N_3}\sqrt{\varkappa_3(R)} - \frac{1}{2}\varkappa_3(R) < -B\sqrt{N_1 N_3}\sqrt{\varkappa_3(R)} < 0, \forall\, t > t_1.$$

$$(4.3.48)$$

Let

$$\mathcal{A} = \{(x, y) \in \mathbb{R}^{2n+1} \mid \; \mathfrak{V}(x, y) \leq C \triangleq \max_{\|(x,y)\| \leq 2R} \mathfrak{V}(x, y)\}.$$

From continuity and radially unboundedness of $\mathfrak{V}(\cdot)$, we know that $\mathcal{A} \subset \mathbb{R}^{2n+1}$ is bounded. If $(\eta(t), \mu(t)) \in \mathcal{A}^C$, then $\|(\eta(t), \mu(t))\| > 2R$. Combining the two cases aforementioned, we obtain that if $(\eta(t), \mu(t)) \in \mathcal{A}^c$ for $t > t_1$, then

$$\left. \frac{d\mathfrak{V}(\eta(t), \mu(t))}{dt} \right|_{\text{along (4.3.13)}} < 0.$$

Thus there exists a $t_2 > t_1$ such that $(\eta(t), \mu(t))$ lies in a bounded set $\mathcal{A}$ for any $t > t_2$, which means that $(\eta(t), \mu(t))$ is uniformly bounded for $t > t_2$.

Step 2. *Convergence of $\eta(t)$ to zero.* Since $(\eta(t), \mu(t))$ is uniformly bounded for $t > t_2$, it follows from the continuity of $g_i(\cdot)$, $\nabla V(\cdot)$, and (4.3.28) that there exists $B_1 > 0$ such that

$$|x_{n+1}(t)| \left| \frac{\partial V(\eta(t))}{\partial \eta_{n+1}} \right| + \sum_{i=1}^{n} \frac{(n+1-i)\dot{r}(t)|\eta_i(t)|}{r(t)} \left| \frac{\partial V(\eta(t))}{\partial \eta_i} \right|$$

$$\leq B_1 + N_0 \sup_{t \in [0,\infty)} \left| \frac{b(t) - b_0}{b_0} \right| r(t) W(\eta(t)), \forall\, t > t_2. \tag{4.3.49}$$

Taking (4.3.40) and (4.3.49) into account and finding the derivative of $V(\eta(t))$ along the solution of (4.3.13) gives

$$\left. \frac{dV(\eta(t))}{dt} \right|_{\text{along (4.3.13)}} = \sum_{i=1}^{n+1} \dot{\eta}_i(t) \frac{\partial V(\eta(t))}{\partial \eta_i}$$

$$= r(t) \left(\sum_{i=1}^{n} (\eta_{i+1}(t) - g_i(\eta_1(t))) \frac{\partial V(\eta(t))}{\partial \eta_i} - g_{n+1}(\eta_1(t)) \frac{\partial V(\eta(t))}{\partial \eta_{n+1}} \right) \tag{4.3.50}$$

$$+ \sum_{i=1}^{n} \frac{(n+1-i)\dot{r}(t)\eta_i(t)}{r(t)} \frac{\partial V(\eta(t))}{\partial \eta_i} + \dot{x}_{n+1}(t) \frac{\partial V(\eta(t))}{\partial \eta_{n+1}}$$

$$\leq -\Delta r(t) W(\eta(t)) + B_1, \forall\, t > t_2.$$

For any given $\sigma > 0$, find $t_{1\sigma} > t_2$ so that $r(t) > 2B_1/(\Delta \kappa_3(\kappa_2^{-1}(\sigma)))$ for all $t \geq t_{1\sigma}$. If $V(\eta(t)) > \sigma$, then

$$W(\eta(t)) \geq \kappa_3(\|\eta(t)\|) \geq \kappa_3(\kappa_2^{-1}(V(\eta(t)))) \geq \kappa_3(\kappa_2^{-1}(\sigma)).$$

This together with (4.3.50) yields

$$\left. \frac{dV(\eta(t))}{dt} \right|_{\text{along (4.3.13)}} \leq -\Delta \frac{2B_1}{\Delta \kappa_3(\kappa_2^{-1}(\sigma))} \kappa_3(\kappa_2^{-1}(\sigma)) + B_1 = -B_1 < 0, \forall\, t \geq t_{1\sigma}.$$

$$\tag{4.3.51}$$

Therefore there exists a $t_{\sigma_2} > t_{\sigma_1}$ such that $\mathcal{V}(\eta(t)) \leq \sigma$ for all $t > t_{\sigma_2}$. This shows that $\mathcal{V}(\eta(t)) \to 0$ as $t \to \infty$. Since $\|\eta(t)\| \leq \kappa_1^{-1}(\mathcal{V}(\eta(t)))$, we finally obtain $\lim_{t\to\infty} \eta(t) = 0$.

Step 3. *Convergence of $\mu(t)$ to zero.* From Step 1, $\mu(t)$ is uniformly bounded for $t > t_2$. This together with the continuity of $\nabla V(\cdot)$ shows that there exists a constant $C_1 > 0$ such that $(Ln + 1)|\partial V(\mu(t))/\partial \mu_{n+1}| < C_1$. With (4.3.26), we can find the derivative of $V(\mu(t))$ along the solution of (4.3.13) to satisfy

$$
\left. \frac{dV(\mu(t))}{dt} \right|_{\text{along (4.3.13)}} = \sum_{i=1}^{n} \dot{\mu}_i(t) \frac{\partial V(\mu(t))}{\partial \mu_i}
$$

$$
= \sum_{i=1}^{n-1} \mu_{i+1}(t) \frac{\partial V(\mu(t))}{\partial \mu_i} + [u_0(\hat{x}_1(t) - z_1(t), \ldots, \hat{x}_n(t) \tag{4.3.52}
$$

$$
- z_n(t)) + \eta_{n+1}(t)] \frac{\partial V(\mu(t))}{\partial \mu_n}
$$

$$
\leq -W(\mu(t)) + (Ln + 1)\|\eta(t)\| \left| \frac{\partial V(\mu(t))}{\partial \mu_n} \right| \leq -W(\mu(t)) + C_1\|\eta(t)\|.
$$

For any $\sigma > 0$, by Step 2, one can find a $t_{\sigma 3} > t_2$ such that $\|\eta(t)\| < 1/(2C_1)\varkappa_3(\varkappa_2^{-1}(\sigma))$ for all $t > t_{\sigma 3}$. It then follows by (4.3.52) that for $V(\mu(t)) > \sigma$ with $t > t_{\sigma 3}$,

$$
\left. \frac{dV(\mu(t))}{dt} \right|_{\text{along (4.3.13)}} \leq -W(\mu(t)) + C_1\|\eta(t)\| < -\varkappa_3(\varkappa_2^{-1}(V(\mu(t)))) + C_1\|\eta(t)\|
$$

$$
\leq -\varkappa_3(\varkappa_2^{-1}(\sigma)) + C_1\|\eta(t)\| \leq -\frac{1}{2}\varkappa_3(\varkappa_2^{-1}(\sigma)) < 0. \tag{4.3.53}
$$

Therefore one can find a $t_{\sigma 4} > t_{\sigma 3}$ such that $V(\mu(t)) < \sigma$ for $t > t_{\sigma 4}$. This together with $\|\mu(t)\| \leq \varkappa_1^{-1}(V(\mu(t)))$ gives $\lim_{t\to\infty} \mu(t) = 0$. Part (I) is thus proved.

Part II: $r(t)$ satisfying (4.3.10). In this case, the error should be separated in different intervals $[0, \frac{1}{a} \ln r_0)$ and $[\frac{1}{a} \ln r_0, \infty)$, respectively. In $[0, \frac{1}{a} \ln r_0)$, the error equation is the same as (4.3.13), while in $[\frac{1}{a} \ln r_0, \infty)$, the errors satisfy

$$
\begin{cases}
\dot{\eta}_1(t) = r_0(\eta_2(t) - g_1(\eta_1(t))), \\
\quad \vdots \\
\dot{\eta}_n(t) = r_0(\eta_{n+1}(t) - g_n(\eta_1(t))), \\
\dot{\eta}_{n+1}(t) = -r_0 g_{n+1}(\eta_1(t)) + \dot{x}_{n+1}(t), \\
\dot{\mu}(t) = A_n \mu(t) + B_n[u_0(\hat{x}_1(t) - z_1(t), \ldots, \hat{x}_n(t) - z_n(t)) + \eta_{n+1}(t)],
\end{cases} \tag{4.3.54}
$$

where A_n and B_n are the same as in (4.3.2). Since $u_0(\cdot)$ and $g_i(\cdot)$ are continuous and $r(t)$ is also continuous in $[0, \ln r_0/a]$, Equation (4.3.54) admits a continuous solution in $[0, \ln r_0/a]$.

Similarly, by the continuity of $u_0(\cdot)$ and $g_i(\cdot)$, Equation (4.3.54) admits a continuous solution as well in $[\ln\ r_0/a, \infty)$.

Let $\mathfrak{V}(\cdot)$ and $\mathfrak{W}(\cdot)$ be defined as in (4.3.29). Finding the derivatives of $\mathfrak{V}(\eta(t), \mu(t))$ along the solution of (4.3.54) gives

$$\frac{d\mathfrak{V}(\eta(t), \mu(t))}{dt}\bigg|_{\text{along (4.3.54)}} = r_0\left(\sum_{i=1}^{n}(\eta_{i+1}(t) - g_i(\eta_1(t)))\frac{\partial V(\eta(t))}{\partial \eta_i} - g_{n+1}(\eta_1(t))\frac{\partial V(\eta(t))}{\partial \eta_{n+1}}\right)$$

$$+ \sum_{i=1}^{n-1}\mu_{i+1}(t)\frac{\partial V(\mu(t))}{\partial \mu_i} + [u_0(\hat{x}_1(t) - z_1(t), \ldots, \hat{x}_n(t) - z_n(t)) + \eta_{n+1}(t)]\frac{\partial V(\mu(t))}{\partial \mu_{n+1}}$$

$$+ \dot{\hat{x}}_{n+1}(t)\frac{\partial V(\eta(t))}{\partial \eta_{n+1}}. \tag{4.3.55}$$

By (4.3.28) and (4.3.10), there exists $B > 0$ such that for all $t > (\ln r_0)/a$,

$$|\dot{\hat{x}}_{n+1}(t)| \leq B\left(1 + \|\eta(t)\| + \|\mu(t)\| + \sum_{i=1}^{n}|g_i(\eta_1(t))|\right) + \left|\frac{b(t) - b_0}{b_0}\right|r_0|g_{n+1}(\eta_1(t))|. \tag{4.3.56}$$

Let

$$r_1 = \max\left\{\frac{4B(n+1)N_1}{\Delta}, \frac{2(B + (n+1)L)^2 N_1 N_2}{\Delta}\right\}, \tag{4.3.57}$$

where Δ is defined as in (4.3.40). By Assumptions 4.3.2 and 4.3.4, (4.3.55), (4.3.56), (4.3.40), (4.3.26), and (4.3.57), for any $r_0 > r_1$ and $t > \ln r_0/a$,

$$\frac{d\mathfrak{V}(\eta(t), \mu(t))}{dt}\bigg|_{\text{along (4.3.54)}} \leq -\Delta r_0 \mathcal{W}(\eta(t)) + B\sqrt{N_1}\sqrt{\mathcal{W}(\eta(t))} - W(\mu(t))$$

$$+ B(n+1)N_1\mathcal{W}(\eta(t)) + (B + (n+1)L)\sqrt{N_1 N_2}\sqrt{\mathcal{W}(\eta(t))}\sqrt{W(\mu(t))}$$

$$\leq -\frac{\Delta r_0 \mathcal{W}(\eta(t))}{2} + B\sqrt{N_1}\sqrt{\mathcal{W}(\eta(t))} - \frac{W(\mu(t))}{2}. \tag{4.3.58}$$

As in Part I, the remainder of the proof is split into the following three steps.

Step 1. *The boundedness of* $\|(\eta(t), \mu(t))\|$. Let

$$R = \max\left\{\kappa_3^{-1}\left(\left(\frac{4BN_1^{1/2}}{\Delta r_0}\right)^2\right), \varkappa_3^{-1}((4B)^2(N_1 N_3))\right\}. \tag{4.3.59}$$

If $\|(\eta(t), \mu(t))\| > 2R$, then there are two cases. When $\|\eta(t)\| > R$ and $r > r_1$,

$$\frac{d\mathfrak{V}(\eta(t), \mu(t))}{dt}\bigg|_{\text{along (4.3.54)}} \leq -\frac{\Delta}{2}r_0\mathcal{W}(\eta(t)) + B\sqrt{N_1}\sqrt{\mathcal{W}(\eta(t))}$$

$$\leq \sqrt{\mathcal{W}(\eta(t))}\left(-\frac{\Delta}{2}r_0\sqrt{\kappa_3(\|\eta(t)\|)} + B\sqrt{N_1}\right) \leq -B\sqrt{N_1}\sqrt{\kappa_3(R)} < 0. \tag{4.3.60}$$

When $\|\eta(t)\| < R$, $\|\mu(t)\| > R$. In this case, by (4.3.9), it follows that

$$\left.\frac{d\mathfrak{V}(\eta(t), \mu(t))}{dt}\right|_{\text{along } (4.3.13)} \leq B\sqrt{N_1}\sqrt{\mathcal{W}(\eta(t))} - \frac{1}{2}W(\mu(t))$$

$$\leq B\sqrt{N_1 N_3}\sqrt{\kappa_4(\|\eta(t)\|)} - \frac{1}{2}\varkappa_3(\|\mu(t)\|) \leq B\sqrt{N_1 N_3}\sqrt{\varkappa_3(R)} - \frac{1}{2}\varkappa_3(R)$$

$$\leq -B\sqrt{N_1 N_3}\sqrt{\varkappa_3(R)} < 0, \forall\, r_0 > r_1. \tag{4.3.61}$$

From aforementioned facts, there exists a $t_1 > \ln r_0/a$ such that $\|(\eta(t), \mu(t))\| \leq 2R$ for all $t > t_1$.

Step 2. *Convergence of $\eta(t)$.* Since $(\eta(t), \mu(t))$ is uniformly bounded for $t > t_1$, it follows from the continuity of $g_i(\cdot)$, $\nabla V(\cdot)$, and (4.3.28) that there exists a $B_1 > 0$ such that for all $t > t_1$,

$$|\dot{x}_{n+1}(t)|\left|\frac{\partial V(\eta(t))}{\partial \eta_{n+1}}\right| \leq B_1 + N_0 \sup_{t \in [0,\infty)}\left|\frac{b(t) - b_0}{b_0}\right| r_0 W(\eta(t)). \tag{4.3.62}$$

By (4.3.62) and Assumption 4.3.2, we find the derivative of $V(\eta(t))$ along the solution of (4.3.54) to be

$$\left.\frac{dV(\eta(t))}{dt}\right|_{\text{along } (4.3.13)} = r_0\left(\sum_{i=1}^{n}(\eta_{i+1}(t) - g_i(\eta_1(t)))\frac{\partial V(\eta(t))}{\partial \eta_i} - g_{n+1}(\eta_1(t))\frac{\partial V(\eta(t))}{\partial \eta_{n+1}}\right)$$

$$+ \dot{x}_{n+1}(t)\frac{\partial V(\eta(t))}{\partial \eta_{n+1}} \leq -\Delta r_0 W(\eta(t)) + B_1. \tag{4.3.63}$$

For any given $\sigma > 0$, let

$$r^* = \max\left\{r_1, \frac{2B_1}{\Delta\kappa_3(\sigma)}\right\}. \tag{4.3.64}$$

For any $r_0 > r^*$, if $\|\eta(t)\| > \sigma$, then $\mathcal{W}(\eta(t)) \geq \kappa_3(\eta(t)) \geq \kappa_3(\sigma)$. It follows that

$$\left.\frac{dV(\eta(t))}{dt}\right|_{\text{along } (4.3.13)} \leq -B_1 < 0. \tag{4.3.65}$$

Thus there exists a $t_\sigma > t_1$ such that $\|\eta(t)\| < \sigma$ for all $t > t_\sigma$.

Step 3. *Convergence of $\mu(t)$.* From Step 1, $\mu(t)$ is uniformly bounded for $t > t_1$. This together with the continuity of $\nabla V(\cdot)$ shows that there exists a constant $C_1 > 0$ such that $(Ln + 1)|\partial V(\mu(t))/\partial\mu_n| < C_1$ for all $t > t_1$. Hence the derivative of $V(\mu(t))$ along the solution of (4.3.54) satisfies

$$\left.\frac{dV(\mu(t))}{dt}\right|_{\text{along } (4.3.54)} = \sum_{i=1}^{n-1}\mu_{i+1}(t)\frac{\partial V(\mu(t))}{\partial\mu_i}$$

$$+ [u_0(\hat{x}_1(t) - z_1(t), \ldots, \hat{x}_n(t) - z_n(t)) + \eta_{n+1}(t)]\frac{\partial V(\mu(t))}{\partial\mu_{n+1}} \tag{4.3.66}$$

$$\leq -W(\mu(t)) + C_1\|\eta(t)\|.$$

For any $\sigma > 0$, by Step 2, one can find $r_\sigma > 0$ and $t_{\sigma 1} > t_1$ such that $\|\eta(t)\| < 1/(2C_1)\varkappa_3(\sigma)$ for all $r_0 > r_\sigma$ and $t > t_{\sigma 1}$. It then follows that for any $t > t_{\sigma 1}$, if $\|\mu(t)\| > \sigma$, then

$$\left. \frac{dV(\mu(t))}{dt} \right|_{\text{along (4.3.54)}} \leq -W(\mu(t)) + C_1\|\eta(t)\| \leq -\varkappa_3(\sigma)$$

$$+ C_1\|\eta(t)\| \leq -\frac{1}{2}\varkappa_3(\sigma) < 0. \tag{4.3.67}$$

Therefore one can find a $t_{\sigma 2} > t_{\sigma 1}$ such that $\|\mu(t)\| < \sigma$ for $t > t_{\sigma 2}$.

As before, if in system (4.3.1) there is something known in "total disturbance", the ESO should make use of this information as much as possible. In what follows, we discuss the case where $f(\cdot)$ satisfies

$$f(t, x, \zeta, w) = \tilde{f}(x) + \overline{f}(t, \zeta, w) \tag{4.3.68}$$

and $\tilde{f}(\cdot)$ in (4.3.68) is known. In this case, the ESO for (4.3.1) is designed as follows:

$$\begin{cases} \dot{\hat{x}}_1(t) = \hat{x}_2(t) + \frac{1}{r^{n-1}(t)}g_1(r^n(t)\delta(t)), \\ \vdots \\ \dot{\hat{x}}_n(t) = \hat{x}_{n+1}(t) + g_n(r^n(t)\delta(t)) + \tilde{f}(\hat{x}_1(t), \ldots, \hat{x}_n(t)) + b_0 u(t), \\ \dot{\hat{x}}_{n+1}(t) = r(t)g_{n+1}(r^n(t)\delta(t)), \end{cases} \tag{4.3.69}$$

where the total disturbance in this case becomes

$$x_{n+1}(t) \triangleq \overline{f}(t, \zeta(t), w(t)) + (b(t) - b_0)u(t). \tag{4.3.70}$$

To deal with this problem, we need some conditions directly about $\tilde{f}(\cdot), \overline{f}(\cdot)$, and $f_0(\cdot)$.

Assumption 4.3.6

- The $\tilde{f}(\cdot)$ is Hölder continuous, that is,

$$|\tilde{f}(\tau) - \tilde{f}(\hat{\tau})| \leq L\|\tau - \hat{\tau}\|^\beta, \forall \tau, \hat{\tau} \in \mathbb{R}^n, \beta > 0. \tag{4.3.71}$$

- There exist constant $K > 0$ and function $\varpi_1 \in C(\mathbb{R}, [0, \infty))$ such that

$$\sum_{i=1}^m \left|\frac{\partial \overline{f}(t, \zeta, w)}{\partial \zeta_i}\right| + \left|\frac{\partial \overline{f}(t, \zeta, w)}{\partial w}\right| + \left|\frac{\partial \overline{f}(t, \zeta, w)}{\partial t}\right| + \|f_0(t, x, \zeta, w)\| \leq K + \varpi_1(w),$$

for all $(t, x, \zeta, w) \in \mathbb{R}^{n+m+2}$.

The functions $g_i(\cdot)$ in (4.3.69) are supposed to satisfy the following assumption.

Assumption 4.3.7 The nonlinear functions $g_i(\cdot) : \mathbb{R} \to \mathbb{R}$ are Hölder continuous, and there exist positive constants $\overline{R}, N > 0$, and radially unbounded, positive definite functions $V \in C^1(\mathbb{R}^{n+1}, [0, \infty)), W \in C(\mathbb{R}^{n+1}, [0, \infty))$ such that the first condition of Assumption 4.3.2 is satisfied, and

- $\displaystyle \max_{i=1,\ldots,n}\left\{\left|\iota_i\frac{\partial \mathcal{V}(\iota)}{\partial \iota_i}\right|\right\}\leq N\mathcal{W}(\iota),\ \iota\in\mathbb{R}^{n+1};$

- $\displaystyle \left|\frac{\partial \mathcal{V}(\iota)}{\partial \iota_{n+1}}\right|+\|\iota\|^\beta\left|\frac{\partial \mathcal{V}(\iota)}{\partial \iota_{n+1}}\right|\leq N\mathcal{W}(\iota),\forall\ \|\iota\|\geq \overline{R},\ \overline{R}>0.$

If the gain function $r(t)$ is chosen as (4.3.10), then Assumption 4.3.7 can be replaced by the following Assumption 4.3.8, which is slightly weaker than Assumption 4.3.7.

Assumption 4.3.8 The nonlinear functions $g_i:\mathbb{R}\to\mathbb{R}$ are Hölder continuous and there exist positive constants $\overline{R},N>0$, and radially unbounded, positive definite functions $\mathcal{V}\in C^1(\mathbb{R}^{n+1},[0,\infty))$, $\mathcal{W}\in C(\mathbb{R}^{n+1},[0,\infty))$ such that the first condition of Assumption 4.3.7 is satisfied and

- $\displaystyle \left|\frac{\partial \mathcal{V}(\iota)}{\partial \iota_{n+1}}\right|+\|\iota\|^\beta\left|\frac{\partial \mathcal{V}(\iota)}{\partial \iota_{n+1}}\right|\leq N\mathcal{W}(\iota),\forall\ \|\iota\|\geq \overline{R},\overline{R}>0.$

The feedback control in this case is changed accordingly into

$$u(t)=\frac{1}{b_0}(u_0(\hat{x}_1(t)-z_1(t),\ldots,\hat{x}_n(t)-z_n(t))$$
$$-\tilde{f}(\hat{x}_1(t),\ldots,\hat{x}_n(t))+z_{n+1}(t)-\hat{x}_{n+1}(t)),\tag{4.3.72}$$

where $u_0\in C^1(\mathbb{R}^n,\mathbb{R})$ is a nonlinear function satisfying Assumption 4.3.4.

Theorem 4.3.2 *Suppose that Assumptions 4.3.6 and 4.3.4 are satisfied,* $\sup_{t\in[0,\infty)}\max\{|w(t)|,|\dot{w}(t)|\}<\infty$, *and* $b(t)\equiv b_0$. *Then the closed loop of system (4.3.1) with ESO (4.3.69) and the feedback control (4.3.72) has the following convergence:*

(i) *If the gain function satisfies Assumption 4.3.3 and* $g_i(\cdot)$ *satisfy Assumption 4.3.7, then for* $i=1,\ldots,n+1,j=1,\ldots,n$,

$$\lim_{t\to\infty}|x_i(t)-\hat{x}_i(t)|=0,\ \lim_{t\to\infty}|x_j(t)-z_j(t)|=0.\tag{4.3.73}$$

(ii) *If the gain function* $r(t)$ *satisfies (4.3.10) and* $g_i(\cdot)$ *satisfy Assumption 4.3.8, then for any given* $\sigma>0$, $i=1,2,\ldots,n+1$, $j=1,2,\ldots,n$, *there exists an* $r^*>0$ *such that for all* $r_0>r^*$,

$$|x_i(t)-\hat{x}_i(t)|<\sigma,|x_j(t)-z_j(t)|<\sigma,t>t_0,$$

where t_0 *is an* r_0-*dependent constant.*

Proof. Here we only present the proof in the case of $r(t)$ satisfying Assumption 4.3.3. The convergence for $r(t)$ satisfying (4.3.10) can be obtained analogously.

Let $\eta_i(t)$ be defined in (4.3.12) with $x_{n+1}(t) = \overline{f}(t, \zeta(t), w(t))$. A straightforward computation shows that

$$
\begin{cases}
\dot{\eta}_1(t) = r(t)(\eta_2(t) - g_1(\eta_1(t))) + \dfrac{n\dot{r}(t)}{r(t)}\eta_1(t), \\
\quad\vdots \\
\dot{\eta}_n(t) = r(t)(\eta_{n+1}(t) - g_n(\eta_1(t))) \\
\qquad + \dfrac{\dot{r}(t)}{r(r)}\eta_n(t) + r(t)[\tilde{f}(x_1(t), \ldots, x_n(t)) - \tilde{f}(\hat{x}_1(t), \ldots, \hat{x}_n(t))], \\
\dot{\eta}_{n+1}(t) = -r(t)g_{n+1}(t)(\eta_1(t)) + \dot{x}_{n+1}(t).
\end{cases}
\tag{4.3.74}
$$

From the second condition of Assumption 4.3.6, there exists a positive constant B such that

$$
|\dot{x}_{n+1}(t)| = \left| \frac{d\overline{f}(t, \zeta(t), w(t))}{dt} \right| < B, \; t > 0.
\tag{4.3.75}
$$

By the first condition of Assumption 4.3.6 and (4.3.12),

$$
|\tilde{f}(x_1(t), \ldots, x_n(t)) - \tilde{f}(\hat{x}_1(t), \ldots, \hat{x}_n(t))| \leq L \left\| \left(\frac{\eta_1(t)}{r^n(t)}, \ldots, \frac{\eta_n(t)}{r(t)} \right) \right\|
$$

$$
\leq \frac{L}{r^\beta(t)} \|(\eta_1(t), \ldots, \eta_n(t))\|^\beta.
\tag{4.3.76}
$$

Let $\mathcal{V}(\cdot)$ and $\mathcal{W}(\cdot)$ be the Lyapunov functions satisfying Assumption 4.3.7. Then there exists the class $\mathcal{K}_\infty$ functions $\kappa_i(\cdot), i = 1, 2, 3, 4$ satisfying (4.3.9). By Assumption 4.3.7, (4.3.75), and (4.3.76), we can find the derivative of $\mathcal{V}(\eta(t))$ along the error equation (4.3.74) to obtain that for any $t > t_1$, if $\|\eta(t)\| > \overline{R}$, then

$$
\left. \frac{d\mathcal{V}(\eta(t))}{dt} \right|_{\text{along (4.3.74)}} = r(t) \left(\sum_{i=1}^{n} (\eta_{i+1}(t) - g_i(\eta_1(t))) \frac{\partial \mathcal{V}(\eta(t))}{\partial \eta_i} - g_{n+1}(\eta_1(t)) \frac{\partial \mathcal{V}(\eta(t))}{\partial \eta_{n+1}} \right)
$$

$$
+ \sum_{i=1}^{n} \frac{(n + 1 - i)\dot{r}(t)\eta_i(t)}{r(t)} \frac{\partial \mathcal{V}(\eta(t))}{\partial \eta_i}
$$

$$
+ r(t)[\tilde{f}(x_1(t), \ldots, x_n(t)) - \tilde{f}(\hat{x}_1(t), \ldots, \hat{x}_n(t))]\frac{\partial \mathcal{V}(\eta(t))}{\partial \eta_n} + \dot{x}_{n+1}(t)\frac{\partial \mathcal{V}(\eta(t))}{\partial \eta_{n+1}}
$$

$$
\leq -r(t)\mathcal{W}(\eta(t)) + LN \; r^{1-\beta}(t)\mathcal{W}(\eta(t)) + n(n + 1)MN\mathcal{W}(t) + BN\mathcal{W}(\eta(t)). \tag{4.3.77}
$$

By Assumption 4.3.3, there exists a $t_1 > 0$ such that for any $t > t_1$, $\dot{r}(t)/r(t) < 2M$ and

$$
r(t) > \max\{2n(n + 1)NM + 2BN, (2LN)^{1/\beta}\}.
\tag{4.3.78}
$$

By (4.3.9), (4.3.77), and (4.3.78), for any $t > t_1$, if $\|\mu(t)\| > \overline{R}$, then

$$
\left. \frac{d\mathcal{V}(\eta(t))}{dt} \right|_{\text{along (4.3.74)}} \leq -((n(n + 1)MN + BN))\kappa_3(\overline{R}) < 0,
\tag{4.3.79}
$$

which shows that the solution $\eta(t)$ of (4.3.74) is uniformly bounded for all $t \geq t_2$ for some $t_2 > t_1$. Therefore, one can find a constant $C > 0$ such that

$$\left.\frac{d\mathcal{V}(\eta(t))}{dt}\right|_{\text{along (4.3.74)}} \leq -\frac{1}{2}r(t)\mathcal{W}(\eta(t)) + C, \forall\, t > t_2. \tag{4.3.80}$$

Similar to the proof of Theorem 4.3.1, $\lim_{t\to\infty} \|\eta(t)\| = 0$.

Let $\mu(t)$ be defined in (4.3.12). Similar to Step 3 of *Part I* in the proof of Theorem 4.3.1, we can also show that $\lim_{t\to\infty} \|\mu(t)\| = 0$. The details are omitted.

To end this section, we present some numerical simulations for illustration.

Example 4.3.1 *Consider the following control system:*

$$\begin{cases} \dot{x}_1(t) = x_2(t), \\ \dot{x}_2(t) = f(t, x(t), \zeta(t), w(t)) + b(t)u(t), \\ \dot{\zeta}(t) = f_0(t, x_1(t), x_2(t), \zeta(t), w(t)), \\ y(t) = x_1(t), \end{cases} \tag{4.3.81}$$

where $f(\cdot)$ and $f_0(\cdot)$ are unknown system functions, $w(t)$ is the unknown external disturbance, control parameter $b(t)$ is also unknown but its nominal value b_0 is given, and $y(t)$ is the output. The control purpose is to design an output feedback control so that $y(t)$ tracks reference signal $v(t)$.

In the numerical simulation, we set

$$\begin{aligned} &f_0(t, x_1, x_2, \zeta, w) = a_1(t)x_1 + \sin x_2 + \cos\zeta + w, \\ &f(t, x_1, x_2, \zeta, w) = a_2(t)x_2 + \sin(x_1 + x_2) + \zeta + w, \\ &a_1(t) = 1 + \sin t, a_2(t) = 1 + \cos t, \\ &b(t) = 10 + 0.1\sin t, b_0 = 10, w(t) = 1 + \cos t + \sin 2t, v(t) = \sin t. \end{aligned} \tag{4.3.82}$$

It is easily to verify that Assumption 4.3.1 is satisfied.

To approximate the state $(x_1(t), x_2(t))$ and the extended state

$$x_3(t) \triangleq f(t, x_1(t), x_2(t), \zeta(t), w(t)) + (b(t) - b_0)u(t) \tag{4.3.83}$$

of system (4.3.1) by output $y(t) = x_1(t)$, the ESO is designed as (4.3.3) with $n = 2$, $g_1(s) = 6s + \phi(s), g_2(s) = 11s, g_3(s) = 6s$, and $s \in \mathbb{R}$, where

$$\phi(s) = \begin{cases} -\dfrac{1}{4\pi}, & s \in (-\infty, -\pi/2), \\[2mm] \dfrac{\sin s}{4\pi}, & s \in (-\pi/2, \pi/2), \\[2mm] \dfrac{1}{4\pi}, & s \in (\pi/2, \infty). \end{cases} \tag{4.3.84}$$

It is easy to verify that $g_i(\cdot)$ satisfy Assumption 4.3.2 with

$$\mathcal{V}(\iota) = 1.7\iota_1^2 + 0.7\iota_2^2 + 1.5333\iota_3^2 - \iota_1\iota_2 - 1.4\iota_1\iota_3 - \iota_2\iota_3,$$

$$\mathcal{W}(\iota) = \frac{1}{2}(\iota_1^2 + \iota_2^2 + \iota_3^2), \iota = (\iota_1, \iota_2, \iota_3) \in \mathbb{R}^3$$

$$\kappa_1(s) = 0.09s^2, \kappa_2(s) = 2.33s^2, \kappa_3(s) = \kappa_4(s) = 0.5s^2, s \geq 0.$$

The gain parameter is chosen as (4.3.10) with $a = 5, r_0 = 200$.

For notation simplicity, we use directly $z_1(t) = \sin t, z_2(t) = \cos t, z_3(t) = -\sin t$ as the target states. In the feedback control (4.3.6), $u_0(\tau_1, \tau_2)$ is chosen as

$$u_0(\tau_1, \tau_2) = -2\tau_1 - 4\tau_2 - \phi(\tau_1). \tag{4.3.85}$$

Define the Lyapunov functions $V(\tau_1, \tau_2), W(\tau_1, \tau_2)$, and the class $\mathcal{K}_\infty$ functions $\varkappa_i(s), i = 1, 2, 3, 4$, in Assumption 4.3.4 as

$$V(\tau_1, \tau_2) = 1.375\tau_1^2 + 0.1875\tau_2^2 + 0.5\tau_1\tau_2,$$

$$W(\tau_1, \tau_2) = 0.5\tau_1^2 + 0.5\tau_2^2, \tag{4.3.86}$$

$$\varkappa_1(s) = 0.13s^2, \varkappa_2(s) = 1.43s^2, \varkappa_3(s) = \kappa_4(s) = 0.5s^2, s \geq 0.$$

One can also easily check that Assumptions 4.3.4 and 4.3.5 are satisfied.

The numerical results are plotted in Figure 4.3.1. From Figure 4.3.1, we can see that: (a) $\hat{x}_i(t)(i = 1, 2, 3)$ of (4.3.3) converge to $x_1(t)$ and $x_2(t)$ of system (4.3.1), and its total disturbance given in (4.3.83) in a relative short time. The most remarkable fact is that peaking value disappears. (b) Under the feedback control, the output $x_1(t)$ and its derivative $x_2(t)$ track reference signal $\sin t$ and its derivative $\cos t$ satisfactorily.

To compare the time-varying ESO over a constant high-gain ESO, we give numerical simulations for system (4.3.81) by constant high-gain ESO, which is a special case of (4.3.3), by setting $r(t) = r_0$. Other functions and parameters are the same as that in Figure 4.3.1. The results are plotted in Figure 4.3.2. It is seen that the constant gain ESO can rapidly track the states $x_1(t)$ and $x_2(t)$, and total disturbance $x_3(t)$. However, very large peaking values in $\hat{x}_2(t)$ around 250 and $\hat{x}_3(t)$ around 25,000 are observed. However, in sharp contrast to Figure 4.3.2, in Figure 4.3.1 the peaking values for $\hat{x}_2(t)$ and $\hat{x}_3(t)$ are completely negligible.

Finally, if in (4.3.81),

$$f(t, x_1, x_2, \zeta, w) = \tilde{f}(x_1, x_2) + \overline{f}(t, \zeta, w),$$

$$w(t) = 1 + \cos t + \sin 2t, \overline{f}(t, \zeta, w) = \cos(t + \zeta) + w, \tag{4.3.87}$$

$$\tilde{f}(x_1, x_2) = [x_1]^\beta \triangleq \text{sign}(x_1)|x_1|^\beta, 0 < \beta < 1,$$

then we can easily check that $f(\cdot)$ does not satisfy Assumption 4.3.1, but satisfies Assumption 4.3.6.

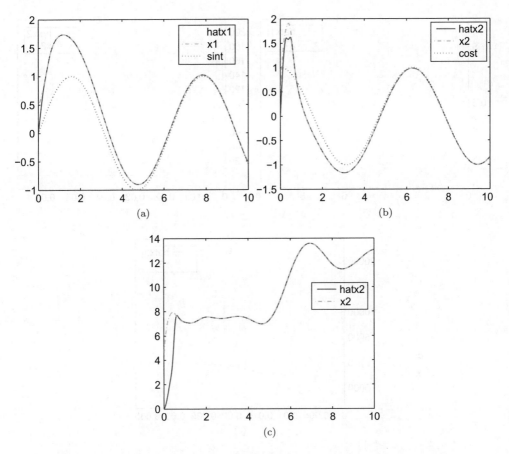

Figure 4.3.1 Numerical simulation for system (4.3.81) with $f(\cdot)$ chosen as (4.3.82).

Suppose that $\tilde{f}(\cdot)$ is known. The total disturbance is

$$x_3(t) \triangleq \overline{f}(t, \zeta(t), w(t)). \tag{4.3.88}$$

Then the ESO can be designed as (4.3.69) with $g_1(x) = [x]^{\theta}$, $g_2(x) = [x]^{2\theta-1}$, $g_3(x) = [x]^{3\theta-2}$, $x \in \mathbb{R}$, $\theta = 0.8$, $\beta = 0.1$. Similar to (3.2.31), (3.2.32), and (3.2.33), we can prove that $g_i(\cdot)$ satisfy Assumption 4.3.8. The gain parameter is the same as that in (4.3.10) with $a = 5, r_0 = 200$. The feedback control is designed as (4.3.72) with $u_0(x_1, x_2) = -2x_1 - 4x_2 - \phi(x_1) - [\hat{x}_1]^{\beta}$. We can verify that all conditions of Theorem 4.3.2 are satisfied. The numerical results are plotted in Figure 4.3.3. From Figure 4.3.3, we see that the states and the total disturbance are convergent very satisfactorily and there is no peaking value problem. Meanwhile, the system state $x_1(t)$ and its derivative $x_2(t)$ track very well the reference signal $\sin t$ and its derivative $\cos t$.

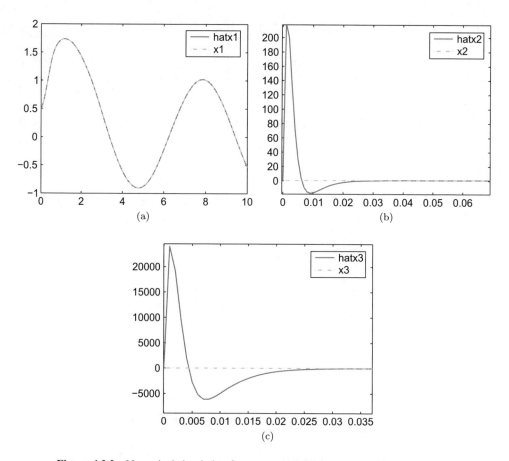

Figure 4.3.2 Numerical simulation for system (4.3.81) by constant high-gain ESO.

4.4 Nonlinear ADRC for MIMO Systems with Vast Uncertainty

In this section, the system we are concerned with is the following canonical form of the ADRC introduced in Chapter 1:

$$
\begin{cases}
\dot{x}^i(t) = A_{n_i} x^i(t) + B_{n_i}[f_i(x(t), \xi(t), w_i(t)) \\
\qquad\qquad + \displaystyle\sum_{j=1}^{m} a_{ij}(x(t), \xi(t), w(t)) u_j(t)], \\
y_i(t) = C_i x^i(t), i = 1, 2, \dots, m, \\
\dot{\xi}(t) = F_0(x(t), \xi(t), w(t)),
\end{cases}
\tag{4.4.1}
$$

where $u(t) \in \mathbb{R}^m$, $\xi(t) \in \mathbb{R}^s$, $x(t) = (x^1(t), x^2(t), \dots, x^m(t)) \in \mathbb{R}^n$, $n = n_1 + \cdots + n_m$, $F_0(\cdot), f_i(\cdot), a_{ij}(\cdot), w_i(t)$ are C^1- functions with their arguments, respectively, the external

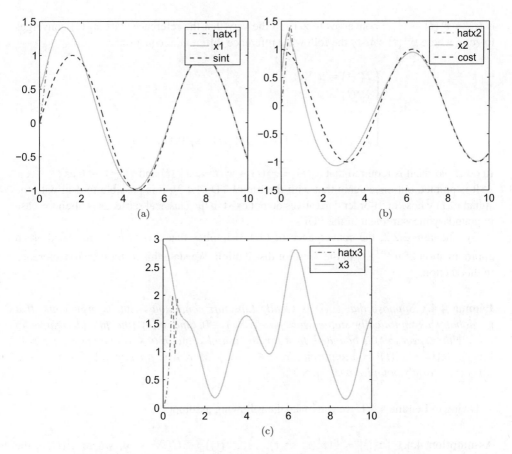

Figure 4.3.3   Numerical results of (4.3.81) for $f(\cdot)$ to be chosen as (4.3.87).

disturbance $w(t) = (w_1(t), w_2(t), \ldots, w_m(t))$ satisfies $\sup_{t \in [0, \infty)} \|(w(t), \dot{w}(t))\| < \infty$, and

$$
A_{n_i} = \begin{pmatrix} 0 & I_{(n_i-1) \times (n_i-1)} \\ 0 & 0 \end{pmatrix}_{n_i \times n_i}, \quad B_{n_i} = (0, \ldots, 0, 1)^\top_{n_i \times 1}, \quad C_{n_i} = (1, 0, \ldots, 0)_{1 \times n_i}.
$$

$$(4.4.2)$$

In general, the unmodeled dynamics terms $f_i(\cdot)$ are unknown, the control parameters $a_{ij}(\cdot)$ have some uncertainties, and the external disturbance $w(t)$ is completely unknown.

Let the reference input signals be $v_i(t)$. We then construct a tracking differentiator (TD) with input $v_i(t)$ and output $z^i(t)$ as follows:

$$
\text{TD: } \dot{z}^i(t) = A_{n_i+1} z^i(t) + B_{n_i+1} \rho^{n_i+1} \psi_i \left(z_1^i(t) - v_i(t), \frac{z_2^i(t)}{\rho}, \ldots, \frac{z_{n_i+1}^i(t)}{\rho^{n_i}} \right),
$$

$$
i = 1, 2, \ldots, m. \qquad (4.4.3)
$$

The control objective of the ADRC is to make the output $x_1^i(t)$ track the measured signal $v_i(t)$ and the $x_j^i(t)$ track $z_j^i(t)$, $1 \le j \le n_i, 1 \le i \le m$. Moreover, each error

$e_j^i(t) = x_j^i(t) - z_j^j(t)$ converges to zero in the way that the reference state $x_j^{i*}(t)$ converges to zero, where $x_j^{i*}(t)$ satisfy the following reference differential equation:

$$\text{Ref:} \quad \begin{cases} \dot{x}_1^{i*}(t) = x_2^{i*}(t), \\ \dot{x}_2^{i*}(t) = x_3^{i*}(t), \\ \quad \vdots \\ \dot{x}_{n_i}^{i*}(t) = \phi_i(x_1^{i*}(t), \dots, x_{n_i}^{i*}(t)), \phi_i(0, 0, \dots, 0) = 0. \end{cases} \tag{4.4.4}$$

In other words, it is required that $e_j^i(t) = x_j^i(t) - z_j^j(t) \approx x_j^{i*}(t)$ and $x_j^{i*}(t) \to 0$ as $t \to \infty$.

It is pointed out once again that with the use of TD (4.4.3), it is possible to deal with the signal $v_i(t)$ whose high-order derivatives do not exist in the classical sense, as is often the case in boundary measurement of the PDEs.

By Theorem 2.2.2, the convergence of (4.4.3) implies that $z_j^i(t)$ can be regarded as an approximation of $v_i^{(j-1)}(t)$ in the sense of distribution. We state this result here for reference in this section.

Lemma 4.4.1 *Suppose that $\psi_i(\cdot)$ is locally Lipschitz continuous with its arguments. If it is globally asymptotically stable with $\rho = 1, v(t) \equiv 0$, and $\dot{v}_i(t)$ (the first derivative of $v_i(t)$ with respect to t) is bounded, then for any initial value of (4.4.3) and constant $\tau > 0$, $\lim_{\rho \to \infty} |z_1^i(t) - v_i(t)| = 0$ uniformly for $t \in [\tau, \infty)$. Moreover, for any $j, 1 \le j \le n_i + 1$, $z_j^i(t)$ is uniformly bounded over $[0, \infty)$.*

Owing to Lemma 4.4.1, we can make the following assumption.

Assumption 4.4.1 $\|z(t)\| = \|(z^1(t), z^2(t), \dots, z^m(t))\| < C_1, \forall\, t > 0$, where $z^i(t)$ is the solution of (4.4.3), $z^i(t) = (z_1^i(t), z_2^i(t), \dots, z_{n_i}^i(t))$, and C_1 is a ρ-dependent positive constant.

The ESO is designed as follows:

$$\text{ESO:} \quad \begin{cases} \dot{\hat{x}}_1^i(t) = \hat{x}_2^i(t) + \varepsilon^{n_i - 1} g_1^i(e_1^i(t)), \\ \dot{\hat{x}}_2^i(t) = \hat{x}_3^i(t) + \varepsilon^{n_i - 2} g_2^i(e_1^i(t)), \\ \quad \vdots \\ \dot{\hat{x}}_{n_i}^i(t) = \hat{x}_{n_i+1}^i(t) + g_{n_i}^i(e_1^i(t)) + u_i^*(t), \\ \dot{\hat{x}}_{i,n_i+1}^i(t) = \frac{1}{\varepsilon} g_{n_i+1}^i(e_1^i(t)), i = 1, 2, \dots, m, \end{cases} \tag{4.4.5}$$

where $e_1^i(t) = (x_1^i(t) - \hat{x}_1^i(t))/\varepsilon^{n_i}$ and $g_j^i(\cdot)$ are possibly some nonlinear functions. The function of ESO is used to estimate, in real time, both the state and the total disturbance in the ith subsystem by $\hat{x}_{i,n_i+1}(t)$ in (4.4.5).

The convergence of ESO (4.4.5) itself (without feedback), like many other observers in nonlinear systems, is an independent issue, which is presented in Chapter 3.

To show the convergence of ADRC, we need the following assumptions on ESO (4.4.5) (Assumption 4.4.2) and reference system (4.4.4) (Assumption 4.4.3).

Assumption 4.4.2 For every $i \leq m$, $|g_j^i(r)| \leq \Lambda_j^i r$ for all $r \in \mathbb{R}$, there exist constants $\lambda_{11}^i, \lambda_{12}^i, \lambda_{13}^i, \lambda_{14}^i$, β_1^i, and positive definite continuous differentiable functions $V_1^i, W_1^i : \mathbb{R}^{n_i+1} \to \mathbb{R}$ such that

- $\lambda_{11}^i \|y\|^2 \leq V_1^i(y) \leq \lambda_{12}^i \|y\|^2$, $\lambda_{13}^i \|y\|^2 \leq W_1^i(y) \leq \lambda_{14}^i \|y\|^2$,

- $\sum_{j=1}^{n_i} (y_{j+1} - g_j^i(y_1)) \dfrac{\partial V_1^i(y)}{\partial y_j} - g_{n_i+1}^i(y_1) \dfrac{\partial V_1^i(y)}{\partial y_{n_i+1}} \leq -W_1^i(y),$

- $\max \left\{ \left| \dfrac{\partial V_1^i(y)}{\partial y_{n_i}} \right|, \left| \dfrac{\partial V_1^i(y)}{\partial y_{n_i+1}} \right| \right\} \leq \beta_1^i \|y\|, \forall\, y \in \mathbb{R}^{n_i+1}.$

Assumption 4.4.3 For every $1 \leq i \leq m$, $\phi_i(\cdot)$ is globally Lipschitz continuous with the Lipschitz constant L_i: $|\phi_i(x) - \phi_i(y)| \leq L_i \|x - y\|$ for all $x, y \in \mathbb{R}^{n_i}$. There also exist constants $\lambda_{21}^i, \lambda_{22}^i, \lambda_{23}^i, \lambda_{24}^i$, β_2^i, and positive definite continuous differentiable functions $V_2^i, W_2^i : \mathbb{R}^{n_i} \to \mathbb{R}$ such that

- $\lambda_{21}^i \|y\|^2 \leq V_2^i(y) \leq \lambda_{22}^i \|y\|^2$, $\lambda_{23}^i \|y\|^2 \leq W_2^i(y) \leq \lambda_{24}^i \|y\|^2$,

- $\sum_{j=1}^{n_i-1} y_{j+1} \dfrac{\partial V_2^i(y)}{\partial y_j} + \phi_i(y_1, y_2, \ldots, y_{n_i}) \dfrac{\partial V_2^i(y)}{\partial y_{n_i}} \leq -W_2^i(y),$

- $\left| \dfrac{\partial V_2^i(y)}{\partial y_{n_i}} \right| \leq \beta_2^i \|y\|, \forall\, y = (y_1, y_2, \ldots, y_{n_i}) \in \mathbb{R}^{n_i}.$

Throughout the section, the following notation will be used without specifying every time:

$$\begin{cases} \tilde{x}^i(t) = (\hat{x}_1^i(t), \hat{x}_2^i(t), \ldots, \hat{x}_{n_i}^i(t))^\top, \hat{x}^i(t) = (\hat{x}_1^i(t), \hat{x}_2^i(t), \ldots, \hat{x}_{n_i+1}^i(t))^\top, \\ \tilde{x}(t) = (\tilde{x}^{1\top}(t), \ldots, \tilde{x}^{m\top}(t))^\top, \\ e_j^i(t) = \dfrac{x_j^i(t) - \hat{x}_j^i(t)}{\varepsilon^{n_i+1-j}}, \quad 1 \leq j \leq n_i + 1, 1 \leq i \leq m, \\ e^i(t) = (e_1^i(t), \ldots, e_{n_i+1}^i(t))^\top, e(t) = (e^{1\top}(t), \ldots, e^{m\top}(t))^\top, \\ \eta(t) = x(t) - z(t), \eta^i(t) = x^i(t) - z^i(t), \end{cases} \quad (4.4.6)$$

$$V_1 : \mathbb{R}^{2n+m} \to \mathbb{R}, \quad V_1(e) = \sum_{i=1}^m V_1^i(e^i), V_2 : \mathbb{R}^{2n} \to \mathbb{R}, \quad V_2(\eta) = \sum_{i=1}^m V_2^i(\eta^i), \quad (4.4.7)$$

and $u_i^*(t)$ and $x_{n_i+1}^i(t)$ will be specified later in different cases.

4.4.1   Semi-Global ADRC for MIMO Systems with Uncertainty

In this section, we assume that the initial values of system (4.4.1) lie in a compact set. This information is used to construct a saturated feedback control to avoid peaking value problems caused by the high gain in the ESO.

Assumption 4.4.4 There are constants C_2 and C_3 such that $\|x(0)\| < C_2$, $\|(w(t), \dot{w}(t))\| < C_3$.

Let $C_1^* = \max_{\{y \in \mathbb{R}^n : \|y\| \le C_1 + C_2\}} V_2(y)$. The following assumption is to guarantee the input-to-state stable for zero dynamics.

Assumption 4.4.5 There exist positive definite functions $V_0, W_0 : \mathbb{R}^s \to \mathbb{R}$ such that $L_{F_0} V_0(\xi) \le -W_0(\xi)$ for all $\xi : \|\xi\| > \chi(x, w)$, where $\chi(\cdot) : \mathbb{R}^{n+m} \to \mathbb{R}$ is a class $\mathcal{K}_\infty$ function and $L_{F_0} V_0(\xi)$ denotes the Lie derivative along the zero dynamics in system (4.4.1).

Set

$$\max \left\{ \sup_{\|x\| \le C_1 + (C_1^*+1)/(\min \lambda_{23}^i)+1, \|w\| \le C_3,} |\chi(x, w)|, \|\xi(0)\| \right\} \le C_4,$$

$$M_1 \ge 2 \left(1 + M_2 + C_1 \right.$$

$$\left. + \max_{1 \le i \le m} \sup_{\|x\| \le C_1 + (C_1^*+1)/(\min \lambda_{23}^i)+1, \|w\| \le C_3, \|\xi\| \le C_4} |f_i(x, \xi, w_i)| \right), \tag{4.4.8}$$

$$M_2 \ge \max_{\|x\| \le C_1 + (C_1^*+1)/(\min \lambda_{23}^i)+1} |\phi_i(x)|.$$

The following assumption is for the control parameters.

Assumption 4.4.6 For each $a_{ij}(x, \xi, w_i)$, there exists a nominal parameter function $b_{ij}(x)$ such that

(i) The matrix with entries $b_{ij}(x)$ are globally invertible with the inverse matrix given by

$$\begin{pmatrix} b_{11}^*(x) & b_{12}^*(x) & \cdots & b_{1m}^*(x) \\ b_{21}^*(x) & b_{22}^*(x) & \cdots & b_{2m}^*(x) \\ \vdots & \vdots & \ddots & \vdots \\ b_{m1}^*(x) & b_{m2}^*(x) & \cdots & b_{mm}^*(x) \end{pmatrix} = \begin{pmatrix} b_{11}(x) & b_{12}(x) & \cdots & b_{1m}(x) \\ b_{21}(x) & b_{22}(x) & \cdots & b_{2m}(x) \\ \vdots & \vdots & \ddots & \vdots \\ b_{m1}(x) & b_{m2}(x) & \cdots & b_{mm}(x) \end{pmatrix}^{-1}. \tag{4.4.9}$$

(ii) For every $1 \le i, j \le m$, $b_{ij}(x)$ and $b_{ij}^*(x)$, all partial derivatives of $b_{ij}(x)$ and $b_{ij}^*(x)$ with respect to their arguments are globally bounded.

(iii)

$$\vartheta = \max_{1 \le i \le m} \sup_{\|x\| \le C_1 + (C_1^*+1)/(\min \lambda_{23}^i)+1, \|\xi\| \le C_4, \|w\| \le C_3, \nu \in \mathbb{R}^n}$$

$$|a_{ij}(x, \xi, w) - b_{ij}(x)||b_{ij}^*(\nu)| \tag{4.4.10}$$

$$< \min_{1 \le i \le m} \left\{ \frac{1}{2}, \lambda_{13}^i \left(m \beta_1^i \Lambda_{n_i+1}^i \left(M_1 + \frac{1}{2} \right) \right)^{-1} \right\}.$$

Let $\mathrm{sat}_M(r)$ be an odd continuous differentiable saturated function (4.1.73). The feedback control is designed as

$$\text{ADRC(S):} \quad \begin{cases} u_i^*(t) = -\mathrm{sat}_{M_1}(\hat{x}_{n_i+1}^i(t)) + \mathrm{sat}_{M_2}(\phi_i(\tilde{x}^i(t) - z^i(t))) + z_{n_i+1}^i(t), \\ u_i(t) = \displaystyle\sum_{k=1}^{m} b_{ik}^*(\tilde{x}(t))u_k^*(t). \end{cases}$$

(4.4.11)

The roles played by the different terms in control design (4.4.11) are as follows: $\hat{x}_{n_i+1}^i(t)$ is to compensate the total disturbance; $x_{n_i+1}^i(t) = f_i(x(t), \xi(t), w_i(t)) + \sum_{j=1}^{m}(a_{ij}(x(t), \xi(t), w_i(t)) - b_{ij}(x(t)))u_j(t)$, $\phi_i(\tilde{x}^i(t) - z^i(t)) + z_{n_i+1}^i(t)$ is to guarantee the output tracking; and $\hat{x}_{n_i+1}^i(t)$, $\phi_i(\hat{x}^i(t) - z^i(t))$ are bounded by using $\mathrm{sat}_{M_1}(r), \mathrm{sat}_{M_2}(r)$, respectively, to limit the peaking value in the control signal. Since both the cancelation and estimation are processing online, the control signal in the ADRC does not need to be unnecessarily large. This means that the ADRC would spend less energy in control in order to cancel the effect of the disturbance.

Under feedback (4.4.11), the closed-loop of system (4.4.1) and ESO (4.4.5) is rewritten as

$$\begin{cases} \dot{x}^i(t) = A_{n_i}x^i(t) + B_{n_i}[f_i(x(t), \xi(t), w_i(t)) \\ \qquad\qquad + \displaystyle\sum_{j=1}^{m} a_{ij}(x(t), \xi(t), w(t))u_j(t)], \\ \dot{\xi}(t) = F_0(x(t), \xi(t), w(t)), \\ \dot{\hat{x}}^i(t) = A_{n_i+1}\hat{x}^i(t) + \begin{pmatrix} B_{n_i} \\ 0 \end{pmatrix} u_i^*(t) + \begin{pmatrix} \varepsilon^{n_i-1}g_1^i(e_1^i(t)) \\ \vdots \\ \frac{1}{\varepsilon}g_{n_i+1}^i(e_1^i(t)) \end{pmatrix}, \\ u_i^*(t) = -\mathrm{sat}_{M_1}(\hat{x}_{n_i+1}^i(t)) + \mathrm{sat}_{M_2}(\phi_i(\tilde{x}^i(t) - z^i(t))) + z_{n_i+1}^i(t), \\ u_i(t) = \displaystyle\sum_{k=1}^{m} b_{ik}^*(\tilde{x}(t))u_k^*(t). \end{cases}$$

(4.4.12)

Theorem 4.4.1 *Assume that Assumptions 4.4.1 to 4.4.6 are satisfied. Let the ε-dependent solution of (4.4.12) be $(x(t), \hat{x}(t))$. Then for any $\sigma > 0$, there exists an $\varepsilon_0 > 0$ such that for any $\varepsilon \in (0, \varepsilon_0)$, there exists an ε-independent constant $t_0 > 0$, such that*

$$|\tilde{x}(t) - x(t)| \leq \sigma \text{ for all } t > t_0$$

(4.4.13)

and

$$\overline{\lim}_{t\to\infty} \|x(t) - z(t)\| \leq \sigma.$$

(4.4.14)

The proof of Theorem 4.4.1 is based on the boundedness of the solution stated in the following Lemma 4.4.2.

Lemma 4.4.2 *Assume that Assumptions 4.4.1, 4.4.2, 4.4.4, and 4.4.5 are satisfied. Let* $\Omega_0 = \{y | V_2(y) \le C_1^*\}$ *and* $\Omega_1 = \{y | V_2(y) \le C_1^* + 1\}$. *Then there exists an* $\varepsilon_1 > 0$ *for any* $\varepsilon \in (0, \varepsilon_1)$, *and* $t \in [0, \infty)$, $\eta(t) \in \Omega_1$.

Proof. Firstly, we see that for any $\varepsilon > 0$,

$$
\begin{cases}
|\eta_j^i(t)| \le |\eta_j^i(0)| + |\eta_{j+1}^i(t)|t, 1 \le j \le n_i - 1, 1 \le i \le m, \\
|\eta_{n_i}^i(t)| \le |\eta_{n_i}^i(0)| + [C_1 + M_1 + mM_1^*(C_1 + M_1 + M_2)]t, \quad \eta(t) \in \Omega_1,
\end{cases}
\tag{4.4.15}
$$

where

$$
M_1^* = \max_{1 \le i,j \le m} \sup_{\|x\| \le C_1 + (C_1^*+1)/(\min \lambda_{23}^i)+1, \|w\| \le \bar{C}_3, \|\xi\| \le \bar{C}_4} |a_{ij}(x, \xi, w)|.
$$

Next, by the iteration process, we can show that all terms on the right-hand side of (4.4.15) are ε-independent. Since $\|\eta(0)\| < C_1 + C_2, \eta(0) \in \Omega_0$, there exists an ε-independent constant $t_0 > 0$ such that $\eta(t) \in \Omega_0$ for all $t \in [0, t_0]$.

We then suppose that Lemma 4.4.2 is not true to obtain a contradiction. Since then for any $\varepsilon > 0$, there exists an $\varepsilon^* \in (0, 1)$ and $t^* \in (0, \infty)$ such that

$$
\eta(t^*) \in \mathbb{R}^n - \Omega_1.
\tag{4.4.16}
$$

Since for any $t \in [0, t_0], \eta(t, \varepsilon^*) \in \Omega_0$, and $\eta(t, \varepsilon)$ is continuous in t, there exists a $t_1 \in (t_0, t_2)$ such that

$$
\begin{cases}
\eta(t_1) \in \partial \Omega_0 \text{ or } V_2(\eta(t_1)) = C_1^*, \\
\|\eta(t_2)\| \in \Omega_1 - \Omega_0 \text{ or } C_1^* < V_2(\eta(t_2)) \le C_1^* + 1, \\
\eta(t) \in \Omega_1 - \Omega_0^\circ, \forall\, t \in [t_1, t_2] \text{ or } C_1^* \le V_2(\eta(t)) \le C_1^* + 1, \\
\eta(t) \in \Omega_1, \forall\, t \in [0, t_2].
\end{cases}
\tag{4.4.17}
$$

By (4.4.1) and (4.4.6), it follows that the error $e^i(t)$ in this case satisfies

$$
\varepsilon \dot{e}^i(t) = A_{n_i+1} e^i(t) + \Delta_{i1}(t) \begin{pmatrix} B_{n_i} \\ 0 \end{pmatrix} + \varepsilon \Delta_{i2}(t) B_{n_i+1} - \begin{pmatrix} g_1^i(e_1^i(t)) \\ \vdots \\ g_{n_i+1}^i(e_1^i(t)) \end{pmatrix}, 1 \le i \le m,
\tag{4.4.18}
$$

where

$$
\Delta_{i1}(t) = \sum_{j=1}^m \Big(b_{ij}(x(t)) - b_{ij}(\tilde{x}(t)) \Big) u_j(t),
$$

$$
\Delta_{i2}(t) = \frac{d}{dt} \left(f_i(x(t), \xi(t), w_i(t)) + \sum_{j=1}^m \Big(a_{ij}(x(t), \xi(t), w_i(t)) - b_{ij}(x(t)) \Big) u_j(t) \right) \Bigg|_{\text{along (4.4.12)}}.
\tag{4.4.19}
$$

Since all derivatives of $b_{ij}(x(t))$ are globally bounded, there exists a constant $N_0 > 0$ such that $|\Delta_{i1}(x(t))| \le \varepsilon N_0 \|e(t)\|$.

We define two vector fields by

$$F_i(x^i) = \begin{pmatrix} x_2^i \\ x_3^i \\ \vdots \\ x_{n_i+1}^i + f_i(x,\xi,w_i) + \sum_{j=1}^{m} a_{ij}(x,\xi,w_j)u_j - u_i^* \end{pmatrix}, \qquad (4.4.20)$$

$$F(x) = (F_1(x^1)^\top, F_2(x^2)^\top, \ldots, F_m(x^m)^\top)^\top; \qquad (4.4.21)$$

$$\hat{F}_i(\tilde{x}^i) = \begin{pmatrix} \hat{x}_2^i + \varepsilon^{n_i-1} g_1^i(e_1^i) \\ \hat{x}_3^i + \varepsilon^{n_i-2} g_2^i(e_1^i) \\ \vdots \\ \hat{x}_{n_i+1}^i + g_{n_i}^i(e_1^i) + u_i^* \end{pmatrix}, \qquad (4.4.22)$$

$$\hat{F}(\tilde{x}) = (\hat{F}_1(\tilde{x}^1)^\top, \hat{F}_2(\tilde{x}^2)^\top, \ldots, \hat{F}_m(\tilde{x}^m)^\top)^\top.$$

Considering the derivative of $x_{n_i+1}^i(t)$ with respect to t in the interval $[t_1, t_2]$, we have

$$\Delta_{i2}(t) = \frac{d}{dt}\left(f_i(x(t),\xi(t),w_i(t)) + \sum_{j=1}^{m} \left(a_{ij}(x(t),\xi(t),w_j(t)) - b_{ij}(x(t)) \right) u_j(t) \right)$$

$$= \left[L_{F(x)} f_i(x(t),\xi(t),w_i(t)) + L_{F_0(\xi)} f_i(x(t),\xi(t),w_i(t)) + \frac{\partial f_i}{\partial w_i} \dot{w}_i(t) \right]$$

$$+ \frac{d}{dt}\left(\sum_{j,l=1}^{m} \left(a_{ij}(x(t),\xi(t),w_j(t)) - b_{ij}(x(t)) \right) b_{jl}^*(\tilde{x}(t)) u_l^*(t) \right)$$

$$= \left[L_{F(x)} f_i(x(t),\xi(t),w_i(t)) + L_{F_0(\xi)} f_i(x(t),\xi(t),w_i(t)) + \frac{\partial f_i}{\partial w_i} \dot{w}_i(t) \right]$$

$$+ \sum_{j,l=1}^{m} \left(L_{F(x)} \left(a_{ij}(x(t),\xi(t),w_j(t)) - b_{ij}(x(t)) \right) + L_{F_0(\xi)} a_{ij}(x(t),\xi(t),w_j(t)) \right.$$

$$\left. + \frac{\partial a_{ij}}{\partial w_i} \dot{w}_i(t) \right) b_{jl}^*(\tilde{x}) u_l^*(t)$$

$$+ \sum_{j,l=1}^{m} \left(a_{ij}(x(t),\xi(t),w_j(t)) - b_{ij}(x(t)) \right) L_{\hat{F}(\tilde{x})}(b_{jl}^*(\tilde{x}(t)) u_l^*(t)$$

$$+ \sum_{j,l=1}^{m} \left(a_{ij}(x(t),\xi(t),w_j(t)) - b_{ij}(x(t)) \right) b_{jl}^*(\tilde{x}(t)) \left(-\frac{1}{\varepsilon} h_{M_1}^i(\hat{x}_{n_i+1}^i(t)) g_{n_i+1}^i(e_1^i(t)) \right.$$

$$\left. + L_{\hat{F}_i(\tilde{x}^i)} \mathrm{sat}_{M_2}(\phi_i(\tilde{x}^i(t) - z^i(t))) - \sum_{s=1}^{n_i} z_{s+1}(t) \frac{\partial \mathrm{sat}_{M_2} \circ \phi_i}{\partial y_s}(\tilde{x}^i(t) - z^i(t)) + \dot{z}_{n_i+1}^i(t) \right).$$

$$(4.4.23)$$

From the assumptions that all $\|(w(t), \dot{w}(t))\|$, $\|x(t)\|$, $\|\xi(t)\|$, $\|z(t)\|$, and $|z^i_{n_i+1}(t)|$ are bounded in $[t_1, t_2]$, we conclude that there exists a positive ε-independent number N_i such that for all $t \in [t_1, t_2]$,

$$
\left| L_{F(x)} f_i(x(t), \xi(t), w_i(t)) + L_{F_0(\xi)} f_i(x(t), \xi(t), w_i(t)) + \frac{\partial f_i}{\partial w_i} \dot{w}_i(t) \right.
$$

$$
+ \sum_{j,l=1}^{m} \left(L_{F(x)} \left(a_{ij}(x(t), \xi(t), w_j(t)) - b_{ij}(x(t)) \right) + L_{F_0(\xi)} a_{ij}(x(t), \xi(t), w_j(t)) + \frac{\partial a_{ij}}{\partial w_i} \dot{w}_i(t) \right)
$$

$$
\left. \times b^*_{jl}(\tilde{x}(t)) u^*_l(t) \right| \leq N_1, \tag{4.4.24}
$$

$$
\left| \sum_{j,l=1}^{m} \left(a_{ij}(x(t), \xi(t), w_j(t)) - b_{ij}(x(t)) \right) L_{\hat{F}(\tilde{x})} b^*_{jl}(\tilde{x}(t)) u^*_l(t) \right| \leq N_2 \|e(t)\| + N_3, \tag{4.4.25}
$$

$$
\left| \sum_{j,l=1}^{m} \left(a_{ij}(x(t), \xi(t), w_j(t)) - b_{ij}(x(t)) \right) b^*_{jl}(\tilde{x}(t)) \left(-\frac{1}{\varepsilon} h_{M_1}(\hat{x}^i_{n_i+1}(t)) g^i_{n_i+1}(e^i_1(t)) \right. \right.
$$

$$
\left. \left. + L_{\hat{F}_i(\tilde{x}^i)} \mathrm{sat}_{M_2}(\phi_i(\tilde{x}^i(t) - z^i(t))) - \sum_{s=1}^{n_i} z_{s+1}(t) \frac{\partial \mathrm{sat}_{M_2} \circ \phi_i}{\partial y_s}(\tilde{x}^i(t) - z^i(t)) + \dot{z}^i_{n_i+1}(t) \right) \right|
$$

$$
\leq \frac{N}{\varepsilon} \|e^i(t)\| + N_4 \|e(t)\| + N_5, \tag{4.4.26}
$$

where

$$
N = \Lambda^i_{n_i+1} \left(M_1 + \frac{1}{2} \right)
$$

$$
\times \max_{1 \leq i \leq m} \sup_{\|x\| \leq C_1 + (C^*_1 + 1)/(\min \lambda^i_{23}) + 1, \|\xi\| \leq C_4, \|w\| \leq C_3, \nu \in \mathbb{R}^n} \tag{4.4.27}
$$

$$
\sum_{j,l=1}^{m} |a_{ij}(x, \xi, w) - b_{ij}(x)||b^*_{jl}(\nu)| = \Lambda^i_{n_i+1} \left(M_1 + \frac{1}{2} \right) \vartheta.
$$

Finding the derivative of $V_1(e(t))$ along system (4.4.18) with respect to t shows that for any $0 < \varepsilon < \min_{1 \leq i \leq m} (\lambda^i_{13} - N\beta^i_1)/((N_0 + N_2 + N_4) \max_{1 \leq i \leq m} \beta^i_1)$ and $t \in [0, t_2]$,

$$
\left. \frac{dV_1(e(t))}{dt} \right|_{\text{along (4.4.18)}}
$$

$$
\leq \sum_{i=1}^{m} \left\{ -\frac{1}{\varepsilon} W^i_1(e^i(t)) + \beta^i_1 \|e^i(t)\|(N_0\|e(t)\| + N_1 + N_2\|e(t)\| + N_3 + N_5 \right.
$$

$$
\left. + N_4\|e(t)\| + \frac{N}{\varepsilon}\|e^i(t)\|) \right\}
$$

$$\leq -\left(\frac{1}{\varepsilon} \min_{1\leq i\leq m} (\lambda_{13}^i - N\beta_1^i) - (N_0 + N_2 + N_4) \cdot \max_{1\leq i\leq m} (\beta_1^i)\right) \|e(t)\|^2 \qquad (4.4.28)$$

$$+ m \max \beta_1^i (N_1 + N_3 + N_5)\|e(t)\|$$

$$\leq -\frac{1}{\max\{\lambda_{12}^i\}} \left(\frac{1}{\varepsilon} \min_{1\leq i\leq m} (\lambda_{13}^i - N\beta_1^i) - (N_0 + N_2 + N_4)\right.$$

$$\left. \times \max_{1\leq i\leq m} (\beta_1^i)\right) V_1(e(t)) + \frac{m \max \beta_1^i (N_1 + N_3 + N_5)}{\sqrt{\lambda_{12}^i}} \sqrt{V_1(e(t))}.$$

Hence, for any $0 < \varepsilon < \min_{1\leq i\leq m}(\lambda_{13}^i - N\beta_1^i)/((N_0 + N_2 + N_4)\max_{1\leq i\leq m} \beta_1^i)$ and $t \in [0, t_2]$, one has

$$\frac{d}{dt}\sqrt{V_1(e(t))} \leq -\left(\frac{\Pi_1}{\varepsilon} - \Pi_2\right)\sqrt{V_1(e(t))} + \Pi_3, \qquad (4.4.29)$$

where

$$\Pi_1 = \frac{\min(\lambda_{13}^i - N\beta_1^i)}{\max\{\lambda_{12}^i\}}, \Pi_2 = \frac{(N_0 + N_2 + N_4)\max(\beta_1^i)}{\max\{\lambda_{12}^i\}},$$

$$\Pi_3 = \frac{m \max\beta_1^i(N_1 + N_3 + N_5)}{\sqrt{\lambda_{12}^i}}. \qquad (4.4.30)$$

Therefore, for every $0 < \varepsilon < \min_{1\leq i\leq m}(\lambda_{13}^i - N\beta_1^i)/((N_0 + N_2 + N_4)\max_{1\leq i\leq m} \beta_1^i)$ and $t \in [0, t_2]$, we have

$$\|e(t)\| \leq \frac{1}{\sqrt{\lambda_{11}^i}} \sqrt{V_1(e(t))} \leq \frac{1}{\sqrt{\lambda_{11}^i}} \left[e^{(-\Pi_1/\varepsilon+\Pi_2)t}\sqrt{V_1(e(0))}\right.$$

$$\left. +\Pi_3 \int_0^t e^{(-\Pi_1/\varepsilon+\Pi_2)(t-s)}ds\right]. \qquad (4.4.31)$$

Passing to the limit as $\varepsilon \to 0$ yields, for any $t \in [t_1, t_2]$, that

$$e^{(-\Pi_1/\varepsilon+\Pi_2)t}\sqrt{V_1(e(0))} \leq \frac{1}{\sqrt{\min\{\lambda_{11}^i\}}} e^{(-\Pi_1/\varepsilon+\Pi_2)t}$$

$$\times \sum_{i=1}^m \left\|\left(\frac{e_{i1}(t)}{\varepsilon^{n_i+1}}, \frac{e_{i2}(t)}{\varepsilon^{n_i}}, \ldots, e_{i(n_i+1)}(t)\right)\right\| \to 0. \qquad (4.4.32)$$

Hence, for any $\sigma \in (0, \min\{1/2, \lambda_{23}^i(C_1 + C_2)/(mN_6)\})$, there exists an $\varepsilon_1 \in (0, 1)$ such that $\|e(t)\| \leq \sigma$ for all $\varepsilon \in (0, \varepsilon_1)$ and $t \in [t_1, t_2]$, where $N_6 = \max_{1\leq i\leq m}\{\beta_2^i(1 + \hat{L}_i)\}$ and $\hat{L}_i$ is the Lipschitz constant of $\phi_i(\cdot)$.

Notice that for any $0 < \varepsilon < \varepsilon_1$ and $t \in [t_1, t_2]$, $\eta \in \Omega_1$, $\|e\| \le \sigma$,

$$\|\tilde{x}^i(t) - z^i(t)\| \le \|x(t) - \tilde{x}^i(t)\| + \|x^i(t) - z^i(t)\| \le (C_1^* + 1)/\min\lambda_{23}^i + 1,$$
$$|\phi_i(\tilde{x}^i(t) - z^i(t))| \le M_2,$$

$$|\hat{x}_{n_i+1}^i(t)| \le |e_{n_i+1}^i(t)| + |x_{n_i+1}^i(t)| \le |e_{n_i+1}^i(t)| + \left| f_i(x(t), \xi(t), w_i(t)) \right.$$

$$\left. + \sum_{j=1}^m (a_{ij}(x(t), \xi(t), w_i(t)) - b_{ij}(x(t)))u_i(t) \right|$$

$$\le |e_{n_i+1}^i(t)| + |f_i(x(t), \xi(t), w_i(t))| + \vartheta(M_1 + M_2 + C_2)$$

$$\le 1 + M_2 + C_1 + |f_i(x(t), \xi(t), w_i(t))| + \vartheta M_1 \le M_1. \qquad (4.4.33)$$

Thus $u_i^*(t)$ in (4.4.12) takes the form: $u_i^*(t) = \hat{x}_{n_i+1}^i(t) + \phi_i(\tilde{x}^i(t) - z^i(t)) + z_{n_i+1}^i(t)$ for all $t \in [t_1, t_2]$. With this $u_i^*(t)$, the derivative of $V_2(\eta(t))$ along system (4.4.12) with respect to t in interval $[t_1, t_2]$, satisfies

$$\left.\frac{dV_2(\eta(t))}{dt}\right|_{\text{along } (4.4.12)} = \sum_{i=1}^m (-W_2^i(\eta^i(t)) + N_6 \sigma \|\eta^i(t)\|)$$

$$\le - \min_{1 \le i \le m} \{\lambda_{23}^i\} \|\eta(t)\|^2 + m N_6 \|e(t)\| \|\eta(t)\| < 0, \qquad (4.4.34)$$

which contradicts (4.4.17). This completes the proof of the lemma. $\square$

Proof of Theorem 4.4.1 From Lemma 4.4.2, $\eta(t) \in \Omega_1$ for all $\varepsilon \in (0, \varepsilon_1)$ and $t \in (0, \infty)$, it follows that (4.4.43) holds true for all $t \in [0, \infty)$. Therefore, (4.4.34) and (4.4.28) also hold true for any $\varepsilon \in (0, \varepsilon_1)$ and $t \in [0, \infty)$.

For any $\sigma > 0$, it follows from (4.4.34) that there exists a $\sigma_1 \in (0, \sigma/2)$ such that $\overline{\lim}_{t \to \infty} \|\eta(t)\| \le \sigma/2$ provided that $\|e(t)\| \le \sigma_1$. From (4.4.28), for any $\tau > 0$ and this determined $\sigma_1 > 0$, there exists an $\varepsilon_0 \in (1, \varepsilon_1)$ such that $\|x(t) - \hat{x}(t)\| \le \sigma_1$ for any $\varepsilon \in (0, \varepsilon_0)$, $t > \tau$. This completes the proof of the theorem. $\square$

Remark 4.4.1 *From Theorem 4.4.1, we can conclude Theorem 4.2.4, where the output stabilization for a class of SISO systems with linear ESO is used. This is just to let $m = 1$ in (4.4.1) and $g_i(r) = r$ in (4.4.5).*

4.4.2   Global ADRC for MIMO Systems with Uncertainty

In this subsection, we develop a semi-global convergence for nonlinear ADRC. The advantage of this result, as mentioned in previous chapters, is that the peaking problem can be effectively alleviated by introducing the saturation function in the control. However, the saturation function depends on the bound of initial values. When this bound is not available, we need the global convergence. The price for global convergence is probably the peaking value problem, and more restricted assumptions as well.

Assumption 4.4.7 For every $1 \leq i \leq m$, all partial derivatives of $f_i(\cdot)$ are bounded over $\mathbb{R}^{n+m}$ where $n = n_1 + \cdots + n_m$.

Assumption 4.4.8 For every $1 \leq i, j \leq m$, $a_{ij}(x, \xi, w_i) = a_{ij}(w_i)$ and there exist constant nominal parameters b_{ij} such that the matrix with entry b_{ij} is invertible:

$$
\begin{pmatrix}
b_{11}^* & b_{12}^* & \cdots & b_{1m}^* \\
b_{21}^* & b_{22}^* & \cdots & b_{2m}^* \\
\vdots & \vdots & \ddots & \vdots \\
b_{m1}^* & b_{m2}^* & \cdots & b_{mm}^*
\end{pmatrix}
=
\begin{pmatrix}
b_{11} & b_{12} & \cdots & b_{1m} \\
b_{21} & b_{22} & \cdots & b_{2m} \\
\vdots & \vdots & \ddots & \vdots \\
b_{m1} & b_{m2} & \cdots & b_{mm}
\end{pmatrix}^{-1}.
$$

Moreover,

$$
\min\{\lambda_{13}^i\} - \sqrt{m} \sum_{i,k,l=1}^{m} \beta_1^i \sup_{t \in [0,\infty)} |a_{ik}(w_i(t)) - b_{ik}| b_{kl}^* \Lambda_{n_l+1}^l > 0, \ \forall \, t \in [0, \infty). \quad (4.4.35)
$$

Assumption 4.4.9 For zero dynamics, there exist positive constants K_1 and K_2 such that $\|F_0(x, \xi, w)\| \leq K_1 + K_2(\|x\| + \|w\|)$ for all $x \in \mathbb{R}^n, \xi \in \mathbb{R}, w \in \mathbb{R}$.

The observer-based feedback control is then designed as

$$
\text{ADRC(G):} \quad
\begin{cases}
u_i^*(t) = \phi_i(\tilde{x}^i(t) - z^i(t)) + z_{n_i+1}^i(t) - \hat{x}_{n_i+1}^i(t), \\
u_i(t) = \sum_{j=1}^{m} b_{ij}^* u_j^*(t).
\end{cases}
\quad (4.4.36)
$$

It is seen that in feedback control (4.4.36), $\hat{x}_{n_i+1}^i(t)$ is used to compensate for the uncertainty $x_{n_i+1}^i(t) = f_i(x(t), \xi(t), w_i(t)) + \sum_{j=1}^{m}(a_{ij}(w_i(t)) - b_{ij})u_j(t)$ and $\phi_i(\tilde{x}^i(t) - z^i(t)) + z_{n_i+1}^i(t)$ is used to guarantee the output tracking.

The closed loop of system (4.4.1) under the ESO (4.4.5) and ADRC (4.4.36) becomes

$$
\text{Closed loop:}
\begin{cases}
\dot{x}^i(t) = A_{n_i} x^i(t) + B_{n_i}\left(f_i(x(t), \xi(t), w_i(t)) + \sum_{k=1}^{m} a_{ik} u_k(t) \right), \\[2mm]
\dot{\hat{x}}^i(t) = A_{n_i+1}\hat{x}^i(t) +
\begin{pmatrix}
\varepsilon^{n_i-1} g_1^i(e_1^i(t)) \\
\vdots \\
\frac{1}{\varepsilon} g_{n_i+1}^i(e_1^i(t))
\end{pmatrix}
+
\begin{pmatrix}
B_{n_i} \\
0
\end{pmatrix}
u_i^*(t), \\[2mm]
u_i(t) = \sum_{k=1}^{m} b_{ik}^* u_k^*(t), \quad u_i^*(t) = \phi_i(\tilde{x}^i(t) - z^i(t)) + z_{n_i+1}^i(t) - \hat{x}_{n_i+1}^i(t).
\end{cases}
$$

$$(4.4.37)$$

Theorem 4.4.2 *Assume that Assumptions 4.4.1 and 4.4.3 and 4.4.7 to 4.4.9 are satisfied. Let $(x(t), \hat{x}(t))$ be the ε-dependent solution of (4.4.37). Then there exists a constant $\varepsilon_0 > 0$ such that for any $\varepsilon \in (0, \varepsilon_0)$, there exists an ε, and the initial value dependent constant $t_\varepsilon > 0$ such that for all $t > t_\varepsilon$,*

$$
|x_j^i(t) - \hat{x}_j^i(t)| \leq \Gamma_1 \varepsilon^{n_i+2-j}, \quad 1 \leq j \leq n_i + 1, \quad 1 \leq i \leq m, \quad (4.4.38)
$$

and

$$\|x_j^i(t) - z_j^i(t)\| \le \Gamma_2 \varepsilon, \quad 1 \le j \le n_i, \quad 1 \le i \le m, \tag{4.4.39}$$

where Γ_1 and Γ_2 are constants independent of ε and initial value. However, they depend on the bounds of $\|z^i(t)\|$ and $\|(w(t), \dot{w}(t))\|$.

Proof. Using the notation of $\eta^i(t)$ and $e^i(t)$ in (4.4.6), we obtain the error equation as follows:

$$\begin{cases} \dot{\eta}^i(t) = A_{n_i}\eta^i(t) + B_{n_i}[\phi_i(\eta^i(t)) + e_{n_i+1}^i(t) + (\phi_i(\tilde{x}^i(t) - z^i(t)) - \phi_i(x^i(t) - z^i(t)))], \\ \\ \varepsilon\dot{e}^i(t) = A_{n_i+1}e^i(t) + \varepsilon\overline{\Delta}_i(t)B_{n_i+1} - \begin{pmatrix} g_1^i(e_1^i(t)) \\ \vdots \\ g_{n_i+1}^i(e_1^i(t)) \end{pmatrix}. \end{cases} \tag{4.4.40}$$

Let

$$\overline{\Delta}_i(t) = \frac{d\left[f_i(x(t), \xi(t), w_i(t)) + \sum_{k=1}^m a_{ik}(w_i(t))u_k(t) - u_i^*(t) \right]}{dt}\Bigg|_{\text{along (4.4.37)}}$$

$$= \frac{d}{dt}\Bigg|_{\text{along (4.4.37)}} \Bigg[f_i(x(t), \xi(t), w_i(t))$$

$$+ \sum_{k,l=1}^m (a_{ik}(w_i(t)) - b_{ik})b_{kl}^* \left(\phi_l(\tilde{x}^l(t) - z^l(t)) + z_{n_l+1}^l(t) - \hat{x}_{n_l+1}^l(t) \right) \Bigg]. \tag{4.4.41}$$

A straightforward computation shows that

$$\overline{\Delta}_i(t) = \sum_{s=1}^m \sum_{j=1}^{n_s-1} x_{j+1}^s(t)\frac{\partial f_i(x(t), \xi(t), w_i(t))}{\partial x_j^s(t)}$$

$$+ \sum_{s,k,l=1}^m b_{ik}b_{kl}^*(\phi_l(\tilde{x}^l(t) - z^l(t)) + z_{n_l+1}^l(t) - \hat{x}_{n_l+1}^l(t))\frac{\partial f_i(x(t), \xi(t), w_i(t))}{\partial x_{n_s}^s}$$

$$+ \dot{w}_i(t)\frac{\partial f_i(x(t), \xi(t), w_i(t))}{\partial w_i} + L_{F_0(\xi)}f_i(x(t), \xi(t), w_i(t))$$

$$+ \sum_{k,l=1}^m (a_{ik} - b_{ik})b_{kl}^* \left\{ \sum_{j=1}^{n_l} (\tilde{x}_{j+1}^l(t) - z_{j+1}^l(t) - \varepsilon^{n_l-j}g_j^l(e_1^l(t)))\frac{\partial\phi_l(\tilde{x}^l(t) - z^l(t))}{\partial y_j} \right\}$$

$$+ \sum_{k,l=1}^m (a_{ik} - b_{ik})b_{kl}^* \left\{ \dot{z}_{n_l+1}^l(t) - \frac{1}{\varepsilon}g_{n_l+1}^l(e_1^l(t)) \right\}$$

$$+ \sum_{k,l=1}^m \dot{a}_{ik}(w_i(t))\dot{w}_i(t)b_{kl}^*(\phi_l(\tilde{x}^l(t) - z^l(t)) + z_{n_l+1}^l(t) - \hat{x}_{n_l+1}^l(t)). \tag{4.4.42}$$

It follows that

$$|\bar{\Delta}_i(t)| \leq \Xi_0^i + \Xi_1^i\|e(t)\| + \Xi_2^j\|\eta(t)\| + \frac{\Xi^i}{\varepsilon}\|e(t)\|,$$

$$\Xi^i = \sqrt{m}\sum_{k,l=1}^{m}\sup_{t\in[0,\infty)}|a_{ik}(w_i(t)) - b_{ik}|b_{kl}^*\Lambda_{n_l+1}^l, \tag{4.4.43}$$

where Ξ_0^i, Ξ_1^i, and Ξ_2^i are ε-independent positive constants.

Construct a Lyapunov function $V : \mathbb{R}^{2n_1+\cdots+2n_m+m} \to \mathbb{R}$ for error system (4.4.40) as

$$V(\eta^1,\ldots,\eta^m,e^1,\ldots,e^m) = \sum_{i=1}^{m}[V_1^i(e^i) + V_2^i(\eta^i)]. \tag{4.4.44}$$

The derivative of $V(\eta(t), e(t))$ along the solution of (4.4.40) is computed as

$$\left.\frac{dV(\eta(t), e(t))}{dt}\right|_{\text{along (4.4.40)}} = \sum_{i=1}^{m}\left\{\frac{1}{\varepsilon}\left[\sum_{j=1}^{n_i}(e_{j+1}^i(t) - g_j^i(e_1^i(t)))\frac{\partial V_1^i(e^i(t))}{\partial e_j^i}\right.\right.$$

$$\left. -g_{n_i+1}^i(e_1^i(t))\frac{\partial V_1^i(e^i(t))}{\partial e_{n+1}^i(t)}\right]$$

$$+\bar{\Delta}_i(t)\frac{\partial V_1^i(e^i(t))}{\partial e_{n_i+1}^i} + \sum_{j=1}^{n_i-1}\eta_{j+1}^i(t)\frac{\partial V_2^i(\eta^i(t))}{\partial \eta_j^i}$$

$$+\left\{\phi_i(\eta^i(t)) + e_{n_i+1}^i(t)\right.$$

$$\left.\left. + [\phi_i(\tilde{x}^i(t) - z^i(t)) - \phi_i(x^i(t) - z^i(t))]\right\}\frac{\partial V_2^i(\eta^i(t))}{\partial x_{n_i}^i(t)}\right\}. \tag{4.4.45}$$

It follows from Assumptions 4.4.2 and 4.4.3 that

$$\left.\frac{dV(\eta(t), e(t))}{dt}\right|_{\text{along (4.4.40)}} \leq \sum_{i=1}^{m}\left\{-\frac{1}{\varepsilon}W_1^i(e^i(t)) + \beta_1^i\|e^i(t)\|(\Xi_0^i + \Xi_1^i\|e(t)\| + \Xi_2^i\|\eta(t)\|\right.$$

$$\left. + \frac{\Xi^i}{\varepsilon}\|e(t)\|) - W_2^i(\eta^i(t)) + \beta_2^i(L_i + 1)\|e^i(t)\|\|\eta^i(t)\|\right\}. \tag{4.4.46}$$

This together with Assumptions 4.4.2 and 4.4.3 gives

$$\left.\frac{dV(\eta(t), e(t))}{dt}\right|_{\text{along (4.4.40)}} \leq \sum_{i=1}^{m}\left\{-\frac{\lambda_{13}^i}{\varepsilon}\|e^i(t)\|^2 + \beta_1^i\|e^i(t)\|(\Xi_0^i + \Xi_1^i\|e(t)\| + \Xi_2^i\|\eta(t)\|\right.$$

$$\left. + \frac{\Xi^i}{\varepsilon}\|e(t)\|) - \lambda_{23}^i\|\eta^i(t)\|^2 + \beta_2^i(L_i + 1)\|e^i(t)\|\|\eta^i(t)\|\right\}$$

$$
\leq - \left(\frac{1}{\varepsilon} \left(\min\{\lambda_{13}^i\} - \sum_{i=1}^m \beta_1^i \Xi^i \right) - \sum_{i=1}^m \beta_1^i \Xi_1^i \right) \|e(t)\|^2
$$

$$
+ \left(\sum_{i=1}^m \beta_1^i \Xi_0^i \right) \|e(t)\| - \min\{\lambda_{23}^i\} \|\eta(t)\|^2 + \sum_{i=1}^m \beta_2^i (L_i + 1) \|e(t)\| \|\eta(t)\|.
$$

(4.4.47)

For notational simplicity, we denote

$$
\Pi_1 = \min\{\lambda_{13}^i\} - \sum_{i=1}^m \beta_1^i \Xi^i, \quad \Pi_2 = \sum_{i=1}^m \beta_1^i \Xi_1^i,
$$

$$
\Pi_3 = \sum_{i=1}^m \beta_1^i \Xi_0^i, \quad \Pi_4 = \sum_{i=1}^m \beta_2^i (L_i + 1), \quad \lambda = \min\{\lambda_{23}^i\},
$$

(4.4.48)

and rewrite inequality (4.4.47) as

$$
\left. \frac{dV(\eta(t), e(t))}{dt} \right|_{\text{along (4.4.40)}} \leq - \left(\frac{\Pi_1}{\varepsilon} - \Pi_2 \right) \|e(t)\|^2 + \Pi_3 \|e(t)\| - \lambda \|\eta(t)\|^2 + \Pi_4 \|e(t)\| \|\eta(t)\|.
$$

(4.4.49)

Let $\varepsilon_1 = \Pi_1/(2\Pi_2)$. For every $\varepsilon \in (0, \varepsilon_1)$, $\Pi_2 = \Pi_1/(2\varepsilon_1) \leq \Pi_1/(2\varepsilon)$, and

$$
\Pi_4 \|e(t)\| \|\eta(t)\| = \sqrt{\frac{\Pi_1}{4\varepsilon}} \|e\| \sqrt{\frac{4\varepsilon}{\Pi_1}} \Pi_2 \|\eta(t)\| \leq \frac{\Pi_1}{4\varepsilon} \|e(t)\|^2 + \frac{4\varepsilon \Pi_2^2}{\Pi_1} \|\eta(t)\|^2.
$$
(4.4.50)

Hence (4.4.49) can be estimated further as

$$
\left. \frac{dV(\eta(t), e(t))}{dt} \right|_{\text{along (4.4.40)}} \leq - \frac{\Pi_1}{4\varepsilon} \|e(t)\|^2 + \Pi_3 \|e(t)\| - \left(\lambda - 4\varepsilon \frac{\Pi_2^2}{\Pi_1} \right) \|\eta(t)\|^2.
$$
(4.4.51)

Now we show that the solution of (4.4.40) is bounded when ε is sufficiently small. To this purpose, let

$$
R = \max \left\{ 2, \frac{2\Pi_3}{\lambda} \right\}, \quad \varepsilon_0 = \min \left\{ 1, \varepsilon_1, \frac{\Pi_1}{4\Pi_3}, \frac{\lambda \Pi_1}{8\Pi_2^2} \right\}.
$$
(4.4.52)

For any $\varepsilon \in (0, \varepsilon_0)$ and $\|(e(t), \eta(t))\| \geq R$, we consider the derivative of $V(\eta(t), e(t))$ along the solution of (4.4.40) by two different cases.

Case 1: $\|e(t)\| \geq R/2$. In this case, $\|e(t)\| \geq 1$. From a direct computation from (4.4.51), with the definition of ε_0 in (4.4.52)

$$
\left. \frac{dV(\eta(t), e(t))}{dt} \right|_{\text{along (4.4.40)}} \leq - \frac{\Pi_1}{4\varepsilon} \|e(t)\|^2 + \Pi_3 \|e(t)\|
$$

$$
\leq - \left(\frac{\Pi_1}{4\varepsilon} - \Pi_3 \right) \|e(t)\|^2 \leq - \left(\frac{\Pi_1}{4\varepsilon} - \Pi_3 \right) < 0.
$$

(4.4.53)

Case 2: $\|e(t)\| < R/2$. In this case, from $\|\eta(t)\| + \|e(t)\| \geq \|(e(t), \eta(t))\|$, $\|\eta(t)\| \geq R/2$. By (4.4.46) and the definition of ε_0 in (4.4.52), we have

$$\frac{dV(\eta(t), e(t))}{dt}\bigg|_{\text{along }(4.4.40)} \leq \Pi_3\|e(t)\| - \left(\lambda - 4\varepsilon\frac{\Pi_2^2}{\Pi_1}\right)\|\eta(t)\|^2$$

$$\leq -\left(\lambda - 4\varepsilon\frac{\Pi_2^2}{\Pi_1}\right)R^2 + \Pi_3 R \leq -\left(\frac{\lambda}{2}R - \Pi_3\right)R \leq 0.$$

$$(4.4.54)$$

Summarizing these two cases, we obtain that, for each $\varepsilon \in (0, \varepsilon_0)$, there exist an ε, and $\tau_\varepsilon > 0$ depending on the initial value such that $\|(e(t), \eta(t))\| \leq R$ for all $t \in (T_\varepsilon, \infty)$. This together with (4.4.42) shows that $|\bar{\Delta}_i(t)| \leq M_i + (\Xi_i/\varepsilon)\|e(t)\|$ for all $t \in (T_\varepsilon, \infty)$, where M_i is an R-dependent constant.

Finding the derivative of $V_1(e(t))$ along the solution of (4.4.40) with respect to t gives, for any $t > \tau_\varepsilon$, that

$$\frac{dV_1(e(t))}{dt}\bigg|_{\text{along }(4.4.40)} = \frac{1}{\varepsilon}\sum_{i=1}^{m}\left\{\sum_{j=1}^{n_i}(e_{j+1}^i(t) - g_j^i(e_1^i(t)))\frac{\partial V_1^i(e^i(t))}{\partial e_j^i}\right.$$

$$\left. + (\varepsilon\bar{\Delta}_i(t) - g_{n_i+1}^i(e_1^i(t)))\frac{\partial V_1^i(e^i(t))}{\partial e_{n_i+1}^i}\right\}$$

$$(4.4.55)$$

$$\leq -\frac{\Pi_1}{\varepsilon}\|e^i(t)\|^2 + \sum_{i=1}^{m}M_i\beta_1^i\|e^i(t)\|$$

$$\leq -\frac{\Pi_1}{\varepsilon\max\{\lambda_{12}^i\}}V_1(e(t)) + \frac{\sum_{i=1}^{m}M_i\beta_1^i}{\sqrt{\min\{\lambda_{11}^i\}}}\sqrt{V_1(e(t))}.$$

Hence

$$\frac{d}{dt}\sqrt{V_1(e(t))}\bigg|_{\text{along }(4.4.40)} \leq -\frac{\Pi_1}{2\varepsilon\max\{\lambda_{12}^i\}}\sqrt{V_1(e(t))} + \frac{\sum_{i=1}^{m}M_i\beta_1^i}{2\sqrt{\min\{\lambda_{11}^i\}}}$$

$$(4.4.56)$$

for all $t > \tau_\varepsilon$.

By comparison principle of ordinary differential equations, we obtain immediately, for all $t > \tau_\varepsilon$, that

$$\sqrt{V_1(e(t))} \leq e^{-\frac{\Pi_1}{2\varepsilon\max\{\lambda_{12}^i\}}(t-\tau_\varepsilon)} + \frac{\sum_{i=1}^{m}M_i\beta_1^i}{2\sqrt{\min\{\lambda_{11}^i\}}}\int_{\tau_\varepsilon}^{t}e^{-\frac{\Pi_1}{2\varepsilon\max\{\lambda_{12}^i\}}(t-s)}\,ds.$$

$$(4.4.57)$$

It is seen that the first term of the right-hand side of the above inequality is convergent to zero as $t \to \infty$, so we may assume that it is less than ε as $t > t_\varepsilon$ for some $t_\varepsilon > 0$. For the second term, we have

$$\left| \int_{\tau_\varepsilon}^{t} e^{-\Pi_1/(2\varepsilon \max\{\lambda_{12}^i\})(t-s)} ds \right| \le \frac{2\max\{\lambda_{12}^i\}}{\Pi_1} \varepsilon. \tag{4.4.58}$$

These together with Assumption 4.4.2 show that there exists a positive constant $\Gamma_1 > 0$ such that

$$|e_j^i(t)| \le \sqrt{\frac{V_1(e(t))}{\min\{\lambda_{11}^i\}}} \le \Gamma_1 \varepsilon, t > t_\varepsilon. \tag{4.4.59}$$

The inequality (4.4.38) then follows by taking (4.4.6) into account.

Finding the derivative of $V_2(\eta(t))$ along the solution of (4.4.40) with respect to t gives, for all $t > t_\varepsilon$, that

$$
\begin{aligned}
\left. \frac{dV_2(\eta(t))}{dt} \right|_{\text{along (4.4.40)}} &= \sum_{i=1}^{m} \left\{ \sum_{j=1}^{n_i-1} \eta_{j+1}^i(t) \frac{\partial V_2^i(\eta^i(t))}{\partial \eta_j^i} \right. \\
&\quad + \{\phi_i(\eta^i(t)) + e_{n_i+1}^i(t) + [\phi_i(\hat{x}^i(t) - z^i(t)) - \phi_i(x^i(t) - z^i(t))]\} \\
&\quad \left. \times \frac{\partial V_2^i(\eta^i(t))}{\partial x_{n_i}^i} \right\} \\
&\le \sum_{i=1}^{m} \{-W_2^i(\eta^i(t)) + \beta_2^i(L_i+1)\|e^i(t)\|\|\eta^i(t)\|\} \\
&\le \sum_{i=1}^{m} \{-W_2^i(\eta^i(t)) + \beta_2^i(L_i+1)\Gamma_1\varepsilon\|\eta^i(t)\|\}.
\end{aligned}
\tag{4.4.60}
$$

By Assumption 4.4.3, we have, for any $t > t_\varepsilon$, that

$$\left. \frac{dV_2(\eta^i(t))}{dt} \right|_{\text{along (4.4.40)}} \le -\min\{\lambda_{23}^i/\lambda_{22}^i\}V_2(\eta) + \frac{\sum_{i=1}^{m} \beta_2^i(L_i+1)\Gamma_1\varepsilon}{\sqrt{\min\{\lambda_{21}^i\}}}\sqrt{V_2(\eta(t))}. \tag{4.4.61}$$

It follows that for all $t > t_\epsilon$,

$$\frac{d}{dt}\sqrt{V_2(\eta(t))}\Big|_{\text{along (4.4.40)}} \le -\frac{\min\{\lambda_{23}^i/\lambda_{22}^i\}}{2}\sqrt{V_2(\eta(t))} + \frac{\sum_{i=1}^{m} \beta_2^i(L_i+1)\Gamma_1\epsilon}{2\sqrt{\min\{\lambda_{21}^i\}}}. \tag{4.4.62}$$

Applying the comparison principle of ordinary differential equations again, we obtain, for all $t > t_\varepsilon$, that

$$
\begin{aligned}
\sqrt{V(\eta(t))} &\le e^{-\min\{\lambda_{23}^i/\lambda_{22}^i\}/2(t-t_\varepsilon)}\sqrt{V(\eta(t_\varepsilon))} + \frac{\sum_{i=1}^{m} \beta_2^i(L_i+1)\Gamma_1\varepsilon}{2\sqrt{\min\{\lambda_{21}^i\}}} \\
&\quad \times \int_{t_\varepsilon}^{t} e^{-\min\{\lambda_{23}^i/\lambda_{22}^i\}/2(t-s)} ds.
\end{aligned}
\tag{4.4.63}
$$

By (4.4.6), we finally obtain that there exist $T_\varepsilon > t_\varepsilon$ and $\Gamma_2 > 0$ such that (4.4.38) holds true. This completes the proof of the theorem.

4.4.3 Global ADRC for MIMO Systems with External Disturbance Only

In this subsection, a special case of ADRC is considered where the functions in dynamics are known in the sense that for any $1 \leq i \leq m$, $f_i(x, \xi, w_i) = \tilde{f}_i(x) + \overline{f}(\xi, w_i)$, where $\tilde{f}_i(\cdot)$ is known but $\overline{f}(\xi, w_i)$ is unknown. In other words, the uncertainty comes from external distur-bance, zero dynamics, and parameter mismatch in control only. In this case, we try to make use of, as before, the known information in the design of ESO.

In this spirit, for each output $y_i(t) = x_1^i(t)(i = 1, 2, \ldots, m)$, the ESO is designed below as ESO(f) to estimate $x_j^i(t)(j = 1, 2, \ldots, n_i)$ and $x_{n_i+1}^i(t) = \overline{f}_i(\xi(t), w_i(t)) + \sum_{k=1}^m a_{ik}(w_i(t))u_i(t) - u_i^*(t)$:

$$\text{ESO(f):} \quad \begin{cases} \dot{\hat{x}}_1^i(t) = \hat{x}_2^i(t) + \varepsilon^{n_i-1}g_1^i(e_1^i(t)), \\ \dot{\hat{x}}_2^i(t) = \hat{x}_3^i(t) + \varepsilon^{n_i-2}g_2^i(e_1^i(t)), \\ \quad \vdots \\ \dot{\hat{x}}_{n_i}^i(t) = \hat{x}_{n_i+1}^i(t) + g_{n_i}^i(e_1^i(t)) + \tilde{f}(\tilde{x}(t)) + u_i^*(t), \\ \dot{\hat{x}}_{n_i+1}^i = \frac{1}{\varepsilon}g_{n_i+1}^i(e_1^i(t)), \quad i = 1, 2, \ldots, m, \end{cases} \quad (4.4.64)$$

and the ESO-based feedback control is designed as

$$\text{ADRC(f):} \quad \begin{cases} u_i^*(t) = -\tilde{f}(\tilde{x}(t)) + \phi_i(\tilde{x}^i(t) - z^i(t)) + z_{n_i+1}^i(t) - \hat{x}_{n_i+1}^i(t), \\ u_i(t) = \sum_{i=1}^m b_{ij}^* u_j(t), \end{cases} \quad (4.4.65)$$

where b_{ij}^* are the same as that in Assumption 4.4.8.

The closed-loop system is now composed of system (4.4.1), ESO(f) (4.4.64), and ADRC(f) (4.4.65):

$$\text{Closed loop(f):} \quad \begin{cases} \dot{x}^i(t) = A_{n_i} x^i(t) + B_{n_i}\left(f_i(x(t), \xi(t), w_i(t)) + \sum_{k=1}^m a_{ik} u_k(t)\right), \\ \dot{\hat{x}}^i(t) = A_{n_i+1}\hat{x}^i(t) + \begin{pmatrix} \varepsilon^{n_i-1}g_1^i(e_1^i(t)) \\ \vdots \\ \frac{1}{\varepsilon}g_{n_i+1}^i(e_1^i(t)) \end{pmatrix} + \begin{pmatrix} B_{n_i} \\ 0 \end{pmatrix}(\tilde{f}(\tilde{x}(t)) + u_i^*(t)) \\ u_i(t) = \sum_{k=1}^m b_{ik}^* u_k^*(t), \\ u_i^*(t) = -\tilde{f}(\tilde{x}(t)) + \phi_i(\tilde{x}^i(t) - z^i(t)) + z_{(n_i+1)}^i(t) - \hat{x}_{(n_i+1)}^i(t). \end{cases}$$
$$(4.4.66)$$

Assumption 4.4.10 All partial derivatives of $\tilde{f}_i(\cdot)$ and $\overline{f}_i(\cdot)$ are bounded by a constant $\tilde{L}_i$.

Theorem 4.4.3 *Let $x_j^i(t)(1 \leq j \leq n_i, 1 \leq i \leq m)$ and $\hat{x}_j^i(t)(1 \leq j \leq n_i + 1, 1 \leq i \leq m)$ be the solutions of the closed-loop system (4.4.66) and let $x_{n_i+1}^i(t) = \overline{f}(\xi(t), w_i(t)) + \sum_{k=1}^m a_{ik}(w_i(t))u_i(t) - u_i^*(t)$ be the extended state. Assume Assumptions 4.4.1 to 4.4.3 and 4.4.8 to 4.4.10 are satisfied. In addition, we assume that (4.4.35) in Assumption 4.4.8 is replaced by*

$$\min\{\lambda_{13}^i\} - \sum_{i=1}^m \beta_1^i \tilde{L}_i - \sqrt{m} \sum_{i,k,l=1}^m \beta_1^i \sup_{t \in [0,\infty)} |a_{ik}(w_i(t)) - b_{ik}| b_{kl}^* \Lambda_{n_l+1}^l > 0. \quad (4.4.67)$$

Then there is a constant $\varepsilon_0 > 0$ such that for any $\varepsilon \in (0, \varepsilon_0)$, there exists a $t_\varepsilon > 0$ such that for all $t > t_\varepsilon$,

$$|x_j^i(t) - \hat{x}_j^i(t)| \leq \Gamma_1 \varepsilon^{n_i+2-j}, \quad 1 \leq j \leq n_i + 1, \quad 1 \leq i \leq m, \quad (4.4.68)$$

and

$$|x_j^i(t) - z_j^i(t)| \leq \Gamma_2 \varepsilon, \quad i = 1, 2, \ldots, n, \quad (4.4.69)$$

where Γ_1, Γ_2, and ε are the initial value independent constants (again they depend on the bounds of $\|z^i(t)\|$ and $\|(w(t), \dot{w}(t))\|$).

Proof. Using the notation of $\eta^i(t)$ and $e^i(t)$ in (4.4.6), the error equation in this case is

$$\begin{cases} \dot{\eta}^i(t) = A_{n_i}\eta^i(t) + B_{n_i}[\phi_i(\tilde{x}^i(t) - z^i(t)) + e_{n_i+1}^i(t) + \tilde{f}_i(x(t)) - \tilde{f}_i(\tilde{x}(t))], \\ \varepsilon \dot{e}^i(t) = A_{n_i+1}e^i + \begin{pmatrix} B_{n_i} \\ 0 \end{pmatrix}(\tilde{f}_i(x(t)) - \tilde{f}_i(\tilde{x}(t))) + \varepsilon \hbar_i B_{n_i+1} - \begin{pmatrix} g_1^i(e_1^i(t)) \\ \vdots \\ g_{n_i+1}^i(e_1^i(t)), \end{pmatrix} \end{cases}$$

$$\quad (4.4.70)$$

where

$$\hbar_i(t) = \frac{d\left[\overline{f}(\xi(t), w_i(t)) + \sum_{k=1}^m a_{ik}(w_i(t))u_k(t) - u_i^*(t) \right]}{dt} \Bigg|_{\text{along (4.4.66)}}$$

$$= \frac{d}{dt}\Bigg|_{\text{along (4.4.66)}} \left[\overline{f}(\xi(t), w_i(t)) + \sum_{k,l=1}^m (a_{ik}(w_i(t)) - b_{ik})b_{kl}^*(\phi_l(\tilde{x}^l(t) - z^l(t)) \right.$$

$$\left. + z_{n_l+1}^l(t) - \hat{x}_{n_l+1}^l(t)) \right].$$

$$\quad (4.4.71)$$

A direct computation shows that

$$\hbar_i(t) = L_{F_0(\xi)}\overline{f}_i(\xi(t), w_i(t)) + \dot{w}_i(t)\frac{\partial \overline{f}_i(\xi(t), w_i(t))}{\partial w_i}$$

$$+ \sum_{k,l=1}^{m}(a_{ik}(w_i(t)) - b_{ik})b_{kl}^* \left\{ \sum_{j=1}^{n_l}\left(\tilde{x}_{j+1}^l(t) - z_{j+1}^l(t) - \varepsilon^{n_l-j}g_j^l(e_1^l(t))\right)\frac{\partial\phi_l(\tilde{x}^l(t) - z^l(t))}{\partial y_j} \right\}$$

$$+ \sum_{k,l=1}^{m}(a_{ik}(w_i(t)) - b_{ik})b_{kl}^* \left\{ \dot{z}_{n_l+1}^l(t) - \frac{1}{\varepsilon}g_{n_l+1}^l(e_1^l(t)) \right\}$$

$$+ \sum_{k,l=1}^{m}\dot{a}_{i,k}(w_i(t))\dot{w}_i(t)b_{kl}^*(\phi_l(\tilde{x}^l(t) - z^l(t)) + z_{n_l+1}^l(t) - \hat{x}_{n_l+1}^l(t)). \tag{4.4.72}$$

It follows that

$$|\hbar_i(t)| \le \Theta_0^i + \Theta_1^i\|e(t)\| + \Theta_2^j\|\eta(t)\| + \frac{\Theta^i}{\varepsilon}\|e(t)\|,$$

$$\Theta^i(t) = \sqrt{m}\sum_{k,l=1}^{m}\sup_{t\in[0,\infty)}|a_{ik}(w_i(t)) - b_{ik}|b_{kl}^*\Lambda_{n_l+1}^l, \tag{4.4.73}$$

for some ε-independent positive constants Θ_0^i, Θ_1^i, and Θ_2^i.

Let $V(\eta, e)$ be defined by (4.4.44). Find the derivative of $V(\eta(t), e(t))$ along the solution of (4.4.70) to give

$$\frac{dV(\eta(t), e(t))}{dt}\bigg|_{\text{along (4.4.70)}} = \sum_{i=1}^{m}\left\{\frac{1}{\varepsilon}\left[\sum_{j=1}^{n_i}(e_{j+1}^i(t) - g_j^i(e_1^i(t)))\frac{\partial V_1^i(e^i(t))}{\partial e_j^i}\right.\right.$$

$$- g_{n_i+1}^i(e_1^i(t))\frac{\partial V_1^i(e^i(t))}{\partial e_{n+1}^i}$$

$$+(\tilde{f}_i(x(t)) - \tilde{f}_i(\tilde{x}(t)))\frac{\partial V_1^i(e^i(t))}{\partial e_{n_i}^i}\Bigg] + \hbar_i\frac{\partial V_1^i(e^i(t))}{\partial e_{n_i+1}^i} + \sum_{j=1}^{n_i-1}\eta_{j+1}^i(t)\frac{\partial V_2^i(\eta^i(t))}{\partial \eta_j^i}$$

$$+ \left\{\phi_i(\eta^i(t)) + e_{n_i+1}^i(t) + [\phi_i(\hat{x}^i(t) - z^i(t)) - \phi_i(x^i(t) - z^i(t))]\right.$$

$$+(\tilde{f}_i(x(t)) - \tilde{f}_i(\tilde{x}(t)))\Bigg\}\frac{\partial V_2^i(\eta^i(t))}{\partial x_{n_i}^i}\Bigg\}. \tag{4.4.74}$$

By Assumptions 4.4.2 and 4.4.3, we obtain

$$\frac{dV(\eta(t), e(t))}{dt}\bigg|_{\text{along (4.4.70)}} \le \sum_{i=1}^{m}\left\{-\frac{1}{\varepsilon}W_1^i(e^i(t)) + \frac{\tilde{L}_i\beta_1^i}{\varepsilon}\|e(t)\|^2\right.$$

$$+ \beta_1^i \|e^i(t)\| \left(\Theta_0^i + \Theta_1^i \|e(t)\| + \Theta_2^i \|\eta(t)\| + \frac{\Theta^i}{\varepsilon} \|e(t)\| \right)$$

$$\left. - W_2^i(\eta^i(t)) + \beta_2^i(L_i + \tilde{L}_i + 1)\|e(t)\|\|\eta^i(t)\| \right\} . \quad (4.4.75)$$

This together with Assumptions 4.4.2 and 4.4.3 again gives

$$\frac{dV(\eta(t), e(t))}{dt} \bigg|_{\text{along } (4.4.70)} \leq \sum_{i=1}^{m} \left\{ -\frac{\lambda_{13}^i}{\varepsilon} \|e^i(t)\|^2 + \frac{\tilde{L}_i \beta_1^i}{\varepsilon} \|e(t)\|^2 \right.$$

$$+ \beta_1^i \|e^i(t)\| \left(\Theta_0^i + \Theta_1^i \|e(t)\| + \Theta_2^i \|\eta(t)\| + \frac{\Theta^i}{\varepsilon} \|e(t)\| \right)$$

$$\left. - \lambda_{23}^i \|\eta^i(t)\|^2 + \beta_2^i(L_i + \tilde{L}_i + 1)\|e(t)\|\|\eta^i(t)\| \right\}$$

$$\leq - \left(\frac{1}{\varepsilon} \left(\min\{\lambda_{13}^i\} - \sum_{i=1}^{m} \beta_1^i \Theta^i - \sum_{i=1}^{m} \beta_1^i \tilde{L}_i \right) - \sum_{i=1}^{m} \beta_1^i \Theta_1^i \right) \|e(t)\|^2$$

$$+ \left(\sum_{i=1}^{m} \beta_1^i \Theta_0^i \right) \|e(t)\| - \min\{\lambda_{23}^i\}\|\eta(t)\|^2 + \sum_{i=1}^{m} \beta_2^i(L_i + \tilde{L}_i + 1)\|e(t)\|\|\eta(t)\|.$$

$$(4.4.76)$$

For simplicity, we introduce the following symbols to represent the parameters in (4.4.76):

$$\$_1 = \min\{\lambda_{13}^i\} - \sum_{i=1}^{m} \beta_1^i \Theta^i - \sum_{i=1}^{m} \beta_1^i \tilde{L}_i, \quad \$_2 = \sum_{i=1}^{m} \beta_1^i \Theta_1^i,$$

$$(4.4.77)$$

$$\$_3 = \sum_{i=1}^{m} \beta_1^i \Theta_0^i, \quad \$_4 = \sum_{i=1}^{m} \beta_2^i(L_i + \tilde{L}_i + 1), \quad \lambda = \min\{\lambda_{23}^i\},$$

and rewrite inequality (4.4.76) as follows:

$$\frac{dV(\eta(t), e(t))}{dt} \bigg|_{\text{along } (4.4.70)} \leq - \left(\frac{\$_1}{\varepsilon} - \$_2 \right) \|e(t)\|^2 + \$_3\|e(t)\| - \lambda\|\eta(t)\|^2 + \$_4\|e(t)\|\|\eta(t)\|.$$

$$(4.4.78)$$

It is seen that (4.4.78) and (4.4.49) are quite similar. In fact, if set $\Pi_i = \$_i$, then (4.4.78) is just (4.4.49). Therefore, the boundedness of the solution to (4.4.70) can be obtained following the corresponding part of proof of Theorem 4.4.2 where for any $\varepsilon \in (0, \varepsilon_0)$, there is a $\tau_\varepsilon > 0$ such that $\|(e(t), \eta(t))\| \leq \bar{R}$ for all $t \in (T_\varepsilon, \infty)$. This together with (4.4.72) yields that $|\hbar_i(t)| \leq \bar{M}_i + (\Theta_i/\varepsilon)\|e(t)\|$ for all $t \in (T_\varepsilon, \infty)$, where $\bar{M}_i$ is an $\bar{R}$-dependent positive constant.

Finding the derivative of $V_1(e(t))$ along the solution of (4.4.70) with respect to t gives, for all $t > \tau_\varepsilon$, that

$$\frac{dV_1(e(t))}{dt}\bigg|_{\text{along }(4.4.70)} = \frac{1}{\varepsilon}\sum_{i=1}^{m}\left\{\sum_{j=1}^{n_i}\left(e_{j+1}^i(t) - g_j^i(e_1^i(t))\right)\frac{\partial V_1^i(e^i(t))}{\partial e_j^i}\right.$$

$$+(\tilde{f}(x(t)) - \tilde{f}(\tilde{x}(t)))\frac{\partial V_1^i(e^i(t))}{\partial e_{n_i}^i} + \left(\varepsilon\hbar_i(t) - g_{n_i+1}^i(e_1^i(t))\right)\frac{\partial V_1^i(e^i(t))}{\partial e_{n_i+1}^i}\right\}$$

$$(4.4.79)$$

$$\leq -\frac{\Pi_1}{\varepsilon}\|e(t)\|^2 + \sum_{i=1}^{m}\bar{M}_i\beta_1^i\|e^i(t)\|$$

$$\leq -\frac{\Pi_1}{\varepsilon\max\{\lambda_{12}^i\}}V_1(e(t)) + \frac{\sum_{i=1}^{m}\bar{M}_i\beta_1^i}{\sqrt{\min\{\lambda_{11}^i\}}}\sqrt{V_1(e(t))}.$$

It follows that

$$\frac{d}{dt}\sqrt{V_1(e(t))}\bigg|_{\text{along }(4.4.70)} \leq -\frac{\Pi_1}{2\varepsilon\max\{\lambda_{12}^i\}}\sqrt{V_1(e(t))} + \frac{\sum_{i=1}^{m}\tilde{M}_i\beta_1^i}{2\sqrt{\min\{\lambda_{11}^i\}}}, \quad \forall\, t > \tau_\varepsilon.$$

$$(4.4.80)$$

Finding the derivative of $V_2(\eta(t))$ along the solution of (4.4.70) with respect to t gives

$$\frac{dV_2(\eta(t))}{dt}\bigg|_{\text{along }(4.4.70)} = \sum_{i=1}^{m}\left\{\sum_{j=1}^{n_i-1}\eta_{j+1}^i(t)\frac{\partial V_2^i(\eta^i(t))}{\partial \eta_j^i} + \left\{\varphi_i(\eta^i(t)) + e_{n_i+1}^i(t)\right.\right.$$

$$+(\tilde{f}_i(x(t)) - \tilde{f}(\tilde{x}(t))) + [\varphi_i(\hat{x}^i(t) - z^i(t)) - \varphi_i(x^i(t) - z^i(t))]\bigg\}\frac{\partial V_2^i(\eta^i(t))}{\partial x_{n_i}^i}\right\}$$

$$\leq \sum_{i=1}^{m}\left\{-W_2^i(\eta^i(t)) + \beta_2^i(L_i + \tilde{L}_i + 1)\|e^i(t)\|\|\eta^i(t)\|\right\}$$

$$\leq \sum_{i=1}^{m}\left\{-W_2^i(\eta^i(t)) + \beta_2^i(L_i + \tilde{L}_i + 1)\Gamma_1\varepsilon\|\eta^i(t)\|\right\}, \forall\, t > t_\varepsilon.$$

$$(4.4.81)$$

By Assumption 4.4.3, we have

$$\frac{dV_2(\eta(t))}{dt}\bigg|_{\text{along }(4.4.70)} \leq -\min\{\lambda_{23}^i/\lambda_{22}^i\}V_2(\eta(t))$$

$$(4.4.82)$$

$$+ \frac{\sum_{i=1}^{m}\beta_2^i(L_i + \tilde{L}_i + 1)\Gamma_1\varepsilon}{\sqrt{\min\{\lambda_{21}^i\}}}\sqrt{V_2(\eta(t))}, \forall\, t > t_\varepsilon.$$

It follows that

$$\left.\frac{d}{dt}\sqrt{V_2(\eta(t))}\right|_{\text{along (4.4.70)}} \leq -\frac{\min\{\lambda_{23}^i/\lambda_{22}^i\}}{2}\sqrt{V_2(\eta(t))}$$

$$+ \frac{\sum\limits_{i=1}^{m}\beta_2^i(L_i + \tilde{L}_i + 1)\Gamma_1\varepsilon}{2\sqrt{\min\{\lambda_{21}^i\}}}, \forall\, t > t_\varepsilon. \tag{4.4.83}$$

It is seen that (4.4.80) and (4.4.83) are very similar to (4.4.56) and (4.4.62), respectively. Using similar arguments, we obtain Theorem 4.4.3. The details are omitted.

4.4.4　Numerical Simulations

Example 4.4.1 *Consider the following MIMO system:*

$$\begin{cases} \dot{x}_1^1(t) = x_2^1(t), \quad \dot{x}_2^1(t) = f_1(x(t), \zeta(t), w_1(t)) + a_{11}(t)u_1(t) + a_{12}(t)u_2(t), \\ \dot{x}_1^2(t) = x_2^2(t), \quad \dot{x}_2^2(t) = f_2(x(t), \zeta(t), w_2(t)) + a_{21}(t)u_1(t) + a_{22}(t)u_2(t), \\ \dot{\zeta}(t) = x_2^1(t) + x_1^2(t) + \sin(\zeta(t)) + \sin t, \\ y_1(t) = x_1^1(t), \quad y_2(t) = x_1^2(t), \end{cases} \tag{4.4.84}$$

where $(y_1(t), y_2(t))$ is the output and $u_1(t)$ and $u_2(t)$ are inputs, and

$$f_1(x_1^1, x_2^1, x_1^2, x_2^2, \zeta, w_1) = x_1^1 + x_1^2 + \zeta + \sin(x_2^1 + x_2^2)w_1,$$

$$f_2(x_1^1, x_2^1, x_1^2, x_2^2, \zeta, w_2) = x_2^1 + x_2^2 + \zeta + \cos(x_1^1 + x_1^2)w_2, \tag{4.4.85}$$

$$a_{11}(t) = 1 + \frac{1}{10}\sin t, \ a_{12}(t) = 1 + \frac{1}{10}\cos t, \ a_{21}(t) = 1 + \frac{1}{10}2^{-t}, \ a_{22}(t) = -1$$

are unknown functions.

Suppose that the external disturbances $w_1(t)$ and $w_2(t)$ and the reference signals $v^1(t)$ and $v^2(t)$ are as follows:

$$w_1(t) = 1 + \sin t, w_2(t) = 2^{-t}\cos t, v^1(t) = \sin t, v^2(t) = \cos t. \tag{4.4.86}$$

Let $\phi_1 = \phi_2 = \phi : \mathbb{R}^2 \to \mathbb{R}$ be defined by $\phi(r_1, r_2) = -9r_1 - 6r_2$. The objective is to design an ESO-based feedback control so that $x_1^i(t) - z_1^i(t)$ and $x_2^i(t) - z_2^i(t)$ converge to zero as time goes to infinity in the way of the following global asymptotical stable system converging to zero:

$$\begin{cases} \dot{x}_1^*(t) = x_2^*(t), \\ \dot{x}_2^*(t) = \phi_i(x_1^*(t), x_2^*(t)), \end{cases} \tag{4.4.87}$$

where $z_1^i(t)$, $z_2^i(t)$, and $z_3^i(t)$ are the states of the tracking differentiator (TD) to estimate the derivatives of $v_i(t)$. For simplicity, we use the same TD for $v_1(t)$ and $v_2(t)$ as follows:

$$\begin{cases} \dot{z}_1^i(t) = z_2^i(t), \\ \dot{z}_2^i(t) = z_3^i(t), \\ \dot{z}_3^i(t) = -\rho^3(z_1^i(t) - v_i(t)) - 3\rho^2 z_2^i(t) - 3\rho z_3^i(t), i = 1, 2. \end{cases} \tag{4.4.88}$$

Since the most of functions in (4.4.84) are unknown, the ESO design relies on very little information of the system, and the total disturbance should be estimated. Here we need the approximate values b_{ij} of a_{ij}:

$$b_{11} = b_{12} = b_{13} = 1, \quad b_{22} = -1, \tag{4.4.89}$$

and b_{ij}^* is found to be

$$\begin{pmatrix} b_{11}^* & b_{12}^* \\ b_{21}^* & b_{22}^* \end{pmatrix} = \begin{pmatrix} 1 & 1 \\ 1 & -1 \end{pmatrix}^{-1} = \begin{pmatrix} \dfrac{1}{2} & \dfrac{1}{2} \\ \dfrac{1}{2} & -\dfrac{1}{2} \end{pmatrix}. \tag{4.4.90}$$

By Theorem 4.4.2, we design a nonlinear ESO for system (4.4.39) as

$$\begin{cases} \dot{\hat{x}}_1^1(t) = \hat{x}_2^1(t) + \dfrac{6}{\varepsilon}(y_1(t) - \hat{x}_1^1(t)) - \varepsilon\Phi\left(\dfrac{y_1(t) - \hat{x}_1^1(t)}{\varepsilon^2}\right), \\ \dot{\hat{x}}_2^1(t) = \hat{x}_3^1(t) + \dfrac{11}{\varepsilon^2}(y_1(t) - \hat{x}_1^1(t)) + u_1^*(t), \\ \dot{\hat{x}}_3^1(t) = \dfrac{6}{\varepsilon^3}(y_1(t) - \hat{x}_1^1(t)), \\ \dot{\hat{x}}_1^2(t) = \hat{x}_2^2(t) + \dfrac{6}{\varepsilon}(y_2(t) - \hat{x}_1^2(t)), \\ \dot{\hat{x}}_2^2(t) = \hat{x}_3^2(t) + \dfrac{11}{\varepsilon^2}(y_2(t) - \hat{x}_1^2(t)) + u_2^*(t), \\ \dot{\hat{x}}_3^2(t) = \dfrac{6}{\varepsilon^3}(y_2(t) - \hat{x}_1^2(t)), \end{cases} \tag{4.4.91}$$
$$u_1^*(t) = \phi_1(\hat{x}_1^1(t) - z_1^1(t), \hat{x}_2^1(t) - z_2^1(t)) + z_3^1(t) - \hat{x}_3^1(t),$$
$$u_2^*(t) = \phi_2(\hat{x}_1^2(t) - z_1^2(t), \hat{x}_2^2(t) - z_2^2(t)) + z_3^2(t) - \hat{x}_3^2(t),$$

where $\Phi : \mathbb{R} \to \mathbb{R}$ is given by

$$\Phi(r) = \begin{cases} -\dfrac{1}{4}, & r \in \left(-\infty, -\dfrac{\pi}{2}\right), \\ \dfrac{1}{4}\sin r, & r \in \left(-\dfrac{\pi}{2}, \dfrac{\pi}{2}\right), \\ \dfrac{1}{4}, & r \in \left(\dfrac{\pi}{2}, -\infty\right). \end{cases} \tag{4.4.92}$$

The ADRC for this example is the ESO-based feedback given by

$$u_1(t) = \frac{1}{2}(u_1^*(t) + u_2^*(t)), \quad u_2(t) = \frac{1}{2}(u_1^*(t) - u_2^*(t)). \tag{4.4.93}$$

We take the initial values and parameters as follows:

$$x(0) = (0.5, 0.5, 1, 1), \quad \hat{x}(0) = (0, 0, 0, 0, 0, 0), \quad z(0) = (1, 1, 1, 1, 1, 1), \quad \rho = 50,$$

$$\varepsilon = 0.05, h = 0.001,$$

where h is the integration step. Using the Euler method, the numerical results for system (4.4.84) to (4.4.86) under (4.4.88), (4.4.87), (4.4.91), and (4.4.93) are plotted in Figure 4.4.1.

Figures 4.4.1(a), 4.4.1(b), 4.4.1(d), and 4.4.1(e) indicate that for every $i, j = 1, 2$, $\hat{x}_j^i(t)$ tracks $x_j^i(t)$, $z_j^i(t)$ tracks $v_i^{(j-1)}(t)$, and $x_j^i(t)$ tracks $v_i^{(j-1)}(t)$ very satisfactorily. In addition, from Figures 4.4.1(c) and 4.4.1(f), we see that $\hat{x}_3^i(t)$ tracks satisfactorily the extended state (total disturbance) $x_3^i(t) = f_i(\cdot) + a_{i1}u_1(t) + a_{i2}u_2(t) - u_i^*(t)$.

The peaking phenomenon of the system states $x_2^1(t)$ and $x_2^2(t)$ plotted in Figures 4.4.1(b) and 4.4.1(e) shown by the dotted curve and the extended states $x_3^1(t)$ and $x_3^2(t)$ in Figures 4.4.1(c) and 4.4.1(f) shown by the dot-dash-dot curve are observed.

To overcome the peaking problem, we use the saturated feedback control in Theorem 4.4.2 by replacing $u_1^*(t)$ and $u_2^*(t)$ with

$$u_1^*(t) = \mathrm{sat}_{20}(\phi_1(\hat{x}_1^1(t) - z_1^1(t), \hat{x}_2^1(t) - z_2^1(t))) + z_3^1(t) - \mathrm{sat}_{20}(\hat{x}_3^1(t)),$$
$$u_2^*(t) = \mathrm{sat}_{20}(\phi_2(\hat{x}_1^2(t) - z_1^2(t), \hat{x}_2^2(t) - z_2^2(t))) + z_3^2(t) - \mathrm{sat}_{20}(\hat{x}_3^2(t)). \tag{4.4.94}$$

For simplicity, we use the exact values of $v_i^{(j-1)}(t)$ instead of $z_j^i(t)$. Under the same parameters as that in Figure 4.4.1, the numerical results under control (4.4.94) are plotted in Figure 4.4.2. It is seen that the peaking values of the states $x_2^1(t)$ and $x_2^2(t)$ plotted in Figures 4.4.2(b) and 4.4.2(e) shown by the dotted curve and the extended states $x_3^1(t)$ and $x_3^2(t)$ plotted in Figures 4.4.2(c) and 4.4.2(f) shown by the dot-dash-dot curve are reduced significantly.

4.5 IMP Versus ADRC

The internal model principle (IMP) deals with the following general output regulation problem:

$$\begin{cases} \dot{x}(t) = Ax(t) + Bu(t) + Pw(t), \\ \dot{w}(t) = Sw(t), \\ e(t) = Cx(t) - Qw(t), \end{cases} \tag{4.5.1}$$

where $x(t)$ is the state, $w(t)$ the external signal, and B the control matrix, C the observation matrix. The control purpose is to design an error-based feedback control so that the error $e(t) \to 0$ as $t \to \infty$ and, at the same time, all internal systems are stable.

Let us start from a simple example.

Example 4.5.1 *Consider a one-dimensional system as follows:*

$$\dot{x}(t) = x(t) + a \sin \omega t + u(t), \tag{4.5.2}$$

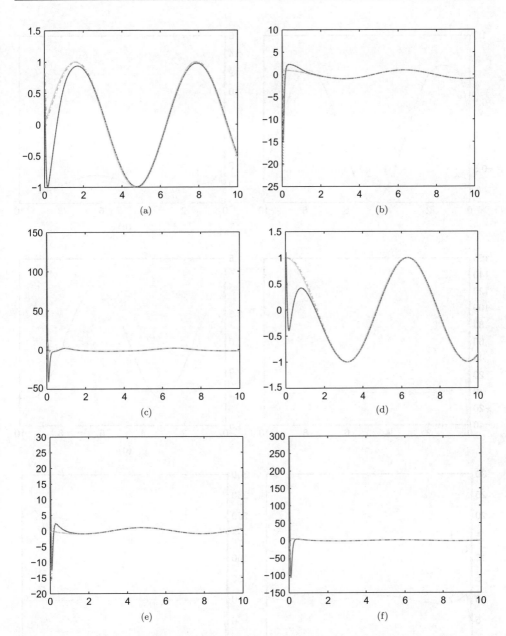

Figure 4.4.1 Numerical results of ADRC for system (4.4.39), (4.4.88), (4.4.91), and (4.4.93) with total disturbance.

where $a \sin \omega t$ is the external disturbance with a being unknown and ω known. Now we design an feedback control to stabilize the system.

Notice that Laplace transform of $a \sin \omega t$ is $\frac{a}{s^2+\omega^2}$, which has two unstable poles $s = \pm i\omega$ located on the imaginary axis. The IMP claims that in the feedback control, we need to design

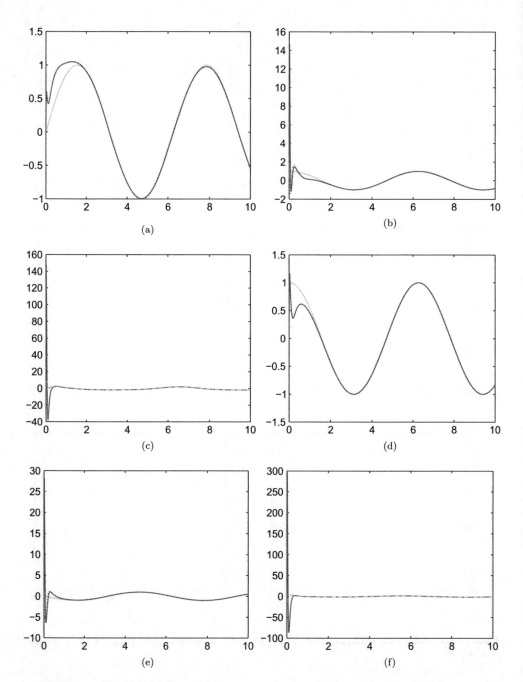

Figure 4.4.2 Numerical results of ADRC for system (4.4.39), (4.4.88), (4.4.91), and (4.4.94) with total disturbance using saturated feedback control.

a loop that has the same unstable poles as the external disturbance, that is,

$$\hat{u}(s) = \left[\frac{C_1}{s^2 + \omega^2} - C_2 \right] \hat{x}(s), \qquad (4.5.3)$$

where $\hat{\cdot}$ denotes the Laplace transform. In the feedback control (4.5.3), the first term on the right-hand side, which has the same unstable poles as the external disturbance, is used to cope with the disturbance, and the second term is the stabilizer. In the state space, (4.5.3) is

$$u(t) = C_1 \int_0^t \sin \omega(t - \tau) x(\tau) d\tau - C_2 x(t). \qquad (4.5.4)$$

Substitute (4.5.4) into (4.5.2) to obtain the closed loop:

$$\dot{x}(t) = (1 - C_2)x(t) + a \sin \omega t + C_1 \int_0^t \sin \omega(t - \tau) x(\tau) d\tau. \qquad (4.5.5)$$

Take the Laplace transform for (4.5.5) to obtain

$$\hat{x}(s) = -\frac{s^2 + \omega^2}{(s^2 + \omega^2)(s - 1 + C_2) - C_1} x(0) + \frac{a}{(s^2 + \omega^2)(s - 1 + C_2) - C_1}, \qquad (4.5.6)$$

where C_1 and C_2 are chosen so that the denominator of the right-hand side of (4.5.6) is stable, this is, all roots are located on the left complex plane. By the Routh–Hurwitz criterion, this is equivalent to

$$C_1 > 0 \text{ and } C_2 > 1 + \frac{1}{\omega^2}. \qquad (4.5.7)$$

Under condition (4.5.8), by the terminal value theorem of Laplace transform, we have

$$\lim_{t \to \infty} x(t) = \lim_{s \to 0} s\hat{x}(s) = 0. \qquad (4.5.8)$$

Proposition 4.5.1 *Suppose that (A, B) is stabilizable and the pair*

$$(C, -Q), \begin{pmatrix} A & P \\ 0 & S \end{pmatrix} \qquad (4.5.9)$$

is detectable. Then the output regulation is solvable if and only if the linear matrix equations

$$\begin{cases} \tilde{M}_1 S = A\tilde{M}_1 + P + B\tilde{M}_2, \\ 0 = C\tilde{M}_1 - Q, \end{cases} \qquad (4.5.10)$$

have solutions $\tilde{M}_1$ and $\tilde{M}_2$.

It is well known that when the output regulation problem (4.5.1) is solvable, then the observer-based feedback $u(t)$ is given by

$$u(t) = \tilde{K}(\hat{x}(t) - \tilde{M}_1 \hat{w}(t)) + \tilde{M}_2 \hat{w}(t), \qquad (4.5.11)$$

where $(\hat{x}(t), \hat{w}(t))$ is the observer for $(x(t), w(t))$:

$$\begin{pmatrix} \dot{\hat{x}}(t) \\ \dot{\hat{w}}(t) \end{pmatrix} = \begin{pmatrix} A & P \\ 0 & S \end{pmatrix} \begin{pmatrix} \hat{x}(t) \\ \hat{w}(t) \end{pmatrix} + \begin{pmatrix} \tilde{N}_1 \\ \tilde{N}_2 \end{pmatrix} (C\hat{x}(t) - Q\hat{w}(t) - e(t)) + \begin{pmatrix} B \\ 0 \end{pmatrix} u(t). \quad (4.5.12)$$

In other words, in IMP, the external disturbance $w(t)$ is also estimated. This is the same as ADRC. By (4.5.11), the first identity of (4.5.10) is what is called the "internal model".

The closed-loop system, with error states $\tilde{x}(t) = \hat{x}(t) - x(t)$ and $\tilde{w}(t) = \hat{w}(t) - w(t)$, is

$$\begin{cases} \dot{x}(t) = (A + B\tilde{K})x(t) + Pw(t) - BK\tilde{M}_1 w(t) + B\tilde{M}_2 w(t) + BK\tilde{x}(t) \\ \qquad + B(\tilde{M}_2 - K\tilde{M}_1)\tilde{w}(t), \\ \begin{pmatrix} \dot{\tilde{x}}(t) \\ \dot{\tilde{w}}(t) \end{pmatrix} = \begin{pmatrix} A + \tilde{N}_1 C & P - \tilde{N}_1 Q \\ \tilde{N}_2 C & S - \tilde{N}_2 Q \end{pmatrix} \begin{pmatrix} \tilde{x}(t) \\ \tilde{w}(t) \end{pmatrix}, \\ \dot{w}(t) = Sw(t), \end{cases} \quad (4.5.13)$$

where $\tilde{K}$ is chosen so that $A + B\tilde{K}$ is Hurwitz, $\tilde{M}_1$ and $\tilde{M}_2$ satisfy the matrix equations (4.5.10), and the matrices $\tilde{N}_1$ and $\tilde{N}_2$ are chosen so that the internal systems are stable, in other words,

$$\begin{pmatrix} A + \tilde{N}_1 C & P - \tilde{N}_1 Q \\ \tilde{N}_2 C & S - \tilde{N}_2 Q \end{pmatrix} \quad (4.5.14)$$

is Hurwitz, which is equivalent to the fact that (4.5.9) is detectable. It should be pointed out that if $w(t)$ is unbounded, the control (4.5.11) may be unbounded.

Example 4.5.2 *We apply (4.5.11) and (4.5.12) to solve stabilization of our example (4.5.2). The system is rewritten in the form of (4.5.1):*

$$\begin{cases} \dot{x}(t) = x(t) + u(t) + d(t), \\ \dot{d}(t) = wd(t). \end{cases} \quad (4.5.15)$$

By (4.5.11), the feedback control can be designed as

$$u(t) = -k(z_1(t) - \Pi z_2(t)) + \Gamma z_2(t), k > 1, \quad (4.5.16)$$

where the parameters Π and Γ are obtained from the following equations:

$$\Pi w = \Pi + 1 + \Gamma, \; 0 = \Pi. \quad (4.5.17)$$

Obviously

$$\Pi = 0, \quad \Gamma = -1. \quad (4.5.18)$$

Therefore

$$u(t) = -kz_1(t) - z_2(t), \quad (4.5.19)$$

where $z_1(t)$ and $z_2(t)$ are estimates of $x(t)$ and $d(t)$, respectively, by the observer, a kind of ESO actually:

$$\frac{d}{dt}\begin{pmatrix} z_1(t) \\ z_2(t) \end{pmatrix} = \begin{pmatrix} 1 & 1 \\ 0 & w \end{pmatrix}\begin{pmatrix} z_1(t) \\ z_2(t) \end{pmatrix} + \begin{pmatrix} l_1 \\ l_2 \end{pmatrix}(z_1(t) - x(t)) + \begin{pmatrix} 1 \\ 0 \end{pmatrix}u(t). \tag{4.5.20}$$

Let

$$\tilde{z}_1(t) = z_1(t) - x(t), \quad \tilde{z}_2(t) = z_2(t) - d(t). \tag{4.5.21}$$

We have the error equation:

$$\frac{d}{dt}\begin{pmatrix} \tilde{z}_1(t) \\ \tilde{z}_2(t) \end{pmatrix} = \begin{pmatrix} 1+l_1 & 1 \\ l_2 & w \end{pmatrix}\begin{pmatrix} \tilde{z}_1(t) \\ \tilde{z}_2(t) \end{pmatrix} \tag{4.5.22}$$

Choose l_1 and l_2 so that the following matrix

$$\begin{pmatrix} 1+l_1 & 1 \\ l_2 & w \end{pmatrix}$$

is Hurwitz. Then $z_1(t) \to x(t)$ and $z_2(t) \to d(t)$ as $t \to \infty$. In this way, the disturbance $d(t)$ is canceled by the observer state $-z_2(t)$.

From this example, we see the IMP principle and its limitation. The unstable poles w should be known.

We show that ADRC can be used to solve the output regulation for a class of MIMO systems in a very different way. Consider the following system:

$$\begin{cases} \dot{x}(t) = Ax(t) + Bu(t) + Pw(t), \\ e(t) = y(t) - Qw(t), y(t) = Cx(t), \end{cases} \tag{4.5.23}$$

where $x(t) \in \mathbb{R}^l, y(t), u(t), w(t) \in \mathbb{R}^m$, A is an $l \times l$ matrix, B is $l \times m$, and P is $l \times m$. Notice that, different to (4.5.1), here we do not need the known dynamic of the disturbance $w(t)$.

Definition 4.5.1 *We say that the output regulation problem (4.5.23) is solvable by ADRC if there is an output feedback control so that for any given $\sigma > 0$, there exists a $t_0 > 0$ such that $\|e(t)\| \leq \sigma$ for all $t \geq t_0$. Meanwhile, all internal systems including control are bounded.*

In order to apply ADRC to solve the regulation problem (4.5.23), we need TD to recover all derivatives of each $Q_i w(t)$ up to $r_i + 1$, where Q_i denotes the ith row of Q and the ESO to estimate the state and the external disturbance by the output $y(t)$. As mentioned before, TD is actually an independent link of ADRC.

For simplicity and comparison with IMP, where all loops are linear, here we also use linear TD for all $1 \leq i \leq m$, a special case of (4.4.3) as follows:

LTD: $\dot{z}^i(t) = A_{r_i+2}z^i(t) + B_{r_i+2}\rho^{r_i+2}$

$$\times \left(d_{i1}(z_1^i(t) - Q_iw(t)), d_{i2}\frac{z_2^i(t)}{\rho}, \dots, d_{i(r_i+2)}\frac{z_{i(r_i+2)(t)}^i}{\rho^{r_i+1}} \right). \tag{4.5.24}$$

The ESO we used here is also linear one, which is a special case of (4.4.5):

$$\text{LESO: } \hat{\ddot{x}}^i(t) = A_{r_i+2}\hat{x}^i(t) + \begin{pmatrix} B_{r_i+1} \\ 0 \end{pmatrix} u_i^*(t) + \begin{pmatrix} \frac{k_{i1}}{\varepsilon}(c_i x(t) - \hat{x}_1^i(t)) \\ \vdots \\ \frac{k_{i(r_i+2)}}{\varepsilon^{r_i+2}}(c_i x - \hat{x}_1^i(t)) \end{pmatrix}. \qquad (4.5.25)$$

In addition, the control $u^*(t)$ also takes the linear form of the following:

$$u_i^*(t) = -\hat{x}_{r_i+2}^i(t) + z_{r_i+2}^i(t) + \sum_{j=1}^{r_i+1} h_{ij}(\hat{x}_j^i(t) - z_j^i(t)), \qquad (4.5.26)$$

where constants d_{ij}, k_{ij}, and h_{ij} are to be specified in Proposition 4.5.2 below.

Proposition 4.5.2 *Assume that the following matrices are Hurwitz:*

$$D_i = \begin{pmatrix} 0 & I_{r_i+1} \\ d_{i1} & \cdots & d_{i(r_i+2)} \end{pmatrix}, \quad K_i = \begin{pmatrix} -k_{i1} & 1 & \cdots & 0 \\ \vdots & \vdots & \ddots & \vdots \\ -k_{i(r_i+2)} & 0 & \cdots & 0 \end{pmatrix},$$

$$H_i = \begin{pmatrix} 0 & I_{r_i} \\ h_{i1} & \cdots & h_{i(r_i+1)} \end{pmatrix}. \qquad (4.5.27)$$

The disturbance is assumed to satisfy $\|(w(t), \dot{w}(t))\| < \infty$, *and there exists a matrix* P^f *such that* $P = BP^f$. *Suppose that the triple* (A, B, C) *is decoupling with relative degree* $\{r_1, r_2, \ldots, r_m\}$ *that is equivalent to the following matrix is invertible:*

$$E = \begin{pmatrix} c_1 A^{r_1} B \\ c_2 A^{r_2} B \\ \vdots \\ c_m A^{r_m} B \end{pmatrix}, \qquad (4.5.28)$$

where c_i *denotes the ith row of C. The output regulation can be solved by ADRC under control* $u(t) = E^{-1} u^*(t)$ *if one of the following two conditions is satisfied:*

(i) $n = r_1 + r_2 + \cdots + r_m = l$ *and the following matrix* T_1 *is invertible:*

$$T_1 = \begin{pmatrix} c_1 A \\ \vdots \\ c_1 A^{r_i} \\ \vdots \\ c_m A^{r_m} \end{pmatrix}_{n \times l} \cdot \qquad (4.5.29)$$

(ii) $n < l$ and there exists an $(l - n) \times l$ matrix T_0 such that the following matrix T_2 is invertible:

$$T_2 = \begin{pmatrix} T_1 \\ T_0 \end{pmatrix}_{l \times l}, \tag{4.5.30}$$

$T_0 A T_2^{-1} = (\tilde{A}_{(l-n) \times n}, \bar{A}_{(l-n) \times (l-n)})$ *where $\tilde{A}$ is Hurwitz, and $T_0 B = 0$.*

Proof. By assumption, the triple (A, B, C) has relative degree $\{r_1, \ldots, r_m\}$, so

$$c_i A^k B = 0, \quad \forall \ 0 \le k \le r_i - 1, \quad c_i A^{r_i} B \ne 0. \tag{4.5.31}$$

Let

$$\bar{x}_j^i(t) = c_i A^{j-1} x(t), \quad j = 1, \ldots, r_i + 1 \quad, i = 1, 2, \ldots, m. \tag{4.5.32}$$

For $i = 1, 2, \ldots, m$, finding the derivative of $\bar{x}_j^i(t)$, we obtain

$$\begin{cases} \dot{\bar{x}}_j^i(t) = c_i A^j x(t) + c_i A^{j-1} B u(t) + c_i A^{j-1} B P^f w(t) = c_i A^j x(t) = \bar{x}_{j+1}^i(t), \\ \qquad\qquad\qquad\qquad j = 1, 2, \ldots, r_i, \\ \dot{\bar{x}}_{r_i+1}^i(t) = c_i A^{r_i+1} x(t) + c_i A^{r_i} B u(t) + c_i A^{r_i} P w(t). \end{cases} \tag{4.5.33}$$

(i) $r_1 + r_2 + \ldots + r_m + m = l$. In this case, under the coordinate transformation $\bar{x}(t) = T_1 x(t)$, system (4.5.23) is transformed into

$$\begin{cases} \dot{\bar{x}}_1^i(t) = \bar{x}_2^i(t), \\ \dot{\bar{x}}_2^i(t) = \bar{x}_3^i(t), \\ \qquad \vdots \\ \dot{\bar{x}}_{r_i+1}^i(t) = c_i A^{r_i+1} T_1^{-1} \bar{x}(t) + c_i A^{r_i} B u(t) + c_i A^{r_i} P w(t). \end{cases} \tag{4.5.34}$$

It is obvious that (4.5.34) has the form of (4.3.84) without zero dynamics.

(ii) $n < l$. In this case, let $\bar{x}(t) = T_1 x(t)$ and $\xi(t) = T_0 x(t)$. Then

$$\begin{cases} \dot{\bar{x}}_1^i(t) = \bar{x}_2^i(t), \\ \dot{\bar{x}}_2^i(t) = \bar{x}_3^i(t), \\ \qquad \vdots \\ \dot{\bar{x}}_{r_i+1}^i(t) = c_i A^{r_i+1} T_2^{-1} \begin{pmatrix} \bar{x}(t) \\ \xi(t) \end{pmatrix} + c_i A^{r_i} B u(t) + c_i A^{r_i} P w(t), \\ \dot{\xi}(t) = T_0 A T_2^{-1} \begin{pmatrix} \bar{x}(t) \\ \xi(t) \end{pmatrix} = \bar{A}\xi(t) + \tilde{A}\bar{x}(t), \end{cases} \tag{4.5.35}$$

which is also of the form (4.4.1).

Now consider the zero dynamics in (4.5.35): $\dot{\xi}(t) = \bar{A}\xi(t) + \tilde{A}\bar{x}(t)$. Since $\bar{A}$ is Hurwitz, there exists a solution $\hat{P}$ to the Lyapunov equation:

$$\hat{P}\bar{A} + \bar{A}^\top \hat{P} = -I_{(l-n)\times(l-n)}$$

We claim that the zero dynamics is input-to-state stable. Actually, let the Lyapunov function $V_0 : \mathbb{R}^{l-n} \to \mathbb{R}$ be $V_0(\xi) = \langle \hat{P}\xi, \xi \rangle$ and let $\chi(\bar{x}, w) = 2\lambda_{\max}(\hat{P}\tilde{A})\|\bar{x}\|^2$, where $\lambda_{\max}(\hat{P}\tilde{A})$ denotes the maximum eigenvalue of $(\hat{P}\tilde{A})(\hat{P}\tilde{A})^\top$.

Finding the derivative of $V_0(\xi(t))$ along the zero dynamics gives

$$\frac{dV_0(\xi(t))}{dt} = \xi^\top(t)\bar{A}^\top \hat{P}\xi(t) + \bar{x}^\top(t)\tilde{A}^\top \hat{P}\xi(t) + \xi^\top(t)\hat{P}\bar{A}\xi(t) + \xi^\top(t)\hat{P}\tilde{A}\bar{x}(t)$$

$$\leq -\|\xi(t)\|^2 + 2\sqrt{\lambda_{\max}(\hat{P}\tilde{A})}\|\xi(t)\|\|\bar{x}(t)\|$$

$$\leq -\frac{1}{2}\|\xi(t)\|^2 + \chi(\bar{x}(t)).$$

Therefore, the zero dynamics is input-to-state stable.

Since all dynamics functions in (4.5.34) or (4.5.35) are linear, they are C^1 and globally Lipschitz continuous. All conditions for dynamics required in Theorem 4.4.2 are satisfied for systems (4.5.34) and (4.5.35). Meanwhile, since matrixes D_i, K_i, and H_i are Hurwitz, all assumptions for LESO (4.5.25) and feedback control in Theorem 4.4.2 are satisfied for systems (4.5.34) and (4.5.35). It then follows directly from Theorem 4.4.2 that for any $\sigma > 0$, there exist $\rho_0 > 0$, $\varepsilon_0 > 0$, and ε-dependent $t_\varepsilon > 0$, such that for every $\rho > \rho_0$, $\varepsilon \in (0, \varepsilon_0)$, $\|e(t)\| < \sigma$ for systems (4.5.34) and (4.5.35). Moreover, since all TD, ESO, and ADRC are convergent, all internal systems of systems (4.5.34) and (4.5.35) are bounded. The result then follows from the equivalence among systems (4.5.34), (4.5.35), and (4.5.23) in the two different cases, respectively. □

We can compare the IMP over ADRC for the class of linear systems discussed in Proposition 4.5.1 as follows:

- The IMP requires known dynamic S of the exosystem, but ADRC does not.
- In the design of the IMP, when the orders of internal and exosystem are high, it is very difficult to choose the corresponding matrices in (4.5.14), while ADRC does not need these and is relatively easy to design.
- The IMP pursues disturbance injection while the ADRC pursues disturbance attenuation.
- In general, two approaches deal with different classes of systems.

In order to have a more direct comparison of IMP and ADRC, we use a concrete example, which can be dealt with by two approaches.

Example 4.5.3 *Consider the following MIMO system:*

$$\begin{cases} \dot{x}(t) = Ax(t) + Bu(t) + Pw(t), y(t) = Cx(t), \\ \dot{w}(t) = Sw(t), \\ e(t) = y(t) - Qw(t), \end{cases} \quad (4.5.36)$$

where

$$A = \begin{pmatrix} 0 & 1 & 0 \\ 1 & 1 & 1 \\ 1 & -1 & 1 \end{pmatrix}, \quad B = \begin{pmatrix} 0 & 0 \\ 1 & 1 \\ 1 & -1 \end{pmatrix}, \quad C = \begin{pmatrix} 1 & 0 & 0 \\ 0 & 0 & 1 \end{pmatrix},$$

$$P = \begin{pmatrix} 0 & 0 \\ 1 & 0 \\ 0 & 1 \end{pmatrix}, \quad Q = \begin{pmatrix} 0 & 0 \\ 0 & 1 \end{pmatrix}, \quad S = \begin{pmatrix} 0 & 1 \\ -1 & 0 \end{pmatrix}.$$

$$(4.5.37)$$

A direct verification shows that the following matrix $\tilde{K}$ makes $A + B\tilde{K}$ Hurwitz:

$$\tilde{K} = \begin{pmatrix} -5/2 & -4 & -2 \\ 5/2 & 4 & 2 \end{pmatrix}. \tag{4.5.38}$$

Solving the matrix equation (4.5.14), we obtain the solutions as follows:

$$\tilde{M}_1 = \begin{pmatrix} 0 & 0 \\ 0 & 0 \\ 0 & 1 \end{pmatrix}, \quad \tilde{M}_2 = \begin{pmatrix} -1 & -3/2 \\ 0 & 1/2 \end{pmatrix}. \tag{4.5.39}$$

Furthermore, we find that the matrices N_1 and N_2 below make the matrix in (4.5.14) Hurwitz.

$$\tilde{N}_1 = \begin{pmatrix} -7 & 0 \\ -22 & 0 \\ -53/5 & 0 \end{pmatrix}, \quad \tilde{N}_2 = \begin{pmatrix} -12/5 & 0 \\ 4/5 & 0 \end{pmatrix}. \tag{4.5.40}$$

Choose

$$x(0) = (0,0,0), \quad \tilde{x}(0) = (0.1, 0.5, 0.5), \quad w(0) = (0,1), \quad \tilde{w}_1(0) = (1,0), \quad h = 0.001,$$

where h is the integration step. The numerical results for system (4.5.13) with specified matrices in (4.5.37), (4.5.38), (4.5.39), and (4.5.40) are plotted in Figure 4.5.1. These show the whole process of applying IMP to system (4.5.36).

Now let us look at the design of ADRC for system (4.5.36). Firstly, the TD given below is used to recover the derivatives of $Q_1 w(t)$ and $Q_2 w(t)$, where $Q = (Q_1, Q_2)^\top$,

$$\begin{cases} \dot{z}_1^1(t) = z_2^1(t), \\ \dot{z}_2^1(t) = z_3^1(t), \\ \dot{z}_3^1(t) = -\rho^3(z_1^1(t) - Q_1 w(t)) - 3\rho^2 z_2^1(t) - 3\rho z_3^1(t), \\ \dot{z}_1^2(t) = z_2^2(t), \\ \dot{z}_2^2(t) = -2\rho^2(z_1^2(t) - Q_2 w(t)) - \rho z_2^2(t). \end{cases} \tag{4.5.41}$$

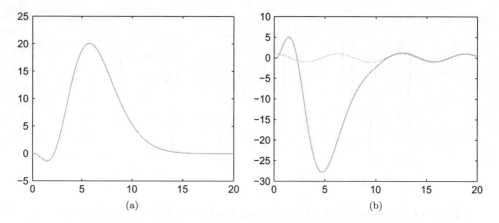

Figure 4.5.1   Numerical results of IMP(4.5.13) for Example (4.5.3).

The ESO is designed below by injection of two outputs $c_1 x(t)$ and $c_2 x(t)$, $(c_1, c_2)^\top = C$ into system (4.5.36):

$$
\begin{cases}
\dot{\hat{x}}_1^1(t) = \hat{x}_2^1(t) + \dfrac{6}{\varepsilon}(c_1 x(t) - \hat{x}_1^1(t)), \\[2mm]
\dot{\hat{x}}_2^1(t) = \hat{x}_3^1(t) + \dfrac{11}{\varepsilon^2}(c_1 x(t) - \hat{x}_1^1(t)) + u_1(t) + u_2(t), \\[2mm]
\dot{\hat{x}}_3^1(t) = \dfrac{6}{\varepsilon^3}(c_1 x(t) - \hat{x}_1^1(t)), \\[2mm]
\dot{\hat{x}}_1^2(t) = \hat{x}_2^2(t) + \dfrac{2}{\varepsilon}(c_2 x(t) - \hat{x}_1^2(t)) + u_1(t) - u_2(t), \\[2mm]
\dot{\hat{x}}_2^2(t) = \dfrac{1}{\varepsilon^2}(c_2 x(t) - \hat{x}_1^2(t)).
\end{cases}
\tag{4.5.42}
$$

The ESO-based feedback controls are designed by

$$
\begin{cases}
u_1^*(t) = -9(\hat{x}_1^1(t) - z_1^1(t)) - 6(\hat{x}_2^1(t) - z_2^1(t)) + z_3^1(t) - \hat{x}_3^1(t), \\[2mm]
u_2^*(t) = -4(\hat{x}_1^2(t) - z_1^2(t)) + z_2^2(t) - \hat{x}_2^2(t), \\[2mm]
u_1(t) = \dfrac{u_1^*(t) + u_2^*(t)}{2}, \quad u_2(t) = \dfrac{u_1^*(t) - u_2^*(t)}{2}.
\end{cases}
\tag{4.5.43}
$$

The numerical results by ADRC (4.5.41), (4.5.42), and (4.5.43) for system (4.5.36) are plotted in Figure 4.5.2 with initial values $x(0) = (0.5, 0.5, 0.5)$, $\hat{x}^1(0) = (0, 0, 0)$, $\hat{x}^2(0) = (0, 0)$, $z^1(0) = (1, 1, 1)$, $z^2(0) = (1, 1)$, $\rho = 50$, $\varepsilon = 0.005$, and the integration step $h = 0.001$.

Figures 4.5.1 and 4.5.2 witness the validity of both IMP and ADRC for the regulation problem of Example 4.5.3. It is seen from Figures 4.5.1 and 4.5.2 that Figure 4.5.2 takes advantage of fast tracking and less overshooting.

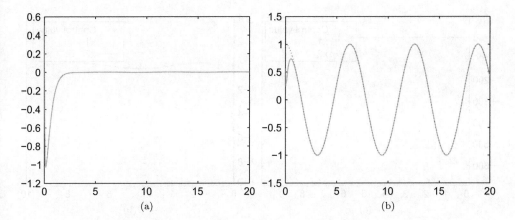

Figure 4.5.2 Numerical results of ADRC for Example (4.5.3).

4.6 HGC and SMC Versus ADRC

In this section, we firstly use a second-order system to illustrate the difference between the high-gain observer (HGO) based high-gain control (HGC) and extended state observer (ESO) based active disturbance rejection control (ADRC). Secondly, we use a first-order system to illustrate the difference of sliding mode control (SMC) with ADRC.

In classical high-gain control, because only the system state is estimated, so there also needs to be high-gain in the feedback loop to suppress the external disturbance. In ADRC, since the total disturbance is estimated by ESO, there is no need the high gain in the feedback loop any more. In the following, we use a simple example to illustrate the difference between ADRC and classical high-gain control numerically. Consider

$$\begin{cases} \dot{x}_1(t) = x_2(t), \\ \dot{x}_2(t) = u(t) + w(t), \end{cases}$$

where $x_1(t) \in \mathbb{R}$ is the output, $u(t) \in \mathbb{R}$ is the control input, and $w(t) \in \mathbb{R}$ is the unknown external disturbance. The control purpose is to stabilize (practically) the system. The high-gain feedback control can be designed as $u(t) = -R(\tilde{x}_1(t) + 2\tilde{x}_2(t))$, where R is the tuning gain parameter to be tuned large enough, and $\tilde{x}_1(t)$ and $\tilde{x}_2(t)$ are coming from the following high-gain observer:

$$\begin{cases} \dot{\tilde{x}}_1(t) = \tilde{x}_2(t) + \tilde{x}_2(t) + 2r_0(x_1(t) - \tilde{x}_1(t)), \\ \dot{\tilde{x}}_2(t) = r_0^2(x_1 - \tilde{x}_1(t)) + u(t), \end{cases}$$

where the high-gain parameter $r_0 = 200$.

Before designing the ADRC, the ESO is given in the following to estimate, in real time, the external disturbance $w(t)$ as well as the system state $(x_1(t), x_2(t))$:

$$\begin{cases} \dot{\hat{x}}_1(t) = \hat{x}_2(t) + 3r(t)(x_1(t) - \hat{x}_1(t)) + u(t), \\ \dot{\hat{x}}_2(t) = \hat{x}_3(t) + 3r^2(t)(x_1(t) - \hat{x}_1(t)), \\ \dot{\hat{x}}_3(t) = r^3(t)(x_1(t) - \hat{x}_1(t)), \end{cases}$$

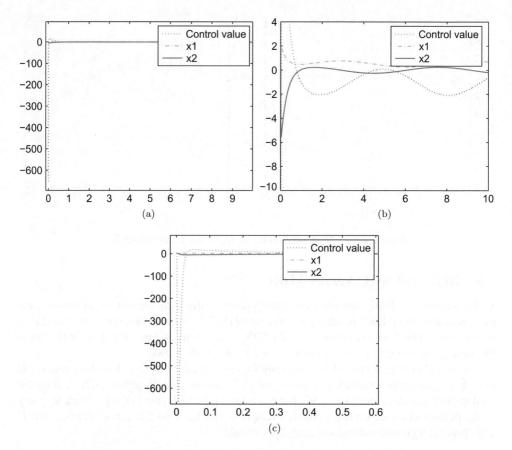

Figure 4.6.1   High-gain observer based high-gain control with $R = 2$.

where $r(t)$ is given as

$$r(t) = \begin{cases} e^{at}, & 0 \le t < \dfrac{1}{a}\ln r_0, \\ r_0, & t \ge \dfrac{1}{a}\ln r_0, \end{cases} \tag{4.6.1}$$

with $a = 2$ and $r_0 = 200$.

The ADRC can be designed as $u(t) = -\hat{x}_3(t) - \hat{x}_1(t) - 2\hat{x}_2(t)$, where $\hat{x}_1(t)$, $\hat{x}_2(t)$, and $\hat{x}_3(t)$ are states of ESO, $\hat{x}_1(t)$ and $\hat{x}_2(t)$ are the estimated values of $x_1(t)$ and $x_2(t)$, respectively, and $\hat{x}_3(t)$ is the estimated value of the external disturbance $w(t)$. Because the external disturbance is estimated and canceled by $x_3(t)$, very different from high-gain control, we do not need the high-gain parameter R in control loop of ADRC.

In numerical simulation, we set $w(t) = 1 + \sin t$. The numerical results of high-gain control with $R = 2$ and its magnifications are plotted in Figure 4.6.1. Figure 4.6.2 plots the numerical results of high-gain control with $R = 20$ and its magnifications.

We can see from Figure 4.6.1 that the high-gain control with small-gain parameter $R = 2$ cannot stabilize the system, as the system state $(x_1(t), x_2(t))$ obviously deviates from the zero

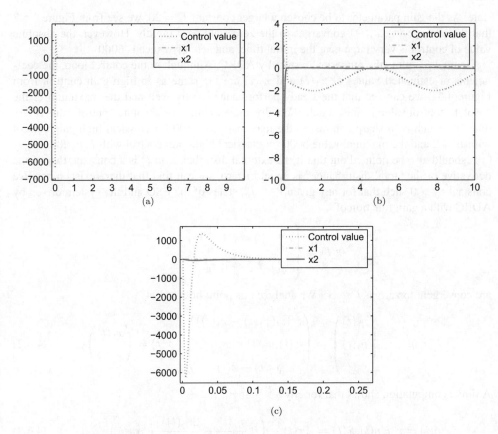

Figure 4.6.2 High-gain observer-based high-gain control with $R = 20$.

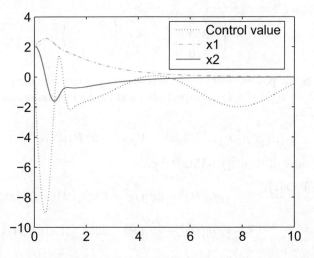

Figure 4.6.3 Extended state observer-based active disturbance rejection control.

state. As the gain parameter to be chosen a larger constant $R = 20$, we see from Figure 4.6.2 that the state $(x_1(t), x_2(t))$ converges to the zero state very rapidly. However, the absolute value of control is very large near the initial time, and actually exceeds 6000.

Figure 4.6.3 plots the numerical results by ADRC. Although in the control loop, the coefficients of estimated values of $x_1(t)$ and $x_2(t)$ are the same as in high-gain control, from Figure 4.6.3 we can see that the steady performance is very well and the maximum of the absolute control value is quite small. Actually, the maximum of absolute control value is less than 10, which is in sharp contrast to the maximal value 600 for classical high-gain control with $R = 2$ and the maximal value 6000 for classical high-gain control with $R = 20$.

It should also be pointed out that if the external disturbance $w(t)$ is a constant, that is, the derivative of the "total disturbance" is equal to zero, we can find that there exists a positive constant $r_0^* > 0$ such that for any given $r_0 > r_0^*$, states of the ESO and the system driven by ADRC with a gain function of

$$r(t) = \begin{cases} e^{at}, & 0 \le t < \dfrac{1}{a}\ln r_0, \\ r_0, & t \ge \dfrac{1}{a}\ln r_0, \end{cases}$$

are convergent to zero as $t \to \infty$. We analyze this point briefly. Let

$$\eta(t) = \begin{pmatrix} \eta_1(t) \\ \eta_2(t) \\ \eta_3(t) \end{pmatrix} = \begin{pmatrix} r^2(t)(x_1(t) - \hat{x}_1(t)) \\ r(t)(x_2(t) - \hat{x}_2(t)) \\ x_3(t) - \hat{x}(t) \end{pmatrix}, \quad x(t) = \begin{pmatrix} x_1(t) \\ x_2(t) \end{pmatrix} \tag{4.6.2}$$

A direct computation shows that for any $t > \frac{1}{a}\ln r_0$,

$$\dot{\eta}(t) = r_0 E\eta(t), \quad \dot{x}(t) = Ax(t) - \left(0, \frac{\eta_1(t)}{r^2(t)} + \frac{2\eta_2(t)}{r(t)} + \eta_3(t)\right)^\top, \tag{4.6.3}$$

where

$$E = \begin{pmatrix} -3 & 1 & 0 \\ -3 & 0 & 1 \\ -1 & 0 & 0 \end{pmatrix}, \quad A = \begin{pmatrix} 0 & 1 \\ -1 & -2 \end{pmatrix}. \tag{4.6.4}$$

Let P_E and P_A be the positive matrix solutions to the Lyapunov equation $E^\top P_E + P_E E = -I_{3\times3}$ and $A^\top P_A + P_A A = -I_{2\times2}$, respectively, and let the Lyapunov functions $V_1(\mu)$ and $V_2(\nu)$ be defined as

$$V_1(\mu) = \mu^\top P_E \mu, \ \mu \in \mathbb{R}^3, \quad V_2(\nu) = \nu^\top P_A \nu, \nu \in \mathbb{R}^2. \tag{4.6.5}$$

Finding the derivative of $V_1(\eta(t)) + V_2(x(t))$ gives

$$\frac{d(V_1(\eta(t)) + V_2(x(t)))}{dt} \le -r_0\|\eta(t)\|^2 - \|x(t)\|^2 + 8\lambda_{\max}(A)\|\eta(t)\|\|x(t)\|$$

$$\le -r_0\|\eta(t)\|^2 - \|x(t)\|^2 + 32(\lambda_{\max}(A))^2\|\eta(t)\|^2 + \frac{1}{2}\|x(t)\|^2$$

$$\le -(r_0 - 32(\lambda_{\max}(A))^2)\|\eta(t)\|^2 - \frac{1}{2}\|x(t)\|^2. \tag{4.6.6}$$

From (4.6.6), we can conclude that $\lim_{t \to \infty}[|x_1(t)| + |x_2(t)|] = 0$ for any given $r_0 > 32(\lambda_{\max}(A))^2$.

While by classical high-gain control, we can only obtain practical stability results with the error estimate:

$$\|(x_1(t), x_2(t))\| \leq \frac{M}{R}, \quad t > t_0$$

where $t_0 > 0$ and M is a w-dependent constant but is independent of R.

In the following, we use a first-order system to illustrate the difference between ADRC and sliding model control (SMC) numerically. The considered system is described by

$$\dot{x}(t) = u(t) + w(t), \tag{4.6.7}$$

where $x(t) \in \mathbb{R}$ is the output, $u(t) \in \mathbb{R}$ is the control input, and $w(t) \in \mathbb{R}$ is the unknown external disturbance. The control purpose is to stabilize the system practically. The sliding mode control (SMC) can be designed as $u(t) = 2 \operatorname{sign}(x(t))$. The ADRC is designed as $u(t) = -\hat{w}(t) - 2x(t)$, where $\hat{w}(t)$ is coming from the following ESO to cancel the disturbance $w(t)$:

$$\begin{cases} \dot{\hat{x}}(t) = \hat{w}(t) + 2r(t)(x(t) - \hat{x}(t)) + u(t), \\ \dot{\hat{w}}(t) = r^2(t)(x(t) - \hat{x}(t)), \end{cases}$$

with $r(t)$ given in (4.3.10) with $a = 2$ and $r_0 = 200$:

$$\varrho(t) = \begin{cases} e^{2t}, & 0 \leq t < \frac{1}{2}\ln r, \\ r, & t \geq \frac{1}{2}\ln r. \end{cases}$$

The numerical results of SMC are plotted in Figure 4.6.4. From Figure 4.6.4, we can see that the steady performance is very good. However, the control $u(t)$ fills almost the whole box and therefore the control energy is very large. Figure 4.6.5 is the numerical results of ADRC. From Figure 4.6.5 we can see that the steady performance is very good and the control energy consumption is smaller than SMC, because $-u(t)$ is almost equal to the external disturbance, while the control $u(t)$ of SMC almost equals $+2$ or -2 along the whole process.

4.7 Applications to PMSMs

The current control for permanent-magnet synchronous motors (PMSM) with parameter uncertainties is described by the following model:

$$\begin{cases} \dot{i}_d(t) = \frac{1}{L}u_d(t) - \frac{R}{L}i_d(t) + \omega_e i_q(t), \\ \dot{i}_q(t) = \frac{1}{L}u_q(t) - \frac{R}{L}i_q(t) - \omega_e i_d(t) - \lambda\omega_e(t), \end{cases} \tag{4.7.1}$$

where $i_d(t)$ and $i_q(t)$ are d-axis and p-axis stator currents respectively, $u_d(t)$ and $u_q(t)$ represent d-axis and p-axis stator voltages respectively, R is the armature resistance, L is the

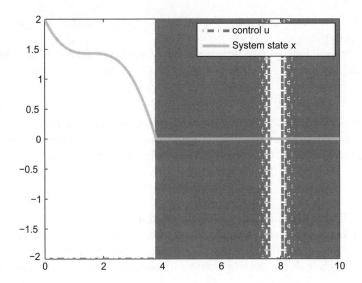

Figure 4.6.4   State of system (4.6.7) driven by sliding mode control.

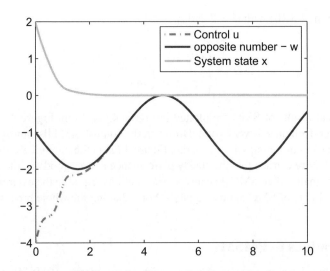

Figure 4.6.5   State of system (4.6.7) driven by ADRC.

armature inductance, λ is the flux linkage of parameter magnet, ω_e is the electrical angular velocity, and $\omega_e(t) = N_r \omega_r(t)$ where $\omega_r(t)$ is the mechanical angular speed of the motor rotor. The motion equation of the motor can be written as

$$J\dot{\omega}_r(t) = \tau_e(t) - \tau_l(t),$$

where $\tau_e(t) = Ki_q(t)$ is the generated motor torque, $\tau_l(t) = B\omega_r(t)$ is the load torque, N_r is the number of pole pairs of the motor, K is the torque constant, and B is the coefficient of viscous friction.

The control purpose is to design control voltages $u_d(t)$ and $u_q(t)$ so that $i_d(t)$ and $i_q(t)$ track the desired current references $i_d^*(t)$ and $i_q^*(t)$, respectively.

In practice, the parameters L, R, λ, J, and B are usually not known exactly, and hence $f_1(i_d, i_q, \omega_e) = -\frac{R}{L}i_d + \omega_e i_q$ and $f_2(i_d, i_q, \omega_e) = -\frac{R}{L}i_q - \omega_e i_d - \lambda\omega_e$ contain some uncertainties. We therefore consider $f_1(\cdot)$ and $f_2(\cdot)$ as the total disturbances that can be estimated by the ESO. If we use Theorem 3.4.1 in Section 3.4 for the control design, then we need to check Assumption 3.4.1 used in Theorem 3.4.1. This is true for some parameters. For instance, if

$$RN_r BJ > \lambda^2 LJ^2 + LN_r^2 K^2, \tag{4.7.2}$$

we can define the Lyapunov function

$$V(t) = \frac{1}{2}[i_d^2(t) + i_q^2(t) + \omega_e^2(t)].$$

Then finding the derivative of $V(t)$ along the solution of (4.7.1), we can obtain

$$\dot{V}(t) = \frac{1}{L}i_d(t)u_d(t) - \frac{R}{L}i_d^2(t) + \omega_e i_q(t)i_d(t) + \frac{1}{L}i_q(t)u_q(t) - \frac{R}{L}i_q^2(t)$$
$$- \omega_e i_q(t)i_d(t) - \lambda\omega_e i_q(t) + \frac{N_r K}{J}i_q(t)\omega_e(t) - \frac{N_r B}{J}\omega_e^2(t)$$
$$\leq \frac{1}{L}i_d(t)u_d(t) - \frac{R}{L}i_d^2(t) + \frac{1}{L}i_q(t)u_q(t) - \frac{R}{L}i_q^2(t) + \left(\frac{\lambda^2 J}{N_r B}i_q^2(t) + \frac{N_r B}{4J}\omega_e^2(t)\right)$$
$$+ \left(\frac{N_r K^2}{BJ}i_q^2(t) + \frac{N_r B}{4J}\omega_e^2(t)\right) - \frac{N_r B}{J}\omega_e^2(t)$$
$$- \frac{R}{2L}i_d^2(t) + \frac{1}{2RL}u_d^2(t) + \frac{1}{L}i_q(t)u_q(t) - \frac{RN_r BJ - \lambda^2 LJ^2 - LN_r^2 K^2}{LN_r BJ}i_q^2(t)$$
$$- \frac{N_r B}{2J}\omega_e^2(t) - \frac{R}{2L}i_d^2(t) + \frac{1}{2RL}u_d^2(t) - \frac{RN_r BJ - \lambda^2 LJ^2 - LN_r^2 K^2}{2LN_r BJ}i_q^2(t)$$
$$+ \frac{N_r BJ}{2L(RN_r BJ - \lambda^2 LJ^2 - LN_r^2 K^2)}u_q^2(t) - \frac{N_r B}{2J}\omega_e^2(t)$$
$$\leq -K_1 V(t) + K_2(u_d^2(t) + u_q^2(t))$$

for some $K_1, K_2 > 0$. From this, we can easily show that $V(t) \leq K_3$ for some $K_3 > 0$ and all $t \geq 0$ provided that $u_d(t)$ and $u_q(t)$ are uniformly bounded, that is, Assumption 3.4.1 is satisfied.

If we only use ESO convergence Theorem 3.4.1 for control design, the inequality (4.7.2) limits applicability of the systems. However, by Theorem 4.4.1 in Section 4.4, this restriction is not needed any more.

We design an ESO for system (4.7.1) as follows:

$$
\begin{cases}
\dot{\hat{x}}_1(t) = \hat{x}_2(t) + \dfrac{1}{\varepsilon}(i_d(t) - \hat{x}_1(t)) + u_d(t), \\[2mm]
\dot{\hat{x}}_2(t) = \dfrac{1}{\varepsilon^2}(i_d(t) - \hat{x}_1(t)), \\[2mm]
\dot{\hat{x}}_3(t) = \hat{x}_4(t) + \dfrac{1}{\varepsilon}(i_q(t) - \hat{x}_3(t)) + u_q(t), \\[2mm]
\dot{\hat{x}}_4(t) = \dfrac{1}{\varepsilon^2}(i_q(t) - \hat{x}_3(t)),
\end{cases}
\tag{4.7.3}
$$

where $\hat{x}_2(t)$ and $\hat{x}_4(t)$ are used to compensate the total disturbances $f_1(i_d(t), \omega_e(t), i_q(t)) = -\frac{R}{L}i_d(t) + \omega_e(t)i_q(t)$, and $f_2(i_d(t), \omega_e(t), i_q(t)) = -\frac{R}{L}i_q(t) - \omega_e(t)i_d(t) - \lambda\omega_e(t)$ respectively.

We design a feedback control as

$$
\begin{cases}
u_d(t) = \mathrm{sat}_{20}(-10(\hat{x}_1(t) - i_d^*(t)) + \dot{i}_d^*(t) - \hat{x}_2(t)), \\[2mm]
u_q(t) = \mathrm{sat}_{20}(-4(\hat{x}_1(t) - i_q^*(t)) + \dot{i}_q^*(t) - \hat{x}_4(t)),
\end{cases}
\tag{4.7.4}
$$

where the saturation function $\mathrm{sat}_M(r)$ is constructed as (4.1.73) to avoid possible damage of peaking value of $\hat{x}_i(t)$ in the beginning of the initial time.

In numerical simulations, let parameters in system function be taken as $L = 1.0$, $R = 2, J = 0.1, N_r = 4, \lambda = 0.001, B = 0.001, K = 3\lambda N_r/2$; the current references are $i_d^*(t) = \sin t, i_q^*(t) = 1$; the tuning parameter $\varepsilon = 0.0002$ in ESO (4.7.3); initial values $i_d(0) = 1, i_q(0) = -0.5, \omega_r(0) = 0, \hat{x}(0) = (0, 0, 0, 0)$; and the integration step $h = 0.0001$. The numerical results of state and "total disturbance" estimates are plotted in Figure 4.7.1 and the $i_d(t), i_d^*(t)$ and $i_q(t), i_q^*(t)$ are plotted in Figure 4.7.2. It is seen from Figure 4.7.1 that the convergence is very fast: the error curve and $i_q^*(t)$ curve are almost coincident. It is seen from Figure 4.7.2 that both $i_d(t)$ and $i_q(t)$ track current references $i_d^*(t)$ and $i_q^*(t)$ and are very satisfactory. More importantly, using a saturated estimation in the feedback loop can avoid effectively the peaking value problem.

4.8 Application to Wave Equation with Uncertainty

In this section we apply ADRC to boundary stabilization for a one-dimensional wave equation with internal anti-damping and boundary disturbance, which is described by

$$
\begin{cases}
u_{tt}(x, t) = u_{xx}(x, t) + 2\lambda(x)u_t(x, t) + \beta(x)u(x, t), \\[2mm]
u(0, t) = 0, \; u_x(1, t) = U(t) + d(t), \\[2mm]
u(x, 0) = u_0(x), \; u_t(x, 0) = u_1(x),
\end{cases}
\tag{4.8.1}
$$

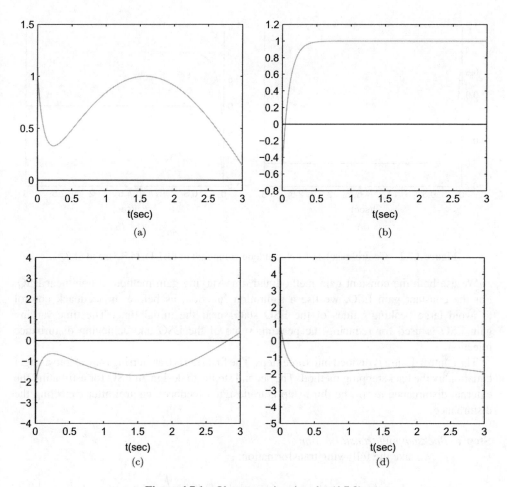

Figure 4.7.1　Observer estimations by (4.7.3).

where $\lambda(x)u_t(x, t)$ is a possible anti-damping term, that is, if $\lambda(x) > 0$ for all $x \in (0, 1)$ and $\beta(x), U(t), d(t)$ are zero, then (4.8.1) is unstable, $U(t)$ is the control input, and $d(t)$ is an external disturbance. The control purpose is to design a feedback to stabilize (4.8.1).

4.8.1　Control Design

In this subsection, firstly, we transfer system (4.8.1) into a target system. Based on this target system, we design a nonlinear ESO to estimate the disturbance and then design a feedback to compensate (cancel) the disturbance.

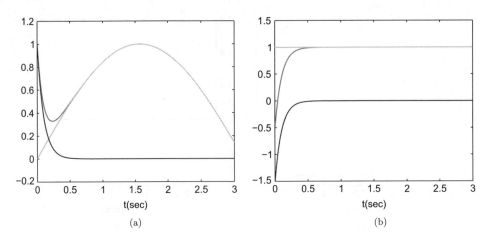

Figure 4.7.2   Control results of a closed loop composed of (4.7.1), (4.7.3), and (4.7.4).

We use both the constant gain method and time-varying gain method in nonlinear ESO. For the constant gain ESO, we use a saturation function, as before, in feedback control to avoid large peaking values of the ESO states near the initial time. The time-varying gain ESO is used for reducing the peaking value of the ESO and achieving disturbance rejection.

The control design is divided into three steps. The first step is transferring (4.8.1) into a target equation by the backstepping method. The second step is to design an ESO for estimating the external disturbance $d(t)$. The third step is to design a feedback control after canceling the disturbance.

Step 1. *Backstepping transformation*

We take the following transformation:

$$w(x,t) = h(x)u(x,t) - \int_0^x K(x,y)u(y,t)dy - \int_0^x S(x,y)u_t(y,t)dy \quad (4.8.2)$$

to transfer (4.8.1) into

$$\begin{cases} w_{tt}(x,t) = w_{xx}(x,t), \\[4pt] w(0,t) = 0, \\[4pt] w_x(1,t) = -w_t(1,t) + U_0(t) + (h(1) - S(1,1))(U(t) + d(t)), \\[4pt] w(x,0) = h(x)u_0(x) - \int_0^x K(x,y)u_0(y)dy + \int_0^x S(x,y)u_1(y)dy, \\[4pt] w_t(x,0) = h(x)u_1(x) - S(x,x)u_0'(x) + S_y(x,x)u_0(x) \\[4pt] \qquad - \int_0^x (\beta(y)S(x,y) + S_{yy}(x,y))u_0(y)dy - \int_0^x (K(x,y) \\[4pt] \qquad + 2\lambda(y)S(x,y))u_1(y)dy, \end{cases} \quad (4.8.3)$$

with

$$U_0(t) = (h'(1) - K(1,1))u(1,t) + (h(1) - S(1,1))u_t(1,t)$$

$$- \int_0^1 (K_x(1,y) + \beta(y)S(1,y) + S_{yy}(1,y)u(y,t)dy \qquad (4.8.4)$$

$$- \int_0^1 (S_x(1,y) + K(1,y) + 2\lambda(y)S(1,y)u_t(y,t)dy,$$

where $K, S \in C^2(\mathbb{I}, \mathbb{R})$ with $\mathbb{I} = \{(x,y)|0 \leq y \leq x \leq 1\}$ are kernel functions and satisfy

$$\begin{cases} K_{xx}(x,y) - K_{yy}(x,y) = 2\lambda(y)S_{yy}(x,y) + \beta(y)K(x,y) + 2(\lambda''(y) \\ \qquad\qquad + \lambda(y)\beta(y))S(x,y) + 4\lambda'(y)S_y(x,y), \\ 2K'(x,x) = -2\lambda(x)S_y(x,x) - 2\lambda'(x)S(x,x) - \beta(x)h(x) + h''(x), \\ K(x,0) = 0, \\ S_{xx}(x,y) - S_{yy}(x,y) = 2\lambda(y)K(x,y) + (4\lambda^2(y) + \beta(y))S(x,y), \\ S'(x,x) = -\lambda(x)h(x), \quad \lambda(x)S(x,x) = -h'(x) \\ S(x,0) = 0. \end{cases}$$

$$(4.8.5)$$

We set $h(0) = 1$ so that when all the coefficients of the original and target systems are the same, we have $w(x,t) = u(x,t)$. By the last three equations in (4.8.5), we obtain

$$h(x) = \cosh\left(\int_0^x \lambda(\tau)d\tau\right), \quad S(x,x) = -\sinh\left(\int_0^x \lambda(\tau)d\tau\right). \qquad (4.8.6)$$

To obtain $K(x,x)$, let

$$f(x) = S_y(x,x), \quad g(x) = K(x,x). \qquad (4.8.7)$$

A straightforward computation shows that

$$\begin{cases} S_x(x,x) = S'(x,x) - f(x), \\ S_{xx}(x,x) - S_{yy}(x,x) = (S_x(x,x) - S_y(x,x))' = (S'(x,x) - 2f(x))'. \end{cases}$$

$$(4.8.8)$$

This together with the second and fourth equations in (4.8.5) gives

$$\begin{cases} 2f'(x) + 2\lambda(x)g(x) = S''(x,x) - (4\lambda^2(x) + \beta(x))S(x,x), \\ 2g'(x) + 2\lambda(x)f(x) = -2\lambda'(x)S(x,x) - \beta(x)h(x) + h''(x), \qquad (4.8.9) \\ f(0) = -\lambda(0), \quad g(0) = 0. \end{cases}$$

Let $G^K(x,y) = K((x+y/2),(x-y/2))$ and $G^S(x,y) = S((x+y)/2,(x-y)/2)$. It follows from (4.8.5) that

$$G^K_{xy}(x,y) = \frac{1}{4}(2\lambda((x-y)/2)(G^S_{xx}(x,y) - 2G^S_{xy}(x,y) + G^S_{yy}(x,y))$$
$$+ \beta((x-y)/2)G^K(x,y) + 4\lambda'((x-y)/2)(G^S_x(x,y) - G^S_y(x,y))$$
$$+ (\lambda''((x-y)/2) + 2\lambda((x-y)/2)\beta((x-y)/2))G^S(x,y)),$$

$$G^K(x,0) = g\left(\frac{x}{2}\right), \quad G^K(x,x) = 0, \tag{4.8.10}$$

$$G^S_{xy}(x,y) = \frac{1}{4}(2\lambda((x-y)/2)(x,y)G^K(x,y)$$
$$+ (4\lambda^2((x-y)/2) + \beta((x-y)/2))G^S(x,y)),$$

$$G^S(x,0) = -\sinh\left(\int_0^{x/2}\lambda(\tau)d\tau\right), \quad G^S(x,x) = 0,$$

where $g(x)$ is the solution of (4.8.9).

Equation (4.8.10) is equivalent to the following integral equation:

$$G^K(x,y) = g\left(\frac{x}{2}\right) - g\left(\frac{y}{2}\right) + \frac{1}{4}\int_y^x\int_0^y \beta((\tau-s)/2)G^K(\tau,s)dsd\tau$$

$$+ \frac{1}{4}\int_y^x\int_0^y \lambda((\tau-s)/2)(G^S_{xx}(\tau,s) - 2G^S_{xy}(\tau,s) + G^S_{yy}(\tau,s))dsd\tau$$

$$+ \frac{1}{4}\int_y^x\int_0^y (4\lambda'((\tau-s)/2)(G^S_x(\tau,s) - G^S_y(\tau,s)))dsd\tau$$

$$+ \frac{1}{4}\int_y^x\int_0^y ((\lambda''((\tau-s)/2) + 2\lambda((\tau-s)/2)\beta((\tau-s)/2))G^S(\tau,s))dsd\tau, \tag{4.8.11}$$

$$G^S(x,y) = -\sinh\left(\int_0^{x/2}\lambda(\tau)d\tau\right) + \sinh\left(\int_0^{y/2}\lambda(\tau)d\tau\right)$$

$$+ \frac{1}{4}\int_x^y\int_0^y (2\lambda((\tau-s)/2)G^K(\tau,s)$$

$$+ (4\lambda^2((\tau-s)/2) + \beta((\tau-s)/2))G^S(\tau,s))dsd\tau.$$

The solution of (4.8.11) can be expressed as the following series:

$$G^S(x,y) = \sum_{n=0}^\infty G^{S\,n}(x,y), \quad G^K(x,y) = \sum_{n=0}^\infty G^{K\,n}(x,y), \tag{4.8.12}$$

where $G^{S\ n}(x,y)$ and $G^{K\ n}(x,y)$ can be obtained iteratively as follows:

$$
\begin{cases}
G^{K\ 0}(x,y) & = g(x/2) - g(y/2), \quad G^{S\ 0}(x,y) = -\sinh\left(\int_0^{x/2} \lambda(\tau)d\tau\right) \\[2mm]
& \quad + \sinh\left(\int_0^{y/2} \lambda(\tau)d\tau\right), \\[2mm]
G^{K\ n+1}(x,y) & = \dfrac{1}{4}\int_y^x \int_0^y \beta((\tau-s)/2)G^{K\ n}(\tau,s)ds\,d\tau \\[2mm]
& \quad + \dfrac{1}{4}\int_y^x \int_0^y \lambda((\tau-s)/2)(G_{xx}^{S\ n}(\tau,s) - 2G_{xy}^{S\ n}(\tau,s) + G_{yy}^{S\ n}(\tau,s))ds\,d\tau \\[2mm]
& \quad + \dfrac{1}{4}\int_y^x \int_0^y (4\lambda'((\tau-s)/2)(G_x^S(\tau,s) - G_y^{S\ n}(\tau,s)))ds\,d\tau \\[2mm]
& \quad + \dfrac{1}{4}\int_y^x \int_0^y ((\lambda''((\tau-s)/2) + 2\lambda((\tau-s)/2)\beta((\tau-s)/2))G^{S\ n}(\tau,s))ds\,d\tau, \\[2mm]
G^{S\ n+1}(x,y) & = \dfrac{1}{4}\int_y^x \int_0^y (\lambda\left(\dfrac{\tau-s}{2}\right)G^{K\ n}(\tau,s) \\[2mm]
& \quad + (4\lambda^2((\tau-s)/2) + \beta((\tau-s)/2))G^{S\ n}(\tau,s))ds\,d\tau.
\end{cases}
$$

$$(4.8.13)$$

Lemma 4.8.1 *The series in (4.8.12) are uniformly convergent in $C^2([0,1]\times[0,1],\mathbb{R})$ and hence there exists kernel functions $K, S \in C^2(\mathbb{I},\mathbb{R})$.*

Proof. Let

$$
b_1(x,y) = 2\lambda\left(\frac{x-y}{2}\right), \quad b_2(x,y) = \beta\left(\frac{x-y}{2}\right), \quad b_4(x,y) = 4\lambda'\left(\frac{x-y}{2}\right),
$$

$$
b_3(x,y) = 2\left(\lambda''\left(\frac{x-y}{2}\right) + \lambda\left(\frac{x-y}{2}\right)\beta\left(\frac{x-y}{2}\right)\right),
$$

$$(4.8.14)$$

$$
b_5(x,y) = 4\lambda^2\left(\frac{x-y}{2}\right) + \beta\left(\frac{x-y}{2}\right).
$$

It follows from (4.8.13) that

$$
G_x^{S\ n+1}(x,y) = \frac{1}{4}\int_0^y (b_1(x,s)G^{K,n}(x,s) + b_5(x,s)G^{S\ n}(x,s))ds,
$$

$$
G_y^{S\ n+1}(x,y) = -\frac{1}{4}\int_0^y (b_1(y,s)G^{K\ n}(y,s) + b_5(y,s)G^{S\ n}(y,s))ds
$$

$$
\quad + \frac{1}{4}\int_y^x (b_1(\tau,y)G^{K\ n}(\tau,y) + b_5(\tau,y)G^{S\ n}(\tau,y))d\tau,
$$

$$
G_{xy}^{S,n+1}(x,y) = \frac{1}{4}(b_1(x,y)G^{K\ n}(x,y) + b_5(x,y)G^{S\ n}(x,y)),
$$

$$G_{xx}^{S,n+1}(x,y) = \frac{1}{4}\int_0^y (b_{1x}(x,s)G^{K,n}(s,x) + b_{5x}(x,s)G^{S\,n}(x,s))ds$$

$$+ \frac{1}{4}\int_0^y (b_1(x,s)G_x^{K,n}(s,x) + b_5(x,s)G_x^{S\,n}(x,s))ds, \qquad (4.8.15)$$

$$G_{yy}^{S,n+1}(x,y) = -\frac{1}{2}(b_1(y,y)G^{K\,n}(y,y) + b_5(y,y)G^{S\,n}(y,y))$$

$$- \frac{1}{4}\int_0^y (b_{1y}(y,s)G^{K\,n}(y,s) + b_{5y}(y,s)G^{S\,n}(y,s))ds$$

$$+ \frac{1}{4}\int_y^x (b_{1y}(\tau,y)G^{K\,n}(\tau,y) + b_{5y}(\tau,y)G^{S\,n}(\tau,y))d\tau$$

$$- \frac{1}{4}\int_0^y (b_1(y,s)G_y^{K\,n}(y,s) + b_5(y,s)G_y^{S\,n}(y,s))ds$$

$$+ \frac{1}{4}\int_y^x (b_1(\tau,y)G_y^{K\,n}(\tau,y) + b_5(\tau,y)G_y^{S\,n}(\tau,y))d\tau,$$

$$G_x^{K\,n+1}(x,y) = \frac{1}{4}\int_0^y b_2(x,s)G^{K\,n}(x,s)ds$$

$$+ \frac{1}{4}\int_0^y b_1(x,s)(G_{xx}^{S\,n}(x,s) - 2G_{xy}^{S\,n}(x,s) + G_{yy}^{S\,n}(x,s)ds$$

$$+ \int_0^y (b_3(x,s)G^{S\,n}(x,s) + b_4(x,s)(G_x^{S\,n}(x,s) - G_y^{S\,n}(x,s)ds,$$

$$G_y^{K\,n+1}(x,y) = -\frac{1}{4}\int_0^y b_2(y,s)G^{K\,n}(y,s)d\sigma$$

$$- \frac{1}{4}\int_0^y b_1(y,s)(G_{xx}^{S\,n}(y,s) - 2G_{xy}^{S\,n}(x,s) - G_{yy}^{S\,n}(y,s)ds \qquad (4.8.16)$$

$$- \int_0^y (b_3(y,s)G^{S\,n}(y,s) + b_4(y,s)(G_x^{S\,n}(y,s) - G_y^{S\,n}(y,s)ds$$

$$+ \frac{1}{4}\int_y^x b_2(\tau,y)G^{K\,n}(\tau,y)d\tau$$

$$+ \frac{1}{4}\int_y^x b_1(\tau,y)(G_{xx}^{S\,n}(\tau,y) - 2G_{xy}^{S\,n}(\tau,y) + G_{yy}^{S\,n}(\tau,y)d\tau$$

$$+ \frac{1}{4}\int_y^x (b_3(\tau,y)G^{S\,n}(\tau,y) + b_4(\tau,y)(G_x^{S\,n}(\tau,y) - G_y^{S\,n}(\tau,y)d\tau.$$

Let

$$M_1 = \max\left\{2\left\|g\left(\frac{x}{2}\right)\right\|_{L^\infty}, \quad 2\left\|\lambda(x)\cosh\left(\int_0^{x/2}\lambda(\tau)d\tau\right)\right\|_{L^\infty}, \right.$$

$$\left. \left\|\lambda^2(x)\sinh\left(\int_0^{x/2}\lambda(\tau)d\tau\right)\right\|_{L^\infty}\right\} \qquad (4.8.17)$$

and

$$M_2 = 2(\|b_1\|_{C^1} + \|b_5\|_{C^1} + 4\|b_1\|_{L^\infty} + \|b_2\|_{L^\infty} + \|b_3\|_{L^\infty}), \tag{4.8.18}$$

where $b_j, j = 1, 2, 3, 4, 5$, are given by (4.8.14). By induction, we can prove that

$$\max\left\{|G^{K\ n}(x,y)|, |G^{S\ n}(x,y)|,\ |G_x^{S\ n}(x,y)|,\ |G_y^{S\ n}(x,y)|,\right.$$

$$\left.|G_x^{S\ n}(x,y)|,\ |G_y^{S\ n}(x,y)|\right\} \le M_1 M_2^n \frac{(x+y)^n}{n!}, \tag{4.8.19}$$

$$\max\{|G_{xx}^{S\ n}(x,y)|,\ |G_{xy}^{S\ n}(x,y)|,\ |G_{yy}^{S\ n}(x,y)|\} \le M_1 M_2^n \frac{(x+y)^{n-1}}{(n-1)!}.$$

Hence the series

$$\sum_{n=0}^\infty G^{K\ n}(x,y),\ \sum_{n=0}^\infty G^{S\ n}(x,y),\ \sum_{n=0}^\infty G_x^{S\ n}(x,y),$$

$$\sum_{n=0}^\infty G_y^{S\ n}(x,y),\ \sum_{n=0}^\infty G_{xx}^{S\ n}(x,y),\ \sum_{n=0}^\infty G_{xy}^{S\ n}(x,y),\ \sum_{n=0}^\infty G_{yy}^{S\ n}(x,y)$$

are uniformly convergent and the series

$$G^K(x,y) = \sum_{n=0}^\infty G^{K\ n}(x,y),\quad G^S(x,y) = \sum_{n=0}^\infty G^{S\ n}(x,y),$$

$$G_x^S = \sum_{n=0}^\infty G_x^{S\ n}(x,y),\quad G_y^S = \sum_{n=0}^\infty G_y^{S\ n}(x,y),$$

$$G_{xx}^S = \sum_{n=0}^\infty G_{xx}^{S\ n}(x,y),\quad G_{xy}^S = \sum_{n=0}^\infty G_{xy}^{S\ n}(x,y), G_{yy}^S = \sum_{n=0}^\infty G_{yy}^{S\ n}(x,y)$$

satisfy (4.8.11). It follows from (4.8.12) and (4.8.13) that there exist $K, S \in C^2(\mathbb{I})$ satisfying (4.8.5). This ends the proof of the lemma. $\qquad\square$

The feedback control is designed mainly based on the auxiliary system (4.8.3). The control contains two parts

$$U(t) = U_1(t) + U_2(t), U_1(t) = -\frac{1}{h(1) - S(1,1)} U_0(t). \tag{4.8.20}$$

The first part $U_1(t)$ is to compensate $U_0(t)$ in (4.8.3) and $U_2(t)$ is designed in the following ESO (4.8.23) to compensate the disturbance $d(t)$.

Under the feedback control $U(t) = U_1(t) + U_2(t)$, Equation (4.8.3) now reads as

$$
\begin{cases}
w_{tt}(x,t) = w_{xx}(x,t), \\
w(0,t) = 0, \\
w_x(1,t) = -w_t(1,t) + (h(1) - S(1,1))(U_2(t) + d(t)), \\
w(x,0) = h(x)u_0(x) - \displaystyle\int_0^x K(x,y)u_0(y)dy + \int_0^x S(x,y)u_1(y)dy, \\
w_t(x,0) = h(x)u_1(x) - S(x,x)u_0'(x) + S_y(x,x)u_0(x) \\
\qquad - \displaystyle\int_0^x (\beta(y)S(x,y) + S_{yy}(x,y))u_0(y)dy - \int_0^x (K(x,y) + 2\lambda(y)S(x,y))u_1(y)dy.
\end{cases}
\tag{4.8.21}
$$

Step 2. *Disturbance estimation*

By the boundary conditions in (4.8.21),

$$
\frac{dw(1,t)}{dt} = -w_x(1,t) + (h(1) - S(1,1))(U_2(t) + d(t)).
\tag{4.8.22}
$$

It is seen that (4.8.22) is an ordinary differential equation, so we can design a special ESO to estimate $d(t)$ by $w(1,t)$ as follows:

$$
\begin{cases}
\dot{\hat{z}}(t) = (h(1) - S(1,1))(\hat{d}(t) + U_2(t)) + \alpha_1\rho^\theta(t)[w(1,t) - \hat{z}(t)]^\theta - w_x(1,t), \\
\dot{\hat{d}}(t) = \dfrac{\alpha_2}{h(1) - S(1,1)}\rho^{2\theta}(t)[(w(1,t) - \hat{z}(t))]^{2\theta-1},
\end{cases}
\tag{4.8.23}
$$

where $[\tau]^\theta = \mathrm{sign}(\tau)|\tau|^\theta$ for $\tau \in \mathbb{R}$, α_1 and α_2 are positive constants, $\theta \in (\theta^*, 1]$, θ^* is a positive constant to be chosen sufficiently close to one. There are three types of gain functions $\rho(t)$ that can be used in ESO (4.8.23). The simplest one is constant high gain. To avoid the peaking phenomena, we can make $U_2(t)$ be saturated.

Case 1: $\rho(t) \equiv r, r > 1$ *is a constant.* In this case, $U_2(t)$ is designed as

$$
U_2(t) = -\mathrm{sat}_M(\hat{d}(t)) = -
\begin{cases}
M, & \hat{d}(t) > M, \\
\hat{d}(t), & |\hat{d}(t)| \leq M, \\
-M, & \hat{d}(t) < -M,
\end{cases}
\tag{4.8.24}
$$

where $M \geq \sup_{t \in [0,\infty)} |d(t)| + 1$ is a priori bound on the external disturbance $d(t)$.

The second type of gain $\rho(t)$ is chosen to be a time-varying gain to reduce peaking value without using saturation function.

Case 2: $\rho(0) = 1, \dot{\rho}(t) = a\rho(t)$, if $\rho(t) < r$ *and* $\rho(t) = 0$ if $\rho(t) \geq r$. In this case, $U_2(t) = -\hat{d}(t)$. To pursue the disturbance rejection, the gain function $\rho(t)$ can be chosen in Case 3.

Case 3: $\rho(0) = 1, \dot{\rho}(t) = a\rho(t), a > 0$. In this case, the second part of control can also be designed as in Case 2: $U_2(t) = -\hat{d}(t)$.

Before giving the convergence, we need the following Sobolev spaces:

$$
\mathcal{H} = H_L^1(0,1) \times L^2(0,1), \quad H_L^k(0,1) = \{f \in H^k(0,1) | f(0) = 0\},
\tag{4.8.25}
$$

with the following inner product:

$$\langle (\phi_1, \psi_1)^\top, (\phi_2, \psi_2)^\top \rangle = \int_0^1 (\phi_1'(x)\overline{\phi_2'(x)} + \psi_1(x)\overline{\psi_2(x)})dx,$$

$$\forall\, (\phi_i, \psi_i) \in \mathcal{H}, i = 1, 2. \tag{4.8.26}$$

Theorem 4.8.1 *Let* $(u_0, u_1)^\top \in \mathcal{H}$, $\lambda(\cdot) \in C^2([0,1], \mathbb{R})$, $\beta \in C([0,1], \mathbb{R})$, *and let the external disturbance* $d \in C^1([0,\infty), \mathbb{R})$ *satisfy* $N = \sup_{t \in [0,\infty)} |\dot{d}(t)| < \infty$. *Then the wave equation (4.8.21) admits a unique solution* $(u, u_t) \in \mathcal{H}$. *Moreover,*

(i) *If the gain function* $\rho(t)$ *in ESO (4.8.23) and* $U_2(t)$ *in the feedback loop are chosen as those in Cases 1, and 2, then there exists* $\theta^* \in (0,1)$ *such that for any* $\theta \in (\theta^*, 1]$ *and for any* $\sigma > 0$, *there exists* $r_\sigma > 0$ *such that*

$$|d(t) - \hat{d}(t)| \le \sigma, \quad E(t) = \int_0^1 \left[u_t^2(x,t) + u_x^2(x,t) \right] dx \le \sigma,\; t > t_r, r > r_\sigma, \tag{4.8.27}$$

where t_r *is an* r-*dependent positive constant.*

(ii) *If the gain function* $\rho(t)$ *in ESO (4.8.23) and* $U_2(t)$ *in the feedback loop are chosen as in Case 3, then there exists* $\theta^* \in (0,1)$ *such that for any* $\theta \in (\theta^*, 1]$,

$$\lim_{t \to \infty} |d(t) - \hat{d}(t)| = 0, \quad \lim_{t \to \infty} E(t) = 0. \tag{4.8.28}$$

4.8.2   Proof of Theorem 4.8.1

To prove Theorem 4.8.1, we first introduce some operators. The operator $\mathcal{A}$ and $\mathcal{B}$ are defined as follows:

$$\mathcal{A}(\phi, \psi)^\top = (\psi, \phi'')^\top, \quad \mathcal{B} = (0, \delta(x-1))^\top, \tag{4.8.29}$$

with

$$D(\mathcal{A}) = \{(\phi, \psi)^\top \in \mathcal{H} |\; \phi \in H_L^2(0,1), \quad \psi \in H_L^1(0,1), \quad \phi'(1) = -\psi(1)\}, \tag{4.8.30}$$

where $\delta(\cdot)$ is the Dirac function. A simple computation shows that

$$\mathcal{A}^*(\phi, \psi)^\top = (-\psi, -\phi'')^\top, \quad D(\mathcal{A}^*) = D(\mathcal{A}). \tag{4.8.31}$$

Lemma 4.8.2 *The wave equation (4.8.21) can be rewritten as*

$$\frac{d}{dt} \begin{pmatrix} w(\cdot,t) \\ w_t(\cdot,t) \end{pmatrix} = \mathcal{A} \begin{pmatrix} w(\cdot,t) \\ w_t(\cdot,t) \end{pmatrix} + (h(1) - S(1,1))(U_2(t) + d(t))\mathcal{B}. \tag{4.8.32}$$

The operator $\mathcal{A}$ *generates an exponential stable* C_0-*semigroup* $e^{\mathcal{A}t}$ *on* $\mathcal{H}$ *that is identical to zero after* $t \ge 2$, *and* $\mathcal{B}$ *is admissible to* $e^{\mathcal{A}t}$. *For any* $(u_0, u_1)^\top \in \mathcal{H}$, $U_1, d \in L_{loc}^2(0, \infty)$, *(4.8.3) admits a unique solution, which can be written as*

$$\begin{pmatrix} w(\cdot,t) \\ w_t(\cdot,t) \end{pmatrix} = e^{\mathcal{A}t} \begin{pmatrix} w(\cdot,0) \\ w_t(\cdot,0) \end{pmatrix} + \int_0^t e^{\mathcal{A}(t-s)} \mathcal{B}(h(1) - S(1,1))(d(s) + U_2(s))ds \in C(0, \infty; \mathcal{H}).$$

$$\tag{4.8.33}$$

In addition, there exist positive constants ω and $\overline{M}$ such that $\|e^{At}\| \leq \overline{M}e^{-\omega t}$.

Proof. The fact that $\mathcal{A}$ generates a C_0-semigroup e^{At} on $\mathcal{H}$ and is identical to zero after $t \geq 2$ are well-known facts by the characteristic line method. A direct computation shows that

$$(I - \mathcal{A})^{-1}\begin{pmatrix} \phi(x) \\ \psi(x) \end{pmatrix} = \begin{pmatrix} (\Theta(\phi, \psi))(x) \\ (\Theta(\phi, \psi))(x) - \phi(x) \end{pmatrix}, \quad \forall\, (\phi, \psi)^\top \in D(\mathcal{A}), \tag{4.8.34}$$

with

$$(\Theta(\phi, \psi))(x) = -\frac{1}{2}\left(e^x \int_0^x (\phi(s) + \psi(s))e^{-s}ds - e^{-x}\int_0^1 (\phi(s) + \psi(s))e^s)ds\right). \tag{4.8.35}$$

In the following we shall prove that $\mathcal{B}$ is admissible to e^{At}. This is equivalent to prove that the solution of the following equation

$$\begin{cases} \dot{z}(t) = \mathcal{A}^* z(t), & z(0) = z_0, \\ y(t) = \langle \mathcal{B}, z(t) \rangle = \mathcal{B}^* z(t) \end{cases} \tag{4.8.36}$$

satisfies

$$\int_0^t |y(t)|^2 dt \leq C_T \|z_0\|_{\mathcal{H}}^2, \tag{4.8.37}$$

where C_T is a T-dependent constant. By the definition of $\mathcal{A}^*$, the equation (4.8.36) is equivalent to

$$\begin{cases} \tilde{w}_{tt}(x, t) = \tilde{w}_{xx}(x, t), & x \in (0, 1), \\ \tilde{w}_t(0, t) = \tilde{w}(0, t) = \tilde{w}_x(0, t) = 0, \\ y(t) = \tilde{w}_t(1, t). \end{cases} \tag{4.8.38}$$

Let

$$\tilde{E}(t) = \frac{1}{2}\int_0^1 [\tilde{w}_t^2(x, t) + \tilde{w}_x^2(x, t)]dx. \tag{4.8.39}$$

Then it is computed that $\tilde{E}(t) = \tilde{E}(0)$ for all $t > 0$. Let

$$\tilde{\rho}(t) = \int_0^1 x\tilde{w}_x(x, t)\tilde{w}_t(x, t)dx. \tag{4.8.40}$$

Then $|\tilde{\rho}(t)| \leq \tilde{E}(t)$ and

$$\dot{\tilde{\rho}}(t) = \frac{1}{2}\tilde{w}_t^2(1, t) - \tilde{E}(t). \tag{4.8.41}$$

This yields

$$\tilde{w}_t^2(1, t) = 2\tilde{E}(t) + \dot{\tilde{\rho}}(t). \tag{4.8.42}$$

Therefore,

$$\int_0^t \tilde{w}_t^2(1, t)dt \leq (2T + 2)\tilde{E}(0). \tag{4.8.43}$$

This gives (4.8.37) with $C_T = 2T + 2$.

Since $\mathcal{A}$ generates an exponential stable C_0-semigroup and $\mathcal{B}$ is admissible to e^{At}, it follows that for any initial value $(u_0, u_1) \in \mathcal{H}$, (4.8.3) admits a unique solution provided that $U_2 + d \in L^2_{loc}(0, \infty)$, which can be written as (4.8.33).   □

Proof of Theorem 4.8.1 First, we show the convergence of ESO. Let

$$\begin{cases} \tilde{z}(t) = \rho(t)(w(1,t) - \hat{z}(t)), \\ \tilde{d}(t) = (h(1) - S(1,1))(\hat{d}(t) - d(t)). \end{cases} \tag{4.8.44}$$

It follows that

$$\begin{cases} \dot{\tilde{z}}(t) = \rho(t)(\tilde{d}(t) - \alpha_1[\tilde{z}(t)]^\theta) + \dfrac{\dot{\rho}(t)}{\rho(t)}\tilde{z}(t), \\ \dot{\tilde{d}}(t) = -\alpha_2\rho(t)[\tilde{z}(t)]^{2\theta-1} - (h(1) - S(1,1))\dot{d}(t). \end{cases} \tag{4.8.45}$$

Let matrix $\wedge$ be defined as

$$\wedge = \begin{pmatrix} -\alpha_1 & 1 \\ -\alpha_2 & 0 \end{pmatrix}. \tag{4.8.46}$$

A simple computation shows that the eigenvalues of $\wedge$ are $(-\alpha_1 + \sqrt{\alpha_1^2 - 4\alpha_2})/2$ and $(-\alpha_1 - \sqrt{\alpha_1^2 - 4\alpha_2})/2$. Therefore, for any $\alpha_1 > 0$ and $\alpha_2 > 0$, two eigenvalues have negative real parts. Hence $\wedge$ is Hurwitz and there exists a positive definite matrix P satisfying the Lyapunov equation:

$$P\wedge + \wedge^\top P = -I_{2\times 2}. \tag{4.8.47}$$

Since the matrix $\wedge$ given in (4.8.46) is Hurwitz, it follows from Theorem 1.3.8 that there exists $\theta_1^* \in \left(\frac{1}{2}, 1\right)$ such that for all $\vartheta \in (\vartheta_1^*, 1)$, the following system

$$\begin{cases} \dot{\eta}_1(t) = \eta_2(t) - \alpha_1[\eta_1(t)]^\theta, \\ \dot{\eta}_2(t) = -\alpha_2[\eta_1(t)]^{2\theta-1} \end{cases} \tag{4.8.48}$$

is finite-time stable. By Theorem 1.3.8, there exists a positive definite, radially unbounded function $V : \mathbb{R}^2 \to \mathbb{R}$ such that $V(\eta_1, \eta_2)$ is homogeneous of degree $\gamma > 1$ with the weights $\{1, \theta\}$, and the derivative of $V(\eta_1(t), \eta_2(t))$ along the system (4.8.48) is given by

$$\left.\dfrac{dV(\eta_1(t), \eta_2(t))}{dt}\right|_{\text{along } (4.8.48)}$$

$$= (\eta_2(t) - \alpha_1[\eta_1(t)]^\theta)\dfrac{\partial V(\eta_1(t), \eta_2(t))}{\partial \eta_1} - \alpha_2[\eta_2(t)]^{2\theta-1}\dfrac{\partial V(\eta_1(t), \eta_2(t))}{\partial \eta_2}, \tag{4.8.49}$$

which is negative definite and is homogeneous of degree $\gamma + \theta - 1$. From the homogeneity of $V(\eta_1, \eta_2)$, we see that $|\partial V(\eta_1, \eta_2)/\partial \eta_2|$ is homogeneous of degree $\gamma - \theta$. By Property 1.3.3, there exist positive constants $b_1, b_2, b_3, b_4 > 0$ such that

$$\left|\dfrac{\partial V(\eta_1, \eta_2)}{\partial \eta_i}\right| \le c_1(V(\eta_1, \eta_2))^{(\gamma-((i-1)\theta-(i-2)))/\gamma}, \quad |\eta_i| \le c_2(V(\eta_1, \eta_2))^{((i-1)\theta-(i-2))/\gamma},$$

$$i = 1, 2, \tag{4.8.50}$$

and

$$-c_3(V(\eta_1(t),\eta_2(t)))^{(\gamma-(1-\theta))/\gamma} \leq \left.\frac{dV(\eta_1(t),\eta_2(t))}{dt}\right|_{\text{along (4.8.48)}}$$

$$\leq -c_4(V(\eta_1(t),\eta_2(t)))^{(\gamma-(1-\theta))/\gamma}. \qquad (4.8.51)$$

For $\theta = 1$, let

$$V(\eta_1,\eta_2) = (\eta_1,\eta_2)^\top P(\eta_1,\eta_2). \qquad (4.8.52)$$

It follows from the Lyapunov equation (4.8.47) that

$$\left.\frac{dV(\eta_1(t),\eta_2(t))}{dt}\right|_{\text{along (4.8.48)}} = -(\eta_1^2(t)+\eta_2^2(t)). \qquad (4.8.53)$$

It is easy to verify that when $\theta = 1$, the Lyapunov function $V(\eta_1,\eta_2)$ defined in (4.8.52) satisfies (4.8.50) and (4.8.51). Finding the derivatives of the Lyapunov function $V(\tilde{z}(t),\tilde{d}(t))$ along the error equation (4.8.45) to give

$$\left.\frac{dV(\tilde{z}(t),\tilde{d}(t))}{dt}\right|_{\text{along (4.8.45)}} = \rho(t)\left((\tilde{d}(t)-\alpha_1[\tilde{z}(t)]^\theta)\frac{\partial V(\tilde{z}(t),\tilde{d}(t))}{\partial \tilde{d}}\right.$$

$$\left. -\alpha_2[\tilde{z}(t)]^{2\theta-1}\frac{\partial V(\tilde{z}(t),\tilde{d}(t))}{\partial \tilde{d}}\right) + \frac{\dot\rho(t)}{\rho(t)}\tilde{z}(t)\frac{\partial V(\tilde{z}(t),\tilde{d}(t))}{\partial \tilde{z}} - \dot{d}(t)\frac{\partial V(\tilde{z}(t),\tilde{d}(t))}{\partial \tilde{d}}.$$

$$(4.8.54)$$

Proof of claim (i) of Theorem 4.8.1 In this case, $\rho(t) \equiv r(t)$, or $\rho(0)=1$, $\dot\rho(t)=a\rho(t)$ if $\rho(t) < r$, and $\rho(t) = 0$ if $\rho(t) \geq r$. Obviously, thre are two types of gain functions $\rho(t) \equiv r$ for $t \geq \ln r/a$. This together with (4.8.50) and (4.8.51) gives

$$\left.\frac{dV(\tilde{z}(t),\tilde{d}(t))}{dt}\right|_{\text{along (4.8.45)}}$$

$$= r\left((\tilde{d}(t)-\alpha_1[\tilde{z}(t)]^\theta)\frac{\partial V(\tilde{z}(t),\tilde{d}(t))}{\partial \tilde{d}} - \alpha_2[\tilde{z}(t)]^{2\theta-1}\frac{\partial V(\tilde{z}(t),\tilde{d}(t))}{\partial \tilde{d}}\right) - \dot{d}(t)\frac{\partial V(\tilde{z}(t),\tilde{d}(t))}{\partial \tilde{d}}$$

$$(4.8.55)$$

$$\leq -c_4 r(V(\tilde{z}(t),\tilde{d}(t)))^{(\gamma-(1-\theta))/\gamma} + c_1 N(V(\tilde{z}(t),\tilde{d}(t)))^{(\gamma-\theta)/\gamma}.$$

By $\theta \in (1/2,1]$, $\gamma-(1-\theta) > \gamma-\theta$, it follows that if $V(\tilde{z}(t),\tilde{d}(t)) \geq 1$ then

$$\left.\frac{dV(\tilde{z}(t),\tilde{d}(t))}{dt}\right|_{\text{along (4.8.45)}} \leq -(c_4 r - c_1 N)(V(\tilde{z}(t),\tilde{d}(t)))^{(\gamma-(1-\theta))/\gamma} < 0, \quad r > \frac{2c_1 N}{c_4}.$$

$$(4.8.56)$$

Therefore for any $r > (2c_1 N)/c_4$, there exists $t_{r1} > \ln r/a$ such that for all $t > t_{r1}$ $V(\tilde{z}(t),\tilde{d}(t)) < 1$. It then follows that

$$\left.\frac{dV(\tilde{z}(t),\tilde{d}(t))}{dt}\right|_{\text{along (4.8.45)}} \leq -c_4 r(V(\tilde{z}(t),\tilde{d}(t)))^{(\gamma-(1-\theta))/\gamma} + c_1 N, \quad r > \frac{2c_1 N}{c_4}, \quad t > t_{r1}.$$

$$(4.8.57)$$

For any $\sigma \in (0,1)$, if $V(\tilde{z}(t), \tilde{d}(t)) > \sigma^{\gamma/\theta}$, then

$$\frac{dV(\tilde{z}(t), \tilde{d}(t))}{dt}\bigg|_{\text{along (4.8.45)}} \leq -c_4 r\sigma^{(\gamma-(1-\theta))/\theta} + c_1 N < 0, \quad r > \frac{2c_1 N}{c_4}\sigma^{-\frac{\gamma-(1-\theta)}{\theta}}, \quad t > t_{r1}.$$

$$(4.8.58)$$

Therefore there exists $t_{r2} > t_{r1}$ such that $V(\tilde{z}(t), \tilde{d}(t)) < \sigma^{\gamma/\theta}$. This together with (4.8.50) yields

$$|\tilde{d}(t)| < \sigma, \quad \forall\, t > t_{2r}, \quad r > r^* \triangleq \frac{2c_1 N}{c_4}\sigma^{-(\gamma-(1-\theta))/\theta}. \tag{4.8.59}$$

Hence

$$|\hat{d}(t)| \leq |d(t)| + \sigma \leq M, \operatorname{sat}_M(\hat{d}(t)) = \hat{d}(t),$$

$$|(h(1) - S(1,1))(U_1(t) + d(t))| = |\tilde{d}(t)| < \sigma, \quad \forall\, t > t_{2r}, \quad r > r^*. \tag{4.8.60}$$

By (4.8.33),

$$\begin{pmatrix} w(\cdot,t) \\ w_t(\cdot,t) \end{pmatrix} = e^{\mathcal{A}t}\begin{pmatrix} u_0 \\ u_1 \end{pmatrix} + e^{\mathcal{A}(t-t_{r2})}\int_0^{t_{r2}} e^{\mathcal{A}(t_0-s)}\mathcal{B}(d(s) + U_1(s))ds$$

$$+ \int_{t_{r2}}^t e^{\mathcal{A}(t-s)}\mathcal{B}(d(s) + U_1(s))ds. \tag{4.8.61}$$

The admissibility of $\mathcal{B}$ implies that

$$\left\|\int_0^t e^{\mathcal{A}(t-s)}\mathcal{B}(d(s) + U_1(s))ds\right\|^2 \leq t^2 C_t \|d + U_1\|_{L^\infty(0,t)}^2. \tag{4.8.62}$$

It follows from Proposition 2.5 in [136] that

$$\left\|\int_{t_0}^t e^{\mathcal{A}(t-s)}\mathcal{B}(d(s) + U_1(s))ds\right\| = \left\|\int_0^t e^{\mathcal{A}(t-s)}\mathcal{B}(0\diamond_{t_0}(d+U_1))(s)ds\right\|$$

$$\leq M_4 \|d + U_1\|_{L^\infty(t_{2r},\infty)} \tag{4.8.63}$$

with

$$(\mu \diamond_\tau \nu) = \begin{cases} \mu(t), & 0 \leq t \leq \tau, \\ \nu(t), & t > \tau. \end{cases} \tag{4.8.64}$$

By (4.8.60), (4.8.61), (4.8.62), and (4.8.62), we have

$$\|(w(\cdot,t), w_t(\cdot,t))^\top\| \leq \overline{M}(\|(u_0, u_1)^\top\| + t_{r2}^2 C_{t_{r2}}\|d + U_1\|_{L^\infty(0,t_{r2})}^2 e^{\omega t_{r2}})e^{-\omega t} + M_4\sigma. \tag{4.8.65}$$

From $\lim_{t\to\infty} e^{-\omega t} = 0$, there exists $t_r > t_{r2}$ such that $e^{-\omega t} < \sigma, \forall\, t > t_r$. This yields

$$\|(w(\cdot,t), w_t(\cdot,t))^\top\| < (1 + M_4)\sigma. \tag{4.8.66}$$

The remainder of the proof in this case can be accomplished by the transformation from $w(x,t)$ to $u(x,t)$. Considering (4.8.73), $\tilde{h} \in C([0,1], \mathbb{R})$, $\tilde{K}, \tilde{S} \in C^2(\mathbb{I}, \mathbb{R})$, it is easy to obtain that there exists constant M_5 such that

$$\|(u(\cdot,t), u_t(\cdot,t))^\top\| \leq M_5\|(w(\cdot,t), w_t(\cdot,t))^\top\|. \tag{4.8.67}$$

This together with (4.8.66) gives (i) of Theorem 4.8.1.

Proof of claim (ii) of Theorem 4.8.1 In this case, $\theta = 1$, $\dot{\rho}(t)/\rho(t) = a$. It follows from (4.8.54) that

$$
\left.\frac{dV(\tilde{z}(t), \tilde{d}(t))}{dt}\right|_{\text{along (4.8.45)}} = \rho(t)\left((\tilde{d}(t) - \alpha_1 \tilde{z}(t))\frac{\partial V(\tilde{z}(t), \tilde{d}(t))}{\partial \tilde{d}} - \alpha_2 \tilde{z}(t)\frac{\partial V(\tilde{z}(t), \tilde{d}(t))}{\partial \tilde{d}}\right)
$$

$$
+ a\tilde{z}(t)\frac{\partial V(\tilde{z}(t), \tilde{d}(t))}{\partial \tilde{z}} - \dot{d}(t)\frac{\partial V(\tilde{z}(t), \tilde{d}(t))}{\partial \tilde{d}} \tag{4.8.68}
$$

$$
\leq -c_4 \rho(t) V(\tilde{z}(t), \tilde{d}(t)) + ac_1 c_2 V(\tilde{z}(t), \tilde{d}(t)) + c_1 N \sqrt{V(\tilde{z}(t), \tilde{d}(t))}.
$$

For any $\sigma \in (0, 1)$, there exists $t_1 > 0$ such that for any $t > t_1$,

$$
\rho(t) > \max\left\{\frac{2ac_1 c_2}{c_4}, \frac{4c_1 N}{c_4 \sqrt{\sigma}}\right\}. \tag{4.8.69}
$$

It follows that if $V(\tilde{z}(t), \tilde{d}(t)) \geq \sigma$, then

$$
\left.\frac{dV(\tilde{z}(t), \tilde{d}(t))}{dt}\right|_{\text{along (4.8.45)}} \leq -c_1 N \sqrt{\sigma} < 0, \; t > t_1. \tag{4.8.70}
$$

Therefore there exists $t_2 > t_1$ such that for any $t > t_2$, $V(\tilde{z}(t), \tilde{d}(t)) < \sigma$. This gives $\lim_{t \to \infty} V(\tilde{z}(t), \tilde{d}(t)) = 0$. By (4.8.50), $\lim_{t \to \infty} \tilde{d}(t) = 0$.

Similar to the proof of claim (i), in this case,

$$
\|(w(\cdot, t), w_t(\cdot, t))^\top\| \leq \overline{M}(\|(u_0, u_1)^\top\| + t_2^2 C_{t_2}\|d + U_1\|_{L^\infty(0, t_2)}^2 e^{\omega t_2})e^{-\omega t} + M_4 \sigma. \tag{4.8.71}
$$

Hence there exists $t_3 > t_2$ such that

$$
\|(w(\cdot, t), w_t(\cdot, t))^\top\| \leq (1 + M_4)\sigma, \; t > t_3. \tag{4.8.72}
$$

This means $\lim_{t \to \infty} \|(w(\cdot, t), w_t(\cdot, t))^\top\| = 0$. Let

$$
u(x, t) = \tilde{h}(x)w(x, t) - \int_0^x \tilde{K}(x, y)w(y, t)dy - \int_0^x \tilde{S}(x, y)w_t(y, t)dy, \tag{4.8.73}
$$

where $\tilde{h}(x)$, $\tilde{K}(x, y)$, and $\tilde{S}(x, y)$ are the solution of the following equation:

$$
\begin{cases}
\tilde{K}_{xx}(x, y) - \tilde{K}_{yy}(x, y) = -2\lambda(x)\tilde{S}_{yy}(x, y) - \beta(x)\tilde{K}(x, y), \\
2\tilde{K}'(x, x) = 2\lambda(x)\tilde{S}_y(x, x) + \beta(x)\tilde{h}(x) + \tilde{h}''(x), \\
K(x, 0) = 0, \\
\tilde{S}_{xx}(x, y) - \tilde{S}_{yy}(x, y) = -2\lambda(x)\tilde{K}(x, y) - \beta(x)\tilde{S}(x, y), \\
\tilde{S}'(x, x) = \lambda(x)\tilde{h}(x), \quad \lambda(x)\tilde{S}(x, x) = \tilde{h}'(x), \\
\tilde{S}(x, 0) = 0.
\end{cases} \tag{4.8.74}
$$

Similar to (4.8.5), we can obtain

$$\tilde{h}(x) = \cosh\left(\int_0^x \lambda(\tau)d\tau\right), \quad \tilde{S}(x,x) = \sinh\left(\int_0^x \lambda(\tau)d\tau\right), \tag{4.8.75}$$

where $\tilde{K}(x,y)$ and $\tilde{S}(x,y)$ are obtained by substitution:

$$\tilde{K}\left(\frac{x+y}{2}, \frac{x-y}{2}\right) = \tilde{G}^K(x,y) = \sum_{n=1}^{\infty} \tilde{G}^{K\,n}(x,y),$$

$$\tilde{S}\left(\frac{x+y}{2}, \frac{x-y}{2}\right) = \tilde{G}^S(x,y) = \sum_{i=1}^{\infty} \tilde{G}^{S\,n}(x,y), \tag{4.8.76}$$

$$\begin{cases} \tilde{G}^{K\,0}(x,y) & = \tilde{g}\left(\frac{x}{2}\right) - \tilde{g}\left(\frac{y}{2}\right), \\[2mm] \tilde{G}^{S\,0}(x,y) & = \sinh\left(\int_0^{x/2} \lambda(\tau)d\tau\right) + \sinh\left(\int_0^{y/2} \lambda(\tau)d\tau\right), \\[2mm] \tilde{G}^{K\,n+1}(x,y) & = -\frac{1}{4}\int_y^x \int_0^y \beta\left(\frac{\tau+s}{2}\right) \tilde{G}^{K\,n}(\tau,s)ds\,d\tau \\[2mm] & \quad -\frac{1}{2}\int_y^x \int_0^y \lambda\left(\frac{\tau+s}{2}\right)(\tilde{G}^{S\,n}_{xx}(\tau,s) - 2\tilde{G}^{S\,n}_{xy}(\tau,s) + \tilde{G}^{S\,n}_{yy}(\tau,s))ds\,d\tau, \\[2mm] \tilde{G}^{S\,n+1}(x,y) & = -\frac{1}{4}\int_y^x \int_0^y \left(2\lambda\left(\frac{\tau+s}{2}\right)\tilde{G}^{K\,n}(\tau,s) + \beta\left(\frac{\tau+s}{2}\right)\tilde{G}^{S\,n}(\tau,s)\right)ds\,d\tau, \end{cases} \tag{4.8.77}$$

and $\tilde{g}(x)$ comes from the following equation:

$$\begin{cases} 2\tilde{f}'(x) + 2\lambda(x)\tilde{g}(x) = \tilde{S}''(x) + \beta(x)\tilde{S}(x,x), \\ 2\tilde{g}'(x) - 2\lambda(x)\tilde{f}(x) = \beta(x)\tilde{h}(x) + \tilde{h}''(x), \\ \tilde{f}(0) = \lambda(0), \quad \tilde{g}(0) = 0. \end{cases} \tag{4.8.78}$$

Similarly with Lemma 4.8.1, if $\lambda \in C^2([0,1], \mathbb{R})$, $\beta \in C^0([0,1], \mathbb{R})$, then the series defined in (4.8.77) is uniformly convergent in $[0,1] \times [0,1]$ and hence there exist kernel functions $\tilde{K}, \tilde{S} \in C^2(\mathbb{I}, \mathbb{R})$. Therefore, $\lim_{t\to\infty} \|(u(\cdot,t), u_t(\cdot,t))^\top\|_{H^1\times L^2} = 0$. This completes the proof of the theorem. $\qquad\square$

Finally, we give a numerical simulation to illustrate the main result of this section. In (4.8.1), let $\lambda(x) \equiv 1, \beta(x) \equiv 0$ and in ESO (4.8.23) and let the parameters be $\theta = 0.9, \alpha_1 = \alpha_2 = 1$, and $\rho = 50$. The initial values are chosen as $u(x,0) = 2\sin x + x, u_t(x,0) = x$. Firstly, we use the constant gain $\rho \equiv 0$ and the control $U_2(t)$ in (4.8.21) to be set as $U_2(t) = -(1/h(1))\text{sat}_4(\hat{d}(t))$. By the finite difference method, the numerical results of wave equation (4.8.1) are plotted in Figure 4.8.1. The numerical results of the disturbance estimate are plotted in Figure 4.8.2. We can see from Figure 4.8.1 that the states are convergent to

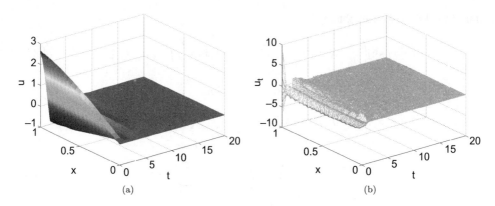

Figure 4.8.1   The numerical result for the state of (4.8.3) under constant gain ESO.

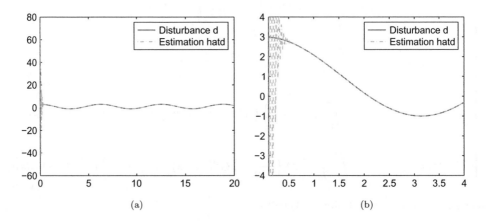

Figure 4.8.2   The numerical result of constant gain ESO.

zero very well. Also it is seen from Figure 4.8.2 that the ESO tracks the external disturbance very well.

Next, we use the time-varying gain $\rho(0) = 1$, $\dot{\rho}(t) = 2\rho(t)$, if $\rho(t) < 50$ and $\rho(t) = 0$ if $\rho(t) \geq 50$. By the finite difference method again, the numerical results of wave equation (4.8.1) are plotted in Figure 4.8.3. The numerical results of the disturbance estimate are plotted in Figure 4.8.4. We can see from Figure 4.8.3 that the states are convergent to zero very satisfactorily. Also it is seen from Figure 4.8.4 that the ESO tracks the external disturbance very well.

Comparing Figures 4.8.2 and 4.8.4, we can see that in Figure 4.8.4, the time-varying gain ESO has a much smaller peaking value than the constant gain ESO in Figure 4.8.2. For accuracy, the former is only 10% of the later. However, the time-varying gain ESO takes a slightly longer time to track the disturbance.

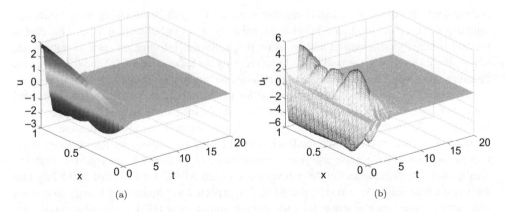

Figure 4.8.3   The numerical result of the state of (4.8.3) under time-varying gain ESO.

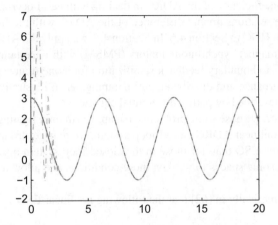

Figure 4.8.4   The numerical result of time-varying gain ESO.

4.9   Summary and Open Problems

In this chapter, we analyze the active disturbance rejection control closed loop (ADRC) composed of the tracking differentiator (TD), extended state observer (ESO) and ESO-based output feedback control, where the TD is used to recover the derivative of the reference signal and the ESO is used to recover the system state and the total disturbance. The disturbance is compensated for in the feedback loop.

In Section 4.1, we discuss a simple case where the control system is a single-input single-output (SISO) system with vast uncertainty. The linear ESO is used for the state and total disturbance estimation and the functions in the feedback control are also linear. Two feedback control design methods in this section are proposed. The first design method presented in Subsections 4.1.1 and 4.1.2 is a global design method that takes advantage of a simple design and guarantees the system output tracking, regardless of the initial values. However, this method may come up with the peaking value problem if the tuning parameter and the derivative of the total disturbance are large. By using the bound of the initial state and saturated method, we design a semi-global feedback control in Subsection 4.1.3. This design method is slightly complicated and relies on the additional information such as initial state bound, but it can reduce the peaking value effectively. In Subsections 4.1.2 and 4.1.3 we also discuss some special cases where the uncertainties do not contain system dynamics.

In Section 4.2, we give a design principle for general nonlinear ADRC feedback control with a nonlinear ESO. In addition, both global and semi-global nonlinear ADRC are analyzed. In Section 4.3, we use a time-varying gain ESO. This method can also reduce the possible peaking value and, moreover, the feedback control design is simpler than the saturated function method and does not use additional information. In Section 4.4, we establish both the semi-global convergence and global convergence of the ADRC for a kind of MIMO system with vast uncertainty. As a result, the ADRC is expected to require less energy in control compared to other control strategies such as high-gain control (HGC) or sliding mode control (SMC). If the uncertainty is the external disturbance only, the ADRC method also has the advantage of needing less information for the external disturbance than the internal model principle (IMP). The efficiency of the ADRC in dealing with vast uncertainty is demonstrated by simulations. The relationship and difference of the ADRC with the IMP are illustrated in Section 4.5 and with HGC in Section 4.6. In Section 4.7, we apply the ADRC to a current control for permanent-magnet synchronous motors (PMSM) with uncertainties. In Section 4.8, we apply the ADRC to boundary feedback stabilization for a one-dimensional wave equation with boundary disturbance and distributed anti-damping, as an illustrative application of the ADRC to systems described by partial differential equations.

To end this chapter, we pose some problems that are worth investigating. In the proof of the convergence for a nonlinear ADRC closed-loop system, we require strong conditions on nonlinear functions in the ESO than for those in the closed-loop system discussed in Chapter 3. Naturally, there are some spaces to weaken these conditions and allow a larger class of nonlinear functions.

It is indicated numerically in [62] that the following nonlinear ESO

$$
\begin{cases}
\dot{\hat{x}}_1(t) = \hat{x}_2(t) - \beta_1(\hat{x}_1(t) - x_1(t)), \\
\dot{\hat{x}}_2(t) = \hat{x}_3(t) - \beta_2\text{fal}(\hat{x}_1(t) - x_1(t), 1/2, \delta), \\
\dot{\hat{x}}_3(t) = -\beta_3\text{fal}(\hat{x}_1(t) - x_1(t), 1/4, \delta),
\end{cases}
\tag{4.9.1}
$$

is very effective in the ADRC, where

$$
\text{fal}(e, \alpha, \delta) = \begin{cases} \dfrac{e}{\delta^{\alpha-1}}, & |e| \le \delta, \\ |e|^\alpha \text{sign}(e), & |e| > \delta, \end{cases} \qquad \delta > 0, \ \alpha > 0,
\tag{4.9.2}
$$

and

$$
\beta_1 = \frac{1}{h}, \quad \beta_2 = \frac{1}{3h^2}, \quad \beta_3 = \frac{2}{8^2 h^3}, \quad \beta_4 = \frac{5}{13^3 h^4}, \ \ldots.
\tag{4.9.3}
$$

However, the convergence is lacking for this case.

Additional attention should be paid to the nonlinear systems that are not in canonical form. Systems like triangular structure as follows:

$$
\begin{cases}
\dot{x}_1(t) & = f_1(x_1(t), x_2(t), w_1(t)), \\
& \vdots \\
\dot{x}_n(t) & = f_n(x_1(t), \ldots, x_n(t), \xi(t), w_n(t)) + b(x(t), \xi(t), w_n(t))u(t), \\
\dot{\xi}(t) & = F(x_1(t), \ldots, x_n(t), \xi(t), w_1(t), \ldots, w_n(t)), \\
y(t) & = x_1(t),
\end{cases}
$$

where $w_i(t)(1 \leq i \leq n)$ is the external disturbance and $f_i(\cdot)(1 \leq i \leq n)$ is the system function with some uncertainty. The difficulty is dealing with the external disturbances $w_1(t), \ldots, w_{n-1}(t)$ and other uncertainties that are not in the control channel.

4.10   Remarks and Bibliographical Notes

Section 4.2. This part is taken from [159] and is reproduced by permission of Springer. Equations (4.2.3) and (4.8.4) are stronger than (4.0.6) in Definition 4.0.1 for the convergence of ADRC. In Theorem 4.2.1, $\varepsilon_0, \varepsilon_1, \Gamma_1, \Gamma_2, t_\varepsilon$, and $t_{R\varepsilon}$ are all R-dependent and $\varepsilon_0, \varepsilon_1, \Gamma_1$, and Γ_2 are independent of the initial value of (4.2.2); t_ε and $t_{R\varepsilon}$ depend on the initial value of (4.2.2).

It is remarked that it is not necessary to assume that the functions $g_i(\cdot)$ $(1 \leq i \leq n+1)$ are bounded by linear functions, as required in Assumption 4.2.2, if we want to have the convergence result of [54] for ESO (4.0.4). This additional assumption for $g_i(\cdot)$ can be considered as a sufficient condition that guarantees the separation principle. However, the assumption of [54] on the system itself for the convergence of the ESO is stronger than Assumption A1 in the sense that in [54], the solution of (4.0.1) is assumed to be bounded.

In [35], a linear extended high-gain observer is used to establish the semi-global convergence of stabilization for the system (4.1.72), which is a special case of the nonlinear extended state observer discussed in this paper with $v(t) \equiv 0$. Since the results here are global convergence rather than the semi-global convergence only, the conditions for system function $f(\cdot)$ and control parameters b in Theorem 4.2.1 and 4.2.3 are slightly stronger than that in [35]. Actually, it is seen that the semi-global convergence claimed in Theorem 4.2.4 requires weaker conditions than those in Theorems 4.2.1 and 4.2.3. Another work that is relevant to this work is [111], where an adaptive linear high-gain observer is used for stabilization with the constant external disturbance only, while the uncertainty in this paper is vast uncertainty that comes not only from the external disturbance but also from the model uncertainty.

Section 4.3. This section is taken from paper [156] and is reproduced by permission of the Elsevier license.

Section 4.5. This part is taken from [53] and is reproduced by permission of the Society for Industrial and Applied Mathematics (SIAM). Proposition 4.5.1 can be found in [33].

Section 4.6. The high-gain observer and control have been studied extensively. A recent work is reviewed in [86]. The observer with time-varying gain (updated gain or dynamic gain) is also used in [8, 111], where the gain is a dynamics determined by some nonlinear function related to the control plant. The major difference between [8] and many others and this section is that there is no estimation for uncertainty in these works. Only in [111], a *constant* unknown nominal control value is estimated on stabilization for an affine nonlinear system.

It is remarked that all $z_j^i(t)$ are ρ-dependent and that $z_j^i(t)$ is regarded as an approximation of the $(i-1)$th order derivative of $v_i(t)$ in the sense of a generalized derivative (see Chapter 2). If all $v_i^{(j-1)}(t)$ exist in the classical sense, we may consider simply $z_j^i(t) = v_i^{(j-1)}(t)$ for $j = 2, 3, \ldots, n_i$. In the latter case, the TD (4.4.3) does not need to be coupled into the ADRC.

Section 4.8 Stabilization for the wave equation is an important issue, which has been studied by many researchers. In paper [119], a wave equation (4.8.1) without the external disturbance is studied. Papers [44, 51] study a one-dimensional wave equation with boundary disturbance.

The ADRC has been successfully applied in stabilization for distributed systems with external disturbances. See, for instance, [42, 43, 44, 45, 46, 47, 48, 49, 50]. In all these works, the control and external disturbance are matched. For stabilization, this is necessary. However, for other control purposes, such as output reference tracking, the control and disturbance are allowed to be in different channels. Such works are just beginning to be written for distributed parameter systems.

5

ADRC for Lower Triangular Nonlinear Systems

In this chapter, we first extend the ESO to lower triangular systems and then use ADRC to stabilize a class of lower triangular systems.

5.1 ESO for Lower Triangular Systems

In this section, we are concerned with convergence of the nonlinear ESO further for the following lower triangular systems:

$$
\begin{cases}
\dot{x}_1(t) = x_2(t) + g_1(u(t), x_1(t)), \\
\dot{x}_2(t) = x_3(t) + g_2(u(t), x_1(t), x_2(t)), \\
\quad \vdots \\
\dot{x}_n(t) = f(t, x(t), w(t)) + g_n(u(t), x(t)), \\
y(t) = x_1(t),
\end{cases}
\tag{5.1.1}
$$

where $g_i \in C(\mathbb{R}^{i+m}, \mathbb{R})$ is a known nonlinear function, $f \in C(\mathbb{R}^{n+s+1}, \mathbb{R})$ is usually an unknown nonlinear function, $x(t) = (x_1(t), x_2(t), \cdots, x_n(t))$ is the state of system, $u \in \mathbb{R}^m$ is the input (control), $y(t)$ is the output (measurement), and $w \in C([0, \infty), \mathbb{R})$ is the external disturbance. The objective of this section is to design a nonlinear ESO for system (5.1.1) to estimate both state $x(t)$ and the "total disturbance" defined by the extended state of system (5.1.1) as follows:

$$
x_{n+1}(t) \triangleq f(t, x(t), w(t)).
\tag{5.1.2}
$$

Active Disturbance Rejection Control for Nonlinear Systems: An Introduction, First Edition.
Bao-Zhu Guo and Zhi-Liang Zhao.
© 2016 John Wiley & Sons Singapore Pte. Ltd. Published 2016 by John Wiley & Sons, Ltd.

We proceed as follows. In the next subsection, we propose a generalized nonlinear constant high-gain ESO. The proof for the practical convergence is presented. We exemplify analytically the proposed ESO by two classes of ESO with an explicit estimation of convergence. In Subsection 5.1.2, we propose a time-varying gain nonlinear ESO for the system (5.1.1). Subsection 5.1.3 presents some numerical simulations for illustration. In particular, the numerical results demonstrate visually the peaking value reduction by the time-varying gain approach.

5.1.1 Constant High-Gain ESO

In this subsection, we design a constant high-gain ESO to recover both the state of system (5.1.1) and its extended state (5.1.2), which reads as follows:

$$\text{ESO:} \begin{cases} \dot{\hat{x}}_1(t) = \hat{x}_2(t) + \dfrac{1}{r^{n-1}}h_1(r^n(y(t) - \hat{x}_1(t))) + g_1(u(t), \hat{x}_1(t)), \\ \quad\vdots \\ \dot{\hat{x}}_n(t) = \hat{x}_{n+1}(t) + h_n(r^n(y(t) - \hat{x}_1(t))) + g_n(u(t), \hat{x}_1(t), \cdots, \hat{x}_n(t)), \\ \dot{\hat{x}}_{n+1}(t) = rh_{n+1}(r^n(y(t) - \hat{x}_1(t))), \end{cases} \tag{5.1.3}$$

where r is the constant high-gain parameter and $h_i \in C(\mathbb{R}, \mathbb{R})$, $i = 1, 2, ..., n+1$ are the design functions.

To achieve convergence of ESO (5.1.3), some mathematical assumptions are required. The following Assumptions 5.1.1 and 5.1.2 are on $g_i(\cdot)$ and $f(\cdot)$ in system (5.1.1).

Assumption 5.1.1 $g_i : \mathbb{R}^{i+1} \to \mathbb{R}$ satisfies

$$|g_i(u, \nu_1, \cdots, \nu_i) - g_i(u, \tilde{\nu}_1, \cdots, \tilde{\nu}_i)| \le \Gamma(u)\|(\nu_1 - \tilde{\nu}_1, \cdots, \nu_i - \tilde{\nu}_i)\|^{\theta_i}, \quad \Gamma \in C(\mathbb{R}^m, \mathbb{R}), \tag{5.1.4}$$

where $\theta_i \in ((n-i)/(n+1-i), 1]$, $i = 1, 2, \cdots, n$.

The condition (5.1.4) means that $g_i(\cdot)$, $i = 1, 2 \cdots, n$, are Hölder continuous. For triangular systems, the widely assumed Lipschitz continuity is just a special case of the Hölder continuity with the exponents $\theta_i = 1$. Systems with appropriate Hölder continuous functions such as weighted homogeneous functions have merits of being finite-time stable, and these kinds of functions can be used for feedback control design.

Assumption 5.1.2 $f \in C^1(\mathbb{R}^{n+2}, \mathbb{R})$ satisfies

$$|f(t, x, w)| + \left|\frac{\partial f(t, x, w)}{\partial t}\right| + \left|\frac{\partial f(t, x, w)}{\partial x_i}\right| + \left|\frac{\partial f(t, x, w)}{\partial w}\right| \le \varpi_1(x) + \varpi_2(w),$$

where $i = 1, 2, \cdots, n$, and $\varpi_1 \in C(\mathbb{R}^n, [0, \infty))$ and $\varpi_2 \in C(\mathbb{R}, [0, \infty))$ are two known functions.

The following Assumption 5.1.3 is on the control input $u(t)$ and external disturbance $w(t)$.

Assumption 5.1.3 $\sup_{t \in [0, \infty)}(\|w(t)\| + |\dot{w}(t)| + \|u(t)\|) < \infty.$

For many practical systems, since the control is bounded, Assumption 5.1.3 is reasonable when ESO is used in fault diagnosis. For ESO-based feedback control, the boundedness of control should be analyzed separately. However, if the $\Gamma(u)$ in (5.1.4) is constant, then the assumption of boundedness for control can be removed.

Assumption 5.1.4 below is on functions $h_i(\cdot)$ in ESO (5.1.3). It gives a principle of choosing $h_i(\cdot)$.

Assumption 5.1.4 All $h_i \in C(\mathbb{R}, \mathbb{R})$ satisfy the following Lyapunov conditions: There exist positive constants $R, N > 0$, and continuous, radially unbounded, positive definite functions $\mathcal{V}, \mathcal{W} \in C(\mathbb{R}^{n+1}, [0, \infty))$ such that

1. $\displaystyle\sum_{i=1}^{n}(\nu_{i+1} - h_i(\nu_1))\frac{\partial \mathcal{V}(\nu)}{\partial \nu_i} - h_{n+1}(\nu_1)\frac{\partial \mathcal{V}(\nu)}{\partial \nu_n} \le -\mathcal{W}(\nu), \forall\, \nu = (\nu_1, \nu_2, \cdots, \nu_{n+1}) \in \mathbb{R}^{n+1};$

2. $\displaystyle\max_{i=1,\cdots,n}\left\{\|(\nu_1, \cdots, \nu_i)\|^{\theta_i}\left|\frac{\partial \mathcal{V}(\nu)}{\partial \nu_i}\right|\right\} \le N\mathcal{W}(\nu), \left|\frac{\partial \mathcal{V}(\nu)}{\partial \nu_{n+1}}\right| \le N\mathcal{W}(\nu), \nu \in \mathbb{R}^{n+1}, \|\nu\| \ge R.$

Assumption 5.1.4 guarantees that the zero equilibrium of the following system

$$\dot{\nu}(t) = (\nu_2(t) - h_1(\nu_1(t)), ..., \nu_{n+1}(t) - h_n(\nu_1(t)), -h_{n+1}(\nu_1(t)))^\top, \nu \in \mathbb{R}^{n+1}$$

is asymptotically stable.

Theorem 5.1.1 *Assume that Assumptions A1 to A4 are satisfied and the solution of (5.1.1) is globally bounded. Then the states of ESO (5.1.3) converge practically to the states and extended state of system (5.1.1): For any $\sigma > 0$, there exists a positive constant $r_0 > 0$ such that*

$$|\hat{x}_i(t) - x_i(t)| < \sigma, \quad \forall\, t > t_r, \quad r > r_0, \quad i = 1, 2, ..., n+1, \tag{5.1.5}$$

where t_r is an r-dependent constant.

Proof. Set

$$\eta_i(t) = r^{n+1-i}(x_i(t) - \hat{x}_i(t)), \quad i = 1, 2, ..., n+1, \quad \eta(t) = (\eta_1(t), ..., \eta_{n+1}(t))^\top, \tag{5.1.6}$$

where $x_i(t)$ is the solution of (5.1.1) and $\hat{x}_i(t)$ the solution of ESO (5.1.3).

A straightforward computation shows that $\eta_i(t)$ satisfies

$$\begin{cases} \dot{\eta}_1(t) = r(\eta_2(t) - h_1(\eta_1(t))) + r^n(g_1(u(t), x_1(t)) - g_1(u(t), \hat{x}_1(t))), \\ \quad\vdots \\ \dot{\eta}_n(t) = r(\eta_{n+1}(t) - h_n(\eta_1(t))) \\ \qquad\quad + r(g_n(u(t), x_1(t), \cdots, x_n(t)) - g_n(u(t), \hat{x}_1(t), \cdots, \hat{x}_n(t))), \\ \dot{\eta}_{n+1}(t) = -rh_{n+1}(\eta_1(t)) + \dot{x}_{n+1}(t). \end{cases} \tag{5.1.7}$$

By Assumption 5.1.1, substitute (5.1.6) into (5.1.7) to obtain

$$r^{n+1-i}|g_i(u(t), x_1(t), \cdots, x_i(t)) - g_i(u(t), \hat{x}_1(t), \cdots, \hat{x}_i(t))|$$
$$\le \Gamma(u(t))r^{n+1-i}\|(\eta_1(t)/r^n, \cdots, \eta_i(t)/r^{n+1-i})\|^{\theta_i} \tag{5.1.8}$$
$$\le \Gamma(u(t))r^{(n+1-i)(1-\theta_i)}\|(\eta_1(t), \cdots, \eta_i(t))\|^{\theta_i}, \quad \forall\, r > 1.$$

By Assumption 5.1.2,

$$
\begin{aligned}
|\dot{x}_{n+1}(t)| \leq (\varpi_1(x(t)) + \varpi_2(w(t))) & \left(1 + \dot{w}(t) + \sum_{i=1}^{n-1} |x_{i+1}(t)| \right. \\
& \left. + \sum_{i=1}^{n} |g_i(u(t), x_1(t), \cdots, x_i(t))| + \varpi_1(x(t)) + \varpi_2(w(t)) \right).
\end{aligned}
\tag{5.1.9}
$$

Let $\mathcal{V} : \mathbb{R}^{n+1} \to \mathbb{R}$ be the Lyapnuov function satisfying Assumption 5.1.4. Finding the derivative of $\mathcal{V}(\eta(t))$ along the solution of (5.1.7) yields

$$
\begin{aligned}
\frac{d\mathcal{V}(\eta(t))}{dt}\Big|_{\text{along (5.1.7)}} = & \sum_{i=1}^{n} (r(\eta_{i+1}(t) - h_i(\eta_1(t))) \\
& + r^{n+1-i}[g_i(u(t), x_1(t), \cdots, x_i(t)) - g_i(u(t), \hat{x}_1(t), \cdots, \hat{x}_i(t))]) \frac{\partial \mathcal{V}(\eta(t))}{\partial \eta_i} \\
& + (-rh_{n+1}(\eta_1(t)) + \dot{x}_{n+1}(t)) \frac{\partial \mathcal{V}(\eta(t))}{\partial \eta_{n+1}}.
\end{aligned}
\tag{5.1.10}
$$

Equation (5.1.10) can be simplified further by condition 1 of Assumption 5.1.4, (5.1.8), and (5.1.9) that

$$
\begin{aligned}
\frac{d\mathcal{V}(\eta(t))}{dt}\Big|_{\text{along (5.1.7)}} \leq & -r\mathcal{W}(\eta(t)) + |\dot{x}_{n+1}(t)| \left| \frac{\partial \mathcal{V}(\eta(t))}{\partial \eta_{n+1}} \right| \\
& + \sum_{i=1}^{n} \Gamma(u(t)) r^{(n+1-i)(1-\theta_i)} \|(\eta_1(t), \cdots, \eta_i(t))\|^{\theta_i} \left| \frac{\partial \mathcal{V}(\eta(t))}{\partial \eta_i} \right|.
\end{aligned}
\tag{5.1.11}
$$

By Assumption 5.1.2 and condition 2 of Assumption 5.1.4, it follows that

$$
\begin{aligned}
\|(\eta_1(t), \cdots, \eta_i(t))\|^{\theta_i} \left| \frac{\partial \mathcal{V}(\eta(t))}{\partial \eta_i} \right| &\leq N\mathcal{W}(\eta(t)), \\
\left| \frac{\partial \mathcal{V}(\eta(t))}{\partial \eta_{n+1}} \right| &\leq N\mathcal{W}(\eta(t)) \text{ when } \|\eta(t)\| \geq R.
\end{aligned}
\tag{5.1.12}
$$

Set

$$
\Lambda = \max_{1 \leq i \leq n} (n + 1 - i)(1 - \theta_i), \quad N_{11} = \sup_{t \in [0,\infty)} nN\Gamma(u(t)),
$$

$$
\begin{aligned}
N_{12} = N \sup_{t \in [0,\infty)} [\varpi_1(x(t)) + \varpi_2(w(t))] & \left(1 + \dot{w}(t) + \sum_{i=1}^{n-1} |x_{i+1}(t)| \right. \\
& \left. + \sum_{i=1}^{n} |g_i(u(t), x_1(t), \cdots, x_i(t))| + \varpi_1(x(t)) + \varpi_2(w(t)) \right).
\end{aligned}
\tag{5.1.13}
$$

By Assumption 5.1.1, $\Lambda \in (0,1)$. Since all $w(t)$, $\dot{w}(t)$, $u(t)$, and $x(t)$ are uniformly bounded, it follows that $N_{11} < \infty$ and $N_{12} < \infty$. Therefore, if $|\eta(t)| \geq R$ and $r > 1$, then

$$
\frac{d\mathcal{V}(\eta(t))}{dt}\Big|_{\text{along (5.1.7)}} \leq -(r - N_{11}r^\Lambda - N_{12})\mathcal{W}(\eta(t)).
\tag{5.1.14}
$$

Since $\mathcal{V}(\nu)$ and $\mathcal{W}(\nu)$ are radially unbounded and continuous positive definite, it follows from Theorem 1.3.1 that there exist class $\mathcal{K}_\infty$ functions $\kappa_i : [0, \infty) \to [0, \infty)(i = 1, 2, 3, 4)$ such that

$$\kappa_1(\|\nu\|) \le \mathcal{V}(\nu) \le \kappa_2(\|\nu\|), \kappa_3(\|\nu\|) \le \mathcal{W}(\nu) \le \kappa_4(\|\nu\|), \forall \nu \in \mathbb{R}^{n+1}. \tag{5.1.15}$$

If $\mathcal{V}(\eta(r;t)) \ge \kappa_2(R)$, then $\|\eta(t)\| \ge \kappa_2^{-1}(\mathcal{V}(\eta(r;t))) \ge R$ and $\mathcal{W}(\eta(t)) \ge \kappa_3(\|\eta(t)\|) \ge \kappa_3(R)$.

Let

$$r > r_1 \triangleq \max\left\{1, (3N_{11})^{1/(1-\Lambda)}, (N_{12}/N_{11})^{1/\Lambda}\right\}.$$

Then, the derivative of $\mathcal{V}(\eta(t))$ along system (5.1.7) satisfies

$$\frac{d\mathcal{V}(\eta(t))}{dt}\bigg|_{\text{along (5.1.7)}} < -N_{11}\kappa_3(R)r^\Lambda < 0, \quad \forall r > r_0, \quad \|\eta(t)\| \ge R. \tag{5.1.16}$$

Hence, for every $r > r_1$, there exists an r-dependent constant t_{r_1} such that $\mathcal{V}(\eta(t)) \le \kappa_2(R)$ for all $t > t_{r_1}$.

Now, finding the derivative of $\mathcal{V}(\eta(t))$ along the solution of error equation (5.1.7) gives

$$\frac{d\mathcal{V}(\eta(t))}{dt}\bigg|_{\text{along (5.1.7)}} \le -r\mathcal{W}(\eta(t)) + M_{11}r^\Lambda + M_{12}, \quad \forall r > N_0, \quad t > t_{r_1}, \tag{5.1.17}$$

where

$$M_{11} = \sup_{t \in [0,\infty)} \Gamma(u(t)) \sum_{i=1}^n \sup_{\nu \in \{\nu \in \mathbb{R}^{n+1}: \mathcal{V}(\nu) \le \kappa_2(R)\}} \|\nu\|^{\theta_i} \left\|\frac{\partial \mathcal{V}(\nu)}{\partial \nu}\right\|,$$

$$M_{12} = \frac{N_{12}}{N} \sup_{\nu \in \{\nu \in \mathbb{R}^{n+1}: \mathcal{V}(\nu) \le \kappa_2(R)\}} \left|\frac{\partial \mathcal{V}(\nu)}{\partial \nu_{n+1}}\right|.$$

By the radial unboundedness of $\mathcal{V}(\nu)$, $\{\nu \in \mathbb{R}^{n+1} : \mathcal{V}(\nu) \le \kappa_2(R)\} \subset \mathbb{R}^{n+1}$ is bounded. This together with the continuity of $\nabla \mathcal{V}(\nu)$ yields $M_{11} < \infty$ and $M_{12} < \infty$.

For any given $\sigma > 0$, it follows from (5.1.17) and (5.1.15) that if

$$r > r_0 \triangleq \max\left\{r_1, \left(\frac{3M_{11}}{\kappa_3(\sigma)}\right)^{\frac{1}{1-\Lambda}}, \left(\frac{M_{12}}{M_{11}}\right)^{1/\Lambda}\right\}$$

and $\|\eta(t)\| \ge \sigma$, then $\mathcal{W}(\eta(t)) \ge \kappa_3(\sigma)$. Hence,

$$\frac{d\mathcal{V}(\eta(t))}{dt}\bigg|_{\text{along(5.1.7)}} \le -r\kappa_3(\sigma) + 2M_{11}r^\Lambda \le -M_{11}r^\Lambda < 0. \tag{5.1.18}$$

Therefore, there exists $t_r > t_{r_1}$ such that $\|\eta(t)\| < \sigma$ for any $r > r_0$ and all $t > t_r$. By (5.1.6),

$$|x_i(t) - \hat{x}_i(t)| = \frac{|\eta_i(t)|}{r^{n+1-i}} \le \sigma, i = 1, 2, \cdots, n+1. \tag{5.1.19}$$

This completes the proof of the theorem. $\square$

Roughly speaking, the larger the absolute value of the derivative of disturbance is, the larger the constant high gain r needed for practical convergence (5.1.5). In the following, we consider a special case of $f(t, x(t), w(t)) = w(t)$. In other words, the unknown part comes from the external disturbance $w(t)$ only.

Theorem 5.1.2 shows that if $\lim_{t \to \infty} \dot{w}(t) = 0$, the asymptotic convergence can be achieved.

Theorem 5.1.2 *Assume that* $\lim_{t \to \infty} \dot{w}(t) = 0$; *Assumptions 5.1.1 and 5.1.3 are satisfied; the functions* $h_i(\cdot)(i = 1, 2, \cdots, n+1)$ *in ESO (5.1.3) satisfy Assumption 5.1.4; and the first inequality of condition 2 in Assumption 5.1.4 holds on* $\mathbb{R}^{n+1}$. *Then, there exists an* $r_0 > 0$ *such that*

$$\lim_{t \to \infty} |\hat{x}_i(t) - x_i(t)| = 0, \quad r > r_0, \quad i = 1, 2, \cdots, n+1, \quad x_{n+1}(t) = w(t). \quad (5.1.20)$$

Proof. Let $\eta(t)$ and $\eta_i(t)$ for $i = 1, 2, \cdots, n+1$ be defined by (5.1.6). Then it is easy to verify that $\eta_i(t)$ also satisfies (5.1.7) with $x_{n+1}(t) = w(t)$. Let $\mathcal{V}(\nu)$ and $\mathcal{W}(\nu)$ be the Lyapunov functions satisfying Assumption 5.1.4. Finding the derivative of $\mathcal{V}(\eta(t))$ along the error equation (5.1.7) gives

$$\frac{d\mathcal{V}(\eta(t))}{dt}\Big|_{\text{along (5.1.7)}} = \sum_{i=1}^{n} (r(\eta_{i+1}(t) - h_i(\eta_1(t)))$$

$$+ r^{n+1-i}[g_i(u(t), x_1(t), \cdots, x_i(t)) - g_i(u(t), \hat{x}_1(t), \cdots, \hat{x}_i(t))]) \frac{\partial \mathcal{V}(\eta(t))}{\partial \eta_i}$$

$$+ (-rh_{n+1}(\eta_1(t)) + \dot{w}(t)) \frac{\partial \mathcal{V}(\eta(t))}{\partial \eta_{n+1}}. \quad (5.1.21)$$

Similar to the proof of Theorem 5.1.1, there exists $r_1 > 0$ such that for every $r > r_1$, $\mathcal{V}(\eta(t)) \leq \kappa_2(R)$ for all $t > t_{r_1}$, where t_{r_1} is an r-dependent constant. This together with Assumption 5.1.4 shows that the derivative of $\mathcal{V}(\eta(t))$ along the solution of (5.1.7) satisfies

$$\frac{d\mathcal{V}(\eta(t))}{dt}\Big|_{\text{along (5.1.7)}} \leq -(r - N_{21}r^{\Lambda})\mathcal{W}(\eta(t)) + M|\dot{w}(t)|, \quad \forall t > t_{r_1}, \quad (5.1.22)$$

where Λ is given in (5.1.13) and N_{21} and M are positive constants. Let $r_0 = \max\{r_1, (2N_{21})^{1/(1-\Lambda)}\}$. For given $\sigma > 0$, since $\lim_{t \to \infty} \dot{w}(t) = 0$, there exists $t_2 > t_{r_1}$ such that $|\dot{w}(t)| < (r/(4M)) (\kappa_3 \circ \kappa_2^{-1})(\sigma)$ for all $t > t_2$, where $\kappa_2(\cdot)$ and $\kappa_3(\cdot)$ are class $\mathcal{K}_\infty$ functions defined by (5.1.15). Therefore, if $\mathcal{V}(\eta(t)) \geq \sigma$, then

$$\frac{d\mathcal{V}(\eta(t))}{dt}\Big|_{\text{along (5.1.7)}} \leq -\frac{r}{2}\mathcal{W}(\eta(t)) + M|\dot{w}(t)| \leq -\frac{r}{4}(\kappa_3 \circ \kappa_2^{-1})(\sigma) < 0, \forall t > t_2, \ r > r_0. \quad (5.1.23)$$

Then $t_3 > t_2$ can then be found such that $\mathcal{V}(\eta(t)) \leq \sigma$ for all $t > t_3$. In other words, $\lim_{t \to \infty} \mathcal{V}(\eta(t)) = 0$. By (5.1.15), this amounts to $\lim_{t \to \infty} \|\eta(t)\| = 0$. Finally,

$$\lim_{t \to \infty} |x_i(t) - \hat{x}_i(t)| = \lim_{t \to \infty} \frac{|\eta_i(t)|}{r^{n+1-i}} = 0, \quad i = 1, 2, \cdots, n+1.$$

In what follows, we construct two classes of ESO in terms of Theorem 5.1.1. The first one is the LESO. That is, the nonlinear functions $h_i(\cdot)(i = 1, 2, \cdots, n+1)$ in (5.1.3) are linear functions: $h_i(\nu) = \alpha_i\nu, \nu \in \mathbb{R}$, where α_i are constants to be specified. □

Corollary 5.1.1 *Let the matrix*

$$
E = \begin{pmatrix}
-\alpha_1 & 1 & \cdots & 0 \\
\vdots & \vdots & \ddots & \vdots \\
-\alpha_n & 0 & \cdots & 1 \\
-\alpha_{n+1} & 0 & \cdots & 0
\end{pmatrix}
\tag{5.1.24}
$$

be Hurwitz and assume Assumption 5.1.3 and Assumption 5.1.1 with all $\theta_i = 1$. Then, the following assertions are valid.

(i) If $f(\cdot)$ is completely unknown, the solution of system (5.1.1) is globally bounded and Assumption 5.1.2 is satisfied; then the states of linear ESO (5.1.3) converge to the states and extended state $x_{n+1}(t) = f(t, x(t), w(t))$ of system (5.1.1) in the sense of (5.1.5). In addition, there exists $r_0 > 0$ such that

$$
|x_i(t) - \hat{x}_i(t)| \le D_1\left(\frac{1}{r}\right)^{n+2-i}, \quad r > r_0, t > t_r, \quad i = 1, 2, \cdots, n+1.
\tag{5.1.25}
$$

where t_r is an r-dependent constant and D_1 is an r-independent constant.
(ii) If $f(t, x(t), w(t)) = w(t)$ and $\lim\limits_{t\to\infty} \dot{w}(t) = 0$, then (5.1.20) holds true for some constant $r_0 > 0$.

Proof. We only show (i) since the proof for (ii) is similar. Let P be the positive definite matrix solution of the Lyapunov equation $PE + E^\top P = -I_{(n+1)\times(n+1)}$. Define the Lyapunov functions $\mathcal{V}, \mathcal{W} : \mathbb{R}^{n+1} \to \mathbb{R}$ by

$$
\mathcal{V}(\nu) = \langle P\nu, \nu\rangle, \mathcal{W}(\nu) = \langle \nu, \nu\rangle, \quad \forall\, \nu \in \mathbb{R}^{n+1}.
\tag{5.1.26}
$$

Then

$$
\lambda_{\min}(P)\|\nu\|^2 \le \mathcal{V}(\nu) \le \lambda_{\max}(P)\|\nu\|^2,
\tag{5.1.27}
$$

$$
\sum_{i=1}^{n}(\nu_{i+1} - \alpha_i\nu_1)\frac{\partial \mathcal{V}(\nu)}{\partial \nu_i} - \alpha_{n+1}\nu_1\frac{\partial \mathcal{V}(\nu)}{\partial \nu_{n+1}} = -\nu^\top\nu = -\|\nu\|^2 = -\mathcal{W}(\nu),
$$

and

$$
\left\|\frac{\partial \mathcal{V}(\nu)}{\partial \nu}\right\| = \|2\nu^\top P\| \le 2\lambda_{\max}(P)\|\nu\|.
$$

where $\lambda_{\max}(P)$ and $\lambda_{\min}(P)$ are the maximal and minimal eigenvalues of P. Therefore, Assumption 5.1.4 is satisfied. The practical convergence follows from Theorems 5.1.1 and 5.1.2 directly.

Now, we show the estimate (5.1.25). By the Lyapunov function defined in (5.1.26) and considering $\theta_i = 1$, we have

$$\frac{dV(\eta(t))}{dt}\Big|_{\text{along (5.1.7)}} \le -(\overline{M}_1 r)V(\eta(t)) + \overline{M}_2\sqrt{V(\eta(t))}, \quad r > 4n\lambda_{\max}(P)\sup_{t\in[0,\infty)}\Gamma(u(t)),$$

(5.1.28)

where $\overline{M}_1 = 1/(2\lambda_{\max}(P))$, $\overline{M}_2 = 2N_{12}\lambda_{\max}(P)/(N\sqrt{\lambda_{\min}(P)})$. It is easy to verify that if $V(\eta(t)) > (2\overline{M}_2/\overline{M}_1)^2 (1/r)^2$, then

$$\frac{dV(\eta(t))}{dt}\Big|_{\text{along (5.1.7)}} \le -\frac{2\overline{M}_2^2}{\overline{M}_1 r} < 0. \tag{5.1.29}$$

Thus there exists $t_r > 0$ such that $V(\eta(t)) \le (2\overline{M}_2/\overline{M}_1)^2(1/r)^2$ for all $t > t_0$. Hence

$$\|\eta(t)\| \le \sqrt{(1/\lambda_{\min}(P))V(\eta(t))} \le 2\overline{M}_2/(\overline{M}_1\lambda_{\min}(P))(1/r).$$

This leads to estimate (i) by variable transform (5.1.7). □

The second class of ESO is the homogeneous ESO (HESO), that is, the nonlinear functions $h_i(\cdot)$ are chosen as (5.1.31), which satisfy weighted homogeneity.

It can be easily shown that the vector field $F(\nu)$ given in the following is homogeneous of degree $-d$, $d = 1 - \beta$, with respect to weights $\{r_i = (i-1)\beta - (i-2)\}_{i=1}^{n+1}$:

$$F(\nu) = (\nu_2 + \alpha_1 h_1(\nu), \nu_3 + \alpha_2 h_2(\nu), \cdots, \nu_{n+1} + \alpha_n h_n(\nu)\alpha_{n+1}(\nu), h_{n+1}(\nu))^\top, \tag{5.1.30}$$

where $\nu = (\nu_1, \nu_2, \cdots, \nu_{n+1}) \in \mathbb{R}^{n+1}$ and

$$h_i(\nu) = [\nu]^{i\beta-(i-1)} \triangleq \text{sign}(\nu)|\nu|^{i\beta-(i-1)}, \nu \in \mathbb{R}, \beta \in (0,1). \tag{5.1.31}$$

The following Corollary 5.1.2 is on the convergence of ESO (5.1.3) with nonlinear functions $h_i(\cdot)$ given in (5.1.31), which is the homogeneous ESO(HESO).

Corollary 5.1.2 *Assume that Assumptions 5.1.1 and 5.1.2 hold true with $\theta_i \in (0,1]$ and the matrix in (5.1.24) is Hurwitz. Then there exists $\beta^* \in (0,1)$ such that for any $\beta \in (\beta^*, 1)$, the following assertions are valid.*

(i) If the solution of system (5.1.1) is globally bounded and Assumption 5.1.2 is satisfied, then the states of ESO (5.1.3) with weighted homogeneous function $h_i(\cdot)$ defined by (5.1.31) converge to the states and extended state $x_{n+1}(t) = f(t, x(t), w(t))$ of system (5.1.1) in the sense of (5.1.20). Moreover, there exists an $r_0 > 0$ such that for all $r > r_0, i \in \{1, 2, \cdots, n+1\}$,

$$\overline{\lim_{t\to\infty}}|x_i(t) - \hat{x}_i(t)| \le D_2\left(\frac{1}{r}\right)^{n+1-i+\frac{(i-1)\beta-(i-2)}{(n+1)\beta-n}(1-\Lambda)}, \tag{5.1.32}$$

$$r > r_0, t > r_r, i = 1, 2, \cdots, n+1,$$

where $\Lambda = \max(n+1-i)(1-\theta_i)$, t_r is an r-dependent constant, and D_2 is an r-independent.

(ii) If $f(t, x(t), w(t)) = w(t)$ and $\lim_{t \to \infty} \dot{w}(t) = 0$, then (5.1.20) holds true for some constant $r_0 > 0$.

Remark 5.1.1 *If $\theta_i = 1$ in Assumption 5.1.1, then the error estimation of the weighted homogeneous ESO (5.1.3) with (5.1.31) is*

$$|x_i(t) - \hat{x}_i(t)| \le D_2 \left(\frac{1}{r}\right)^{n+1-i+\frac{(i-1)\beta - (i-2)}{(n+1)\beta - n}}.$$

Since $(i-1)\beta - (i-2) > (n+1)\beta - n$, we have

$$n + 1 - i + \frac{(i-1)\beta - (i-2)}{(n+1)\beta - n} > n + 2 - i$$

Compared with the error estimation (5.1.25) by LESO, HESO has a more accurate estimation for system (5.1.1) with Lipschitz continuous functions $g_i(\cdot)$. For instance, for $n = 2$ and $\beta = 0.8$, by HESO, the error between $x_3(t)$ and $\hat{x}_3(t)$ is $D_2/r^{3/2}$, while by LESO, the error is D_1/r, where D_1 and D_2 are r-independent constants. Therefore, the observer error of HESO drops much more rapidly than the observer error of LESO as r is increasing. The numerical simulation also witnesses this point.

Proof of Corollary 5.1.2 For the vector field $F(\nu)$ defined by (5.1.30), it follows from Theorem 1.3.9 that there exists a positive definite, radial unbounded Lyapunov function $\mathcal{V}(\nu)$ such that $\mathcal{V}(\nu)$ is homogeneous of degree $\gamma \ge \max\{d = 1 - \beta, r_i\}$ with weights $\{r_i = (i-1)\beta - (i-2)\}_{i=1}^{n+1}$ and $L_F \mathcal{V}(\nu)$ is negative definite. From the homogeneity of $\mathcal{V}(\nu)$, for any positive constant λ,

$$\mathcal{V}(\lambda^{r_1}\nu_1, \lambda^{r_2}\nu_2, \cdots, \lambda^{r_{n+1}}\nu_{n+1}) = \lambda^{\gamma}\mathcal{V}(\nu_1, \nu_2, \cdots, \nu_{n+1}). \tag{5.1.33}$$

Finding the derivatives of both sides of (5.1.33) with respect to ν_i yields

$$\lambda^{r_i} \frac{\partial \mathcal{V}(\lambda^{r_1}\nu_1, \lambda^{r_2}\nu_2, \cdots, \lambda^{r_{n+1}}\nu_{n+1})}{\partial \lambda^{r_i}\nu_i} = \lambda^{\gamma} \frac{\partial \mathcal{V}(\nu_1, \nu_2, \cdots, \nu_{n+1})}{\partial \nu_i}. \tag{5.1.34}$$

This shows that $\partial \mathcal{V}(\nu)/\partial \nu_i$ is homogeneous of degree $\gamma - r_i$ with respect to weights $\{r_i\}_{i=1}^{n+1}$. Moreover, the Lie derivative of $\mathcal{V}(\nu)$ along the vector field $F(\nu)$ satisfies

$$L_F \mathcal{V}(\lambda^{r_1}\nu_1, \lambda^{r_2}\nu_2, \cdots, \lambda^{r_{n+1}}\nu_{n+1})$$

$$= \sum_{i=1}^{n+1} \frac{\partial \mathcal{V}(\lambda^{r_1}\nu_1, \lambda^{r_2}\nu_2, \cdots, \lambda^{r_{n+1}}\nu_{n+1})}{\partial \lambda^{r_i}\nu_i} F_i(\lambda^{r_1}\nu_1, \lambda^{r_2}\nu_2, \cdots, \lambda^{r_{n+1}}\nu_{n+1})$$

$$= \lambda^{\gamma - d} \sum_{i=1}^{n+1} \frac{\partial \mathcal{V}(\nu_1, \nu_2, \cdots, \nu_{n+1})}{\partial \nu_i} F_i(\nu_1, \nu_2, \cdots, \nu_{n+1}) \tag{5.1.35}$$

$$= \lambda^{\gamma - d} L_F \mathcal{V}(\nu_1, \nu_2, \cdots, \nu_{n+1}).$$

Therefore, $L_F \mathcal{V}(\nu)$ is homogeneous of degree $\gamma - d$ with respect to weights $\{r_i\}_{i=1}^{n+1}$. Since $|\nu_i|$ is a function of $(\nu_1, \nu_2, \cdots, \nu_{n+1})$ and is homogeneous of degree r_i with weights $\{r_i\}_{i=1}^{n+1}$, it follows from Property 1.3.3 that

$$\left|\frac{\partial \mathcal{V}(\nu)}{\partial \nu_i}\right| \le b_1(\mathcal{V}(\nu))^{(\gamma-r_i)/\gamma}, \quad \nu \in \mathbb{R}^{n+1}, \quad i = 1, 2, \cdots, n+1, \tag{5.1.36}$$

$$L_F \mathcal{V}(\nu) \le -b_2(\mathcal{V}(\nu))^{(\gamma-d)/\gamma}, \quad \forall \nu \in \mathbb{R}^{n+1}, \tag{5.1.37}$$

$$|\nu_i| \le b_3(\mathcal{V}(\nu_1, \cdots, \nu_{n+1}))^{r_i/\gamma}, \quad i = 1, 2, \cdots, n+1 \tag{5.1.38}$$

where $b_1, b_2,$ and b_3 are positive constants. Let

$$\mathcal{W}(\nu) \triangleq b_2(\mathcal{V}(\nu))^{(\gamma-d)/\gamma}, \quad \nu \in \mathbb{R}^{n+1}.$$

By (5.1.37), we obtain condition 1 of Assumption 5.1.4. For continuous positive definite function $\mathcal{V}(\nu)$, there exists a class $\mathcal{K}_\infty$ function $\kappa_h(\cdot)$ such that $\kappa_h(\|\nu\|) \le \mathcal{W}(\nu)$, $\nu \in \mathbb{R}^{n+1}$. Let $R = \kappa_h^{-1}(1)$. For any $\nu \in \mathbb{R}^{n+1}$, if $\|\nu\| > R$, then $\mathcal{V}(\nu) > 1$. For $\beta \ge \max\{n/(n+1), (i-1+\theta_i)/i\}$, we have $\gamma - r_{n+1} \le \gamma - d$ and $\gamma + \theta_i - r_i \le \gamma - d$. This, together with (5.1.36) and (5.1.38), gives

$$\left|\frac{\partial \mathcal{V}(\nu)}{\partial \nu_{n+1}}\right| \le b_1(\mathcal{V}(\nu))^{(\gamma-r_{n+1})/\gamma} \le \frac{b_2}{b_1}\mathcal{W}(\nu), \nu \in \mathbb{R}^{n+1}, \|\nu\| > R \tag{5.1.39}$$

and

$$\|(\nu_1, \cdots, \nu_i)\|^{\theta_i}\left\|\frac{\partial \mathcal{V}(\nu)}{\partial \nu_i}\right\| \le (|\nu_1| + \cdots + |\nu_i|)^{\theta_i}\left\|\frac{\partial \mathcal{V}(\nu)}{\partial \nu_i}\right\| \le (ib_3)^{\theta_i}(\mathcal{V}(\nu))^{(\theta_i+\gamma-r_i)/\gamma}$$

$$\le \frac{(ib_3)^{\theta_i}}{b_1}\mathcal{W}(\nu). \tag{5.1.40}$$

Therefore, all conditions of Assumption 5.1.4 are satisfied. The practical convergence then follows from Theorem 5.1.1.

Now we only need to show (5.1.32) since the proof for (ii) is similar. Using the notation $\eta(t)$ defined by (5.1.6), and practical convergence, there exists an $r_0 > 0$ such that

$$\mathcal{V}(\eta(t)) < 1, \quad \forall t > t_{r_1}, \quad r > r_0, \tag{5.1.41}$$

where t_{r_1} is an r-dependent positive constant. By inequalities (5.1.36) to (5.1.38),

$$\left.\frac{d\mathcal{V}(\eta(t))}{dt}\right|_{\text{along (5.1.7)}} \le -b_2 r(\mathcal{V}(\eta(t)))^{(\gamma-d)/\gamma} + \overline{N}_{11}(\mathcal{V}(\eta(t)))^{(\gamma-r_{n+1})/\gamma}$$

$$+ r^\Lambda \overline{N}_{12} \sum_{i=1}^n (\mathcal{V}(\eta(t)))^{\frac{\gamma+\theta_i r_i - r_i}{\gamma}}, \tag{5.1.42}$$

where

$$\overline{N}_{11} = b_1 N_{12}/N, \quad \overline{N}_{12} = nb_1 \sup_{t\in[0,\infty)} \Gamma(u(t)) \max_{i=1,\cdots,n} b_3^{\theta_i}.$$

Let $\beta^* = \max\{\beta_1^*, (n - \min\{\theta_1, \cdots, \theta_n\})/n, n/(n+1), (i-1+\theta_i)/i\}$. Since $\beta \in (\beta^*, 1)$, $r_i = (i-1)\beta - (i-2)$, it follows that $r_{n+1} \ge 1 - \min\{\theta_i\} \ge r_i(1-\theta_i)$, and hence $(\gamma - r_{n+1})/\gamma \leq (\gamma + \theta_i r_i - r_i)/\gamma$. Now, suppose that $r > r_0$, $t > t_{1r}$. Then (5.1.42) can be rewritten as

$$\left.\frac{d\mathcal{V}(\eta(t))}{dt}\right|_{\text{along (5.1.7)}} \leq -rb_2(\mathcal{V}(\eta(t)))^{(\gamma-d)/\gamma} + r^\Lambda(\overline{N}_{11} + n\overline{N}_{12})(\mathcal{V}(\eta(t)))^{(\gamma-r_{n+1})/\gamma}.$$

(5.1.43)

If

$$\mathcal{V}(\eta(t)) \geq ((N_{11} + N_{12})/b_2)^{\gamma/(r_{n+1}-d)}(1/r)^{\frac{\gamma(1-\Lambda)}{r_{n+1}-d}},$$

then

$$\left.\frac{d\mathcal{V}(\eta(t))}{dt}\right|_{\text{along (5.1.7)}} \leq -r^\Lambda(\overline{N}_{11} + n\overline{N}_{12})(\mathcal{V}(\eta(t)))^{(\gamma-r_{n+1})/\gamma} < 0.$$

Thus there exists $t_r > t_{1r}$ such that $\mathcal{V}(\eta(t)) \leq ((N_{11}+N_{12})/b_2)^{\gamma/(r_{n+1}-d)}(1/r)^{\gamma(1-\Lambda)/(r_{n+1}-d)}$ for all $t > t_r$; (5.1.32) then follows from inequality (5.1.38). $\qquad\square$

5.1.2 Time-Varying Gain ESO

In the previous subsection, the constant high-gain ESO (5.1.3) is designed to attenuate the vast uncertainty for system (5.1.1). The advantage of high gain lies in its fast convergence and filtering function for high-frequency noise. However, as mentioned before, firstly, the constant high gain may cause peaking value problem in the initial stage by different initial values of the system and ESO. The second point is that the derivative of external disturbance must be uniformly bounded. The third problem is that in most situations, only practical convergence is expected unless the derivative of disturbance satisfies additional conditions like $\lim_{t\to\infty} \dot{w}(t) = 0$ in Theorem 5.1.2. To solve all these problems aforementioned, we design once again a time-varying gain ESO where the time-varying gain increases slowly in the initial stage to reach its maximal value. Our aim is to shrink the error between the states of the system and ESO first and reduce subsequently the peaking value. The mechanism of time-varying gain design and constant high gain for the peaking value problem is also analyzed. The time-varying gain is designed as follows:

$$\begin{cases} \dot{\hat{x}}_1(t) = \hat{x}_2(t) + \dfrac{1}{\rho^{n-1}(t)}h_1(\rho^n(t)(y(t) - \hat{x}_1(t))) + g_1(u(t), \hat{x}_1(t)), \\[2mm] \dot{\hat{x}}_2(t) = \hat{x}_3(t) + \dfrac{1}{\rho^{n-2}(t)}h_2(\rho^n(t)(y(t) - \hat{x}_1(t))) + g_2(u(t), \hat{x}_1(t), \hat{x}_2(t)), \\[1mm] \quad\vdots \\[1mm] \dot{\hat{x}}_n(t) = \hat{x}_{n+1}(t) + h_n(\rho^n(t)(y(t) - \hat{x}_1(t))) + g_n(u(t), \hat{x}_1(t), \cdots, \hat{x}_n(t)), \\[1mm] \dot{\hat{x}}_{n+1}(t) = \rho(t)h_{n+1}(\rho^n(t)(y(t) - \hat{x}_1(t))). \end{cases}$$

(5.1.44)

In both systems (5.1.1) and (5.1.44), we first take the time-varying gain $\rho : \mathbb{R} \to \mathbb{R}$ to grow gradually from a small value to its maximal value to reduce the peaking value:

$$\varrho(t) = \begin{cases} e^{at}, & 0 \leq t < \dfrac{1}{a}\ln r, \\[2mm] r, & t \geq \dfrac{1}{a}\ln r. \end{cases}$$

(5.1.45)

Theorem 5.1.3 *Let the gain function $\varrho(t)$ be chosen as in (5.1.45) and assume that Assumptions 5.1.1, 5.1.3, and 5.1.4 are satisfied.*

(i) If $f(\cdot)$ is completely unknown and each solution of (5.1.1) is globally bounded, then under Assumption 5.1.2, for any given $\sigma > 0$, there exists an $r_0 > 0$ such that

$$|\hat{x}_i(t) - x_i(t)| < \sigma, \quad \forall\, t > t_r, r > r_0, \quad i = 1, 2, \cdots, n+1, \tag{5.1.46}$$

where $\hat{x}_i(t)$ is the solution of (5.1.44), $x_i(t)$ is the solution of system (5.1.1), $x_{n+1}(t)$ is the total disturbance (5.1.2), and t_r is an r-dependant constant.

(ii) If $f(t, x(t), w(t)) = w(t)$, $\lim\limits_{t\to\infty} \dot{w}(t) = 0$, and the first inequality of condition 2 in Assumption 5.1.4 holds on $\mathbb{R}^{n+1}$, then there exists an $r_0 > 0$ such that

$$\lim_{t\to\infty} |x_i(t) - \hat{x}_i(t)| = 0, \quad i = 1, 2, \cdots, n+1, \quad \forall\, r > r_0. \tag{5.1.47}$$

The proof of Theorem 5.1.3 is the same as the proofs of Theorems 5.1.1 and 5.1.2 because when $t > \ln r/a$, the time-varying gain ESO (5.1.44) is reduced to ESO (5.1.3).

Remark 5.1.2 *Replacing r by $\rho(t)$ in ESO (5.1.3) with linear $h_i(\cdot)$ or homogeneous $h_i(\cdot)$, we obtain the corresponding Corollaries 5.1.1 and 5.1.2, respectively.*

Theorem 5.1.1 gives only practical convergence (disturbance attenuation) and the derivative of disturbance is required to be bounded. In what follows, we show that the asymptotic stability can be achieved by a properly chosen gain function. To this purpose, we propose that Assumption 5.1.5 replaces Assumption 5.1.3, where the boundedness of the derivative of external disturbance is relaxed to allow exponential growth.

Assumption 5.1.5 The control input is bounded on $\mathbb{R}^m$ and there exist positive constants b, B_1, and B_2 such that $|w(t)| + |\dot{w}(t)| \le B_1 + B_2 e^{bt}$.

Assumption 5.1.2 is replaced by the following Assumption 5.1.6.

Assumption 5.1.6 In system (5.1.1), the partial derivatives of f with respect to $x_i (i = 1, \cdots, n)$ and $w(t)$ are bounded on $\mathbb{R}^{n+2}$ and there exist function $\varpi \in C(\mathbb{R}^n, [0, \infty))$ and positive constant B_3 such that

$$|f(t, x, w)| + \left| \frac{\partial f(t, x, w)}{\partial t} \right| \le \varpi(x) + B_3 |w|, \quad \forall\, (t, x, w) \in \mathbb{R}^{n+2}.$$

Theorem 5.1.4 *Assume that $g_i(\cdot)$ in system (5.1.1) satisfy Assumption 5.1.1, $u(t)$ and $w(t)$ satisfy Assumption 5.1.5. Assume that $h_i(\cdot)$ in ESO (5.1.44) satisfy Assumption 5.1.4 and let $\varrho(t) = e^{at}$, $a > b$. Then the following assertions are valid.*

(i) If $f(\cdot)$ is completely unknown yet satisfies a priori Assumption 5.1.6 and the solution of system (5.1.1) is globally bounded, then

$$\lim_{t\to\infty} |x_i(t) - \hat{x}_i(t)| = 0, \quad i = 1, 2, \cdots, n+1, \tag{5.1.48}$$

where $\hat{x}_i(t)$ *is the solution of ESO (5.1.44),* $x_i(t)$ *is the solution of system (5.1.1), and* $x_{n+1}(t) = f(t, x(t), w(t))$ *is the total disturbance.*

(ii) If $f(t, x(t), w(t)) = w(t)$, *then*

$$\lim_{t\to\infty} |x_i(t) - \hat{x}_i(t)| = 0, \quad i = 1, 2, \cdots, n+1, \qquad (5.1.49)$$

where $\hat{x}_i(t)$ *is the solution of ESO (5.1.44),* $x_i(t)$ *is the solution of system (5.1.1), and* $x_{n+1}(t) = w(t)$.

Proof. We only give a proof of assertion (i) since the proof for (ii) is very similar. Let

$$\eta_i(t) = (\rho(t))^{n+1-i}(x_i(t) - \hat{x}_i(t)), \quad i = 1, 2, \cdots, n+1, \quad \eta(t) = (\eta_1(t), \cdots, \eta_{n+1}(t))^\top. \qquad (5.1.50)$$

A straightforward computation shows that $\eta_i(t)$ satisfies the following differential equation:

$$\begin{cases} \dot{\eta}_1(t) = \rho(t)(\eta_2(t) - h_1(\eta_1(t))) + \dfrac{n\dot{\rho}(t)}{\rho(t)}\eta_1(t) \\ \qquad + (\rho(t))^n(g_1(u(t), x_1(t)) - g_1(u(t), \hat{x}_1(t))), \\ \qquad \vdots \\ \dot{\eta}_n(t) = \rho(t)(\eta_{n+1}(t) - h_n(\eta_1(t))) + \dfrac{\dot{\rho}(t)}{\rho(t)}\eta_n(t), \\ \qquad + \rho(t)(g_n(u(t), x_1(t), \cdots, x_n(t)) - g_n(u(t), \hat{x}_1(t), \cdots, \hat{x}_n(t))), \\ \dot{\eta}_{n+1}(t) = -\rho(t)h_{n+1}(\eta_1(t)) + \dfrac{d}{dt}f(t, x(t), w(t)). \end{cases} \qquad (5.1.51)$$

By Assumption 5.1.1 and (5.1.50),

$$(\rho(t))^{n+1-i}|g_i(u(t), x_1(t), \cdots, x_i(t)) - g_i(u(t), \hat{x}_1(t), \cdots, \hat{x}_i(t))|$$
$$\leq \Gamma(u(t))(\rho(t))^{(n+1-i)(1-\theta_i)}\|(\eta_1(t), \cdots, \eta_i(t))\|^{\theta_i}. \qquad (5.1.52)$$

By Assumptions 5.1.1, 5.1.6, 5.1.5, and the boundedness of solution of (5.1.1),

$$\left|\frac{df(t, x(t), w(t))}{dt}\right| \leq \left|\frac{\partial f(t, x(t), w(t))}{\partial t}\right| + \left|\dot{w}(t)\frac{\partial f(t, x(t), w(t))}{\partial w}\right|$$
$$+ \left|\sum_{i=1}^{n-1}(x_{i+1}(t) + g_i(u(t), x_1(t), \cdots, x_i(t)))\frac{\partial f(t, x(t), w(t))}{\partial x_i}\right|$$
$$+ \left|(f(t, x(t), w(t)) + g_n(u(t), x(t)))\frac{\partial f(t, x(t), w(t))}{\partial x_{n+1}}\right| \leq \Pi_1 + \Pi_2 e^{bt}, \qquad (5.1.53)$$

where Π_1 and Π_2 are positive constants.

Let $\mathcal{V}(\eta)$ be the Lyapnuov function satisfying Assumption 5.1.4. Finding the derivative of $\mathcal{V}(\eta(t))$ along the error equation (5.1.51) gives

$$\left.\frac{d\mathcal{V}(\eta(t))}{dt}\right|_{\text{along (5.1.51)}} = \sum_{i=1}^{n}\left(\rho(t)(\eta_{i+1}(t) - h_i(\eta_1(t))) + \frac{(n+1-i)\dot{\rho}(t)}{\rho(t)}\eta_i(t)\right.$$
$$\left. + (\rho(t))^{n+1-i}[g_i(u(t), x_1(t), \cdots, x_i(t)) - g_i(u(t), \hat{x}_1(t), \cdots, \hat{x}_i(t))]\right)\frac{\partial\mathcal{V}(\eta(t))}{\partial\eta_i}$$
$$+ \left(-\rho(t)h_{n+1}(\eta_1(t)) + \frac{df(t, x(t), w(t))}{dt}\right)\frac{\partial\mathcal{V}(\eta(t))}{\partial\eta_{n+1}}. \qquad (5.1.54)$$

If $\|\eta(t)\| \geq R$, it follows from Assumption 5.1.4, (5.1.52), and (5.1.53) that

$$\left. \frac{d\mathcal{V}(\eta(t))}{dt} \right|_{\text{along } (5.1.51)} \leq - \left(\rho(t) - N_{11}(\rho(t))^\Lambda - \frac{n(n+1)N}{2a} - \Pi_1 - \Pi_2 e^{bt} \right) \mathcal{W}(\eta(t)),$$
(5.1.55)

where N_{11} and Λ are defined in (5.1.13). Since $\varrho(t) = e^{at}, a > b$, there exists a $t_1 > 0$ such that

$$\rho(t) - N_{11}(\rho(t))^\Lambda - \frac{n(n+1)N}{2a} - \Pi_1 - \Pi_2 e^{bt} > 1, \quad \forall\, t > t_1.$$

Hence, there exists $t_2 > t_1$ such that

$$\left. \frac{d\mathcal{V}(\eta(t))}{dt} \right|_{\text{along } (5.1.51)} \leq -\mathcal{W}(\eta(t)) \leq -\kappa_3(\|\eta(t)\|) \leq -\kappa_3(R) < 0, \quad \forall\, t > t_2. \quad (5.1.56)$$

Therefore, one can find $t_3 > t_2$ such that $\|\eta(t)\| \leq R$ for all $t > t_3$. Let

$$M_{31} = \sup_{t \in [0,\infty)} \Gamma(u(t)) \sum_{i=1}^n R^{\theta_i} \sup_{\|\nu\| \leq R} \left| \frac{\partial \mathcal{V}(\nu)}{\partial \nu_i} \right|, \quad M_{32} = \Pi_2 \sup_{\|\nu\| \leq R} \left| \frac{\partial \mathcal{V}(\nu)}{\partial \nu_{n+1}} \right|,$$

$$M_{33} = \left(\frac{an(n+1)N}{2} + \Pi_1 \right) \sup_{\|\nu\| \leq R} \left\| \frac{\partial \mathcal{V}(\nu)}{\partial \nu} \right\|, \quad \nu = (\nu_1, \cdots, \nu_{n+1}) \in \mathbb{R}^{n+1}.$$
(5.1.57)

Then

$$\left. \frac{d\mathcal{V}(\eta(t))}{dt} \right|_{\text{along } (5.1.51)} \leq e^{at}(-\mathcal{W}(\eta(t)) + M_{31}e^{(\Lambda-1)at} + M_{32}e^{(b-a)t} + M_{33}e^{-at}), \quad \forall\, t > t_3.$$
(5.1.58)

Furthermore, since for any $\sigma > 0$,

$$\lim_{t \to \infty} (M_{31}e^{(\Lambda-1)at} + M_{32}e^{(b-a)t} + M_{33}e^{-at}) = 0,$$

there exists $t_4 > t_3$ such that

$$M_{31}e^{(\Lambda-1)at} + M_{32}e^{(b-a)t} + M_{33}e^{-at} < \frac{1}{2}(\kappa_3 \circ \kappa_2^{-1})(\sigma), \quad \forall\, t > t_4,$$

where $\kappa_2(\cdot)$ and $\kappa_3(\cdot)$ are class $\mathcal{K}_\infty$ functions given in (5.1.15). Hence for any $t > t_4$, if $\mathcal{V}(\eta(t)) > \sigma$, then

$$\left. \frac{d\mathcal{V}(\eta(t))}{dt} \right|_{\text{along } (5.1.51)} \leq -\frac{1}{2}(\kappa_3 \circ \kappa_2^{-1})(\sigma) < 0. \quad (5.1.59)$$

Therefore, one can find $t_5 > t_4$ such that $\mathcal{V}(\eta(t)) < \sigma$ for all $t > t_5$. This amounts to $\lim_{t \to \infty} \mathcal{V}(\eta(t)) = 0$. The assertion (i) then follows from (5.1.15) and (5.1.50). $\quad\square$

To end this section, we give a brief discussion on the notorious peaking-value problem in the high-gain method, both by constant high gain and time-varying gain. To bring the object into focus, we take $g_i(\cdot) = 0, i = 1, 2, \cdots, n$ in system (5.1.1) and consider the linear ESO

(5.1.3) as that in Corollary 5.1.1 with the Hurwitz matrix E given by (5.1.24). Suppose that E has $n + 1$ different negative real eigenvalues $\lambda_1, \cdots, \lambda_{n+1}$. Firstly, by the constant high gain $\varrho(t) \equiv r$, the solution of ESO (5.1.3) is

$$\hat{x}_i(t) = \frac{1}{r^{n+1-i}} \eta_i(t) + x_i(t), \tag{5.1.60}$$

where

$$\begin{pmatrix} \eta_1(t) \\ \eta_2(t) \\ \vdots \\ \eta_{n+1}(t) \end{pmatrix} = e^{-rEt} \begin{pmatrix} r^n(x_1(0) - \hat{x}_1(0)) \\ r^{n-1}(x_2(0) - \hat{x}_2(0)) \\ \vdots \\ x_{n+1}(0) - \hat{x}_{n+1}(0) \end{pmatrix} + \int_0^t e^{-rE(t-s)} \begin{pmatrix} 0 \\ 0 \\ \vdots \\ \dot{x}_{n+1}(s) \end{pmatrix} ds. \tag{5.1.61}$$

The peaking value is mainly caused by a large initial value of $\eta(t) = (\eta_1(t), \cdots, \eta_{n+1}(t))$:

$$\eta_i(t) = \sum_{j=1}^{n+1} \sum_{l=1}^{n+1} d_{ij}^l e^{rt\lambda_l} r^{n+1-i} (x_i(0) - \hat{x}_i(0)) + \sum_{l=1}^{n+1} \int_0^t \dot{x}_{n+1}(s) d_{i(n+1)}^l e^{rt\lambda_l} ds, \tag{5.1.62}$$

where d_{ij}^l are real numbers determined by the matrix E. It is seen that the peaking value occurs only at $t = 0$ since for any $a > 0$, $\eta_i(t) \to 0$ as $r \to \infty$ uniformly in $t \in [a, \infty)$. On the other hand, in the initial time stage, however, $e^{rt\lambda_i}$ is very close to 1. This is the reason behind the peaking value problem by constant high gain. Actually, the peaking values for $\hat{x}_2(t), \cdots, \hat{x}_{n+1}(t)$ are on the order of $r, r^2, \cdots, r^n$, respectively. The larger r is, the larger the peaking values are.

Next, when we apply the time-varying gain and let the gain be relatively small in the initial stage, the initial value of error $\eta(t)$ is

$$(\rho(0)^n (x_1(0) - \hat{x}_1(0)), \rho^{n-1}(0)(x_2(0) - \hat{x}_2(0)), \cdots, x_{n+1}(0) - \hat{x}_{n+1}(0))^\top, \tag{5.1.63}$$

which is also small. Actually, if $\rho(0) = 1$, the initial value of error $\eta(t)$ is

$$((x_1(0) - \hat{x}_1(0)), (x_2(0) - \hat{x}_2(0)), \cdots, x_{n+1}(0) - \hat{x}_{n+1}(0))^\top. \tag{5.1.64}$$

Since the gain function $\rho(t)$ is small in the initial stage, when $\|\eta(t)\|$ increases with increasing eigenvalues to some given value, $\|\eta(t)\|$ stops increasing at some value that is determined by the system functions and the external disturbances, but does not rely on the maximal value of $\rho(t)$. Actually, let $\mathcal{V} : \mathbb{R}^{n+1} \to [0, \infty)$ be $\mathcal{V}(v) = \langle Pv, v \rangle$, $v \in \mathbb{R}^{n+1}$ and let the gain function $\rho(t)$ be chosen as in (5.1.45) or an exponential function; we can then see from the proof of Theorem 5.1.4 that the derivative of $\mathcal{V}(\eta(t))$ along the error equation (5.1.51) satisfies

$$\left. \frac{d\mathcal{V}(\eta(t))}{dt} \right|_{\text{along (5.1.51)}} \leq -(\varrho(t) - N_{12})\langle \eta(t), \eta(t) \rangle, \tag{5.1.65}$$

where N_{12} is the upper bound of the derivative of total disturbance, which is given by (5.1.13). When ϱ increases to N_{12}, then $\mathcal{V}(\eta(t))$ stops increasing. This together with

$$\|\eta(t)\| \leq \frac{1}{\lambda_{\max}(P)} \mathcal{V}(\eta(t)) \tag{5.1.66}$$

shows that $\|\eta(t)\|$ does not increase any more although $\rho(t)$ increases continuously to a large number r or ∞. If $N_{12} \leq 1$, $\mathcal{V}(\eta(t))$ decreases from the beginning. It follows from (5.1.66) and (5.1.60) that the peaking values become much smaller.

5.1.3 Numerical Simulation

In this section, we present some examples and numerical simulations for illustration.

Example 5.1.1 *Consider the following uncertain nonlinear system:*

$$\begin{cases} \dot{x}_1(t) = x_2(t) + g_1(u(t), x_1(t)), \\ \dot{x}_2(t) = f(t, x(t), w(t)) + g_2(u(t), x_1(t), x_2(t)), \end{cases} \tag{5.1.67}$$

where $x(t) = (x_1(t), x_2(t))$ *is the state,* $g_1(u(t), x_1(t)) = u(t) \sin x_1(t)$, $g_2(u(t), x_1(t), x_2(t)) = u(t) \sin x_2(t)$ *are known functions,* $u(t) = 1 + \sin t$ *is the control input, and* $y(t) = x_1(t)$ *is the output. The total disturbance* $x_3(t) \triangleq f(t, x(t), w(t))$ *is completely unknown.*

According to Corollary 5.1.1, we can design the following LESO for system (5.1.67):

$$\begin{cases} \dot{\hat{x}}_1(t) = \hat{x}_2(t) + r\alpha_1((y(t) - \hat{x}_1(t))) + g_1(u(t), \hat{x}_1(t)), \\ \dot{\hat{x}}_2(t) = \hat{x}_3(t) + r^2\alpha_2((y(t) - \hat{x}_1(t))) + g_2(u(t), \hat{x}_1(t), \hat{x}_2(t)), \\ \dot{\hat{x}}_3(t) = r^3\alpha_3((y(t) - \hat{x}_1(t))), \end{cases} \tag{5.1.68}$$

where $\alpha_1 = \alpha_2 = 3, \alpha_3 = 1$. In the numerical simulation, we take the external disturbance $w(t)$ and nonlinear function $f(t, x, w)$ in system (5.1.67) as, respectively,

$$w(t) = \sin(2t + 1), \quad f(t, x, w) = \sin t - 2x_1 - 4x_2 + w + \cos(x_1 + x_2 + w).$$

System (5.1.67) satisfies Assumptions 5.1.1, 5.1.2, and 5.1.3. This is because the matrix of the linear main part of system (5.1.67) is

$$\begin{pmatrix} 0 & 1 \\ -2 & -4 \end{pmatrix},$$

which is Hurwitz. Since the control and external disturbance are uniformly bounded, the solution of system (5.1.67) is bounded as well. The eigenvalues of the matrix

$$E = \begin{pmatrix} -\alpha_1 & 1 & 0 \\ -\alpha_2 & 0 & 1 \\ -\alpha_3 & 0 & 0 \end{pmatrix} = \begin{pmatrix} -3 & 1 & 0 \\ -3 & 0 & 1 \\ -1 & 0 & 0 \end{pmatrix}$$

are identical to be $\{-1\}$. This shows that all conditions of Corollary 5.1.1 are satisfied.

We first take the tuning-gain parameter $r = 10$, the discrete step $\Delta t = 0.001$ for time, and the initial states of system (5.1.67) and LESO (5.1.68) to be $(1, 1)$ and $(0, 0, 0)$, respectively.

The numerical results are plotted in Figure 5.1.1, where Figure 5.1.1(a) is on $x_1(t)$ and $\hat{x}_1(t)$, Figure 5.1.1(b) is on $x_2(t)$ and $\hat{x}_2(t)$, and Figure 5.1.1(c) is on $x_3(t)$ and $\hat{x}_3(t)$.

According to Corollary 5.1.2, an HESO for system (5.1.67) can be designed as follows:

$$
\begin{cases}
\dot{\hat{x}}_1(t) = \hat{x}_2(t) + \dfrac{1}{r^{n-1}}[r^n(y(t) - \hat{x}_1(t))]^\beta + u(t)\sin\left([\hat{x}_1(t)]^{2/3}\right), \\[2mm]
\dot{\hat{x}}_2(t) = \hat{x}_3(t) + \dfrac{1}{r^{n-2}}[r^n(y(t) - \hat{x}_1(t))]^{2\beta-1} + u(t)\sin([\hat{x}_2(t)]^{1/2}), & (5.1.69) \\[2mm]
\dot{\hat{x}}_3(t) = r[r^2(y(t) - \hat{x}_1(t))]^{3\beta-2}.
\end{cases}
$$

Let $\alpha_1 = 3$, $\alpha_2 = 3$, $\alpha_3 = 1$, and $\beta = 0.8$, with the tuning gain $r = 10$. The numerical results of HESO (5.1.69) for (5.1.67) are plotted in Figure 5.1.2, where Figure 5.1.2(a) is on $x_1(t)$ and $\hat{x}_1(t)$, Figure 5.1.2(b) is on $x_2(t)$ and $\hat{x}_2(t)$, and Figure 5.1.2(c) is on $x_3(t)$ and $\hat{x}_3(t)$.

From Figures 5.1.1 and 5.1.1 we can see that although the tuning gain $r = 10$ is not large, the estimations of the states $x_1(t)$, $x_2(t)$ and the extended state $x_3(t) \triangleq f(x_1(t), x_2(t), w(t)) =$

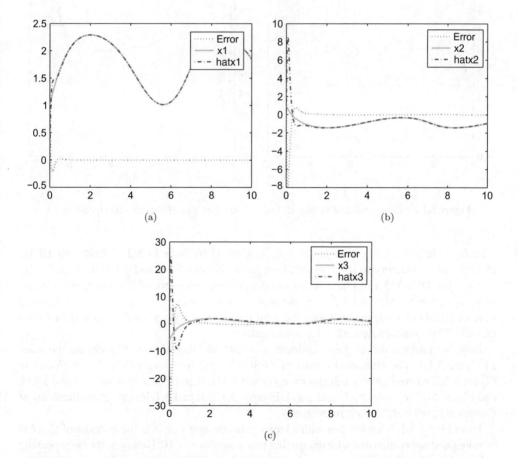

Figure 5.1.1 The numerical results for system (5.1.67) by constant gain LESO (5.1.68) with $r = 10$.

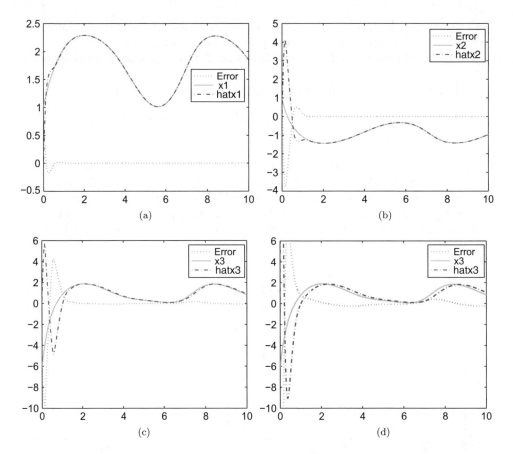

Figure 5.1.2 Observation of system (5.1.67) by constant gain HESO (5.1.69) with $r = 10$.

$-2x_1(t) - 4x_2(t) + w(t) + \cos(x_1(t) + x_2(t) + w(t))$ by both LESO (5.1.68) and HESO (5.1.69) are satisfactory to some extent. Comparing Figures 5.1.1 and 5.1.1, the states $\hat{x}_2(t)$ and $\hat{x}_3(t)$ of HESO (5.1.69) have much smaller peaking values than LESO (5.1.68). We also magnify Figure 5.1.1(c) in 5.1.2(d) by the same scale as Figure 5.1.2(c). We see that the third state of HESO (5.1.69) can estimate the total disturbance with a smaller error than LESO (5.1.68). This confirms Remark 5.1.1 numerically.

Now, we take the tuning gain parameter $r = 200$, and the other parameters are the same as Figure 5.1.1. The numerical results of LESO (5.1.68) for system (5.1.67) are plotted in Figure 5.1.3, where Figure 5.1.3(a) is on $x_1(t)$ and $\hat{x}_1(t)$, Figure 5.1.3(b) is on $x_2(t)$ and $\hat{x}_2(t)$, and Figure 5.1.3(c) is on $x_3(t)$ and $\hat{x}_3(t)$. Figures 5.1.3(d) and 5.1.3(e) are magnifications of Figures 5.1.3(b) and 5.1.3(c), respectively.

From Figure 5.1.3, we see that with a larger gain constant $r = 200$, the estimate of (5.1.68) is more satisfactory than that with the smaller gain constant $r = 10$. However, the large peaking values of $\hat{x}_2(t)$ and $\hat{x}_3(t)$ are observed in the initial stage. The peaking value of $\hat{x}_2(t)$ is near 200 and that of $\hat{x}_3(t)$ is even greater than 10^4.

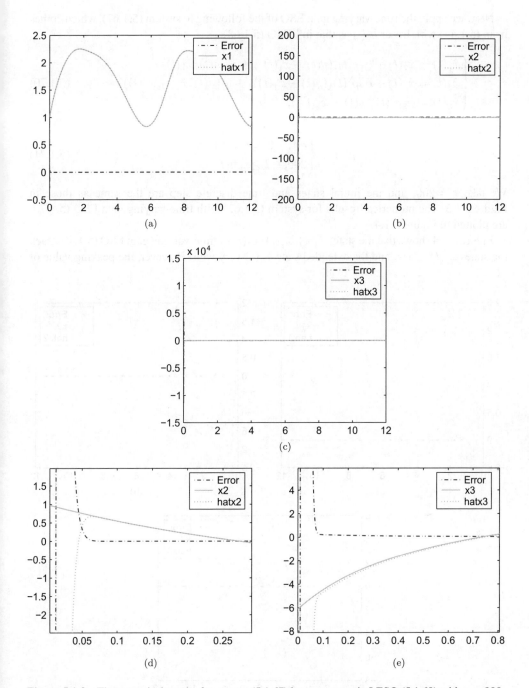

Figure 5.1.3   The numerical results for system (5.1.67) by constant gain LESO (5.1.68) with $r = 200$.

Now, we apply the time-varying gain ESO of the following to system (5.1.67), which comes from (5.1.44) with linear $h_i(\cdot)$ as that in LESO (5.1.68):

$$\begin{cases} \dot{\hat{x}}_1(t) = \hat{x}_2(t) + \alpha_1 \rho(t)((y(t) - \hat{x}_1(t))) + g_1(u(t), \hat{x}_1(t)), \\ \dot{\hat{x}}_2(t) = \hat{x}_3(t) + \alpha_2 \rho^2(t)((y(t) - \hat{x}_1(t))) + g_2(u(t), \hat{x}_1(t), \hat{x}_2(t)), \\ \dot{\hat{x}}_3(t) = \alpha_3 \rho^3(t)((y(t) - \hat{x}_1(t))), \end{cases} \qquad (5.1.70)$$

where

$$\varrho(t) = \begin{cases} e^{2t}, & 0 \le t < \frac{1}{2}\ln r, \\ 200, & t \ge \frac{1}{2}\ln r. \end{cases} \qquad (5.1.71)$$

We take $r = 200$, and the initial states and time discrete step are the same as those in Figure 5.1.3. The numerical results for system (5.1.67) with time-varying gain ESO (5.1.70) are plotted in Figure 5.1.4.

Figure 5.1.4 shows that the states $\hat{x}_1(t), \hat{x}_2(t), \hat{x}_3(t)$ of time-varying gain ESO (5.1.70) track the states $x_1(t), x_2(t)$, and the extended state $x_3(t)$ very well. However, the peaking value of

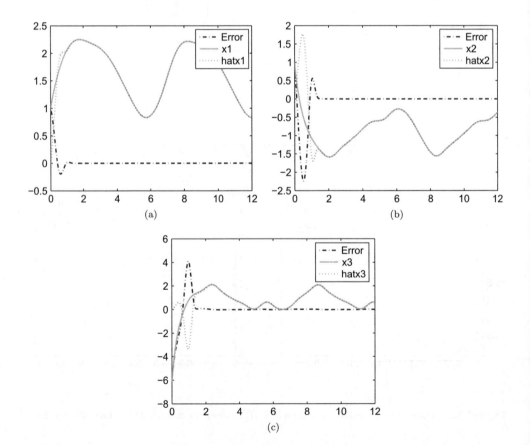

Figure 5.1.4 The numerical results for system (5.1.67) by time-varying LESO (5.1.70).

$\hat{x}_2(t)$ is less than 2 (over 200 by constant high gain) and that of $\hat{x}_3(t)$ is less than 4 (over 10^4 by constant high gain). This shows that the time-varying gain method reduces dramatically the peaking value caused by the constant high gain.

In the following, we consider a simple Hölder continuous system with external disturbance. We consider not only the convergence of the ESO but also the ESO-based feedback control as an application of ESO in control design.

Example 5.1.2 *Consider the following system:*

$$\dot{x}(t) = -[x(t)]^{\frac{1}{2}} + w(t) + u(t), \tag{5.1.72}$$

where $u(t)$ is the control input and $w(t)$ is the external disturbance.

If $w(t) \equiv 0$ and $u(t) \equiv 0$, then system (5.1.72) is finite-time stable. In fact, if $w(t) = u(t) \equiv 0$, the solution of (5.1.72) starting from initial value $x_0 \in \mathbb{R}$ is $x(t; x_0) = \text{sign}(x_0)$ $\left| |x_0|^{\frac{1}{2}} - t/2 \right|^2, t < 2|x_0|^{\frac{1}{2}}$; $x(t; x_0) = 0, t \geq 2|x_0|^{\frac{1}{2}}$. We design an HESO to estimate the external disturbance and design an ESO-based feedback control (ADRC) to (practically) stabilize the system (5.1.72). According to Corollary 5.1.2, the time-varying gain HESO is given as follows:

$$\begin{cases} \dot{\hat{x}}_1(t) = \hat{x}_2(t) + [\rho(t)(y(t) - \hat{x}_1(t))]^\beta - [x(t)]^{\frac{1}{2}} + u(t), \\ \dot{\hat{x}}_2(t) = \rho(t)[\rho(t)(y(t) - \hat{x}_1(t))]^{2\beta-1}, \end{cases} \tag{5.1.73}$$

where $\beta = 0.8$; ρ is given in (5.1.71). Let $u(t) = -\hat{x}_2(t)$, which is used to compensate (cancel) the disturbance $w(t)$. In the numerical simulation the external disturbance is chosen as $w(t) = 1 + \sin t$, and the initial states of (5.1.72) and (5.1.73) are $x(0) = 1$ and $\hat{x}(0) = (0,0)$, respectively. The numerical results with discrete step $\Delta t = 0.001$ are plotted in Figure 5.1.5.

In Figure 5.1.5(a), the maximal value of the gain function ρ is $r = 5$, and in Figure 5.1.5(a), $r = 200$. We can see from Figure 5.1.5 that for large r, $\hat{x}_2(t)$ tracks disturbance $w(t)$ very

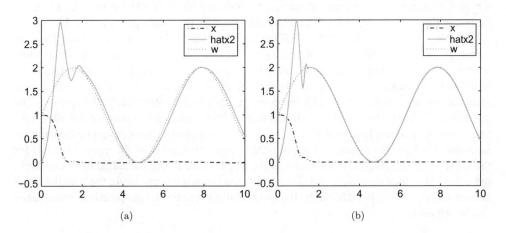

(a) (b)

Figure 5.1.5 Disturbance observation and practical stabilization of system (5.1.72).

well. In addition, under the feedback control $u(t) = -\hat{x}_2(t)$, the state $x(t)$ of system (5.1.72) converges to 0 quite satisfactorily. At the same time, we see that for different r, the maximal value of $\hat{x}_2(t)$ is almost 3 without much change by increasing r. This is coincident with the peaking value reduction of the time-varying gain ESO presented at the end of Subsection 5.1.2.

5.2   Stabilization of Lower Triangular Systems by ADRC

In this section, we consider stabilization for the following lower triangle SISO nonlinear systems with vast uncertainties:

$$\begin{cases} \dot{x}_1(t) = x_2(t) + \phi_1(x_1(t)), \\ \dot{x}_2(t) = x_3(t) + \phi_2(x_1(t), x_2(t)), \\ \quad \vdots \\ \dot{x}_n(t) = f(t, x(t), \zeta(t), w(t)) + b(t, x(t), \zeta(t), w(t))u(t), \\ \dot{\zeta}(t) = f_0(t, x(t), \zeta(t), w(t)), \\ x(0) = (x_{10}, \cdots, x_{n0})^\top, \zeta(0) = (\zeta_{10}, \cdots, \zeta_{s0})^\top, \end{cases} \qquad (5.2.1)$$

where $x(t) = (x_1(t), \cdots, x_n(t)) \in \mathbb{R}^n$ is the state and $\zeta(t) \in \mathbb{R}^s$ is the state of the zero dynamics. The function $\phi_i \in C(\mathbb{R}^i, \mathbb{R})$ is known while $f \in C(\mathbb{R}^{n+s+2}, \mathbb{R})$ and $f_0 \in C(\mathbb{R}^{n+2+s}, \mathbb{R})$ are possibly unknown system functions; $y(t) = x_1(t)$ is the output (measurement); $u(t)$ is the control (input); $w \in C([0, \infty), \mathbb{R})$ is the external disturbance; $b \in C(\mathbb{R}^{n+s+2}, \mathbb{R})$ is the control coefficient with some uncertainty. We assume that there is a known function $b_0 \in C(\mathbb{R}^{n+1}, \mathbb{R})$ serving as a nominal function of $b(\cdot)$. The ESO can estimate the state of the x subsystem in (5.2.1) and the "total disturbance" $x_{n+1}(t)$, which contains an unmodeled system dynamic, external disturbance, and uncertainty caused by the deviation of control parameter $b(\cdot)$ from its nominal value $b_0(\cdot)$:

$$x_{n+1}(t) \triangleq f(t, x(t), \zeta(t), w(t)) + (b(t, x(t), \zeta(t), w(t)) - b_0(t, \hat{x}(t)))u(t), \qquad (5.2.2)$$

where $\hat{x}(t)$ denotes the estimate of state $x(t)$, obtained by ESO. On the basis of ESO, we design an ESO-based output feedback control to stabilize the x-subsystem in (5.2.1).

We design two types of nonlinear ESO for system (5.2.1). The first one is using the constant high-gain ESO. The main problem for constant high-gain ESO is again the peaking-value problem. However, the peaking problem can be significantly reduced through saturation function design by making use of a priori bounds of initial state and external disturbance. The practical convergence of the closed loop under the ESO-based output feedback is proved. The advantage of constant high-gain ESO lies in its easy implementation in practice and allowing a large class of unknown functions $f(\cdot)$ in (5.2.1). However, the feedback design by the constant high-gain ESO is complicated and requires additional information of the a priori bounds for the initial state and external disturbance. If the a priori bounds are estimated to be too small, the closed loop might diverge, and if the estimated bounds are too large, the peaking value becomes large as well. In addition, by the constant high-gain ESO, only practical convergence can be achieved.

To overcome the peaking value problems caused by constant high-gain ESO, we design a second type ESO by using time-varying gain, where the tuning gain is a function of time rather than a constant number. The properly selected gain function can reduce not only the peaking value of ESO dramatically but also simplifies the feedback design without a priori assumption. More importantly, the asymptotic convergence instead of practical convergence can be achieved by the time-varying gain ESO.

On the other hand, the time-varying gain degrades the ability of ESO to filter the high-frequency noise while the constant high-gain ESO does not. A recommended strategy is to apply time-varying gain ESO in the initial stage and then apply constant high-gain ESO after the observer error reaches a reasonable level.

In Subsection 5.2.1, we design a constant high-gain ESO and ESO-based feedback control for system (5.2.1). The practical stability of the closed loop is proved. The asymptotic stability for the closed loop with a special case where the derivative of external disturbance satisfies $\dot{w}(t) \to 0$ as $t \to \infty$ is also discussed. In Subsection 5.2.3, a time-varying gain ESO-based closed loop for system (5.2.1) is proposed. The asymptotic stability is proved. Finally, in Subsection 5.2.4, some numerical simulations are presented for illustration of the convergence and the peaking-value reduction.

5.2.1   ADRC with Constant Gain ESO

The constant gain ESO for (5.2.1) is designed as follows:

$$
\begin{cases}
\dot{\hat{x}}_1(t) = \hat{x}_2(t) + \dfrac{1}{r^{n-1}} g_1(r^n(x_1(t) - \hat{x}_1(t))) + \phi_1(\hat{x}_1(t)), \\
\quad\vdots \\
\dot{\hat{x}}_n(t) = \hat{x}_{n+1}(t) + g_n(r^n(x_1(t) - \hat{x}_1(t))) + b_0(t, \hat{x}_1(t), \cdots, \hat{x}_n(t))u(t), \\
\dot{\hat{x}}_{n+1}(t) = r g_{n+1}(r^n(x_1(t) - \hat{x}_1(t))),
\end{cases}
\tag{5.2.3}
$$

where $g_i \in C(\mathbb{R}, \mathbb{R})$ is a designed function to be specified later and $r \in \mathbb{R}^+$ is the tuning parameter. The main idea of ESO is to choose some appropriate $g_i(\cdot)$ so that when r is large enough, the $\hat{x}_i(t)$ approaches $x_i(t)$ for all $i = 1, 2, \cdots, n + 1$ and sufficiently large t, where $x_{n+1}(t)$ is the total disturbance defined by (5.2.2).

The ESO (5.2.3)-based output feedback control is designed as

$$
u(t) = \frac{\rho u_0(\text{sat}_{M_1}(\rho^{n-1}\hat{x}_1(t)), \text{sat}_{M_2}(\rho^{n-2}\hat{x}_2(t)), \cdots, \text{sat}_{M_n}(\hat{x}_n(t))) - \text{sat}_{M_{n+1}}(\hat{x}_{n+1}(t))}{b_0(t, \hat{x}_1(t), \cdots, \hat{x}_n(t))},
\tag{5.2.4}
$$

where $\rho > 0$ is a constant, $\hat{x}_{n+1}(t)$ is used to compensate (cancel) the total disturbance, and $u_0 : \mathbb{R}^n \to \mathbb{R}$ is chosen so that the following system is globally asymptotically stable:

$$
z^{(n)}(t) = u_0(z(t), \cdots, z^{(n-1)}(t)).
\tag{5.2.5}
$$

The continuous differentiable saturation odd function $\text{sat}_{M_i} : \mathbb{R} \to \mathbb{R}$ is used to avoid the peaking value in control, which is defined by (the counterpart for $t \in (-\infty, 0]$ is obtained by symmetry)

$$\operatorname{sat}_{M_i}(\tau) = \begin{cases} \tau, & 0 \le \tau \le M_i, \\ -\dfrac{1}{2}\tau^2 + (M_i+1)\tau - \dfrac{1}{2}M_i^2, & M_i < \tau \le M_i+1, \\ M_i + \dfrac{1}{2}, & \tau > M_i+1, \end{cases} \tag{5.2.6}$$

where $M_i(1 \le i \le n)$ are constants depending on the bounds of the initial value and external disturbance to be specified in (5.2.10) later. It is obvious that $|\operatorname{sat}_{M_i}(\tau)| \le 1$.

To obtain convergence of the closed-loop system under the ESO-based output feedback control (5.2.4), we need some assumptions.

Assumption 5.2.1 is on functions $\phi_i(\cdot), i = 1, \cdots, n-1, f(\cdot), f_0(\cdot)$, and $b(\cdot)$.

Assumption 5.2.1
- $|\phi_i(x_1, \cdots, x_i) - \phi_i(\hat{x}_1, \cdots, \hat{x}_i)\| \le L|(x_1 - \hat{x}_1, \cdots, x_i - \hat{x}_i)\|, L > 0, \phi_i(0, \cdots, 0) = 0$.
- $f, b \in C^1(\mathbb{R}^{n+2+s}, \mathbb{R}), f_0 \in C^1(\mathbb{R}^{n+2+s}, \mathbb{R}^s)$, and there exists function $\varpi \in C(\mathbb{R}^{n+2+s}, \mathbb{R})$ such that

$$\max\{|f(t, x, \zeta, w)|, \|\nabla f(t, x, \zeta, w)\|, \|b(t, x, \zeta, w)\|, |\nabla b(t, x, \zeta, w)|, \|f_0(t, x, \zeta, w)\|,$$

$$\|\nabla f_0(t, x, \zeta, w)\|\} \le \varpi(x, \zeta, w), \forall\, t \in \mathbb{R}, x \in \mathbb{R}^n, \zeta \in \mathbb{R}^s, w \in \mathbb{R}.$$

- There exist positive definite functions $V_0, W_0 : \mathbb{R}^s \to \mathbb{R}$ such that $L_{f_0}V_0(\zeta) \le -W_0(\zeta)$ for all $\zeta : \|\zeta\| > \chi(x, w)$, where $\chi(\cdot) : \mathbb{R}^{n+1} \to \mathbb{R}$ is a class $\mathcal{K}_\infty$ function, and $L_{f_0}V_0(\zeta)$ denotes the Lie derivative of $V_0(\zeta)$ along the vector field $f_0(\zeta)$.

Assumption 5.2.2 is on nonlinear functions $g_i(\cdot)$ in (5.2.3).

Assumption 5.2.2 $g_i \in C(\mathbb{R}, \mathbb{R}), |g_i(\tau)| \le \Gamma_i|\tau|$ for some $\Gamma_i > 0$ and all $\tau \in \mathbb{R}$, and there exists a continuous, positive definite, and radially unbounded function $\mathcal{V} : \mathbb{R}^{n+1} \to \mathbb{R}$ such that

- $\|e\| \left\| \left(\dfrac{\partial \mathcal{V}(e)}{\partial e_1}, \cdots \dfrac{\partial \mathcal{V}(e)}{\partial e_{n+1}} \right) \right\| \le c_2 \mathcal{V}(e), \left| \dfrac{\partial \mathcal{V}(e)}{\partial e_{n+1}} \right| \le c_3 \mathcal{V}^\theta(e), 0 < \theta < 1;$
- $\displaystyle\sum_{i=1}^{n}(e_{i+1} - g_i(e_1))\dfrac{\partial \mathcal{V}(e)}{\partial \eta_i} - g_{n+1}(e_1)\dfrac{\partial \mathcal{V}(e)}{\partial \eta_{n+1}} \le -c_1 \mathcal{V}(e), e \in \mathbb{R}^{n+1}, c_1, c_2, c_3 > 0;$
- $\mathcal{V}(e) \le \displaystyle\sum_{i=1}^{n}|e_i|^{\mu_i}, i = 1, 2, \cdots, n, \mu_i > 0, e = (e_1, e_2, \cdots, e_{n+1}) \in \mathbb{R}^{n+1}.$

The following Assumption 5.2.3 is for the nonlinear function $u_0(\cdot)$ in feedback control (5.2.4).

Assumption 5.2.3 $u_0 \in C^1(\mathbb{R}^n, \mathbb{R}), u_0(0, \cdots, 0) = 0$, and there exist radially unbounded, positive definite functions $V \in C^1(\mathbb{R}^n, \mathbb{R})$ and $W \in C(\mathbb{R}^n, \mathbb{R})$ such that

$$\sum_{i=1}^{n-1} \iota_{i+1}\dfrac{\partial V(\iota)}{\partial \iota_i} + u_0(\iota)\dfrac{\partial V(\iota)}{\partial \iota_n} \le -W(\iota), \sum_{i=1}^{n-1}\|\iota\|\left| \dfrac{\partial V(\iota)}{\partial \iota_i} \right| \le c_4 W(\iota), \forall\, \iota = (\iota_1, \cdots, \iota_n) \in \mathbb{R}^n.$$

Set

$$\mathcal{A}_1 \triangleq \{\nu \in \mathbb{R}^n : V(\nu) \leq \max_{\iota \in \mathbb{R}^n, \|\iota\| \leq d} V(\iota) + 1\}, \tag{5.2.7}$$

where $d = \left(\sum_{i=1}^n (\rho^{n-i}\beta_i + 1)^2\right)^{1/2}$ and β_i is an upper bound of $|x_{i0}|$, that is, $\beta_i \geq |x_{i0}|$ where x_{i0} is the ith component of the initial state $x(0)$. The continuity and radial unboundedness of $V(\cdot)$ ensures that $\mathcal{A}_1$ is a compact set of $\mathbb{R}^n$.

The following Assumption 5.2.4 is to guarantee that the nominal function $b_0(\cdot)$ is close to $b(\cdot)$.

Assumption 5.2.4 The $b_0 \in C^1(\mathbb{R}^{n+1}, \mathbb{R})$ satisfies $\inf_{(t,y) \in \mathbb{R}^{n+1}} |b_0(t, y)| > c_0 > 0$ and all partial derivatives of $b_0(\cdot)$ are globally bounded. In addition,

$$\wedge \triangleq \sup_{t \in [0,\infty), x \in \mathcal{A}_1, w \in \mathcal{B}, \zeta \in C} |b(t, x, \zeta, w) - b_0(t, x)| \leq \min \left\{\frac{c_0}{2}, \frac{c_0 c_1}{2c_2 \Gamma_{n+1}}\right\}, \tag{5.2.8}$$

where $\mathbb{C} \subset \mathbb{R}^s$ is a compact set defined by

$$\mathbb{C} = \{\varsigma \in \mathbb{R}^s : \|\varsigma\| \leq \max_{\nu \in \mathcal{A}_1, \tau \in \mathcal{B}} \chi(\nu, \tau)\}, \quad \mathcal{B} = [-B, B] \subset \mathbb{R}, B > 0. \tag{5.2.9}$$

The positive constants M_i used in the saturation function in (5.2.4) are chosen so that

$$M_i \geq \sup\{|\iota_i| : (\iota_1, \cdots, \iota_n) \in \mathcal{A}_1\}, i = 1, \cdots, n, \quad M_{n+1} = \frac{c_0 B_1 + \wedge \rho B_2}{c_0 - \wedge}. \tag{5.2.10}$$

Theorem 5.2.1 *Let ρ be chosen so that $\rho \geq L + c_4 + c_4 L + 1$. Suppose that $w(t) \in \mathcal{B}, \dot{w}(t) \in \mathcal{B}$ for all $t \geq 0$ and $\zeta(0) \in \mathbb{C}$. Then under Assumptions 5.2.1–5.2.3, and 5.2.4, the closed-loop system composed of (5.2.1), (5.2.3), and (5.2.4) has the following convergence: For any $a > 0, \sigma > 0$, there exists an $r^* > 0$ such that for all $r > r^*$,*

- *$|x_i(t) - \hat{x}_i(t)| < \sigma$ uniformly for $t \in [a, \infty)$, that is, $\lim_{r \to \infty} |x_i(t) - \hat{x}_i(t)| = 0$ uniformly for $t \in [a, \infty), 1 \leq i \leq n+1$;*
- *$|x_j(t)| < \sigma \ (1 \leq j \leq n)$ uniformly for $t \in [t_r, \infty)$, where t_r is an r-dependent positive constant.*

Proof. Let

$$\xi_i(t) = \rho^{n-i}x_i(t), \eta_i(t) = r^{n+1-i}(x_i(t) - \hat{x}_i(t)). \tag{5.2.11}$$

A direct computation shows that

$$\begin{cases} \dot{\xi}_1(t) = \rho\xi_2(t) + \rho^{n-1}\phi_1(\xi_1(t)/\rho^{n-1}), \\ \dot{\xi}_2(t) = \rho\xi_3(t) + \rho^{n-2}\phi_2(\xi_1(t)/\rho^{n-1}, \xi_2(t)/\rho^{n-2}), \\ \quad \vdots \\ \dot{\xi}_n(t) = \rho u_0(\text{sat}_{M_1}(\rho^{n-1}\hat{x}_1(t)), \cdots, \text{sat}_{M_n}(\hat{x}_n(t))) + x_{n+1}(t) - \text{sat}_{M_{n+1}}(\hat{x}_{n+1}(t)) \end{cases} \tag{5.2.12}$$

and

$$\begin{cases} \dot{\eta}_1(t) = r(\eta_2(t) - g_1(\eta_1(t))) + r^n(\phi_1(x_1(t)) - \phi_1(\hat{x}_1(t))), \\ \quad\vdots \\ \dot{\eta}_{n-1}(t) = r(\eta_n(t) - g_{n-1}(\eta_1(t))) \\ \quad + r^2(\phi_{n-1}(x_1(t), \cdots, x_{n-1}(t)) - \phi_{n-1}(\hat{x}_1(t), \cdots, \hat{x}_{n-1}(t))), \\ \dot{\eta}_n(t) = r[(\eta_{n+1}(t) - g_n(\eta_1(t))) + (b_0(t, x(t)) - b_0(t, \hat{x}(t)))u(t)], \\ \dot{\eta}_{n+1}(t) = -rg_{n+1}(\eta_1(t)) + \dot{x}_{n+1}(t). \end{cases} \quad (5.2.13)$$

Define a compact set $\mathcal{A}_2 \subset \mathbb{R}^n$ by

$$\mathcal{A}_2 \triangleq \{\nu \in \mathbb{R}^n : V(\nu) \le \max_{\iota \in \mathbb{R}^n, \|\iota\| \le d} V(\iota)\}, \quad (5.2.14)$$

where d is the same as in (5.2.7). It is obvious that $\mathcal{A}_2 \subset \mathcal{A}_1$. The remainder of the proof is split into three steps.

Step 1. There exists an $r_0 > 0$ such that $\{\xi(t) : t \in [0, \infty)\} \subset \mathcal{A}_1$ for all $r > r_0$.

Since by (5.2.11), $\xi(0) = (\rho^n x_{10}, \rho^{n-1} x_{20}, \cdots, x_{n0})$ and by (5.2.14), $\xi(0) \in \mathcal{A}_2$, we claim that there exists a $T > 0$ such that $\xi(t) \in \mathcal{A}_2$ for all $t \in [0, T]$, which shows that the possible escape time for $\xi(t)$ from $\mathcal{A}_2$ is larger than T. Let

$$B_1 \triangleq \sup_{(t,x,\zeta,w)\in[0,\infty)\times\mathcal{A}_1\times\mathbb{C}\times\mathcal{B}} \{|f(t, x, \zeta, w)|, |b(t, x, \zeta, w)|, \|\nabla b(t, x, \zeta, w),$$

$$\|\nabla f(t, x, \zeta, w)\|, \|f_0(t, x, \zeta, w)\|\},$$

$$B_2 \triangleq \sup_{|x_i|\le M_i} \{|u_0(x_1, \cdots, x_n)|, \|\nabla u_0(x_1, \cdots, x_n)\|\},$$

$$B_3 \triangleq \sup_{(t,x)\in\mathbb{R}^{n+1}} \{|b_0(t, x)|, \|\nabla b_0(t, x)\|\}, \quad B_4 \triangleq \sup_{x\in\mathcal{A}_1} \|x\|. \quad (5.2.15)$$

By Assumption 5.2.1, $B_i < +\infty$ $(i = 1, 2, 3, 4)$. Before $\xi(t)$ escaping from $\mathcal{A}_2$, $\xi(t) \in \mathcal{A}_2 \subset \mathcal{A}_1$. Noticing $w(t) \in \mathcal{B}$, $\zeta(0) \in \mathbb{C}$, and the input-to-state stable condition on ζ-subsystem in Assumption 5.2.1, if $\xi(t) \in \mathcal{A}_1$, then $\zeta(t) \in \mathbb{C}$. Since $\rho > 1$ and by (5.2.11), if $\xi \in \mathcal{A}_1$, then $x \in \mathcal{A}_1$. By (5.2.12) and (5.2.15), if $\xi \in \mathcal{A}_1$, then

$$\begin{cases} |\xi_n(t)| \le (2B_1 + \rho B_2 + B_3 + M_{n+1})t + |x_{n0}|, \\ |\xi_i(t)| \le \rho(L+1)B_4 t + \rho^{n+1-i}|x_{i0}|. \end{cases} \quad (5.2.16)$$

Let

$$T = \min\left\{\frac{1}{2B_1 + B_2 + \rho B_3 + M_{n+1}}, \frac{1}{(L+1)B_4}\right\}. \quad (5.2.17)$$

By (5.2.16) and (5.2.17), for any $t \in [0, T]$,

$$|\xi_i(t)| \le \rho^{n+1-i}|x_{i0}| + 1, \quad 1 \le i \le n.$$

This gives

$$\{\xi(t) = (\xi_1(t), \cdots, \xi_n(t)) : t \in [0, T]\} \subset \mathcal{A}_2. \quad (5.2.18)$$

We suppose that the conclusion of Step 1 is false and obtain a contradiction. Actually, by continuity of $\xi(t)$ and (5.2.18), there exist $t_2 > t_1 \geq T$ such that

$$\xi(t_1) \in \partial \mathcal{A}_2, \quad \xi(t_2) \in \partial \mathcal{A}_1, \{\xi(t) : t \in (t_1, t_2)\} \in \mathcal{A}_1 - \mathcal{A}_2^\circ, \quad \{\xi(t) : t \in [0, t_2]\} \in \mathcal{A}_1.$$
$$(5.2.19)$$

Finding the derivative of the total disturbance $x_{n+1}(t)$ with respect to t gives

$$\dot{x}_{n+1}(t) = \frac{df(t, x(t), \zeta(t), w(t))}{dt} + \frac{d(b(t, x(t), \zeta(t), w(t)) - b_0(t, \hat{x}(t)))u(t)}{dt}$$

$$= \left.\frac{df(t, x(t), \zeta(t), w(t))}{dt}\right|_{\text{along (5.2.1)}} + \left(\left.\frac{db(t, x(t), \zeta(t), w(t))}{dt}\right|_{\text{along (5.2.1)}}\right.$$

$$\left. - \left.\frac{db_0(t, x(t), \zeta(t), w(t))}{dt}\right|_{\text{along (5.2.3)}}\right)u(t)$$

$$+ (b(t, x(t), \zeta(t), w(t)) - b_0(t, \hat{x}(t)))\left.\frac{du(t, \hat{x}(t))}{dt}\right|_{\text{along (5.2.3)}}.$$
$$(5.2.20)$$

Next, computing the derivative of $f(\cdot)$ along the solution of system (5.2.1) yields

$$\left.\frac{df(t, x(t), \zeta(t), w(t))}{dt}\right|_{\text{along (5.2.1)}}$$

$$= \frac{\partial f(t, x(t), \zeta(t), w(t))}{\partial t} + f_0(t, x(t), \zeta(t), w(t))\frac{\partial f(t, x(t), \zeta(t), w(t))}{\partial \zeta}$$

$$+ \sum_{i=1}^{n-1}(x_{i+1}(t) + \phi_i(x_1(t), \cdots, x_i(t)))\frac{\partial f(t, x(t), \zeta(t), w(t))}{\partial x_i}$$
$$(5.2.21)$$

$$+ \dot{w}(t)\frac{\partial f(f(t, x(t), \zeta(t), w(t))}{\partial w}$$

$$+ (f(t, x(t), \zeta(t), w(t)) + b(t, x(t), \zeta(t), w(t))u(t))\frac{\partial f(t, x(t), \zeta(t), w(t))}{\partial x_n}.$$

From the last expression in (5.2.19), $\xi(t) \in \mathcal{A}_1$ for all $t \in [0, t_2]$. This, together with $w(t) \in \mathcal{B}$ and $\zeta(0) \in \mathcal{C}$, gives $\{\zeta(t) : t \in [0, t_2]\} \subset \mathcal{C}$. It follows from (5.2.15) that for any $t \in [0, t_2]$,

$$\left|\left.\frac{df(t, x(t), \zeta(t), w(t))}{dt}\right|_{\text{along (5.2.1)}}\right|$$
$$(5.2.22)$$

$$\leq B_1\left(1 + B + 2B_1 + n(n-1)(L+1)B_4/2 + \frac{\rho B_2 + M_{n+1}}{c_0}\right).$$

Similarly,

$$\left|\left.\frac{db(t, x(t), \zeta(t), w(t))}{dt}\right|_{\text{along (5.2.1)}}\right|$$
$$(5.2.23)$$

$$\leq B_1\left(1 + B + 2B_1 + n(n-1)(L+1)B_4/2 + \frac{\rho B_2 + M_{n+1}}{c_0}\right), \quad \forall t \in [0, t_2].$$

A direct computation shows that

$$\left.\frac{db_0(t,\hat{x}(t))}{dt}\right|_{\text{along (5.2.3)}} = \frac{\partial b_0(t,\hat{x}(t))}{\partial t}$$

$$+ \sum_{i=1}^{n-1}\left(\hat{x}_{i+1}(t) + \frac{1}{r^{n-i}}g_i(\eta_1(t)) + \phi_i(\hat{x}_1(t),\cdots,\hat{x}_i(t))\right)\frac{\partial b_0(t,\hat{x}(t))}{\partial \hat{x}_i}$$

$$+ (\hat{x}_{n+1}(t) + g_n(\eta_1(t)) + b_0(t,\hat{x}(t))u(t))\frac{\partial b_0(t,\hat{x}(t))}{\partial \hat{x}_n}. \tag{5.2.24}$$

By Assumptions 5.2.1, 5.2.2, and (5.2.15), for every $t \in [0,t_2]$,

$$\left|\left.\frac{db_0(t,\hat{x}(t))}{dt}\right|_{\text{along (5.2.3)}}\right|$$

$$\leq B_3\left(1 + B_1 + n(n-1)(1+L)B_4/2 + (\wedge + B_3)(\rho B_2 + M_{n+1})/c_0\right.$$

$$\left. + |\eta_{n+1}(t)| + \Gamma_n|\eta_1(t)| + \sum_{i=1}^{n-1}\left(|\eta_{i+1}(t)| + L\|(\eta_1(t),\cdots,\eta_i(t))\| + \frac{\Gamma_i}{r^{n-i}}|\eta_1(t)|\right)\right). \tag{5.2.25}$$

Finding the derivative of $u(\cdot)$ along the solution of (5.2.3) yields

$$\left.\frac{du(t,\hat{x}(t))}{dt}\right|_{\text{along (5.2.3)}} = \frac{1}{b_0(t,\hat{x}(t))}\left(\sum_{i=1}^{n-1}\left(\hat{x}_{i+1}(t) + \frac{1}{r^{n-i}}g_i(\eta_1(t))\right.\right.$$

$$+ \phi_i(\hat{x}_1(t),\cdots,\hat{x}_i(t)))\frac{d\text{sat}_{M_i}(\hat{x}_i(t))}{d\hat{x}_i}\frac{\partial u_0(\text{sat}_{M_1}(\hat{x}_1(t)))}{\partial \hat{x}_i},\cdots,$$

$$\text{sat}_{M_n}(\hat{x}_n(t))) + (\hat{x}_{n+1}(t) + g_n(\eta_1(t))$$

$$+ b_0(t,\hat{x}(t))u(t))\frac{d\text{sat}_{M_n}(\hat{x}_n(t))}{d\hat{x}_n}\frac{\partial u_0(\text{sat}_{M_1}(\hat{x}_1(t)))}{\partial \hat{x}_n},$$

$$\cdots,\text{sat}_{M_n}(\hat{x}_n(t))) + rg_{n+1}(r^n(x_1(t) - \hat{x}_1(t)))\frac{d\text{sat}_{M_{n+1}}(\hat{x}_{n+1}(t))}{d\hat{x}_{n+1}}\Bigg)$$

$$- \frac{\rho u_0(\text{sat}_{M_1}(\rho^{n-1}\hat{x}_1(t)),\cdots,\text{sat}_{M_n}(\hat{x}_n(t))) - \text{sat}_{M_{n+1}}(\hat{x}_{n+1}(t))}{b_0^2(t,\hat{x}(t))}$$

$$\left.\frac{db_0(t,\hat{x}(t))}{dt}\right|_{\text{along (5.2.3)}}. \tag{5.2.26}$$

Similarly with (5.2.25), we have

$$\left|\left.\frac{du(t,\hat{x}(t))}{dt}\right|_{\text{along (5.2.3)}}\right|$$

$$\leq \frac{B_2}{c_0}\left(1 + B_1 + n(n-1)(1+L)B_4/2 + (\wedge + B_3)(\rho B_2 + M_{n+1})/c_0\right.$$

$$+ |\eta_{n+1}(t)| + \Gamma_n |\eta_1(t)| + \sum_{i=1}^{n-1} \left(|\eta_{i+1}(t)| + L\|(\eta_1(t), \cdots, \eta_i(t))\| + \frac{\Gamma_i}{r^{n-i}} |\eta_1(t)| \right)$$

$$+ \frac{\Gamma_{n+1} r}{c_0} |\eta_1(t)| + \frac{\rho B_2 + M_{n+1}}{c_0^2} \left| \frac{db_0(t, \hat{x}(t))}{dt} \right|_{\text{along (5.2.3)}}, \quad \forall t \in [0, t_2]. \tag{5.2.27}$$

By (5.2.20), (5.2.22)–(5.2.25), and (5.2.27), there exist positive constants N_1 and N_2 depending on B_i, c_0, L, M_i, and Γ_i such that

$$|\dot{x}_{n+1}(t)| \leq N_1 + N_2 \|\eta(t)\| + \frac{\Gamma_{n+1} \wedge r}{c_0} |\eta_1(t)|, \quad \forall t \in [0, t_2]. \tag{5.2.28}$$

Let $\mathcal{V}(\cdot)$ and $\mathcal{W}(\cdot)$ be the Lyapunov functions satisfying Assumption 5.2.2. The derivative of $\mathcal{V}(\eta(t))$ along the solution of (5.2.13) is

$$\frac{d\mathcal{V}(\eta(t))}{dt} \bigg|_{\text{along (5.2.13)}} = \sum_{i=1}^{n+1} \dot{\eta}_i(t) \frac{\partial \mathcal{V}(\eta(t))}{\partial \eta_i}$$

$$= \sum_{i=1}^{n-1} (r(\eta_{i+1}(t) - g_i(\eta_1(t))) + r^{n+1-i}(\phi_i(x_1(t), \cdots, x_i(t))$$

$$- \phi_i(\hat{x}_1(t), \cdots, \hat{x}_i(t)))) \frac{\partial \mathcal{V}(\eta(t))}{\partial \eta_i}$$

$$+ r[\eta_{n+1}(t) - g_n(\eta_1(t)) + (b_0(t, x(t)) - b_0(t, \hat{x}(t)))u(t)] \frac{\partial \mathcal{V}(\eta(t))}{\partial \eta_n}$$

$$+ (-rg_{n+1}(\eta_1(t)) + \dot{x}_{n+1}(t)) \frac{\partial \mathcal{V}}{\partial \eta_{n+1}}(\eta(t)). \tag{5.2.29}$$

This, together with Assumption 5.2.2 and (5.2.28), gives

$$\frac{d\mathcal{V}(\eta(t))}{dt} \bigg|_{\text{along (5.2.13)}} \leq -c_1 r \mathcal{V}(\eta(t)) + c_2 \left((n-1)L + N_2 + \frac{(\rho B_2 + M_{n+1}) B_3}{c_0} \right) \mathcal{V}(\eta(t))$$

$$+ \frac{c_2 \Gamma_{n+1} \wedge r}{c_0} \mathcal{V}(\eta(t)) + c_2 c_3 N_1 \mathcal{V}^\theta(\eta(t)), \quad \forall t \in [0, t_2]. \tag{5.2.30}$$

Let $r > 4c_2((n-1)L + N_2 + (\rho B_2 + M_{n+1}) B_3/c_0)$. By Assumption 5.2.4, it follows that

$$\frac{d\mathcal{V}(\eta(t))}{dt} \bigg|_{\text{along (5.2.13)}} \leq -\frac{c_1 r}{4} \mathcal{V}(\eta(t)) + c_3 N_1 \mathcal{V}^\theta(\eta(t)), \quad \forall t \in [0, t_2]. \tag{5.2.31}$$

Furthermore, if $\eta(t) \neq 0$, then

$$\frac{d}{dt}(\mathcal{V}^{1-\theta}(\eta(t))) \leq -\frac{c_1(1-\theta)}{4} \mathcal{V}^{1-\theta}(\eta(t)) + c_3 N_1, \quad \forall t \in [0, t_2]. \tag{5.2.32}$$

By comparison principle of ordinary differential equations, we have

$$\mathcal{V}^{1-\theta}(\eta(t)) \leq e^{-\frac{c_1(1-\theta)r}{4}t}\mathcal{V}^{1-\theta}(\eta(0)) + c_3 N_1 \int_0^t e^{-c_1(1-\theta)r/4(t-s)}ds, \quad \forall\, t \in [0, t_2],$$

(5.2.33)

where

$$\eta(0) = (r^n(x_{10} - \hat{x}_{10}), r^{n-1}(x_{20} - \hat{x}_{20}), \cdots, x_{(n+1)0} - \hat{x}_{(n+1)0})$$

(5.2.34)

and $(\hat{x}_{10}, \hat{x}_{20}, \cdots, \hat{x}_{(n+1)0})$ is the initial value of (5.2.3). By Assumption 5.2.2,

$$|\mathcal{V}^{1-\theta}(\eta(0))| \leq \left(\sum_{i=1}^n |r^{n+1-i}(x_{i0} - \hat{x}_{i0})|^{\mu_i} \right)^{1-\theta}.$$

Notice that $t_1 \geq T$ and for any $t \in [t_1, t_2]$,

$$e^{-c_1(1-\theta)r/4t}\mathcal{V}^{1-\theta}(\eta(0)) \leq e^{-\frac{c_1(1-\theta)r}{4}T}\left(\sum_{i=1}^n |r^{n+1-i}(x_{i0} - \hat{x}_{i0})|^{\mu_i} \right)^{1-\theta} \to 0, \quad r \to \infty.$$

(5.2.35)

Since $\mathcal{V}(\cdot)$ is continuous, positive definite, and radially unbounded, by Theorem 1.3.1, there exists the continuous class $\mathcal{K}_\infty$ function $\kappa : [0, \infty) \to [0, \infty)$ such that $\mathcal{V}(\eta) \geq \kappa(\|\eta\|)$ for all $\eta \in \mathbb{R}^{n+1}$. Let

$$\delta = \min\left\{ \frac{1}{2}, \frac{M_i}{2}, \frac{\min_{\nu \in A_l} W(\nu)}{2\rho n B_2 + 3} \right\}.$$

(5.2.36)

By (5.2.35), there exists an $r_1^* > 0$ such that

$$|e^{-\frac{c_1(1-\theta)r}{4}t}\mathcal{V}^{1-\theta}(\eta(0))| \leq \frac{(\kappa(\delta))^{1-\theta}}{2}, \quad \forall\, r > r_1^*, t \in [t_1, t_2].$$

(5.2.37)

The second term on the right-hand side of (5.2.33) satisfies

$$\left| c_3 N_1 \int_0^t e^{-\frac{c_1(1-\theta)r}{4}(t-s)}ds \right| \leq \frac{4c_3 N_1}{c_1(1-\theta)r}.$$

(5.2.38)

By (5.2.35) and (5.2.38), for any $r > r_2^* \triangleq \max\{\rho, r_1^*, (8c_3 N_1)/(c_1(1-\theta)(\kappa(\delta))^{1-\theta})\}$, $\mathcal{V}(\eta(t)) \leq \kappa(\delta)$ uniformly in $t \in [t_1, t_2]$. This, together with (5.2.11), yields

$$|\rho^{n-i}x_i(t) - \rho^{n-i}\hat{x}_i(t)| \leq \|(\rho^{n-1}(x_1(t) - \hat{x}_1(t)), \cdots, x_n(t) - \hat{x}_n(t))\|$$
$$\leq \|(r^n(x_1(t) - \hat{x}_1(t)), \cdots, r(x_n(t) - \hat{x}_n(t)))\|$$

(5.2.39)

$$\leq \|\eta(t)\| \leq \delta, \quad \forall\, t \in [t_1, t_2], r > r_2^*.$$

From (5.2.19), $|x_i(t)| \leq |\rho^{n-i}x_i(t)| = |\xi_i(t)| \leq M_i$ for all $t \in [t_1, t_2]$, $i = 1, 2, \cdots, n$. This, together with (5.2.2) and Assumption 5.2.4, yields

$$|x_{n+1}(t)| \leq B_1 + \wedge\frac{\rho B_2 + M_{n+1}}{c_0} = M_{n+1}, \quad \forall\, t \in [t_1, t_2].$$

If $|\rho^{n-i}\hat{x}_i(t)| \leq M_i$, then $\rho^{n-i}\hat{x}_i(t) - \mathrm{sat}_{M_i}(\rho^{n-i}\hat{x}_i(t)) = 0$, $i = 1, \cdots, n+1$. If $|\rho^{n-i}\hat{x}_i(t)| > M_i$ and $\rho^{n-i}\hat{x}_i(t) > 0$, since $\delta \leq M_i/2$, we have $\rho^{n-i}\hat{x}_i > M_i$. Hence

$$|\rho^{n-i}\hat{x}_i(t) - M_i| = \rho^{n-i}\hat{x}_i(t) - M_i \leq \rho^{n-i}\hat{x}_i(t) - \rho^{n-i}x_i(t) \leq \delta \leq \frac{1}{2}, \quad \forall\, t \in [t_1, t_2].$$
$$(5.2.40)$$

This, together with (5.2.6), gives

$$|\rho^{n-i}\hat{x}_i(t) - \mathrm{sat}_{M_i}(\rho^{n-i}\hat{x}_i(t))| = \rho\hat{x}_i(t) + \frac{1}{2}(\rho\hat{x}_i(t))^2 - (M_i + 1)\rho\hat{x}_i(t) + \frac{1}{2}M_i^2$$
$$= \frac{(\rho\hat{x}_i(t) - M_i)^2}{2} < \frac{\delta^2}{2} < \delta, \quad \forall\, t \in [t_1, t_2].$$
$$(5.2.41)$$

Similarly, (5.2.41) also holds true when $|\rho^{n-i}\hat{x}_i(t)| > M_i$ and $\rho^{n-i}\hat{x}_i(t) < 0$. Therefore $|\rho^{n-i}\hat{x}_i(t) - \mathrm{sat}_{M_i}(\rho^{n-i}\hat{x}_i(t))| \leq \delta$ for all $t \in [t_1, t_2]$.

Let $V(\cdot)$ and $W(\cdot)$ be the Lyapunov functions satisfying Assumption 5.2.3. Finding the derivative of $V(\xi(t))$ along (5.2.12) gives

$$\left.\frac{dV(\xi(t))}{dt}\right|_{\text{along (5.2.12)}} = \sum_{i=1}^{n} \dot{\xi}_i(t)\frac{\partial V(\xi(t))}{\partial \xi_i}$$

$$= \sum_{i=1}^{n-1}(\rho\xi_{i+1}(t) + \rho^{n-i}\phi_i(\xi_1(t)/\rho^{n-1}, \cdots, \xi_i(t)/\rho^{n-i}))\frac{\partial V(\xi(t))}{\partial \xi_i}$$
$$+ (\rho u_0(\mathrm{sat}_{M_1}(\rho^{n-1}\hat{x}_1(t)), \cdots, \mathrm{sat}_{M_n}(\hat{x}_n(t))) + x_{n+1}(t)$$
$$- \mathrm{sat}_{M_1}(\hat{x}_{n+1}(t)))\frac{\partial V(\xi(t))}{\partial \xi_n}.$$
$$(5.2.42)$$

By Assumption 5.2.1, (5.2.39), and (5.2.41),

$$|u_0(\mathrm{sat}_{M_1}(\rho^{n-1}\hat{x}_1(t)), \cdots, \mathrm{sat}_{M_n}(\hat{x}_n(t))) - u_0(\xi(t))|$$
$$\leq |u_0(\mathrm{sat}_{M_1}(\rho^{n-1}\hat{x}_1(t)), \cdots, \mathrm{sat}_{M_n}(\hat{x}_n(t))) - u_0(\rho^{n-1}\hat{x}_1(t), \cdots, \hat{x}_n(t))|$$
$$+ |u_0(\rho^{n-1}\hat{x}_1(t), \cdots, x_n(t)) - u_0(\rho^{n-1}x_1(t), \cdots, x_n(t))| \leq 2nB_2\delta.$$
$$(5.2.43)$$

Let $N_3 = \sup_{\iota \in \mathcal{A}_1}(\partial V(\iota)/\partial\iota_n)$. By Assumption 5.2.3, (5.2.42), (5.2.43), and (5.2.36),

$$\left.\frac{dV(\xi(t))}{dt}\right|_{\text{along (5.2.12)}} = \sum_{i=1}^{n} \dot{\xi}_i(t)\frac{\partial V(\xi(t))}{\partial \xi_i}$$

$$= \sum_{i=1}^{n-1}(\rho\xi_{i+1}(t) + \rho^{n-i}\phi_i(\xi_1(t)/\rho^{n-1}, \cdots, \xi_i(t)/\rho^{n-i}))\frac{\partial V(\xi(t))}{\partial \xi_i}$$
$$+ (\rho u_0(\mathrm{sat}_{M_1}(\rho^{n-1}\hat{x}_1(t)), \cdots, \mathrm{sat}_{M_n}(\hat{x}_n(t))) + x_{n+1}(t)$$
$$- \mathrm{sat}_{M_1}(\hat{x}_{n+1}(t)))\frac{\partial V(\xi(t))}{\partial \xi_n}$$

$$\leq -\rho W(\xi(t)) + L\sum_{i=1}^{n-1}\|\xi(t)\|\left|\frac{\partial V(\xi(t))}{\partial x_i}\right| + 2(\rho n B_2 + 1)N_3\delta$$

$$\leq -W(\xi(t)) + 2(\rho n B_2 + 1)N_3\delta < 0, \quad \forall t \in [t_1, t_2]. \tag{5.2.44}$$

This shows that $V(\xi(t))$ is monotone decreasing in $t \in [t_1, t_2]$. However, by (5.2.19), (5.2.7), and (5.2.14), $V(\xi(t_2)) = V(\xi(t_1)) + 1$, which is a contradiction. Therefore $\{\xi(t) : t \in [0, \infty)\} \subset \mathcal{A}_1$ for all $r > r_2^*$. The claim of Step 1 then follows.

Step 2. $\eta(t) \to 0$ uniformly in $t \in [a, \infty)$ as $r \to \infty$.

By Step 1, $\xi(t) \in \mathcal{A}_1$ for all $t \in [0, \infty)$ as $r > r_2^*$. Similarly to (5.2.28), we can obtain

$$|\dot{x}_{n+1}(t)| \leq N_1 + N_2|\eta(t)| + \frac{\Gamma_{n+1} \wedge r}{c_0}\|\eta_1(t)\|, \quad \forall t \in [0, \infty), r > r_2^*. \tag{5.2.45}$$

Similarly to (5.2.33), (5.2.45) implies that

$$\mathcal{V}^{1-\theta}(\eta(t)) \leq e^{-c_1(1-\theta)r/4t}\mathcal{V}^{1-\theta}(\eta(0)) + c_3 N_1 \int_0^t e^{-c_1(1-\theta)r/4(t-s)}ds, t \in [0, \infty), r > r_2^*. \tag{5.2.46}$$

Since for any $a > 0$, $\sigma > 0$ and $t \in [a, \infty)$,

$$e^{-\frac{c_1(1-\theta)r}{4}t}\mathcal{V}^{1-\theta}(\eta(0)) \leq e^{-c_1(1-\theta)r/4a}\mathcal{V}^{1-\theta}(\eta(0)) \to 0 \text{ as } r \to \infty, \tag{5.2.47}$$

there exists an r_3^* such that when $r > r_3^*$, $e^{-c_1(1-\theta)r/4t}\mathcal{V}^{1-\theta}(\eta(0)) < \sigma/2$ uniformly in $t \in [a, \infty)$. Let

$$r_4^* = \max\left\{r_2^*, \ r_3^*, \ \frac{8c_2 c_3 N_1}{c_1(1-\theta)\sigma^{1-\theta}}\right\}.$$

By (5.2.46) and (5.2.47), $|x_i(t) - \hat{x}_i(t)| \leq \|\eta\| \leq \sigma$ uniformly in $t \in [a, \infty)$ as $r > r_4^*$. This shows that $\lim_{r\to\infty}|x_i(t) - \hat{x}_i(t)| = 0$ uniformly in $t \in [a, \infty)$.

Step 3. $\xi(t) \to 0$ as $t \to \infty$ and $r \to \infty$.

For the positive definite and radially unbounded Lyapunov function $W(\cdot)$, it follows from Theorem 1.3.1 that there exists a class $\mathcal{K}_\infty$ function $\varkappa : [0, \infty) \to [0, \infty)$ such that $\varkappa(\|\nu\|) \leq W(\nu)$ for all $\nu \in \mathbb{R}^n$. For any given $\sigma > 0$, by Step 2, there exists an $r^* > r_4^*$ such that $\|\eta(t)\| < \sigma_1 = \varkappa(\sigma)/(3(\rho n B_2 + 1)N_3)$ uniformly in $t \in [a, \infty)$ as $r > r^*$. Similarly to (5.2.44), we have

$$\left.\frac{dV(\xi(t))}{dt}\right|_{\text{along (5.2.12)}} \leq -W(\xi(t)) + 2(\rho n B_2 + 1)N_3\sigma_1$$
$$\leq -\varkappa(\|\xi(t)\|) + 2(\rho n B_2 + 1)N_3\sigma_1, \quad \forall t \in [a, \infty). \tag{5.2.48}$$

Therefore, if $\|\xi(t)\| \geq \sigma$, then

$$\left.\frac{dV(\xi(t))}{dt}\right|_{\text{along (5.2.12)}} \leq -\varkappa(\sigma) + 2(\rho n B_2 + 1)N_3\sigma_1 = -\frac{1}{3}\varkappa(\sigma) < 0. \tag{5.2.49}$$

Therefore, there exists t_r such that $|\xi(t)| \leq \sigma$ for all $t \in [t_r, \infty)$. By (5.2.11) and $\rho > 1$, we have $\|x_j(t)\| < \sigma$ $(1 \leq j \leq n)$ uniformly in $t \in [t_r, \infty)$. This completes the proof of the theorem.

The simplest example of ADRC satisfying the conditions of Theorem 5.2.1 is the linear ADRC, that is, $g_i(\cdot)$, $i = 1, \cdots, n+1$, in ESO (5.2.3) and $u_0(\cdot)$ in feedback control are linear functions. Let

$$g_i(\tau) = k_i \tau, \quad u_0(y_1, \cdots, y_n) = \alpha_1 y_1 + \cdots + \alpha_n y_n. \tag{5.2.50}$$

Define the matrices K and A as follows:

$$K = \begin{pmatrix} -k_1 & 1 & 0 & \cdots & 0 \\ -k_2 & 0 & 1 & \cdots & 0 \\ \vdots & \vdots & \vdots & \ddots & \vdots \\ -k_n & 0 & 0 & \cdots & 1 \\ -k_{n+1} & 0 & 0 & \cdots & 0 \end{pmatrix}_{(n+1)\times(n+1)}, \qquad A = \begin{pmatrix} 0 & 1 & 0 & \cdots & 0 \\ 0 & 0 & 1 & \cdots & 0 \\ \vdots & \vdots & \vdots & \ddots & \vdots \\ 0 & 0 & 0 & \cdots & 1 \\ \alpha_1 & \alpha_2 & 0 & \cdots & \alpha_n \end{pmatrix}_{n\times n}. \tag{5.2.51}$$

Let $\lambda_{\max}(P)$ and $\lambda_{\min}(P)$ be the maximal and minimal eigenvalues of matrix P that is the unique positive definite matrix solution of the Lyapunov equation $PK + K^\top P = -I_{(n+1)\times(n+1)}$.

Corollary 5.2.1 *Let $\rho = L + 1$. Suppose that $w(t) \in \mathcal{B}$ and $\dot{w}(t) \in \mathcal{B}$ for a compact set $\mathcal{B} \triangleq [-B, B] \subset \mathbb{R}$, and matrices K and A in (5.2.51) are Hurwitz. Then, under Assumptions 5.2.1 and 5.2.4 with $\Gamma_{n+1} = k_{n+1}$ and $\theta = 1/2$, $c_2 = \frac{2\lambda_{\max}(P)}{\lambda_{\min}(P)}$, the closed-loop system composed of (5.2.1), (5.2.3), and (5.2.4) with linear function (5.2.50) has the following convergence: For any $a > 0$, there exists an $r^* > 0$ such that for any $r > r^*$,*

- *$|x_i(t) - \hat{x}_i(t)| \leq \Delta_1 / r^{n+2-i}$ uniformly in $t \in [a, \infty)$, where $\Delta_1 > 0$ is an r-independent constant, that is, $\lim_{r \to \infty} |x_i(t) - \hat{x}_i(t)| = 0$ uniformly in $t \in [a, \infty)$, $1 \leq i \leq n+1$;*
- *$|x_j(t)| < \frac{\Delta_2}{r}$ uniformly in $t \in [t_r, \infty)$, where $\Delta_2 > 0$ is an r-independent constant, t_r is an r and initial value dependent positive constant.*

Proof. Let the Lyapunov functions $\mathcal{V} : \mathbb{R}^{n+1} \to \mathbb{R}$ and $V, W : \mathbb{R}^n \to \mathbb{R}$ be defined as $\mathcal{V}(\eta) = \eta^\top P \eta$ for $\eta \in \mathbb{R}^{n+1}$, $V(\xi) = \xi^\top Q \xi$, $W(\xi) = \|\xi\|^2$ for $\xi \in \mathbb{R}^2$. Let Q be the unique positive definite matrix solution of the Lyapunov equation $QA + A^\top Q = -I_{n\times n}$. Then it is easy to verify that all conditions of Assumptions 5.2.2 and 5.2.3 are satisfied. The results then follow directly from Theorem 5.2.1.

We point out that in Theorem 5.2.1, we have only practical stability by the constant high-gain ESO. Now we indicate a special case where the derivative of the external disturbance converges to 0 as time goes to infinity:

$$\lim_{t \to \infty} \dot{w}(t) = 0. \tag{5.2.52}$$

For this special case, we have the asymptotic stability. To this purpose, we need the following Assumption 5.2.5.

Assumption 5.2.5 There exist continuous functions $\phi_n(\cdot) : \mathbb{R}^n \to \mathbb{R}$, $\bar{f}(\cdot) : \mathbb{R}^n \to \mathbb{R}$ and a positive constant $L_1 > 0$ such that

- $f(t, x, \zeta, w) = \phi_n(x) + \bar{f}(\zeta, w), \ \forall\, t \in [0, \infty), x \in \mathbb{R}^n, \zeta \in \mathbb{R}^s, w \in \mathbb{R};$
- $\phi_i(\cdot), i = 1, 2, \cdots, n$ and $\bar{f}(\cdot)$ are continuous differentiable, $\|\nabla \phi_i(x_1, \cdots, x_i)\| \le L\|(x_1, \cdots, x_i)\|, \phi_i(0, \cdots, 0) = 0, \|\nabla \bar{f}(\zeta, w)\| \le L, x \in \mathbb{R}^n, \zeta \in \mathbb{R}^s, w \in \mathbb{R};$
- $\|f_0(t, x, \zeta, w)\| \le L\|x\|, \ \forall\, t \in [0, \infty), x \in \mathbb{R}^n, \zeta \in \mathbb{R}^s, w \in \mathbb{R};$
- $b(t, x, \zeta, w) \equiv b_0, b_0 \ne 0$ is a constant real number.

For this case, we also allow the nonlinear function $\phi_i(\cdot)$ to be unknown but satisfies some a priori conditions in Assumption 5.2.5. The nonlinear ESO that is independent of $\phi_i(\cdot)$ for this case is designated as follows:

$$
\begin{cases}
\dot{\hat{x}}_1(t) = \hat{x}_2(t) + \dfrac{1}{\rho^{n-1}} g_1(\rho^n(x_1(t) - \hat{x}_1(t))), \\[2mm]
\dot{\hat{x}}_2(t) = \hat{x}_3(t) + \dfrac{1}{\rho^{n-2}} g_2(\rho^n(x_1(t) - \hat{x}_1(t))), \\
\quad \vdots \\
\dot{\hat{x}}_n(t) = \hat{x}_{n+1}(t) + g_n(\rho^n(x_1(t) - \hat{x}_1(t))) + b_0 u(t), \\[2mm]
\dot{\hat{x}}_{n+1}(t) = \rho g_{n+1}(\rho^n(x_1(t) - \hat{x}_1(t))),
\end{cases}
\tag{5.2.53}
$$

where the nonlinear functions $g_i(\cdot)$ are chosen to satisfy the following Assumption 5.2.6.

Assumption 5.2.6 $g_i \in C(\mathbb{R}, \mathbb{R})$, and there exists continuous, positive definite, and radially unbounded functions $\mathcal{V}, \mathcal{W} : \mathbb{R}^{n+1} \to \mathbb{R}$ such that

- $\displaystyle\sum_{i=1}^{n} (e_{i+1} - g_i(e_1)) \frac{\partial \mathcal{V}(e)}{\partial e_i} - g_{n+1}(e_1) \frac{\partial \mathcal{V}(e)}{\partial e_{n+1}} \le -\mathcal{W}(e);$

- $\displaystyle \|e\|^2 + \|\nabla \mathcal{V}(e)\|^2 + \sum_{i=1}^{n+1} \|g_i(e_1)\|^2 \le c_1 \mathcal{W}(e), \quad \left| g_{n+1}(e_1) \frac{\partial \mathcal{V}(e)}{\partial e_{n+1}} \right| \le \Gamma \mathcal{W}(e), \qquad \forall\, e =$

$(e_1, \cdots, e_{n+1})^\top \in \mathbb{R}^{n+1}, c_1 > 0.$

The ESO (5.2.53)-based output feedback control is designed as

$$
u(t) = \frac{u_0(\rho^{n\alpha} \hat{x}_1(t), \rho^{(n-1)\alpha} \hat{x}_2(t), \cdots, \rho^\alpha \hat{x}_n(t)) - \hat{x}_{n+1}(t)}{b_0},
\tag{5.2.54}
$$

where $\alpha \in (n/(n+1), 1)$ and the nonlinear function $u_0 : \mathbb{R}^n \to \mathbb{R}$ is supposed to satisfy the following Assumption 5.2.7.

Assumption 5.2.7 $u_0 \in C^1(\mathbb{R}^n, \mathbb{R}), \|\nabla u_0\| \le L, u_0(0, \cdots, 0) = 0$, and there exist radially unbounded, positive definite functions $V, W \in C^1(\mathbb{R}^n, \mathbb{R})$ such that

- $\displaystyle\sum_{i=1}^{n-1} x_{i+1} \frac{\partial V(x)}{\partial x_i} + u_0(x) \frac{\partial V(x)}{\partial x_n} \le -W(\xi),$
- $\|\xi\|^2 + \|\nabla V(\xi)\|^2 \le c_2 W(\xi), c_2 > 0, \xi = (\xi_1, \cdots, \xi_n) \in \mathbb{R}^n.$

For the high-gain parameter ρ in (5.2.53) and (5.2.54), we need the following Assumption 5.2.8.

Assumption 5.2.8

$$
\rho \ge \max \left\{ 1, 2c_1 L + 4c_1 L^2, 4 \left(nc_2 L + \frac{Lc_2}{2} \right)^{\frac{1}{\alpha}}, \ (\sqrt{c_1 c_2}(1 + nL))^{\frac{2}{1-\alpha}} \right\}.
$$

Theorem 5.2.2 *If* $\lim_{t\to\infty} \dot{w}(t) = 0$ *and Assumptions 5.2.5–5.2.7, and 5.2.8 are satisfied, then the closed-loop system composed of (5.2.1), (5.2.53), and (5.2.54) is convergent in the sense that*

$$\lim_{t\to\infty} |x_i(t) - \hat{x}_i(t)| = 0, i = 1, 2, \cdots, n+1, \lim_{t\to\infty} |x_j(t)| = 0, j = 1, 2, \cdots, n,$$

where the total disturbance $x_{n+1}(t) \triangleq \bar{f}(\zeta(t), w(t))$.

Proof. Let $\eta(t) = (\eta_1(t), \eta_2(t), \cdots, \eta_{n+1}(t))$ and $\xi(t) = (\xi_1(t), \xi_2(t), \cdots, \xi_n(t))$ be defined as

$$\xi_i(t) = \rho^{(n+1-i)\alpha} x_i(t), \ i = 1, 2, \cdots, n, \eta_j(t) = \rho^{n+j-1}(x_j(t) - \hat{x}_j(t)), j = 1, 2, \cdots, n+1.$$
$$(5.2.55)$$

A direct computation shows that $\xi_i(t)$ and $\eta_j(t)$ satisfy

$$\begin{cases} \dot{\xi}_1(t) = \rho^{\alpha}\xi_2(t) + \rho^{n\alpha}\phi_1(\xi_1(t)/\rho^{n\alpha}), \\ \dot{\xi}_2(t) = \rho^{\alpha}\xi_3(t) + \rho^{(n-1)\alpha}\phi_2(\xi_1(t)/\rho^{n\alpha}, \xi_2(t)/\rho^{(n-1)\alpha}), \\ \quad\vdots \\ \dot{\xi}_n(t) = \rho^{\alpha}u_0(\rho^{n\alpha}\hat{x}_1(t), \cdots, \rho^{\alpha}\hat{x}_n(t)) + \rho^{\alpha}\phi_n(\xi_1(t)/\rho^{n\alpha}, \cdots, \xi_n(t)/\rho^{\alpha}) \\ \qquad + \rho^{\alpha}(x_{n+1}(t) - \hat{x}_{n+1}(t)) \end{cases}$$
$$(5.2.56)$$

and

$$\begin{cases} \dot{\eta}_1(t) = \rho(\eta_2(t) - g_1(\eta_1(t))) + \rho^n\phi_1(\xi_1(t)/\rho^{n\alpha}), \\ \quad\vdots \\ \dot{\eta}_{n-1}(t) = \rho(\eta_n(t) - g_{n-1}(\eta_1(t))) \\ \qquad + \rho^2\phi_{n-1}(\xi_1(t)/\rho^{n\alpha}, \cdots, \xi_n(t)/\rho^{2\alpha}), \\ \dot{\eta}_n(t) = \rho(\eta_{n+1}(t) - g_n(\eta_1(t))) + \rho\phi_n(\xi_1(t)/\rho^{n\alpha}, \cdots, \xi_n(t)/\rho^{\alpha}), \\ \dot{\eta}_{n+1}(t) = -\rho g_{n+1}(\eta_1(t)) + \dot{x}_{n+1}(t). \end{cases}$$
$$(5.2.57)$$

Let $\mathcal{V}(\cdot), \mathcal{W}(\cdot), V(\cdot)$, and $W(\cdot)$ be the Lyapunov functions satisfying Assumptions 5.2.6 and 5.2.7, respectively. Define $\mathbb{V}, \mathbb{W} : \mathbb{R}^{n+1} \to [0, \infty)$ as

$$\mathbb{V}(\xi, \eta) = V(\xi) + \mathcal{V}(\eta), \mathbb{W}(\xi, \eta) = W(\xi) + \mathcal{W}(\eta), \xi \in \mathbb{R}^n, \eta \in \mathbb{R}^{n+1}. \qquad (5.2.58)$$

Finding the derivative of $\mathbb{V}(\xi(t), \eta(t))$ along the solutions of the error equations (5.2.56) and (5.2.57) gives

$$\left. \frac{d\mathbb{V}(\xi(t), \eta(t))}{dt} \right|_{\text{along (5.2.56) and (5.2.57)}}$$

$$= \sum_{i=1}^{n} \dot{\xi}_i(t) \frac{\partial V(\xi(t))}{\partial \xi_i} + \sum_{j=1}^{n+1} \dot{\eta}_j(t) \frac{\partial \mathcal{V}(\xi(t))}{\partial \eta_j(t)}$$

$$= \sum_{i=1}^{n-1} \left(\rho^{\alpha}\xi_{i+1}(t) + \rho^{(n+1-i)\alpha}\phi_i\left(\frac{\xi_1(t)}{\rho^{n\alpha}}, \cdots, \frac{\xi_i(t)}{\rho^{(n+1-i)\alpha}}\right) \right) \frac{\partial V(\xi(t))}{\partial \xi_i}$$

$$+ \rho^\alpha u_0(\xi_1(t), \cdots, \xi_n(t)) \frac{\partial V(\xi(t))}{\partial \xi_n} + \left(\rho^\alpha \phi_n \left(\frac{\xi_1(t)}{\rho^{n\alpha}}, \cdots, \frac{\xi_n(t)}{\rho^\alpha} \right) + \rho^\alpha \eta_{n+1}(t) \right.$$

$$+ (u_0(\rho^{n\alpha} \hat{x}_1(t), \cdots, \rho^\alpha \hat{x}_n(t)) - u_0(\rho^{n\alpha} x_1(t), \cdots, \rho^\alpha x_n(t)))) \frac{\partial V(\xi(t))}{\partial \xi_n}$$

$$+ \sum_{i=1}^n \left(\rho(\eta_i(t) - g_i(\eta_1(t))) + \rho^{n+1-i} \phi_i \left(\frac{\xi_1(t)}{\rho^{n\alpha}}, \cdots, \frac{\xi_i(t)}{\rho^{(n+1-i)\alpha}} \right) \right) \frac{\partial V(\xi(t))}{\partial \eta_i}$$

$$- \rho g_{n+1}(\eta_1(t)) \frac{\partial V(e(t))}{\partial \eta_{n+1}}$$

$$+ \left(\dot{w}(t) \frac{\partial \bar{f}(\zeta(t), w(t)) + f_0(t, x(t), \zeta(t), w(t))}{\partial w} \cdot \frac{\partial f(\zeta(t), w(t))}{\partial \zeta} \right) \frac{\partial V(\eta(t))}{\partial \eta_{n+1}}. \tag{5.2.59}$$

By Assumptions 5.2.5, 5.2.6, and 5.2.7, we obtain further that

$$\left. \frac{d\mathbb{V}(\xi(t), \eta(t))}{dt} \right|_{\text{along (5.2.56) and (5.2.57)}}$$

$$\leq -\rho^\alpha W(\xi(t)) - \rho \mathcal{W}(\eta(t))$$

$$+ L \sum_{i=1}^n \|\xi(t)\| \left| \frac{\partial V(\xi(t))}{\partial \xi_i} \right| + L \|\eta(t)\| \left| \frac{\partial V}{\partial \xi_n}(\xi(t)) \right|$$

$$+ \rho^\alpha \|\eta(t)\| \left| \frac{\partial V(\xi(t))}{\partial \xi_n} \right| + L \sum_{i=1}^n \rho^{(n+1-i)(1-\alpha)} \|\xi(t)\| \left| \frac{\partial V(\eta(t))}{\partial \eta_i} \right|$$

$$+ L^2 \|\eta(t)\| \left| \frac{\partial V(\eta(t))}{\partial \eta_{n+1}} \right| + L |\dot{w}(t)| \left| \frac{\partial V}{\partial \eta_{n+1}}(\xi(t)) \right|$$

$$\leq -\rho^\alpha W(\xi(t)) - \rho \mathcal{W}(\eta(t)) + N_1 W(\xi(t)) + N_2 \mathcal{W}(\eta(t))$$

$$+ N_3 \rho^\alpha \sqrt{W(\xi(t))} \sqrt{\mathcal{W}(\eta(t))} + L |\dot{w}(t)| \left| \frac{\partial V}{\partial \eta_{n+1}}(\xi(t)) \right|$$

$$\leq -\frac{\rho^\alpha}{2} W(\xi(t)) - \rho \mathcal{W}(\eta(t)) + N_1 W(\xi(t)) + N_2 \mathcal{W}(\eta(t))$$

$$+ \frac{N_3^2 \rho^\alpha}{2} \mathbb{W}(\xi(t), \eta(t))) + L |\dot{w}(t)| \left| \frac{\partial V(\xi(t))}{\partial \eta_{n+1}} \right|$$

$$\leq -\frac{\rho^\alpha}{4} \mathbb{W}(\xi(t), \eta(t)) + \sqrt{c_2} L |\dot{w}(t)| \sqrt{\mathbb{W}(\xi(t), \eta(t))}, \tag{5.2.60}$$

where

$$N_1 = n c_2 L + \frac{L c_2}{2}, \ N_2 = \frac{c_1 L}{2} + c_1 L^2, \ N_3 = \sqrt{c_1 c_2}(1 + nL).$$

By hypothesis (5.2.52), there exists $t_1 > 0$ such that $|\dot{w}(t)| < \rho^\alpha/(8\sqrt{c_2}L)$ for all $t > t_1$. For $\mathbb{V}(\cdot)$ and $\mathbb{W}(\cdot)$ defined in (5.2.58), there exist continuous class $\mathcal{K}_\infty$ functions $\kappa_i (i = 1, 2)$:

$[0, \infty) \to [0, \infty)$ satisfying

$$\kappa_1(\mathbb{V}(e)) \leq \mathbb{W}(e) \leq \kappa_2(\mathbb{V}(e)), \forall e \in \mathbb{R}^{2n+1}. \tag{5.2.61}$$

This, together with (5.2.60), shows that for any $t > t_1$, if $\mathbb{V}(\xi(t), \eta(t)) \geq \kappa_1^{-1}(1)$, then $\mathbb{W}(\xi(t), \eta(t)) \geq 1$ and

$$\left. \frac{d\mathbb{V}(\xi(t), \eta(t))}{dt} \right|_{\text{along (5.2.56) and (5.2.57)}} \tag{5.2.62}$$

$$\leq -\frac{\rho^\alpha}{4} \mathbb{W}(\xi(t), \eta(t)) + \frac{\rho^\alpha}{8} \sqrt{\mathbb{W}(\xi(t), \eta(t))} \leq -\frac{\rho^\alpha}{8} < 0.$$

Thus there exists $t_2 > t_1$ such that $\mathbb{V}(\eta(t)) \leq \kappa_1^{-1}(1)$ and $\mathbb{W}(\eta(t)) \leq \kappa_2 \circ \kappa_1^{-1}(1)$ for all $t \in (t_2, \infty)$.

For any given $\sigma > 0$, by (5.2.52), there exists $t_3 > t_2$ such that $|\dot{w}(t)| \leq \frac{\rho^\alpha \kappa_1(\sigma)}{8L\sqrt{c_2}\kappa_2 \circ \kappa_1^{-1}(1)}$ for all $t_3 > t_2$. Hence, for $t > t_3$, if $\mathbb{V}(\eta(t)) \geq \sigma$ then

$$\left. \frac{d\mathbb{V}(\xi(t), \eta(t))}{dt} \right|_{\text{along (5.2.56) and (5.2.57)}} \tag{5.2.63}$$

$$\leq -\frac{\rho^\alpha}{4} \mathbb{W}(\xi(t), \eta(t)) + L\sqrt{c_2} \, \kappa_2 \circ \kappa_1^{-1}(1)|\dot{w}(t)| \leq -\frac{\rho^\alpha \kappa_1(\sigma)}{8}.$$

There exists $t_4 > t_3$ such that $\mathbb{V}(\eta(t)) \leq \sigma$ for all $t > t_4$, that is, $\lim_{t \to \infty} \mathbb{V}(\xi(t), \eta(t)) = 0$. Since $\mathbb{V}(\cdot)$ is continuous and positive definite, there exists a class $\mathcal{K}_\infty$ function $\hat{\kappa} : [0, \infty) \to [0, \infty)$ such that $\|(\xi(t), \eta(t))\| \leq \hat{\kappa}(\mathbb{V}(\xi(t), \eta(t)))$. Therefore, $\lim_{t \to \infty} |(\xi(t), \eta(t))\| = 0$. This completes the proof of the theorem.

This also applies to Corollary 5.2.1 for this example, as the linear ESO, that is, the functions $g_i(\cdot)$ and $u_0(\cdot)$ in (5.2.53) and (5.2.54), is chosen as that in (5.2.50), with Hurwitz matrices K and A defined by (5.2.51) that satisfy all the conditions of Theorem 5.2.2.

5.2.2   ADRC with Time-Varying Gain ESO

In this section, we propose a time-varying gain ESO for system (5.2.1) as follows:

$$\begin{cases} \dot{\hat{x}}_1(t) = \hat{x}_2(t) + \dfrac{1}{\varrho^{n-1}(t)} g_1(\varrho^n(t)(x_1(t) - \hat{x}_1(t))) + \phi_1(\hat{x}_1(t)), \\[2mm] \dot{\hat{x}}_2(t) = \hat{x}_3(t) + \dfrac{1}{\varrho^{n-2}(t)} g_2(\varrho^n(t)(x_1(t) - \hat{x}_1(t))) + \phi_2(\hat{x}_1(t), \hat{x}_2(t)), \\[2mm] \quad\vdots \\[2mm] \dot{\hat{x}}_n(t) = \hat{x}_{n+1}(t) + g_n(\varrho^n(t)(x_1(t) - \hat{x}_1(t))) + b_0 u(t), \\[2mm] \dot{\hat{x}}_{n+1}(t) = \varrho(t) g_{n+1}(\varrho^n(t)(x_1(t) - \hat{x}_1(t))), \end{cases} \tag{5.2.64}$$

where $g_i(\cdot)$ are nonlinear functions satisfying Assumption 5.2.6, and $\varrho \in C([0, \infty) \to \mathbb{R}^+)$ is the gain function to be required to satisfy the following Assumption 5.2.9. For simplicity, we assume that b_0 is a constant nominal value of $b(\cdot)$ in this section.

Assumption 5.2.9 $\varrho \in C^1([0, \infty), [0, \infty))$, $\varrho(t) > 0$, $\dot{\varrho}(t) > a > 0$, and $|\dot{\varrho}(t)/\varrho(t)| \leq M$ for all $t \geq 0$, where $a > 0, M > 0$.

The ESO (5.2.64)-based output feedback control is designed as follows:

$$u(t) = \frac{\rho u_0(\rho^{n-1}\hat{x}_1(t), \cdots, \hat{x}_n(t)) - \hat{x}_{n+1}(t)}{b_0}, \tag{5.2.65}$$

where $u_0(\cdot)$ is a nonlinear function satisfying Assumption 5.2.7, $\hat{x}_i(t)$ is the state of ESO (5.2.64), and $\hat{x}_{n+1}(t)$ is used to compensate (cancel) the total disturbance $x_{n+1}(t)$ defined by (5.2.66) below:

$$x_{n+1}(t) \triangleq f(t, x(t), \zeta(t), w(t)) + (b(t, w(t)) - b_0)u(t). \tag{5.2.66}$$

To obtain the convergence, we need the following Assumption 5.2.10 on the nonlinear functions $f(\cdot)$, $f_0(\cdot)$, and $\phi_i(\cdot)$ in system (5.2.1), which is stronger than Assumption 5.2.1.

Assumption 5.2.10 $|f(t, x, \zeta, w)| + \|f_0(t, x, \zeta, w)\| \leq L\|(x, w)\|$; $\|\nabla f(t, x, \zeta, w)\| \leq L\|w\|$, $\|\phi_i(x_1, \cdots, x_i) - \phi_i(\hat{x}_1, \cdots, \hat{x}_i)\| \leq L$, $L > 0$, $\phi_i(0, \cdots, 0) = 0$.

The following Assumption 5.2.11 is the conditions on $b(\cdot)$ and its nominal value b_0.

Assumption 5.2.11 $b(t, x, \zeta, w) = b(t, w)$, $|b(t, w)| + \|\nabla b(t, w)\| \leq L$, and its nominal value $b_0 \in \mathbb{R}$ satisfy $|b(t, w) - b_0| \leq \wedge \triangleq b_0/(2\Gamma)$ for all $(t, w) \in \mathbb{R}^2$.

Theorem 5.2.3 *Let $\rho(t)$ be chosen so that $\rho(t) > \frac{(nL+3)c_2}{2} + 1$. Suppose that $w(t)$ and $\dot{w}(t)$ are uniformly bounded and Assumptions 5.2.10, 5.2.6, 5.2.7, 5.2.11, and 5.2.9 are satisfied. Then, the closed loop composed of (5.2.1), (5.2.64), and (5.2.65) is asymptotically stable:*

$$\lim_{t \to \infty} \|\eta(t)\| = 0, \quad \lim_{t \to \infty} \|\xi(t)\| = 0. \tag{5.2.67}$$

Proof. From Assumptions 5.2.10, 5.2.6, and 5.2.7, we can estimate the derivative of the total disturbance $x_{n+1}(t)$ given in (5.2.66) as

$$|\dot{x}_{n+1}(t)| \leq C(1 + \|\eta(t)\| + \|\xi(t)\|) + \left|\frac{b(t, w(t)) - b_0}{b_0}\right| \varrho(t)|g_{n+1}(\eta_1(t))|, \tag{5.2.68}$$

where C is an r-independent constant. Let

$$\xi_i(t) = \rho^{n-i}x_i(t), \quad \eta_i(t) = \varrho^{n+1-i}(t)(x_i(t) - \hat{x}_i(t)). \tag{5.2.69}$$

A straightforward calculation shows that

$$\begin{cases} \dot{\xi}_1(t) = \rho\xi_2(t) + \rho^{n-1}\phi_1(\xi_1(t)/\rho^{n-1}), \\ \dot{\xi}_2(t) = \rho\xi_3(t) + \rho^{n-2}\phi_2(\xi_1(t)/\rho^{n-1}, \xi_2(t)/\rho^{n-2}), \\ \quad \vdots \\ \dot{\xi}_n(t) = \rho u_0(\rho^{n-1}\hat{x}_1(t), \cdots, \hat{x}_n(t)) + x_{n+1}(t) - \hat{x}_{n+1}(t) \end{cases} \tag{5.2.70}$$

and

$$
\begin{cases}
\dot\eta_1(t) = \varrho(t)(\eta_2(t) - g_1(\eta_1(t))) + \varrho^n(t)(\phi_1(x_1(t)) - \phi_1(\hat{x}_1(t))) + \dfrac{n\dot\varrho(t)}{\varrho(t)}\eta_1(t), \\[2mm]
\quad\vdots \\[2mm]
\dot\eta_{n-1}(t) = \varrho(t)(\eta_n(t) - g_{n-1}(\eta_1(t))) + \dfrac{(n-1)\dot\varrho(t)}{\varrho(t)}\eta_2(t) \\[2mm]
\qquad + \varrho^2(t)(\phi_{n-1}(x_1(t),\cdots,x_{n-1}(t)) - \phi_{n-1}(\hat{x}_1(t),\cdots,\hat{x}_{n-1}(t))), \\[2mm]
\dot\eta_n(t) = \varrho(t)(\eta_{n+1}(t) - g_n(\eta_1(t))) + \dfrac{\dot\varrho(t)}{\varrho(t)}\eta_n(t), \\[2mm]
\dot\eta_{n+1}(t) = -\varrho(t)g_{n+1}(\eta_1(t)) + \dot{x}_{n+1}(t).
\end{cases}
\tag{5.2.71}
$$

Let $\mathcal{V}(\cdot), \mathcal{W}(\cdot)$ and $V(\cdot), W(\cdot)$ be the Lyapunov functions satisfying Assumptions 5.2.6 and 5.2.7, respectively. Let $\mathbb{V}, \mathbb{W} : \mathbb{R}^{2n+1} \to [0,\infty)$ be the same as that in (5.2.58). Finding the derivative of $\mathbb{V}(\xi(t),\eta(t))$ along the solutions of (5.2.70) and (5.2.71) gives

$$
\left.\frac{d\mathbb{V}(\xi(t),\eta(t))}{dt}\right|_{\text{along (5.2.70) and (5.2.71)}}
$$

$$
= \sum_{i=1}^{n-1}\left(\rho\xi_{i+1}(t) + \rho^{n-i}\phi_i\left(\frac{\xi_1(t)}{\rho^{n-1}},\cdots,\frac{\xi_i(t)}{\rho^{(n-i)\alpha}}\right)\right)\frac{\partial V(\xi(t))}{\partial\xi_i}
$$

$$
+ \rho u_0(\xi_1(t),\cdots,\xi_n(t))\frac{\partial V(\xi(t))}{\partial\xi_n} + \eta_{n+1}(t)\frac{\partial V(\xi(t))}{\partial\xi_n}
$$

$$
+ (\rho u_0(\rho^{n-1}\hat{x}_1(t),\cdots,\hat{x}_n(t)) - \rho u_0(\rho^{n-1}x_1(t),\cdots,x_n(t)))\frac{\partial V(\xi(t))}{\partial\xi_n}
$$

$$
+ \sum_{i=1}^{n-1}\left(\varrho(t)(\eta_{i+1}(t) - g_i(\eta_1(t))) + \varrho^{n+1-i}(t)(\phi_i(x_1(t),\cdots,x_i(t)) - \phi_i(\hat{x}_1(t),\cdots,\hat{x}_i(t)))\right)\frac{\partial\mathcal{V}(\eta(t))}{\partial\eta_i}
$$

$$
+ \sum_{i=1}^{n}\frac{(n+1-i)\dot\varrho(t)|\eta_i(t)|}{\varrho(t)}\left|\frac{\partial\mathcal{V}(\eta(t))}{\partial\eta_i}\right|
$$

$$
+ (\varrho(t)(\eta_{n+1}(t) - g_n(\eta_1(t))))\frac{\partial\mathcal{V}(\eta(t))}{\partial\eta_n} + (-\varrho(t)g_{n+1}(\eta_1(t)) + \dot{x}_{n+1}(t))\frac{\partial\mathcal{V}(\eta(t))}{\partial\eta_{n+1}}.
\tag{5.2.72}
$$

By Assumption 5.2.9, there exists a $t_1 > 0$ such that $\varrho(t) > \max\{\rho, 2(Cc_1 + C^2c_1 + nLc_1 + n(n+1)Mc_1)\}$ for all $t > t_1$. This, together with (5.2.72), gives

$$
\left.\frac{d\mathbb{V}(\xi(t),\eta(t))}{dt}\right|_{\text{along (5.2.70) and (5.2.71)}}
$$

$$
\leq -\rho W(\xi(t)) - \varrho(t)\mathcal{W}(\eta(t)) + \sum_{i=1}^{n-1}L\|\xi(t)\|\left|\frac{\partial V(\xi(t))}{\partial\xi_i}\right| + 2\|\eta(t)\|\left|\frac{\partial V(\xi(t))}{\partial\xi_n}\right|
$$

$$
+ \sum_{i=1}^{n-1}L\|\eta(t)\|\left|\frac{\partial\mathcal{V}(\eta(t))}{\partial\eta_i}\right| + \varrho(t)|g_{n+1}(\eta_1(t))|\left|\frac{b(t,w(t)) - b_0}{b_0}\right|\left|\frac{\partial\mathcal{V}(\eta(t))}{\partial\eta_{n+1}}\right|
$$

$$+ C\left(1 + \|\eta(t)\| + \|\xi(t)\| + \sum_{i=1}^{n} |g_i(\eta_1(t))|\right)\left|\frac{\partial \mathcal{V}(\eta(t))}{\partial \eta_{n+1}}\right|$$

$$+ M \sum_{i=1}^{n} (n+1-i)|\eta_i(t)|\left|\frac{\partial \mathcal{V}(\eta(t))}{\partial \eta_i}\right|$$

$$\leq -\rho W(\xi(t)) - \varrho(t)\mathcal{W}(\eta(t)) + \frac{nLc_2}{2}W(\xi(t)) + c_1\mathcal{W}(\eta(t)) + c_2 W(\xi(t)) + \frac{nLc_1}{2}\mathcal{W}(\eta(t))$$

$$+ \varrho(t)\frac{\Lambda\Gamma}{b_0}\mathcal{W}(\eta(t)) + C\sqrt{c_1}\sqrt{\mathcal{W}(\eta(t))} + Cc_1\mathcal{W}(\eta(t)) + \frac{C^2 c_1}{2}\mathcal{W}(\eta(t)) + \frac{c_2}{2}W(\xi(t))$$

$$\leq -\left(\rho - \frac{(nL+3)c_2}{2}\right)W(\xi(t)) + C\sqrt{c_1}\sqrt{\mathcal{W}(\eta(t))}$$

$$- \left(\varrho(t) - \varrho(t)\frac{\Lambda\Gamma}{b_0} + \frac{nL+2+2C+2C^2+n(n+1)M}{2}c_1\right)\mathcal{W}(\eta(t))$$

$$\leq -W(\xi(t)) - \frac{\varrho(t)}{4}\mathcal{W}(\eta(t)) + C\sqrt{c_1}\sqrt{\mathcal{W}(\eta(t))}. \tag{5.2.73}$$

Now we show that $W(\xi(t)) + \mathcal{W}(\eta(t))$ is uniformly ultimately bounded. In fact, if $W(\xi(t)) + \mathcal{W}(\eta(t)) > 16\max\{1, c_1 C^2\}$, then either $\mathcal{W}(\eta(t)) > 8\max\{1, c_1 C^2\}$ or $\mathcal{W}(\eta(t)) \leq 8\max\{1, c_1 C^2\}$ and $W(\xi(t)) > 8\max\{1, c_1 C^2\}$. In the first case, for $t > t_1$,

$$\left.\frac{d\mathbb{V}(\xi(t), \eta(t))}{dt}\right|_{\text{along } (5.2.70) \text{ and } (5.2.71)} \leq -\frac{\varrho(t)}{4}\mathcal{W}(\eta(t)) + C\sqrt{c_1}\sqrt{\mathcal{W}(\eta(t))}$$

$$\leq \sqrt{\mathcal{W}(\eta(t))}\left(-\frac{\sqrt{\mathcal{W}(\eta(t))}}{4} + C\sqrt{c_1}\right) \leq -C^2 c_1. \tag{5.2.74}$$

In the second case,

$$\left.\frac{d\mathbb{V}(\xi(t), \eta(t))}{dt}\right|_{\text{along } (5.2.70) \text{ and } (5.2.71)} \tag{5.2.75}$$

$$\leq -W(\xi(t)) + C\sqrt{c_1}\sqrt{\mathcal{W}(\eta(t))} \leq -(8 - 2\sqrt{2})C^2 c_1, \ t > t_1.$$

Hence, there exists a positive constant $t_2 > t_1$ such that $W(\xi(t)) + \mathcal{W}(\eta(t)) \leq 16\max\{1, c_1 C^2\}$ for all $t > t_2$. This, together with (5.2.68), produces

$$|\dot{x}_{n+1}(t)| \leq D + \frac{\Lambda}{b_0}\varrho(t)|g_{n+1}(\eta_1(t))|, t > t_2, D > 0. \tag{5.2.76}$$

Finding the derivative of $\mathcal{V}(\eta(t))$ along the solution of (5.2.71) yields

$$\left.\frac{d\mathcal{V}(\eta(t))}{dt}\right|_{\text{along } (5.2.71)} \leq -\varrho(t)\mathcal{W}(\eta(t)) + \left(D + \frac{\Lambda}{b_0}\varrho(t)|g_{n+1}(\eta_1(t))|\right)\left|\frac{\partial \mathcal{V}(\eta(t))}{\partial \eta_{n+1}}\right|$$

$$\leq -\frac{\varrho(t)}{2}\mathcal{W}(\eta(t)) + c_1 D\sqrt{\mathcal{W}(\eta(t))}, \quad \forall\, t > t_2. \tag{5.2.77}$$

For positive definite and radial unbounded Lyapunov functions $\mathcal{V}(\cdot)$ and $\mathcal{W}(\cdot)$, there exist class $\mathcal{K}_\infty$ functions $\kappa_{ij}(\cdot)$ $(i, j = 1, 2)$ such that

$$\kappa_{11}(\|\nu\|) \leq \mathcal{V}(\nu) \leq \kappa_{12}(\|\nu\|), \kappa_{21}(\|\nu\|) \leq \mathcal{W}(\nu) \leq \kappa_{22}(\|\nu\|), \nu \in \mathbb{R}^{n+1}. \quad (5.2.78)$$

From Assumption 5.2.9, for any $\sigma > 0$, there exists a positive constant $t_3 > t_2$ such that $\varrho(t) > 4c_1 D(\kappa_{21} \circ \kappa_{12}^{-1} \circ \kappa_{11}(\sigma))^{-1/2}$. This, together with (5.2.78) and (5.2.77), yields that if $\mathcal{V}(\eta(t)) > \sigma$, then

$$\left. \frac{d\mathcal{V}(\eta(t))}{dt} \right|_{\text{along (5.2.70) and (5.2.71)}} \leq -c_1 D \sqrt{\eta(t)} \leq -c_1 D(\kappa_{21} \circ \kappa_{12}^{-1} \circ \kappa_{11}(\sigma))^{-1/2} < 0.$$
$$(5.2.79)$$

Therefore, there exists $t_4 > t_3$ such that $\mathcal{V}(\eta(t)) \leq \kappa_1(\sigma)$ for all $t > t_4$, and hence

$$\|\eta(t)\| \leq \kappa_{11}^{-1}(\mathcal{V}(\eta(t))) \leq \sigma, \quad \forall t \in [t_4, \infty). \quad (5.2.80)$$

This shows that $\lim_{t \to \infty} \|\eta(t)\| = 0$.

Finding the derivative of $V(\xi(t))$ along the solution of (5.2.70) gives

$$\left. \frac{dV(\xi(t))}{dt} \right|_{\text{along (5.2.70)}} \leq -W(\xi(t)) + c_2 \|\eta(t)\| \sqrt{W(\xi(t))}, \quad \forall t > t_3. \quad (5.2.81)$$

By positive definiteness and radial unboundedness of $V(\cdot)$ and $W(\cdot)$, there exist class $\mathcal{K}_\infty$ functions $\tilde{\kappa}_{ij}(\cdot), i, j = 1, 2$ such that

$$\tilde{\kappa}_{11}(\|\nu\|) \leq V(\nu) \leq \tilde{\kappa}_{12}(\|\nu\|), \quad \tilde{\kappa}_{21} \leq W(\nu) \leq \tilde{\kappa}_{22}(\|\nu\|), \nu \in \mathbb{R}^n. \quad (5.2.82)$$

Similar to (5.2.80), there exists $t_4 > t_3$ such that

$$\|\eta(t)\| \leq \sqrt{(\tilde{\kappa}_{21} \circ \tilde{\kappa}_{12}^{-1} \circ \tilde{\kappa}_{11})(\sigma)/(2c_2)}, \quad \forall t > t_4.$$

This, together with (5.2.80), shows that if $V(\xi(t)) \geq \tilde{\kappa}_{11}(\sigma)$, then

$$\left. \frac{dV(\xi(t))}{dt} \right|_{\text{along (5.2.70)}} \leq -\frac{(\tilde{\kappa}_{21} \circ \tilde{\kappa}_{12}^{-1} \circ \tilde{\kappa}_{11})(\sigma)}{2c_2}, \quad \forall t > t_4. \quad (5.2.83)$$

Therefore, there exists a constant $t_5 > 0$ such that $V(\xi(t)) \leq \tilde{\kappa}_{11}(\sigma)$ for all $t > t_5$. By (5.2.82), $\|\xi(t)\| \leq \tilde{\kappa}_{11}^{-1}(V(\xi(t))) \leq \sigma$ for $t > t_5$. This shows that $\lim_{t \to \infty} \|\xi(t)\| = 0$. The result then follows from (5.2.69).

5.3 Numerical Simulations

In this section, we present several numerical simulations for illustration.

Example 5.3.1 *Consider the following system:*

$$\begin{cases} \dot{x}_1(t) = x_2(t) + \sin(x_1(t)), \\ \dot{x}_2(t) = f(t, x_1, x_2, \zeta(t), w(t)) + b(t, x_1(t), x_2(t), \zeta(t), w(t))u(t), \\ \dot{\zeta}(t) = f_0(t, x_1, x_2, \zeta(t), w(t)), \end{cases} \quad (5.3.1)$$

where $f, f_0 \in C(\mathbb{R}^5, \mathbb{R})$, are unknown nonlinear functions and $w(t)$ is the external disturbance.

As in Corollary 5.2.1, we first design a linear ESO to estimate the states $x_1(t), x_2(t)$, and the "total disturbance" $x_3(t) \triangleq f(t, x_1(t), x_2(t), \zeta(t), w(t)) + (b(t, x_1(t), x_2(t), \zeta(t), w(t)) - b_0)u(t)$, where b_0 is a constant nominal value of $b(\cdot)$, as follows:

$$\begin{cases} \dot{\hat{x}}_1(t) = \hat{x}_2(t) + 6r(x_1(t) - \hat{x}_1(t)) + \sin(\hat{x}_1(t)), \\ \dot{\hat{x}}_2(t) = \hat{x}_3(t) + 11r^2(x_1(t) - \hat{x}_1(t)), \\ \dot{\hat{x}}_3(t) = 6r^3(x_1(t) - \hat{x}_1(t)), \end{cases} \quad (5.3.2)$$

A linear ESO (5.3.2)-based output feedback control $u(t)$ is designed to stabilize the x-subsystem of (5.3.1):

$$u(t) = \frac{-8 \, \mathrm{sat}_{10}(\hat{x}_1(t)) - 4 \, \mathrm{sat}_{10}(\hat{x}_2(t)) - \mathrm{sat}_{10}(\hat{x}_3(t))}{b_0}. \quad (5.3.3)$$

We suppose that the functions $f(\cdot), f_0(\cdot)$, and the external disturbance $w(t)$ (it is seen from (5.3.2) and (5.3.3) that these functions are not used to design the ESO (5.3.2) and feedback (5.3.2) for numerical simulation are

$$\begin{cases} f(t, x_1, x_2, \zeta, w) = te^{-t} + x_1^2 + \sin x_2 + \cos \zeta + w, \\ f_0(t, x_1, x_2, \zeta, w) = -(x_1^2 + w^2)\zeta, \quad w(t) = \sin(2t + 1). \end{cases} \quad (5.3.4)$$

The control amplification coefficient is given by

$$b(t, x_1, x_2, \zeta, w) = 2 + \frac{1}{10}\sin(t + x_1 + x_2 + w). \quad (5.3.5)$$

The nominal value for $b(\cdot)$ is chosen as $b_0 = 2$. Set $r = 200$, integration step $h = 0.001$. The numerical results are plotted in Figure 5.3.1 and the local amplifications of Figures 5.3.1(b) and 5.3.1(c) are plotted as Figure 5.3.2. It is seen from Figures 5.3.1(a), 5.3.1(b), 5.3.2(a), and 5.3.2(b) that the stabilization is very satisfactory. Figure 5.3.1(c) shows that the total disturbance estimation is also satisfactory. However, the large peaking values for $\hat{x}_2(t)$ and $\hat{x}_3(t)$ that reach almost 2.5×10^4 in the initial time are observed. However, the saturations of $\hat{x}_i(t), i = 1, 2, 3$, make the control value less than 10, which can be seen from Figure 5.3.2(c). Since we do not apply high gain in the feedback loop like the high-gain control method, there are almost no peaking phenomenon for system states $x_1(t)$ and $x_2(t)$, which is an advantage of the ADRC approach.

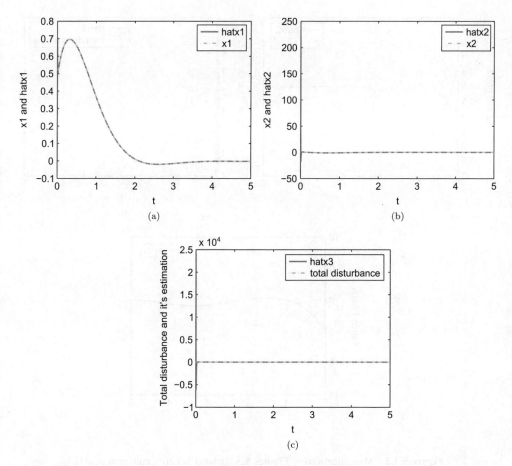

Figure 5.3.1   The numerical results of system (5.3.1) under the saturated control (5.3.3) with constant gain ESO (5.3.2).

Next, we design a nonlinear time-varying gain ESO for system (5.3.1) as

$$\begin{cases} \dot{\hat{x}}_1(t) = \hat{x}_2(t) + 6\varrho(t)(x_1(t) - \hat{x}_1(t)) + \dfrac{1}{\varrho(t)}\Phi\left(\dfrac{x_1(t) - \hat{x}_1(t)}{\varrho^2(t)}\right) + \sin(\hat{x}_1(t)), \\[2mm] \dot{\hat{x}}_2(t) = \hat{x}_3(t) + 11\varrho^2(t)(x_1(t) - \hat{x}_1(t)), \\[2mm] \dot{\hat{x}}_3(t) = 6\varrho^3(t)(x_1(t) - \hat{x}_1(t)), \end{cases}$$

(5.3.6)

where the nonlinear function $\Phi : \mathbb{R} \to \mathbb{R}$ is given by

$$\Phi(\tau) = \begin{cases} \dfrac{1}{4\pi}, & \tau > \pi/2, \\[2mm] \dfrac{\sin\tau}{4\pi}, & -\pi/2 \leq \tau \leq \pi/2, \\[2mm] -\dfrac{1}{4\pi}, & \tau < -\pi/2. \end{cases}$$

(5.3.7)

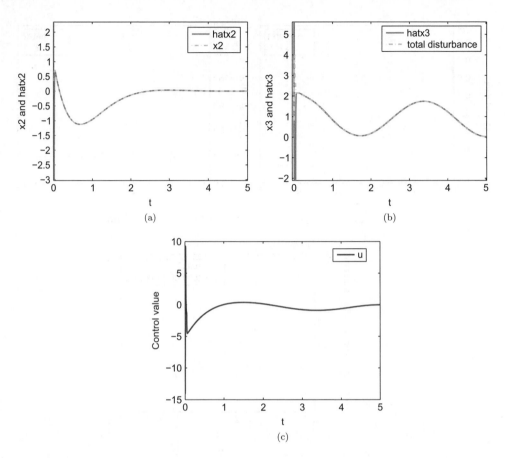

Figure 5.3.2   Magnifications of Figures 5.3.1(b) and 5.3.2(c), and control $u(t)$.

In the following, we use the time-varying gain $\varrho(t) = e^{0.6t}$ for the numerical simulation. A linear ESO (5.3.10)- based output feedback control $u(t)$ is designed as follows:

$$u(t) = \frac{-8\hat{x}_1(t) - 4\hat{x}_1(t) - \hat{x}_3(t)}{b_0}. \tag{5.3.8}$$

The nonlinear function class required by Theorem 5.2.1 is more general than the class required by Theorem 5.2.3, where the nonlinear functions are required to be Lipschitz continuous, which is not satisfied by class (5.3.4). Therefore, we choose the nonlinear functions $f(\cdot)$ and $f_0(\cdot)$ for system (5.3.1) with the time-varying gain ESO (5.3.6) as

$$
\begin{aligned}
f(t, x_1, x_2, \zeta, w) &= te^{-t} + \sin x_2 + \cos \zeta + w, \\
f_0(t, x_1, x_2, \zeta, w) &= -\sin(\zeta(t))|x_1(t)|, \ w(t) = \sin(2t + 1).
\end{aligned}
\tag{5.3.9}
$$

The other functions and parameters are the same as in Figure 5.3.1.

It is seen from Figures 5.3.1(a), 5.3.1(b), and 5.3.1(c) that the stabilization and "total distur-bance" estimate are also very satisfactory. Meanwhile, there are no peaking values for $\hat{x}_2(t)$ and $\hat{x}_3(t)$.

Generally, the large-gain value needs a small integration step. Therefore, as recommended in Section 5.1, in practice, the time-varying gain should be a small value in the beginning and gradually increase to a large constant high gain for which we choose

$$\begin{cases} \varrho(0) = 1, \\ \dot{\varrho}(t) = 5\varrho(t) \text{ if } \varrho(t) < 200, \\ \dot{\varrho}(t) = 0, \text{ otherwise,} \end{cases} \tag{5.3.10}$$

which means that for $t \in [0, \ln 200/5]$, $\varrho(t) = e^{5t}$, and $\varrho(t) = 200$ for every $t > (\ln 200)/5$. Hence the switch time is $\frac{\ln 200}{5}$. With the same other functions and parameters as in Figure 5.3.3, the numerical results are plotted in Figure 5.3.4.

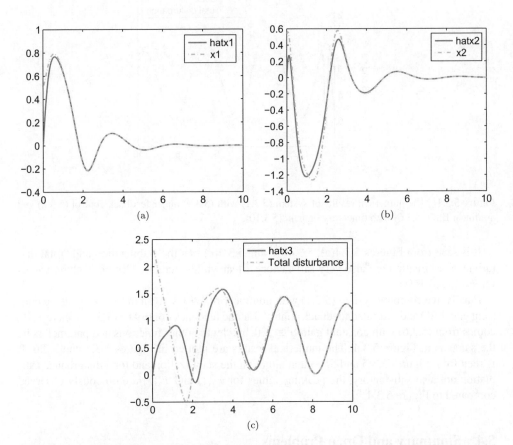

(a)

(b)

(c)

Figure 5.3.3   The numerical results of system (5.3.1) with (5.3.9) under feedback control (5.3.8) and nonlinear ESO (5.3.6) with time-varying gain $\varrho(t) = e^{0.6t}$.

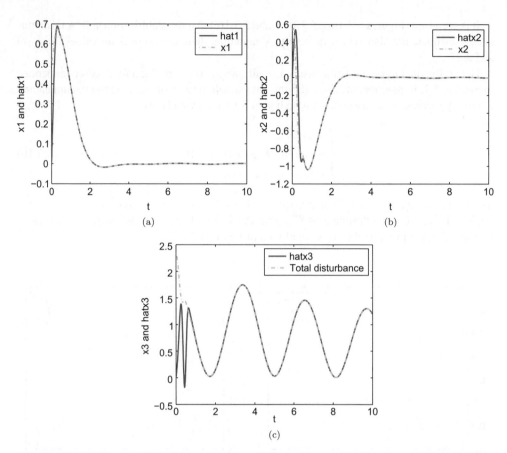

Figure 5.3.4 The numerical results of system (5.3.1) with (5.3.9) under feedback control (5.3.8) and nonlinear ESO (5.3.6) with time-varying gain (5.3.10).

It is seen from Figures 5.3.4(a), 5.3.4(b), and 5.3.4(c) that the stabilization and "total disturbance" estimation are also very satisfactory. Meanwhile, there are also no peaking values for $\hat{x}_2(t)$ and $\hat{x}_3(t)$.

Finally, we compute system (5.3.1) with nonlinear functions given in (5.3.9) by using constant gain ESO and saturated feedback control. Let the feedback control be (5.3.3), where $\hat{x}_i(t)$ comes from (5.3.6) with constant gain $\varrho \equiv 200$, and let the other functions and parameters be the same as in Figure 5.3.4. The numerical results are plotted in Figures 5.3.5 and 5.3.6. It is seen from Figures 5.3.5 and 5.3.6 that although the stabilization and total disturbance estimation are also satisfactory, the peaking values for $\hat{x}_2(t)$ and $\hat{x}_3(t)$ are obviously observed compared to Figure 5.3.4.

5.4   Summary and Open Problems

The aim of this chapter is to extend the extended state observer (ESO) to lower triangular systems with large uncertainty, and to apply the active disturbance rejection

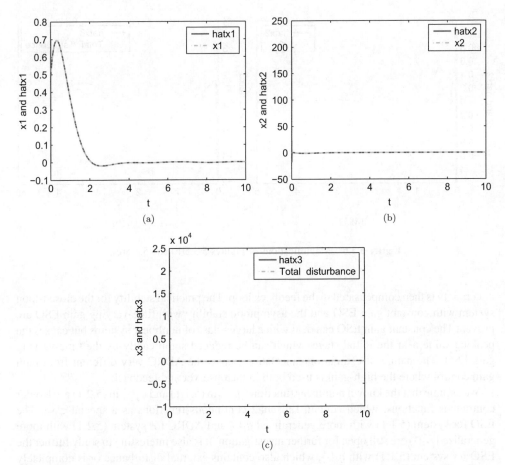

Figure 5.3.5 Stabilization of system (5.3.1) with (5.3.9) under feedback control (5.3.2) and nonlinear ESO (5.3.6) with constant gain.

control (ADRC) to stabilization for lower triangular nonlinear systems with large uncertainties.

In Section 5.1, we propose systematically the design of an extended state observer for a class of lower triangular nonlinear systems with vast uncertainty. The uncertainty (total disturbance) may contain unmodeled system dynamics and external disturbance. The objective of ESO is to estimate, in real time, both states and total disturbance by the output. A constant high-gain approach and time-varying gain approach are adopted in the investigation. The practical convergence by the constant high-gain approach and asymptotical convergence by the time-varying gain approach are presented. In addition, two classes of ESO, namely linear ESO and ESO with homogeneous weighted functions, are constructed. The peaking-value problem caused by the constant high gain is discussed in comparison to the time-varying gain that can reduce dramatically the peaking value. Numerical simulations are presented to validate visually the convergence and peaking-value reduction.

In Section 5.2, the ESO is extended to estimate the state and uncertainty, in real time, simultaneously. The constant gain and the time-varying gain are used in ESO design separately. The

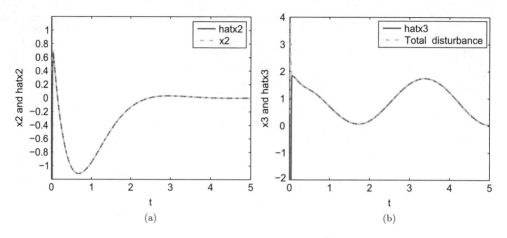

Figure 5.3.6　Magnifications of Figures 5.3.5(b) and 5.3.5(c).

uncertainty is then compensated in the feedback loop. The practical stability for the closed-loop system with constant gain ESO and the asymptotic stability with time-varying gain ESO are proven. The constant gain ESO can deal with a larger class of nonlinear systems but causes the peaking value near the initial stage, which can be reduced significantly by the time-varying gain ESO. The nature of estimation/cancelation makes the ADRC very different from high gain control where the high-gain is used both in the observer and feedback.

We assume that the known nonlinear functions $g_i(\cdot)$ in (5.1.1) and $\phi_i(\cdot)$ in (5.2.1) are Hölder continuous functions, which contain Lipschitz continuous functions as a special case. The ESO for system (5.1.1) with more generalized $g_i(\cdot)$ and ADRC for system (5.2.1) with more generalized $\phi_i(\cdot)$ are still open for further investigation. It is also interesting to study further the ESO for system (5.1.1) with $g_i(\cdot)$, which also contains external disturbance or is completely unknown. The ADRC for system (5.2.1) with $\phi_i(\cdot)$, which also contains external disturbance or is completely unknown is another open problem.

5.5　Remarks and Bibliographical Notes

Section 5.1 This section is taken from [157] and is reproduced by permission of the Elsevier license.

The state observer for lower triangular systems with or without uncertainty has been studied by many researchers and some of them can be found in [3, 5, 103, and 116]. However, in these works, the uncertainty is not estimated.

Section 5.1.1 Theorem 5.1.2 covers the case of constant $w(t)$ considered in [111].

Section 5.1.2 In Theorem 5.1.4, to deal with the disturbance with unbounded derivative and to guarantee the asymptotical convergence, an exponentially growing gain is used. This is not very realistic in practice. The recommended strategy is to use the time-varying gain in the beginning to reduce the peaking value and then use the constant high-gain afterwards. However, Theorem 5.1.4 has its theoretical significance itself. It simply means that, for disturbance with unbounded derivative, we need a large gain in the ESO.

Section 5.2 This is taken from [158] and is reproduced by permission of John Wiley & Sons Ltd license.

It is indicated in [38] that any uniform observable affine SISO nonlinear system can be transformed into the lower triangular form (5.2.1). The state observer design and control for lower triangular systems have been extensively studied, see, for instance, [5, 6, 111, and 116], and the references therein. In [6, and 116], a state observer is designed for an uncertain lower triangular system without estimating the system uncertainty. In [5], an output tracking problem for a lower triangular system is studied by the high-gain approach, where the uncertainty is also not estimated. In [111], an unknown constant in control is estimated on stabilization for lower triangular nonlinear systems. The estimation for uncertainty allows us to design an ESO-based output feedback control without using high gain in the feedback loop, while for the system with general uncertainty, the high-gain dominated method must use (generally) high gain both in the observer and feedback except for some very special cases as shown numerically in [38].

Section 5.2.2: As indicated in Section 5.1, the time-varying gain ESO degrades the ability of ESO to filter high-frequency noise while the constant gain ESO does not. In practical applications, we can use time-varying gain $\varrho(t)$ as follows: (a) given a small initial value $\varrho(0) > 0$; (b) from the constant high-gain, we obtain the convergent high gain value r, which can also be obtained by trial and error experiment for practical systems; (c) our gain function is usually supposed to grow continuously from $\varrho(0) > 0$ to r. For instance, $\dot{\varrho}(t) = a\varrho(t), a > 0$. In this case, we can compute the switching time as $\ln\left(r/\varrho(0)\right)/a$, where a is used to control the convergent speed and the peaking value. The larger a is, the faster the convergence but the larger the peaking, while the smaller a is, the lower the convergence speed and the smaller the peaking.

References

[1] A. Adams and J.J.F. Fournier, *Sobolev Spaces*, 2nd edition, Academic Press, 2003.

[2] J.H. Ahrens and H.K. Khalil, High-gain observers in the presence of measurement noise: a switched- gain approach, *Automatica*, 45(2009), 936–943.

[3] A. Alessandri and A. Rossi, Time-varying increasing-gain observers for nonlinear systems, *Automatica*, 49(2013), 2845–2852.

[4] B. Aloliwi and H.K. Khalil, Adaptive output feedback regulation of a class of nonlinear systems: convergence and robustness, *IEEE Trans. Automat. Control*, 42(97), 1714–1716.

[5] V. Andrieu, L. Praly, and A. Astolfi, High gain observers with updated gain and homogeneous correction terms, *Automatica*, 45(2009), 412–418.

[6] V. Andrieu, L. Praly, and A. Astolfi, Asymptotic tracking of a reference trajectory by output-feedback for a class of non linear systems, *Systems Control Lett.*, 45(2009), 412–418.

[7] D. Angeli and E. Mosca, Lyapunov-based switching supervisory control of nonlinear uncertain systems, *IEEE Trans. Automat. Control*, 47(2002), 500–505.

[8] V. Andrieu, L. Praly, and A. Astolfi, High gain observers with updated gain and homogeneous correction terms, *Automatica*, 45(2009), 412–418.

[9] K. Astrom and T. Hagglund, *PID Controller: Theory, Design, and Tuning*, Instrument Society of America, Research Triangle Park, NC, 1995.

[10] A.N. Atassi and H.K. Khalil, A separation principle for the control of a class of nonlinear systems, *IEEE Trans. Automat. Control*, 46(2001), 742–746.

[11] J. Back and H. Shim, Adding robustness to nominal output-feedback controllers for uncertain nonlinear systems: A nonlinear version of disturbance observer, *Automatica*, 44(2008), 2528–2537.

[12] A. Bacciotti and L. Rosier, *Liapunov Functions and Stability in Control Theory*, Springer-Verlag, Berlin, 2005.

[13] G. Besancon, High-gain observation with disturbance attenuation and application to robust fault detection, *Automatica*, 39(2003), 1095–1102.

[14] G. Besancon, *Nonlinear Observers and Applications*, Springer Verlag, New York, 2007.

[15] S.P. Bhat and D.S. Bernstein, Finite-time stability of continuous autonomous systems, *SIAM J. Control Optim.*, 38(2000), 751–766.

[16] S.P. Bhat and D.S. Bernstein, Geometric homogeneity with applications to finite-time stability, *Math. Control Signals Systems*, 17(2005), 101–127.

[17] S.P. Bhat and D.S. Bernstein, Continuous finite-time stabilization of the translational and rotational double integrators, *IEEE Trans. Automatic Control*, 43(1998), 678–682.

Active Disturbance Rejection Control for Nonlinear Systems: An Introduction, First Edition.
Bao-Zhu Guo and Zhi-Liang Zhao.
© 2016 John Wiley & Sons Singapore Pte. Ltd. Published 2016 by John Wiley & Sons, Ltd.

[18] W.L. Bialkowski, Control of the pulp and paper making process, in: *The Control Handbook* (W.S. Levine, Ed.), IEEE Press, New York, 1996, 1219–1242.

[19] R. Brockett, New issues in the mathematics of control, in: *Mathematics Unlimited-2001 and Beyond*, Springer-Verlag, Berlin, 2001.

[20] A. Chakrabortty and M. Arcak, Time-scale separation redesigns for stabilization and performance recovery of uncertain nonlinear systems, *Automatica*, 46(2009), 34–44.

[21] X. Chen, C.Y. Su, and T. Fukuda, A nonlinear disturbance observer for multivariable systems and its application to magnetic bearing system, *IEEE Trans. Automat. Control*, 49(2004), 569–577.

[22] M. Corless and J. Tu, State/input estimation for a class of uncertain systems, *Automatica*, 34(1998), 757–764.

[23] A.M. Dabroom and H.K. Khalil, Discrete-time implementation of high-gain observers for numerical differentiation, *Internat. J. Control*, 12(1997), 1523–15371.

[24] S. Darkunov, Sliding-mode observers based on equivalent control method, *Proc. the 31st IEEE Conference on Decision and Control*, 1992, 2368–2370.

[25] M. Darouach, M. Zasadzinski, and J.S. Xu, Full-order observers for linear systems with uncertain inputs, *IEEE Trans. Automat. Control*, 39(1994), 606–609.

[26] J. Davila, L. Fridman, and A. Levant, Second-order sliding-modes observer for mechanical systems, *IEEE Trans. Automat. Control*, 50(2005), 1785–1789.

[27] L. Dong, P. Kandula, and Z. Gao, On a robust control system design for an electric power assist steering system, *Proc. the American Control Conference*, 2010, 5356–5361.

[28] P. Dong, G. Ye, and J. Wu, Auto-disturbance rejection controller in the wind energy conversion system, *Proc. the Power Electronics and Motion Control Conference*, 2(2004), 878–881.

[29] L. Dong, Q. Zheng, and Z. Gao, On control system design for the conventional mode of operation of vibrational gyroscopes, *IEEE Sensors Journal*, 8(2008), 1871–1878.

[30] T. Emaru and T. Tsuchiya, Research on estimating smoothed value and differential value by using sliding mode system, *IEEE Trans. Robotics and Automation*, 19(2003), 391–402.

[31] G. Feng, Y. Liu, and L. Huang, A new robust algorithm to improve the dynamic performance on the speed control of induction motor drive, *IEEE Trans. Power Electronics*, 19(2004), 1614–1627.

[32] A.F. Filippov, *Differential Equations with Discontinuous Righthand Sides*, Kluwer Academic Publishers, Dordrecht, 1998.

[33] B. Francis, The linear multivariable regulaor problem, *SIAM J. Control Optim.*, 15(1977), 486–505.

[34] B. Francis and W. Wonham, The internal model principle of control theory, *Automatica*, 12(1976), 457–465.

[35] L.B. Freidovich and H.K. Khalil, Performance recovery of feedback-linearization-based designs, *IEEE Trans. Automat. Control*, 53(2008), 2324–2334.

[36] L. Fridman, Y. Shtessel, C. Edwards, and X.G. Yan, Higher-order sliding-mode observer for state estimation and input reconstruction in nonlinear systems, *Internat. J. Robust Nonlinear Control*, 18(2008), 399–412.

[37] Z. Gao, Scaling and bandwith-parameterization based controller tuning, *Proc. the American Control Conference*, 2003, 4989–4996.

[38] J.P. Gauthier, H. Hammouri, and S. Othman. A simple observer for nonlinear systems application to bioreactors, *IEEE Trans. Automat. Control*, 37(1992), 875–880.

[39] F.J. Goforth and Z.Q. Gao, An active disturbance rejection control solution for hysteresis compensation, *Proc. the American Control Conference*, 2008, 2202–2208.

[40] V. Gourshankar, P. Kudva, and K. Ramar, Reduced order observer for multivariable systems with inaccessible disturbance inputs, *Internat. J. Control*, 25(1977), 311–319.

[41] B.Z. Guo, J.Q. Han, and F.B. Xi, Linear tracking-differentiator and application to online estimation of the frequency of a sinusoidal signal with random noise perturbation, *Internat. J. Systems Sci.*, 33(2002), 351–358.

[42] B.Z. Guo and F.F. Jin, Output feedback stabilization for one-dimensional wave equation subject to boundary disturbance, *IEEE Trans. Automat. Control*, 60(2015), 824–830.

[43] B.Z. Guo and F.F. Jin, The active disturbance rejection and sliding mode control approach to the stabilization of Euler–Bernoulli beam equation with boundary input disturbance, *Automatica*, 49(2013), 2911–2918.

[44] B.Z. Guo and F.F. Jin, Sliding mode and active disturbance rejection control to stabilization of one-dimensional anti-stable wave equations subject to disturbance in boundary input, *IEEE Trans. Automat. Control*, 58(2013), 1269–1274.

[45] B.Z. Guo and J.J. Liu, Sliding mode control and active disturbance rejection control to the stabilization of one-dimensional Schrodinger equation subject to boundary control matched disturbance, *Internat. J. Robust Nonlinear Control*, 24(2014), 2194–2212.

[46] B.Z. Guo, J.J. Liu, A.S. AL-Fhaid, A.M.M. Younas, and A. Asiri, The active disturbance rejection control approach to stabilization of coupled heat and ODE system subject to boundary control matched disturbance, *Internat. J. Control*, 88(2015), 1554–1564.

[47] B.Z. Guo and H.C. Zhou, The active disturbance rejection control to stabilization for multi-dimensional wave equation with boundary control matched disturbance, *IEEE Trans. Automat. Control*, 60(2015), 143–157.

[48] B.Z. Guo, H.C. Zhou, A.S. AL-Fhaid, A.M.M. Younas, and A. Asiri, Stabilization of Euler–Bernoulli beam equation with boundary moment control and disturbance by active disturbance rejection control and sliding model control approach, *J. Dyn. Control Syst.*, 20(2014), 539–558.

[49] B.Z. Guo and H.C. Zhou, Active disturbance rejection control for rejecting boundary disturbance from multi-dimensional Kirchhoff plate via boundary control, *SIAM J. Control Optim.*, 52(2014), 2800–2830.

[50] B.Z. Guo, H.C. Zhou, A.S. AL-Fhaid, A.M.M. Younas, and A. Asiri, Parameter estimation and stabilization for one-dimensional Schrodinger equation with boundary output constant disturbance and non-collocated control, *J. Franklin Inst.*, 352(2015), 2047–2064.

[51] W. Guo and B.Z. Guo, Parameter estimation and non-collocated adaptive stabilization for a wave equation subject to general boundary harmonic disturbance, *IEEE Trans. Automat. Control*, 58(2013), 1631–1643.

[52] B.Z. Guo and Z.L. Zhao, Weak convergence of nonlinear high-gain tracking differentiator, *IEEE Trans. Automat. Control*, 58(2013), 1074–1080.

[53] B.Z. Guo and Z.L. Zhao, On convergence of the nonlinear active disturbance rejection control for MIMO systems, *SIAM J. Control Optim.*, 51(2013), 1727–1757.

[54] B.Z. Guo and Z.L. Zhao, On the convergence of an extended state observer for nonlinear systems with uncertainty, *Systems Control Lett.*, 60(2011), 420–430.

[55] B.Z. Guo and Z.L. Zhao, On convergence of tracking differentiator, *Internat. J. Control*, 84(2011), 693–701.

[56] B.Z. Guo and Z.L. Zhao, On convergence of nonlinear extended state observer for MIMO systems with uncertainty, *IET Control Theory Appl.*, 6(2012), 2375–2386.

[57] B.Z. Guo and Z.L. Zhao, On convergence of nonlinear active disturbance rejection for SISO systems, *Proc. the 24th Chinese Control and Decision Conference*, 2012, 3524–3529.

[58] J.Q. Han, Control theory: model approach or control approach, *J. Systems Sci. Math. Sci.*, 9(4)(1989), 328–335 (in Chinese).

[59] J.Q. Han and W. Wang, Nonlinear tracking-differentiator, *J. Systems Sci. Math. Sci.*, 14(2)(1994), 177–183 (in Chinese).

[60] J.Q. Han, A class of extended state observers for uncertain systems, *Control & Decision*, 10(1)(1995), 85–88 (in Chinese).

[61] J.Q. Han, Auto-disturbance rejection control and applications, *Control & Decision*, 13(1)(1998), 19–23 (in Chinese).

[62] J.Q. Han, *Active Disturbance Rejection Control–The Technique for Extimating and Compensating Uncertainties*, National Defence Industry Press, Beijing, 2008 (in Chinese).

[63] J. Q. Han, From PID to active disturbance rejection control, *IEEE Trans. Ind. Electron.*, 56(2009), 900–906.

[64] J.Q. Han, A new type of controller: NLPID, *Control & Decision*, 6(6)(1994), 401–407 (in Chinese).

[65] J.Q. Han, The improvement of PID control law by using nonlinearity, *Information and Control*, 24(6)(1995), 356–364 (in Chinese).

[66] G. Hillerstrom and J. Sternby, Application of repetitive control to a peristsltic pump, *ASME J. Dynamic Systems, Measurement, and Control*, 116(1994), 786–789.

[67] Y.G. Hong, J. Huang, and Y. Xu, On an output feedback finite-time stable stabilization problem, *Proc. the 38th IEEE Conference on Desion and Control*, 1999, 1302–1307.

[68] Y. Hou, Z. Gao, F. Jiang, and B.T. Boulter, Active disturbance rejection control for web tension regulation, *Proc. the 40th IEEE Conference on Decision and Control*, 2001, 4974–4979.

[69] H. Huang, L. Wu, and J.Q. Han, A new synthesis method for unit coordinated control system in thermal power plant-adrc control scheme, *Proc. the International Conference on Power System Technology*, 2004, 133–138.

[70] L. Huang, *Stability Theory*, Peking University Press, Beijing, 1992 (in Chinese).

[71] Y. Huang and J.Q. Han, A new synthesis method for uncertain systems–the self-stable region approach, *Internat. J. Systems Sci.*, 30(1999), 33–38.

[72] Y. Huang, K. Xu, J. Han, and J. Lam, Flight control design using extended state observer and non-smooth feedback, *Proc. the 40th IEEE Conference on Decision and Control*, 2001, 223–228.

[73] Y. Huang, K. Xu, J. Han, and J. Lam, Extended state observer based technique for control of robot systems, *Proc. the 4th World Congress on Intelligent Control and Automation*, Shanghai, 2002, 2807–2811.

[74] L. Hsu, R. Ortega, and G. Damm, A global convergent frequency estimator, *IEEE Trans. Automatic Control*, 44(1999), 698–713.

[75] Y. Hou, Z. Gao, F. Jiang, and B. Boulter, Active disturbance rejection control for web tension regulation, *Proc. the 40th IEEE Conference on Decision and Control*, 2001, 4974–4979.

[76] Y. Huang, W.C. Xue, and X.X. Yang, Active disturbance rejection control: Methodology, theoretical analysis and applications, *Proc. the 29th Chinese Control Conference*, 2010, 6113–6120.

[77] J. Huang, *Nonlinear Output Regulation: Theorey and Applications*, SIAM, Philadelphia, 2004.

[78] S. Ibrir, Linear time-derivative trackers, *Automatica*, 40(2004), 397–405.

[79] A. Isidori, *Nonlinear Control Systems*, Spinger-Verlag, New York, 1995.

[80] R. E. Kalman and J. F. Bertram, Control System Analysis and Design via the Second Method of Lyapunov, *J. Basic Engrg*, 88(1960), 371–394.

[81] R. Kalman, A new approach to linear filtering and prediction problems, *Trans. ASME, Journal of Basic Engineering*, 82(1960), 35–45.

[82] H. Khalil, High-gain observer in nonlinear feedback control, in: *New Direction in Nonlinear Observer Design*, Lecture Notes in Control and Inform. Sci., 244, Springer, London, 1999, 249–268.

[83] H.K. Khalil, High-gain observers in nonlinear feedback control, *Proc. the International Conference of Control, Automation and Systems*, New Zealand, (2009), 1527–1528.

[84] H. K. Khalil, *Nonlinear Systems*, Prentice Hall, New Jersey, 2002.

[85] A. Koshkouei and A. Zinober, Sliding mode controller observer design for SISO linear systems, *Internat. J. Systems Sci.*, 29(1998), 1363–1373.

[86] H. Khalil and L. Praly, High-gain observers in nonlinear feedback control, *Int. J. Robust. Nonlinear Control*, 24(2014), 993–1015.

[87] A. Krener and A. Isidori, Linearization by output injection and nonlinear observers, *Systems Control Lett.*, 3(1983), 47–52.

[88] A. Krener and A. Respondek, Nonlinear observers with linearizable error dynamics, *SIAM J. Control Optim.*, 23(1985), 197–216.

[89] P. Krishnamurthy, F. Khorrami, and R. S. Chandra, Global high-gain-based observer and backstepping controller for generalized output–feedback canonical form, *IEEE Trans. Automat. Control*, 48(2003), 2277–2284.

[90] P. Kudva, N. Viswanadham, and A. Ramakrishna, Observers for linear systems with unknown inputs, *IEEE Trans. Automat. Control*, 25(1980), 113–115.

[91] J. P. LaSalle and S. Lefschetz, *Stability by Lyapunov's Second Method with Applications*, Academic Press, New York, 1961.

[92] A. M. Letov, *Stability of Nonlinear Control Systems*, English tr., Princeton, 1961.

[93] A. Levant, Robust exact differentiation via sliding mode technique, *Automatica*, 34(1998), 379–384.

[94] A. Levant, Higher order sliding modes, differentiation and output feedbacck control, *Internat. J. Control*, 76(2003), 924–941.

[95] Q. Li and J. Wu, Application of auto-disturbance rejection control for a shunt hybrid active power filter, *Proc. the IEEE International Symposium on Industrial Electronics*, 2005, 627–632.

[96] S. Li and Z. Liu, Adaptive speed control for permanent-magnet synchronous motor system with variations of load inertia, *IEEE Trans. Ind. Electron.*, 56(2009), 3050–3059.

[97] S. Li, X. Yang, and D. Yang, Active disturbance rejection control for high pointing accuracy and rotation speed, *Automatica*, 45(2009), 1856–1860.

[98] X.X. Liao, *Theory and Application of the Stability for Dynamical Systems*, National Defence Industy Press, Beijing, 1999 (in Chinese).

[99] L. Liu, Z. Chen, and J. Huang, Parameter convergence and nomimal internal model with an adaptive output regulation problem, *Automatica*, 45(2009), 1306–1311.

[100] Z. Long, Y. Li, and X. Wang, On maglev train automatic operation control system based on auto-disturbance-rejection control algorithm, *Proc. the 27th Chinese Control Conference*, 2008, 681–685.

[101] A.M. Lyapunov, *The General Problem of the Stability of Motion*, Doctoral dissertation, Univ. Kharkov, 1892 (in Russian).

[102] Y. Ma, Y. Yu, and X. Zhou, Control technology of maximal wind energy capture of VSCF wind power generation, *Proc. the 2010 WASE International Conference on Information Engineering*, 2010, 268–271.

[103] T. Menard, E. Moulay, and W. Perruquetti, A global high-gain finite-time observer, *IEEE Trans. Automat. Control*, 55(2010), 1500–1506.

[104] A. Medvedev and G. Hillerstrom, An external model control system, *Control Theory and Advanced Techenology*, 4(1995), 1643–1665.

[105] Y. Mei and L. Huang, A second-order auto disturbance rejection controller for matrix converter fed induction motor drive, *Proc. the Power Electronics and Motion Control Conference*, 2009, 1964–1967.

[106] R. Miklosovic and Z. Gao, A dynamic decoupling method for controlling high performance turbofan engines, *Proc. the 16th IFAC World Congress*, Czech Republic, 2005, 488–488.

[107] S. Nazrulla and H.K. Khalil, Robust stabilization of non-minimum phase nonlinear systems using extended high-gain observers, *IEEE Trans. Automat. Control*, 56(2011), 802–813.

[108] W. Pan, Y. Zhou, and Y. Han, Design of ship main engine speed controller based on optimal active disturbance rejection technique, *Proc. the IEEE International Conference on Automation and Logistics*, 2010, 528–532.

[109] W. Perruquetti, T. Floquet, and E. Moulay, Finite-time observers: application to secure communication, *IEEE Trans. Automatic Control*, 53(2008), 356–360.

[110] L. Praly, Asymptotic stabilization via output feedback for lower triangular systems with output dependent incremental rate, *IEEE Trans. Automat. Control*, 48(2003), 1103–1108.

[111] L. Praly and Z.P. Jiang, Linear output feedback with dynamic high gain for nonlinear systems, *Systems Control Lett.*, 53(2004), 107–116.

[112] L. Rosier, Homogeneous Lyapunov function for homogeneous continuous vector field, *Systems Control Lett.*, 19(1992), 467–473.

[113] J. Ruan, Z. Li, and F. Zhou, ADRC based ship tracking controller design and simulations, *Proc. the IEEE International Conference on Automation and Logistics*, 2008, 1763–1768.

[114] J. Ruan, X. Rong, and S. Wu, Study on ADRC controller design and simulation of rock drill robot joint hydraulic drive system, *Proc. the 26th Chinese Control Conference*, 2007, 133–136.

[115] J. Ruan, F. Yang, R. Song, and Y. Li, Study on ADRC-based intelligent vehicle lateral locomotion control, *Proc. the 7th World Congress on Intelligent Control and Automation*, 2008, 2619–2624.

[116] Y. Shen and X. Xia, Semi-global finite-time observers for nonlinear systems, *Automatica*, 44(2008), 3152–3156.

[117] G. J. Silva, A. Datta, and S. P. Bhattacharyya, New results on the synthesisi of PID controllers, *IEEE Trans. Automat. Control*, 47(2002), 241–252.

[118] J. Slotine, J. Hedrick, and E. Misawa, On sliding observers for nonlinear systems, *ASME J. Dynam. Sys. Measur. Contr.*, 109(1987), 245–252.

[119] A. Smyshlyaev, E. Cerpa, and M. Krist, Boundary stabilization of a 1-D wave equation with in-domain antidamping, *SIAM J. Control Optim.*, 48(2010), 4014–4031.

[120] R. Song, Y. Li, W. Yu, and F. Wang, Study on ADRC based lateral control for tracked mobile robots on stairs, *Proc.the International Conference on Automation and Logistics*, 2008, 2048–2052.

[121] B. Sun and Z. Gao, A DSP-based active disturbance rejection control design for a 1-kW H-bridge DCCDC power converter, *IEEE Trans. Ind. Electron.*, 52(2005), 1271–1277.

[122] Y. Su, C. Zheng, and B. Duan, Automatic disturbances rejection controller for precise motion control of permanent-magnet synchronous motors, *IEEE Trans. Ind. Electron.*, 52(2005), 814–823.

[123] Y. X. Su, C. H. Zheng, and Dong Sun, A simple nonlinear velocity stimatiom for high-performance motion control, *IEEE Trans. Ind. Electron.*, 52(2005), 1661–1169.

[124] Y. Su, B. Y. Duan, C. H. Zheng, Y. F. Zhang, G. D. Chen, and J. W. Mi, Disturbance-rejection high-precision motion control of a Stewart platform, *IEEE Trans. Control Syst. Technol.*, 12(2005), 364–374.

[125] B. Sun and Z. Gao, A DSP-based active disturbance rejection control design for a 1-kW H-bridge DC-DC power converter, *IEEE Trans. Ind. Electron.*, 52(2005), 1271–1277.

[126] M. Sun, Z. Chen, and Z. Yuan, A practical solution to some problems in flight control, *Proc. the 48th IEEE Conference on Decision and Control and 28th Chinese Control Conference*, 2009, 1482–1487.

[127] K. Sun. and Y.L. Zhao, A new robust algorithm to improve the dynamic performance on the position control of magnet synchronous motor drive, *Proc. the International Conference on Future Computer and Communication*, 2010, 268–272.

[128] A. Teel and L. Praly, Tools for semiglobal stabilization by partial state and output feedback, *SIAM J. Control Optim*, 33(1995), 1443–1488.

[129] M. Tomizuka, K. Chew, and W. Yang, Disturbance rejection through an external model, *ASEM J. Dynamic Systems, Measurement, and Control*, 112(1990), 559–564.

[130] V. Utkin, J. Guldner, and J. Shi, *Sliding Mode Control in Electro-Mechanical Systems*, 2nd edition, Taylor Francis Group, Boca Raton, 2009.

[131] V. Utkin, Slide mode control design principles and applications to electric drives, *IEEE Trans. Industrial Electronics*, 40(1993), 23–36.

[132] B.L. Walcott, M.J. Corless, and S.H. Zak, Comparative study of non-linear state-observation technique, *Internat. J. Control*, 45(1987), 2109–2132.

[133] B. L. Walcott, and S.H. Zak, State observation of nonlinear uncertain dynamical systems, *IEEE Trans. Automat. Control*, 32(1987), 166–170.

[134] X. Wang, Z. Chen, and G. Yang, Finite-time-convergent differentiator based on singular perturbation technique, *IEEE Trans. Automat. Control*, 52(2007), 1731–1737.

[135] J. Wang, L. He and M. Sun, Application of active disturbance rejection control to integrated flight-propulsion control, *Proc. the 22nd Chinese Control and Decision Conference*, 2010, 2565–2569.

[136] G. Weiss, Admissibility of unbounded control operators, *SIAM J.Control Optim.*, 27(1989), 527–545.

[137] D. Wu, Design and analysis of precision active disturbance rejection control for noncircular turning process, *IEEE Trans. Ind. Electron.*, 56(2009), 2746–2753.

[138] D. Wu, K. Chen, and X. Wang, Tracking control and active disturbance rejection with application to noncircular machining, *Int. J. Mach. Tool Manu.*, 47(2007), 2207–2217.

[139] Y. Xia and M. Fu, *Compound Control Methodology for Fight Vehicles*, Springer-Verlag, Berlin, 2013.

[140] Y. Xia, P. Shi, G. Liu, D. Rees, and J. Han, Active disturbance rejection control for uncertain multivariable systems with time-delay, *IET Control Theory Appl.*, 1(2007), 75–81.

[141] X. Xia, Global frequency estimation using adaptive identifiers, *IEEE Trans. Automatic Control*, 47(2002), 1188–1193.

[142] J. Xie, B. Cao, D. Xu, and S. Zhang, Control of regenerative retarding of a vehicle equipped with a new energy recovery retarder, *Proc. the 4th Industrial Electronics and Applications*, 2009, 3177–3180.

[143] H. Xiong, J. Yi, J. and G. Fan, Autolanding of unmanned aerial vehicles based on active disturbance rejection control, *Proc. the IEEE International Conference on Intelligent Computing and Intelligent Systems*, 2009, 772–776.

[144] W. Xue, Y. Huang, and X. Yang, What kinds of system can be used as tracking-differentiator, *Proc. the 29th Chinese Control Conference*, 2010, 6113–6120.

[145] B. Yan, Z. Tian, S. Shi, and Z. Wang, Fault diagnosis for a class of nonlinear systems, *ISA Transactions*, 47(2008), 386–394.

[146] F. Yang, Y. Li, J. Ruan, and Z. Yin, ADRC based study on the anti-braking system of electric vehicles with regenerative braking, *Proc. the 8th World Congress on Intelligent Control and Automation*, 2010, 2588–2593.

[147] C.Z. Yao and J.Q. Han, The application of nonlinear tracking-differentiator in the forcasting of fertility pattern, *Systems Engineering–Theory & Practice*, 2(1996), 57–61 (in Chinese).

[148] J.Y. Yao, Z.X. Jiao, and D.W. Ma, Extended-state-observer-based output feedback nonlinear robust control of hydraulic systems with backstepping, *IEEE Trans. Ind. Electron.*, 61(2014), 6285–6293.

[149] G. Zarikian and A. Serrani, Harmonic disturbance rejection in tracking control of Euler–Lagrange systems: a exernal model approach, *IEEE Trans. Control Systems Technology*, 15(2007), 118–129.

[150] X. X. Yang and Y. Huang, Capability of extended state observer for estimating uncertainties, *Proc. the American Control Conference*, 2009, 3700–3705.

[151] Y. Zhang, L. Dong, and Z. Gao, Load frequency control for multiple-area power systems, *Proc. the American Control Conference*, 2009, 2773–2778.

[152] W.G. Zhang and J.Q. Han, The application of tracking differentiator in allocation of zero, *Acta Automatica Sinica*, 27(5)2001, 724–727 (in Chinese).

[153] W.G. Zhang and J.Q. Han, Continuous-time system identification with the tracking-differentiator, *Control & Decision*, 14(1999), 557–560 (in Chinese).

[154] Q.T. Zhang, Z.F. Tan, and L.D. Guo, A new controller of stabilizing circuit in FOGS, *Proc. International Conference on Mechatronics and Automation*, 2009, 757–761.

[155] L.J. Zhang, X. Qi, and Y.J. Pang, Adaptive output feedback control based on DRFNN for AUV, *Ocean Engineering*, 36(2009), 716–722.

[156] Z.L. Zhao and B.Z. Guo, On active disturbance rejection control for nonlinear systems using time-varying gain, *Eur. J. Control*, 23(2015), 62–70.

[157] Z.L. Zhao and B.Z. Guo, Extended state observer for uncertain lower triangular nonlinear systems, *Systems Control Lett.*, 85(2015), 100–108.

[158] Z.L. Zhao and B.Z. Guo, Active disturbance rejection control approach to stabilization of lower triangular systems with uncertainty, *Internat. J. Robust Nonlinear Control*, 26(2016), 2314–2337.

[159] Z.L. Zhao and B.Z. Guo, On convergence of nonlinear active disturbance rejection control for a class of nonlinear systems, *J. Dyn. Control Syst.*, 22(2016), 385–412.

[160] Q. Zheng, L. Gao, and Z. Gao, On stability analysis of active disturbance rejection control for nonlinear time-varying plants with unknow dynamics, *Proc. IEEE Conference on Decision and Control*, 2007, 3501–3506.

[161] Q. Zheng, L.L. Dong, D.H. Lee, and Z. Gao, Active disturbance rejection control for MEMS gyroscopes, *Proc. the American Control Conference*, 2008, 4425–4430.

[162] Q. Zheng, L.L. Dong, and Z. Gao, Control and rotation rate estimation of vibrational MEMS gyroscopes, *Proc. IEEE Multi-conference on Systems and Control*, 2007, 118–123.

[163] Q. Zheng and Z. Gao, On applications of active disturbance rejection control, *Proc. the 29th Chinese Control Conference*, 2010, 6095–6100.

[164] Q. Zheng, Z. Chen, and Z. Gao, A dynamic decoupling control approach and its applications to chemical processes, *Proc. the American Control Conference*, 2007, 5176–5181.

[165] W. Zhou and Z. Gao, An active disturbance rejection approach to tension and velocity regulations in Web processing lines, *Proc. the IEEE Multi-conference on Systems and Control*, 2007, 842–848.

[166] Q. Zheng and Z. Gao, An energy saving, factory-validated disturbance decoupling control design for extrusion processes, *Proc. the 10th World Congress on Intelligent Control and Automation*, 2012, 2891–2896.

[167] Q.L. Zheng, Z. Gao, and W. Tan, Disturbance rejection in thermal power plants, *Proc. the 30th Chinese Control Conference*, 2011, 6350–6355.

[168] Q. Zhong, Y. Zhang, and J. Yang, Non-linear auto-disturbance rejection control of parallel active power filters, *IET Control Theory Appl.*, 3(2008), 907–916.

[169] Z. Zhu, Y. Xia, and M. Fu, Attitude tracking of rigid spacecraft based on extended state observer, *Proc. the International Symposium on Systems and Control in Aeronautics and Astronautics*, 2010, 621–626.

Index

C_0-semigroup, 279

Hölder continuity, 222

active disturbance rejection control, 2, 155, 263
ADRC, 1, 2, 5, 6, 9, 209, 230, 260, 291
asymptotic convergence, 313
asymptotical stability, 9, 230
asymptotically stable, 4, 19, 23, 60
attracting basin, 23, 31

backstepping method, 272
boundary disturbance, 157, 270
boundary stabilization, 85, 157, 270

canonical form, 9, 16, 228
canonical form of ADRC, 13, 18
class $\mathcal{K}$ function, 21
class $\mathcal{K}_\infty$ function, 61, 105, 211, 226, 232, 295, 300, 314, 320
class $\mathcal{K}_\infty$ function, 21
comparison principle, 320
constant gain, 272
control (input), 3, 4, 312

differential inclusion, 32
distributed control system, 157

eigenvalue, 116, 297
equilibrium, 19, 22, 31, 293
ESO, 3, 5, 93, 209, 230, 291, 293
Euclid norm, 18
extended high-gain observer, 5, 94
extended state, 4, 96, 179, 197, 207, 291, 297
extended state observer, 3, 4, 93, 156, 263
external disturbance, 1, 4, 155, 157, 167, 229, 291, 312
external model principle (EMP), 6

finite-time observer, 4
finite-time stability, 9, 27, 292
finite-time stable, 108, 126
finite-time stable system, 69
finite-time stable tracking differentiator, 77
frequency online estimation, 81
Frobenius theorem, 16

Global ADRC, 238, 245
globally asymptotical stability, 189

Hölder continuity, 292
high frequency noise, 313
high gain control (HGC), 6, 157, 263
high gain observer, 93, 263
high gain tuning parameter, 134
high order tracking differentiator, 57, 64

Active Disturbance Rejection Control for Nonlinear Systems: An Introduction, First Edition.
Bao-Zhu Guo and Zhi-Liang Zhao.
© 2016 John Wiley & Sons Singapore Pte. Ltd. Published 2016 by John Wiley & Sons, Ltd.